PROCEEDINGS OF THE FOURTH COMPTON SYMPOSIUM

PROCEEDINGS OF THE FOURTH COMPTON SYMPOSIUM

Williamsburg, VA   April 1997

PART ONE: The Compton Observatory in Review

EDITORS
Charles D. Dermer
Mark S. Strickman
James D. Kurfess
Naval Research Laboratory

American Institute of Physics

AIP CONFERENCE
PROCEEDINGS 410

Woodbury, New York

Authorization to photocopy items for internal or personal use, beyond the free copying permitted under the 1978 U.S. Copyright Law (see statement below), is granted by the American Institute of Physics for users registered with the Copyright Clearance Center (CCC) Transactional Reporting Service, provided that the base fee of $10.00 per copy is paid directly to CCC, 222 Rosewood Drive, Danvers, MA 01923. For those organizations that have been granted a photocopy license by CCC, a separate system of payment has been arranged. The fee code for users of the Transactional Reporting Service is: 1-56396-659-X/ 97 /$10.00.

© 1997 American Institute of Physics

Individual readers of this volume and nonprofit libraries, acting for them, are permitted to make fair use of the material in it, such as copying an article for use in teaching or research. Permission is granted to quote from this volume in scientific work with the customary acknowledgment of the source. To reprint a figure, table, or other excerpt requires the consent of one of the original authors and notification to AIP. Republication or systematic or multiple reproduction of any material in this volume is permitted only under license from AIP. Address inquiries to Office of Rights and Permissions, 500 Sunnyside Boulevard, Woodbury, NY 11797-2999; phone: 516-576-2268; fax: 516-576-2499; e-mail: rights@aip.org.

L.C. Catalog Card No. 97-77179
ISBN 1-56396-659-X (Set)
ISBN 1-56396-772-3 (Part One)
ISBN 1-56396-773-1 (Part Two)
ISSN 0094-243X
DOE CONF- 9704154

Printed in the United States of America

CONTENTS

Preface xxv
Prologue xxvii

PART ONE—THE COMPTON OBSERVATORY IN REVIEW

COMPTON STATUS AND FUTURE

Status and Future of the Compton Gamma Ray Observatory 3
    N. Gehrels and C. Shrader

SOLAR AND STELLAR GAMMA RAY ASTRONOMY

Solar and Stellar Gamma Ray Observations with Compton 17
    G. H. Share, R. J. Murphy, and J. Ryan

GALACTIC GAMMA RAY ASTRONOMY

Gamma-Ray Pulsars: The Compton Observatory Contribution
to the Study of Isolated Neutron Stars 39
    D. J. Thompson, A. K. Harding, W. Hermsen, and M. P. Ulmer
Recent Results from Observations of Accreting Pulsars 57
    M. H. Finger and T. A. Prince
Low-Mass X-Ray Binaries and Radiopulsars in Binary Systems 75
    M. Tavani and D. Barret
GRO J1744-28: The Bursting Pulsar 96
    C. Kouveliotou and J. van Paradijs
The Soft Gamma-Ray Repeaters 110
    I. A. Smith
Galactic Black Hole Binaries: High-Energy Radiation 122
    J. E. Grove, J. E. Grindlay, B. A. Harmon, X.-M. Hua, D. Kazanas,
    and M. McConnell
Galactic Black Hole Binaries: Multifrequency Connections 141
    S. N. Zhang, I. F. Mirabel, B. A. Harmon, R. A. Kroeger,
    L. F. Rodriguez, R. M. Hjellming, and M. P. Rupen
CGRO Studies of Supernovae and Classical Novae 163
    M. D. Leising
Supernova Remnants and Plerions in the Compton Gamma-Ray
Observatory Era 171
    O. C. de Jager and M. G. Baring
Diffuse Galactic Continuum Radiation 192
    S. D. Hunter, R. L. Kinzer, and A. W. Strong
Galactic $e^+ - e^-$ Annihilation Line Radiation 208
    D. M. Smith, W. R. Purcell, and M. Leventhal

Galactic Gamma-Ray Line Emission from Radioactive Isotopes 218
    R. Diehl and F. X. Timmes
Galactic Nuclear Deexcitation Gamma-Ray Lines......................... 249
    H. Bloemen and A. M. Bykov

EXTRAGALACTIC GAMMA RAY ASTRONOMY

High Energy Emission from Starburst Galaxies and Clusters 271
    Y. Rephaeli and C. D. Dermer
Seyferts and Radio Galaxies... 283
    W. N. Johnson, A. A. Zdziarski, G. M. Madejski, W. S. Paciesas,
    H. Steinle, and Y.-C. Lin
Gamma-Ray Blazars 307
    R. C. Hartman, W. Collmar, C. von Montigny, and C. D. Dermer
Multiwavelength Campaigns... 328
    C. R. Shrader and A. E. Wehrle
The Extragalactic Diffuse Gamma-Ray Emission 344
    P. Sreekumar, F. W. Stecker, and S. C. Kappadath

HIGH ENERGY GAMMA RAY ASTRONOMY

VHE and UHE Gamma-Ray Astronomy in the EGRET Era 361
    T. C. Weekes, F. Aharonian, D. J. Fegan, and T. Kifune

GAMMA RAY MYSTERIES

Pulsar Counterparts of Gamma-Ray Sources 387
    P. A. Caraveo and G. F. Bignami
On the Nature of the Unidentified EGRET Sources 394
    R. Mukherjee, I. A. Grenier, and D. J. Thompson
A Review of Gamma Ray Bursts....................................... 407
    C. Meegan, K. Hurley, A. Connors, B. Dingus, and S. Matz
Gamma-Ray Line Transients 418
    M. J. Harris
Constraints from Undetected Gamma-Ray Sources 436
    C. E. Fichtel and P. Sreekumar

HIGH ENERGY PHYSICS AND ASTROPHYSICS

Gamma Ray Implications for the Origin and the Acceleration
of Cosmic Rays.. 449
    R. Schlickeiser, M. Pohl, R. Ramaty, and J. G. Skibo
Spectral Signatures and Physics of Black Hole Accretion Disks.............. 461
    E. Liang and R. Narayan

Comptonization Processes in Galactic and Extragalactic
High Energy Sources 477
    L. G. Titarchuk
Radiation Processes in Blazars 494
    M. Sikora

COMPTON INSTRUMENTS AND HISTORY

Overview of the Compton Observatory Instruments 509
    J. D. Kurfess, D. L. Bertsch, G. J. Fishman, and V. Schönfelder
The COMPTON Observatory: Reflections on its Origins and History 524
    D. A. Kniffen and N. Gehrels

PART TWO-PAPERS AND PRESENTATIONS

ISOLATED NEUTRON STARS

COMPTEL Gamma-Ray Study of the Crab Nebula 537
    R. D. van der Meulen, H. Bloemen, K. Bennett, W. Hermsen,
    L. Kuiper, R. P. Much, J. Ryan, V. Schönfelder, and A. Strong
5 Years of Crab Pulsar Observations with COMPTEL 542
    R. Much, K. Bennett, C. Winkler, R. Diehl, G. Lichti, V. Schönfelder,
    H. Steinle, A. Strong, M. Varendorff, W. Hermsen, L. Kuiper,
    R. van der Meulen, A. Connors, M. McConnell, J. Ryan, and R. Buccheri
The Infrared to Gamma-Ray Pulse Shape of the Crab Nebula Pulsar 547
    S. S. Eikenberry and G. G. Fazio
Observation of the Crab Pulsar with BeppoSAX: Study of the
Pulse Profile and Phase Resolved Spectroscopy 553
    G. Cusumano, D. Dal Fiume, S. Giarrusso, E. Massaro, T. Mineo,
    L. Nicastro, A. N. Parmar, and A. Segreto
The Spectrum of TeV Gamma Rays from the Crab Nebula 558
    J. P. Finley, S. Biller, P. J. Boyle, J. H. Buckley, A. Burdett,
    J. Bussóns Gordo, D. A. Carter-Lewis, M. A. Catanese, M. F. Cawley,
    D. J. Fegan, J. A. Gaidos, A. M. Hillas, F. Krennrich, R. C. Lamb,
    R. W. Lessard, C. Masterson, J. E. McEnery, G. Mohanty, J. Quinn,
    A. J. Rodgers, H. J. Rose, F. W. Samuelson, G. H. Sembroski,
    R. Srinivasan, T. C. Weekes, M. West, and A. Zweerink
First Stereoscopic Measurements at the Whipple Observatory 563
    F. Krennrich, for the Whipple Collaboration
The "COS-B/EGRET 1997" Geminga Ephemeris 568
    J. R. Mattox, J. P. Halpern, and P. A. Caraveo
On the Accurate Positioning of Geminga 573
    P. A. Caraveo, M. G. Lattanzi, G. Massone, R. Mignani, V. V. Makarov,
    M. A. C. Perryman, and G. F. Bignami
Studies of the Gamma Ray Pulsar Geminga 578
    S. Zhang, T. P. Li, M. Wu, Y. Yu, X. J. Sun, L. M. Song, and F. J. Lu

Timing Analysis of Four Years of COMPTEL Data on PSR B1509-58 583
A. Carramiñana, K. Bennett, W. Hermsen, L. Kuiper, V. Schönfelder,
A. Connors, V. Kaspi, M. Bailes, and R. N. Manchester

**Search for the Pulse of PSR B1823-13 in the COMPTEL
and EGRET Databases** 588
A. Carramiñana, K. T. S. Brazier, K. Bennett, W. Hermsen,
L. Kuiper, V. Schönfelder, and A. Lyne

Very High Energy Observations of PSR B1951+32 592
R. Srinivasan, P. J. Boyle, J. H. Buckley, A. M. Burdett, J. Gordo,
D. A. Carter-Lewis, M. Catanese, M. F. Cawley, E. Colombo, D. J. Fegan,
J. P. Finley, J. A. Gaidos, A. M. Hillas, R. C. Lamb, F. Krennrich,
R. W. Lessard, C. Masterson, J. E. McEnery, G. Mohanty, P. Moriarty,
J. Quinn, A. J. Rodgers, H. J. Rose, F. W. Samuelson, G. H. Sembroski,
T. C. Weekes, and J. Zweerink

A Candidate γ-Ray Pulsar in CTA 1 597
K. T. S. Brazier, O. Reimer, G. Kanbach, and A. Carramiñana

Discovery of the Young, Energetic Radio Pulsar PSR J1105-6107 602
V. M. Kaspi, M. Bailes, R. N. Manchester, B. W. Stappers, J. S. Sandhu,
J. Navarro, and N. D'Amico

RXTE Observation of PSR1706-44 607
A. Ray, A. K. Harding, and M. Strickman

VHE Gamma Rays from PSR B1706-44 612
P. M. Chadwick, M. R. Dickinson, N. A. Dipper, J. Holder,
T. R. Kendall, T. J. L. McComb, K. J. Orford, J. L. Osborne, S. M. Rayner,
I. D. Roberts, S. E. Shaw, and K. E. Turver

RXTE Observations of the Anomalous Pulsar 4U0142+61 617
S. Dieters, C. Wilson, M. Finger, M. Scott, and J. van Paradijs

Search for a Gamma-Ray Pulsar in the SNR RCW103 623
M. Mori and K. Ebisawa

Search for X-ray Pulsation from Rotation-Powered Pulsars with *ASCA* 628
Y. Saito, N. Kawai, T. Kamae, and S. Shibata

Gamma Ray Pulsar Luminosities 633
M. A. McLaughlin, J. M. Cordes, and M. P. Ulmer

A New Class of Radio Quiet Pulsars 638
M. G. Baring and A. K. Harding

The Pulse Profile of γ-ray Pulsars and the Emission Region Geometry 643
E. Massaro and M. Litterio

Geometry of Pulsar X-Ray and Gamma-Ray Pulse Profiles 648
A. K. Harding and A. Muslimov

**A New Method for Statistical Study of Gamma-Ray Phase Curves
of Radio Pulsars** 653
A. Chernenko

**Evidence for Spontaneous Magnetic Field Decay in an Isolated
Neutron Star** 658
J. C. L. Wang

NEUTRON STAR BINARIES

**A Multi-Year Light Curve of Sco X-1 Based on BATSE SD Data
and the Variability States of Sco X-1**.. 665
        B. J. McNamara, T. E. Harrison, P. A. Mason, M. Templeton,
        C. W. Heikkila, T. Buckley, E. Galvan, and A. Silva
**Comparison of the BATSE LAD and SD Light Curves of Sco X-1:
1991–1996**.. 670
        T. E. Harrison, B. J. McNamara, P. A. Mason, and M. Templeton
**Application of the Gabor Transform to BATSE Spectroscopy
Detector Observations of Scorpius X-1**... 675
        P. A. Mason, M. Templeton, B. J. McNamara, T. E. Harrison, E. Galvan,
        and T. Buckley
High-Energy Transient Events From Scorpius X-1 and Cygnus X-1............ 679
        P. A. Mason, B. J. McNamara, and T. E. Harrison
**Low-Energy Line Emission from Cygnus X-2 Observed
by the BeppoSAX LECS**.. 683
        E. Kuulkers, A. N. Parmar, A. Owens, T. Oosterbroek,
        and U. Lammers
BATSE Observations of the Second Outburst of GRO J1744-28............... 687
        P. Woods, C. Kouveliotou, J. van Paradijs, M. S. Briggs, K. Deal,
        C. A. Wilson, B. A. Harmon, G. J. Fishman, W. H. G. Lewin,
        and J. Kommers
Determination of Peak Fluxes and α for Bursts from GRO J1744–28........ 692
        T. E. Strohmayer, K. Jahoda, J. H. Swank, and M. J. Stark
Kilohertz Oscillations in 4U 0614+091 and Other LMXBs...................... 697
        E. C. Ford, P. Kaaret, M. Tavani, D. Barret, P. Bloser, J. Grindlay,
        B. A. Harmon, W. S. Paciesas, and S. N. Zhang
General Relativity and Quasi-Periodic Oscillations.............................. 703
        P. Kaaret and E. C. Ford
Compact Hard X-ray Sources Near Galactic Longitude 20..................... 708
        G. V. Jung, J. D. Kurfess, and W. R. Purcell
Hard and Soft X-ray Observations of Aquila X-1................................ 713
        B. C. Rubin, B. A. Harmon, W. S. Paciesas, C. R. Robinson,
        and S. N. Zhang
**Observation of X-Ray Bursters with the Beppo-SAX
Wide Field Cameras**... 719
        A. Bazzano, M. Cocchi, L. Natalucci, P. Ubertini, J. Heise, J. in 't Zand,
        J. M. Muller, and M. J. S. Smith
**Aperiodic Variability of the X-ray Burster 1E1724-3045:
First Results from RXTE/PCA**... 724
        J.-F. Olive, D. Barret, L. Boirin, J. Grindlay, P. Bloser, J. Swank,
        and A. Smale
**New X-Ray Bursters with the WFCs on Board SAX:
SAX J1750. 8-2900, GS 1826-24 and SLX 1735-269**........................... 729
        A. Bazzano, M. Cocchi, L. Natalucci, P. Ubertini, J. Heise,
        J. in 't Zand, J. M. Muller, and M. J. S. Smith

Kilo-Hertz QPO and X-ray Bursts in 4U 1608-52 in Low Intensity State 734
    W. Yu, S. N. Zhang, B. A. Harmon, W. S. Paciesas, C. R. Robinson,
    J. E. Grindlay, P. Bloser, D. Barret, E. C. Ford, M. Tavani, and P. Kaaret

Long-Term Observations of Her X-1 with BATSE 739
    R. B. Wilson, D. M. Scott, and M. H. Finger

RXTE Spectroscopy of Her X-1 744
    D. E. Gruber, W. A. Heindl, R. E. Rothschild, R. Staubert, M. Kunz,
    and D. M. Scott

Observations of Pulse Evolution in Her X-1 748
    D. M. Scott, R. B. Wilson, M. H. Finger, and D. A. Leahy

Evolution of the Orbital Period of Her X-1: Determination of a New Ephemeris Using RXTE Data 753
    B. Stelzer, R. Staubert, J. Wilms, R. D. Geckeler, D. Gruber,
    and R. Rothschild

The Pulsed Light Curves of Her X-1 as Observed by BeppoSAX 758
    D. Dal Fiume, M. Orlandini, G. Cusumano, S. Del Sordo, M. Feroci,
    F. Frontera, T. Oosterbroek, E. Palazzi, A. N. Parmar, A. Santangelo,
    and A. Segreto

The 35 Day Cycle of Her X-1 and the Coronal Wind Model 763
    S. Schandl, R. Staubert, and M. König

CGRO/EGRET Observations of Centaurus X-3 768
    W. T. Vestrand, P. Sreekumar, and M. Mori

GRO J2058+42 X-Ray Observations 773
    C. A. Wilson, M. H. Finger, B. A. Harmon, R. B. Wilson,
    D. Chakrabarty, and T. Strohmayer

Observation of a Long Term Spin-up Trend in 4U1538-52 778
    B. C. Rubin, M. H. Finger, D. M. Scott, and R. B. Wilson

EGRET Observations of X-R Binaries 783
    B. B. Jones, Y. C. Lin, P. F. Michelson, P. L. Nolan, M. S. E. Roberts,
    and W. F. Tompkins

Observations of Vela X-1 with RXTE 788
    P. Kretschmar, I. Kreykenbohm, R. Staubert, J. Wilms, M. Maisack,
    E. Kendziorra, W. Heindl, D. Gruber, R. Rothschild, and J. E. Grove

BeppoSAX Observation of the X-ray Binary Pulsar Vela X-1 793
    M. Orlandini, D. Dal Fiume, L. Nicastro, S. Giarrusso, A. Segreto,
    S. Piraino, G. Cusumano, S. Del Sordo, M. Guainazzi, and L. Piro

New Radio Observations of Circinus X-1 798
    R. P. Fender

Orbit Determination for the Be/X-Ray Transient EXO 2030+375 803
    M. T. Stollberg, M. H. Finger, R. B. Wilson, D. M. Scott, D. J. Crary,
    and W. S. Paciesas

The Orbital Ephemeris and X-Ray Light Curve of Cyg X-3 808
    S. M. Matz

A Multiwavelength Study of Cygnus X-3 813
    M. L. McCollough, C. R. Robinson, S. N. Zhang, B. A. Harmon,
    W. S. Paciesas, R. M. Hjellming, M. Rupen, A. J. Mioduszewski,
    E. B. Waltman, R. S. Foster, F. D. Ghigo, G. G. Pooley, R. P. Fender,
    and W. Cui

Is There Any Evidence for a Massive Black Hole in Cyg X-3 818
    A. Mitra

Generation of Periodical Gamma Radiation in Binary System
with a Millisecond Pulsar 822
    M. A. Chernyakova and A. F. Illarionov

GALACTIC BLACK HOLE CANDIDATES

The MeV Spectrum of Cygnus X-1 as Observed with COMPTEL 829
    M. McConnell, K. Bennett, H. Bloemen, W. Collmar, W. Hermsen,
    L. Kuiper, R. Much, J. Ryan, V. Schönfelder, H. Steinle, A. Strong,
    and R. van Dijk

Five Years in the Life of Cygnus X-1: BATSE Long-Term Monitoring 834
    W. S. Paciesas, C. R. Robinson, M. L. McCollough, S. N. Zhang,
    B. A. Harmon, and C. A. Wilson

Spectral Evolution of Cyg X-1 During Its 1996 Soft State Transition 839
    S. N. Zhang, W. Cui, B. A. Harmon, and W. S. Paciesas

X-ray and γ-ray Spectra of Cyg X-1 in the Soft State 844
    M. Gierliński, A. A. Zdziarski, T. Dotani, K. Ebisawa, K. Jahoda,
    and W. N. Johnson

RXTE Observation of Cygnus X-1: Spectra and Timing 849
    J. Wilms, J. Dove, M. Nowak, and B. A. Vaughan

Spectral Variability of Cygnus X-1 in the Soft State 854
    W. Focke, J. Swank, B. Phlips, W. Heindl, and W. Cui

Modeling Cygnus X-1 γ_2 Spectra Observed by BATSE 858
    X.-M. Hua, J. C. Ling, and Wm. A. Wheaton

A Model for the High-Energy Emission of Cyg X-1 863
    I. V. Moskalenko, W. Collmar, and V. Schönfelder

A Thermal-Nonthermal Inverse Compton Model for Cyg X-1 868
    A. Crider, E. P. Liang, I. A. Smith, D. Lin, and M. Kusunose

Two Distinct States of Microquasars 1E1740-294 and GRS1758-258 873
    S. N. Zhang, B. A. Harmon, and E. P. Liang

Broad-Band Spectral Modeling of Cyg X-1 and 1E1740.7 878
    S. Sheth, E. Liang, M. Burger, C. Luo, A. Harmon, and S. N. Zhang

Observational Constraints on Annihilation Sites in 1E 1740. 7-2942
and Nova Muscae 881
    I. V. Moskalenko and E. Jourdain

Two-Phase Spectral Modelling of 1E1740.7-2942 887
    O. Vilhu, J. Nevalainen, J. Poutanen, M. Gilfanov, P. Durouchoux,
    M. Vargas, R. Narayan, and A. Esin

Multi-Wavelength Monitoring of GRS 1915+105 892
    R. Bandyopadhyay, P. Martini, E. Gerard, P. A. Charles,
    R. M. Wagner, C. Shrader, T. Shahbaz, and I. F. Mirabel
The Hard X-Ray Spectrum of GRS 1915+105 897
    W. A. Heindl, P. Blanco, D. E. Gruber, M. Pelling,
    R. Rothschild, E. Morgan, and J. H. Swank
OSSE Upper Limit on Positron Annihilation from GRS 1915+105 902
    D. M. Smith, M. Leventhal, L. X. Cheng, J. Tueller, N. Gehrels,
    I. F. Mirabel, L. F. Rodriguez, and W. Purcell
RXTE Observations of GRS 1915+105 907
    J. Greiner, E. H. Morgan, and R. A. Remillard
Near-Infrared Observations of GRS 1915+105 912
    W. A. Mahoney, S. Corbel, Ph. Durouchoux, T. N. Gautier,
    J. C. Higdon, and P. Wallyn
Infrared Observations and Energetic Outburst of GRS 1915+105 917
    S. Chaty and I. F. Mirabel
ASCA Observations of Galactic Jet Systems 922
    T. Kotani, N. Kawai, M. Matsuoka, T. Dotani, H. Inoue, F. Nagase,
    Y. Tanaka, Y. Ueda, K. Yamaoka, W. Brinkmann, K. Ebisawa,
    T. Takeshima, N. E. White, A. Harmon, C. F. Robinson, S. N. Zhang,
    M. Tavani, and R. Foster
BATSE Observations of GX339-4 927
    B. C. Rubin, B. A. Harmon, W. S. Paciesas, C. R. Robinson,
    S. N. Zhang, and G. J. Fishman
Multiwavelength Observations of GX 339-4 932
    I. A. Smith, E. P. Liang, M. Moss, J. Dobrinskaya, R. P. Fender,
    Ph. Durouchoux, S. Corbel, R. Sood, A. V. Filippenko, and D. C. Leonard
Radio Observations of the Black Hole Candidate GX 339-4 937
    S. Corbel, R. P. Fender, Ph. Durouchoux, R. K. Sood, A. K. Tzioumis,
    R. E. Spencer, and D. Campbell-Wilson
Infrared Observations of the Ellipsoidal Light Variation in J0422+32 942
    D. M. Leeber, T. E. Harrison, and B. J. McNamara
Rapid X-ray Variability in GRO J0422+32 (Nova Per 1992) 947
    F. van der Hooft, C. Kouveliotou, J. van Paradijs, D. J. Crary,
    B. C. Rubin, M. H. Finger, B. A. Harmon, M. van der Klis,
    W. H. G. Lewin, and G. J. Fishman
BATSE Observations of Two Hard X-ray Outbursts from 4U 1630-47 952
    P. F. Bloser, J. E. Grindlay, D. Barret, S. N. Zhang, B. A. Harmon,
    G. J. Fishman, and W. S. Paciesas
Hard X-ray Observations of GRS 1009-45 with the SIGMA Telescope 957
    P. Goldoni, M. Vargas, A. Goldwurm, P. Laurent, E. Jourdain,
    J.-P. Roques, V. Borrel, L. Bouchet, M. Revnivtsev, E. Churazov,
    M. Gilfanov, R. Sunyaev, A. Dyachkov, N. Khavenson, N. Tserenin,
    and N. Kuleshova
Relativistic Effects in the X-ray Spectra of the Black Hole Candidate
GS 2023+338 962
    P. T. Życki, C. Done, and D. A. Smith

A Search for Gamma-ray Flares from Black-Hole Candidates
on Time Scales of ~1.5 Hours 967
      R. van Dijk, K. Bennett, H. Bloemen, R. Diehl, W. Hermsen,
      M. McConnell, J. Ryan, and V. Schönfelder

Physical Characteristics of the Spectral States of Galactic Black Holes 972
      J. Poutanen, J. H. Krolik, and F. Ryde

Temporal Characteristics of Compton Reflection from Accretion Disks 977
      W. T. Bridgman, C. D. Dermer, and J. G. Skibo

Temporal and Spectral Properties of Comptonized Radiation 982
      D. Kazanas, X.-M. Hua, and L. Titarchuk

Phase Difference and Coherence as Diagnostics of Accreting Sources 987
      X.-M. Hua, D. Kazanas, and L. Titarchuk

Global Spectra of Transonic Accretion Disks............................. 992
      E. Liang and C. Luo

Horizontal Branch Oscillations from Black Hole Candidates 995
      X. Chen, R. E. Taam, and J. H. Swank

Evolution of the Optically Thick Disk in Nova Muscae.................... 1000
      F. Melia and R. Misra

GALACTIC GAMMA-RAY LINE EMISSION

TGRS Results on the Spatial and Temporal Behavior
of the Galactic Center 511 keV Line.................................. 1007
      B. J. Teegarden, T. L. Cline, N. Gehrels, R. Ramaty, H. Seifert,
      M. Harris, D. Palmer, and K. H. Hurley

A BATSE Measurement of the Galactic Positron Annihilation Line 1012
      D. M. Smith, L. X. Cheng, M. Leventhal, J. Tueller, N. Gehrels,
      and G. Fishman

OSSE Constraints on the Galactic Positron Source Distribution 1017
      P. A. Milne and M. D. Leising

Is Positron Escape Seen in the Late-Time Light Curves
of Type Ia Supernovae?... 1022
      P. A. Milne, L.-S. The, and M. D. Leising

The Origin of the High-Energy Activity at the Galactic Center............. 1027
      F. Yusef-Zadeh, W. Purcell, and E. Gotthelf

The Galactic Center Lobe and its Interpretation 1034
      M. Pohl

Evidence for GeV Emission from the Galactic Center Fountain 1039
      D. H. Hartmann, D. D. Dixon, E. D. Kolaczyk, and J. Samimi

Positron Transport and Annihilation in Expanding Flows:
A Model for the High-Latitude Annihilation Feature..................... 1044
      C. D. Dermer and J. G. Skibo

Issues Concerning the Orion Gamma Ray Line Observations:
Overview and X-Ray Emission 1049
      R. Ramaty, B. Kozlovsky, and V. Tatischeff

**Constraints from Pion Production on the Spectral Hardness
of the Low Energy Cosmic Rays in Orion** 1054
    V. Tatischeff, R. Ramaty, and N. Mandzhavidze
Gamma-Ray Lines from OB Associations at $Z=Z_\odot$ and $Z=2Z_\odot$ 1059
    E. Parizot, J. Paul, and M. Cassé
On the Origin of the Orion Energetic Particles 1064
    E. M. G. Parizot
On the Origin of 3 to 7 MeV γ-Ray Excess in the Direction of Orion 1069
    V. A. Dogiel, M. J. Freyberg, G. E. Morfill, and V. Schönfelder
COMPTEL Spectral Study of the Inner Galaxy 1074
    H. Bloemen, A. M. Bykov, R. Diehl, W. Hermsen, R. van der Meulen,
    V. Schönfelder, and A. W. Strong
OSSE Results on Galactic γ-Ray Line Emission 1079
    M. J. Harris, W. R. Purcell, K. McNaron-Brown, R. J. Murphy,
    J. E. Grove, W. N. Johnson, R. L. Kinzer, J. D. Kurfess, G. H. Share,
    and G. V. Jung
Reassessment of the ^{56}Co emission from SN 1991T 1084
    D. J. Morris, K. Bennett, H. Bloemen, R. Diehl, W. Hermsen,
    G. G. Lichti, M. L. McConnell, J. M. Ryan, and V. Schönfelder
RXTE Observations of Cas A 1089
    R. E. Rothschild, R. E. Lingenfelter, W. A. Heindl, P. R. Blanco,
    M. R. Pelling, D. E. Gruber, G. E. Allen, K. Jahoda, J. H. Swank,
    S. E. Woosley, K. Nomoto, and J. C. Higdon
**Fluctuation Analysis of OSSE Measurements of the
1.275 MeV Line of ^{22}Na** 1094
    M. J. Harris
COMPTEL All-Sky Imaging at 2.2 MeV 1099
    M. McConnell, S. Fletcher, K. Bennett, H. Bloemen, R. Diehl,
    W. Hermsen, J. Ryan, V. Schönfelder, A. Strong, and R. van Dijk
^{26}Al Constraints from the COMPTEL/OSSE/SMM Data 1104
    R. Diehl, M. D. Leising, J. Knödlseder, and U. Oberlack
^{26}Al and the COMPTEL ^{60}Fe Data 1109
    R. Diehl, U. Wessolowski, U. Oberlack, H. Bloemen, R. Georgii,
    A. Iyudin, J. Knödlseder, G. Lichti, W. Hermsen, D. Morris, J. Ryan,
    V. Schönfelder, A. Strong, P. von Ballmoos, and C. Winkler
Models for COMPTEL ^{26}Al Data 1114
    R. Diehl, U. Oberlack, J. Knödlseder, H. Bloemen, W. Hermsen,
    D. Morris, J. Ryan, V. Schönfelder, A. Strong, P. von Ballmoos,
    and C. Winkler
**γ-Ray Emitting Radionuclide Production in A Multidimensional
Supernovae Model** 1119
    G. Bazán and D. Arnett
**Predictions of Gamma-Ray Emission from Classical Novae
and their Detectability by CGRO** 1125
    M. Hernanz, J. Gómez-Gomar, J. José, and J. Isern
New Studies of Nuclear Decay γ-rays From Novae 1130
    S. Starrfield, J. W. Truran, M. C. Wiescher, and W. M. Sparks

SUPERNOVA REMNANTS AND COSMIC RAYS, DIFFUSE GAMMA-RAY CONTINUUM RADIATION, AND UNIDENTIFIED SOURCES

CTA 1 Supernova Remnant: A High Energy Gamma-Ray Source? 1137
    D. Bhattacharya, A. Akyüz, G. Case, D. Dixon, and A. Zych

Constraints on Cosmic-Ray Origin from TeV Gamma-Ray Observations of Supernova Remnants 1142
    R. W. Lessard, P. J. Boyle, S. M. Bradbury, J. H. Buckley,
    A. C. Burdett, J. Bussóns Gordo, D. A. Carter-Lewis, M. Catanese,
    M. F. Cawley, D. J. Fegan, J. P. Finley, J. A. Gaidos, A. M. Hillas,
    F. Krennrich, R. C. Lamb, C. Masterson, J. E. McEnery,
    G. Mohanty, J. Quinn, A. J. Rodgers, H. J. Rose, F. W. Samuelson,
    G. H. Sembroski, R. Srinivasan, T. C. Weekes, and J. Zweerink

Hard X-ray Emission from Cassiopeia A SNR 1147
    L.-S. The, M. D. Leising, D. H. Hartmann, J. D. Kurfess, P. Blanco,
    and D. Bhattacharya

Nonthermal SNR Emission 1152
    S. J. Sturner, J. G. Skibo, C. D. Dermer, and J. R. Mattox

Gamma-Rays from Supernova Remnants: Signatures of Non-Linear Shock Acceleration 1157
    M. G. Baring, D. C. Ellison, S. J. Reynolds, and I. A. Grenier

Modelling Cosmic Rays and Gamma Rays in the Galaxy 1162
    A. W. Strong and I. V. Moskalenko

Production of Beryllium and Boron by Spallation in Supernova Ejecta 1167
    D. Majmudar and J. H. Applegate

Gamma Rays and Cosmic Rays from Supernova Explosions and Young Pulsars in the Past 1172
    L. I. Dorman

Angle Distribution and Time Variation of Gamma Ray Flux from Solar and Stellar Winds, 1. Generation of Flare Energetic Particles 1178
    L. I. Dorman

Angle Distribution and Time Variation of Gamma Ray Flux from Solar and Stellar Winds, 2. Generation of Galactic Cosmic Rays 1183
    L. I. Dorman

Diffuse High-Energy Gamma-Ray Emission in Monoceros 1188
    S. W. Digel, I. A. Grenier, S. D. Hunter, T. M. Dame,
    and P. Thaddeus

Diffuse 50 keV to 10 MeV Gamma-Ray Emission from the Inner Galactic Ridge 1193
    R. L. Kinzer, W. R. Purcell, and J. D. Kurfess

Diffuse Galactic Continuum Emission: Recent Studies using COMPTEL Data 1198
    A. W. Strong, R. Diehl, V. Schönfelder, K. Bennett, M. McConnell,
    and J. Ryan

Galactic Diffuse γ-ray Emission at TeV Energies
and the Ultra-High Energy Cosmic Rays.................................. 1203
    G. A. Medina Tanco and E. M. de Gouveia Dal Pino
The Diffuse Galactic Continuum Observed with EGRET:
Where's the Bump?... 1208
    J. Skibo
The Pulsar Contribution to the Diffuse Galactic γ-ray Emission............ 1213
    M. Pohl, G. Kanbach, S. D. Hunter, and B. B. Jones
The Total Cosmic Diffuse Gamma-Ray Spectrum from 9 to 30 MeV
Measured with COMPTEL 1218
    S. C. Kappadath, J. Ryan, K. Bennett, H. Bloemen, R. Diehl,
    W. Hermsen, M. McConnell, V. Schönfelder, M. Varendorff,
    G. Weidenspointner, and C. Winkler
The Cosmic γ-Ray Background from Supernovae 1223
    K. Watanabe, D. H. Hartmann, M. D. Leising, L.-S. The, G. H. Share,
    and R. L. Kinzer
The γ-Ray and Neutrino Background and Cosmic
Chemical Evolution 1228
    D. H. Hartmann, K. Watanabe, M. D. Leising, L.-S. The,
    and S. E. Woosley
The Contribution of Blazars to the Extragalactic Diffuse γ-ray
Background.. 1233
    A. Mücke and M. Pohl
Absorption of High Energy Gamma Rays by Interactions
with Extragalactic Starlight Photons at High Redshifts.................... 1238
    M. H. Salamon and F. W. Stecker
Further COMPTEL Observations of the Region Around GRO J1753+57:
Are There Several MeV Sources Present? 1243
    O. R. Williams, K. Bennett, R. Much, V. Schönfelder, W. Collmar,
    H. Bloemen, J. J. Blom, W. Hermsen, and J. Ryan
Temporal and Spectral Studies of Unidentified EGRET High
Latitude Sources... 1248
    O. Reimer, D. L. Bertsch, B. L. Dingus, J. A. Esposito, R. C. Hartman,
    S. D. Hunter, B. B. Jones, G. Kanbach, D. A. Kniffen, Y. C. Lin,
    H. A. Mayer-Hasselwander, C. von Montigny, P. L. Nolan, P. Sreekumar,
    D. J. Thompson, and W. F. Tompkins
Discovery of a Non-Blazar Gamma-ray Transient in the Galactic Plane 1253
    M. Tavani, R. Mukherjee, J. R. Mattox, J. Halpern, D. J. Thompson,
    G. Kanbach, W. Hermsen, S. N. Zhang, and R. S. Foster
Searches for Short-Term Variability of EGRET Sources
in the Galactic Anticenter... 1257
    D. J. Thompson, S. D. Bloom, J. A. Esposito, D. A. Kniffen,
    and C. von Montigny
Short Time-scale Gamma-Ray Variability of Blazars and EGRET
Unidentified Sources 1262
    S. D. Bloom, D. J. Thompson, R. C. Hartman, and C. von Montigny

Optical Identification of EGRET Source Counterparts 1267
    A. Carramiñana, J. Guichard, K. T. S. Brazier, G. Kanbach,
    and O. Reimer
Possible Identification of Unidentified EGRET Sources
with Wolf-Rayet Stars 1271
    R. K. Kaul and A. K. Mitra
Accreting Isolated Black Holes and the Unidentified EGRET Sources 1275
    C. D. Dermer

SEYFERT AND RADIO GALAXIES

Multi-Year BATSE Earth Occultation Monitoring of NGC4151 1283
    A. Parsons, N. Gehrels, W. Paciesas, A. Harmon, G. Fishman,
    C. Wilson, and S. N. Zhang
Broad-Band Continuum and Variability of NGC 5548 1288
    P. Magdziarz, O. Blaes, A. A. Zdziarski, W. N. Johnson,
    and D. A. Smith
Detection of a High Energy Break in the Seyfert Galaxy
MCG+8−11−11 1293
    P. Grandi, F. Haardt, G. Ghisellini, J. E. Grove, L. Maraschi,
    and C. M. Urry
Compton Gamma-Ray Observatory Observations of the Nearest
Active Galaxy Centaurus A 1298
    H. Steinle, K. Bennett, H. Bloemen, W. Collmar, R. Diehl,
    W. Hermsen, G. G. Lichti, D. Morris, V. Schönfelder, A. W. Strong,
    and O. R. Williams
An Anisotropic Illumination Model of Seyfert I Galaxies 1303
    P. O. Petrucci, G. Henri, J. Malzac, and E. Jourdain
Scattered Emission and the X−γ Spectra of Seyfert Galaxies 1308
    J. Chiang, C. D. Dermer, and J. G. Skibo
Pair Models Revivified for High Energy Emission of AGNs 1313
    G. Henri and P. O. Petrucci
Big Blue Bump and Transient Active Regions in Seyfert Galaxies 1318
    S. Nayakshin and F. Melia
Magnetic Flares and the Observed $\tau_T \sim 1$ in Seyfert Galaxies 1323
    S. Nayakshin and F. Melia
Physical Constraints for the Active Regions in Seyfert Galaxies 1328
    S. Nayakshin and F. Melia
Are Gamma-ray Bursts related to Active Galactic Nuclei? 1333
    J. Gorosabel and A. J. Castro-Tirado

BLAZARS

Evidence for γ-Ray Flares in 3C 279 and PKS 1622-297 at ∼10 MeV 1341
    W. Collmar, V. Schönfelder, H. Bloemen, J. J. Blom, W. Hermsen,
    M. McConnell, J. G. Stacy, K. Bennett, and O. R. Williams

EGRET Observations of PKS 0528+134 from 1991 to 1997............. 1346
    R. Mukherjee, D. L. Bertsch, S. D. Bloom, B. L. Dingus,
    J. A. Esposito, R. C. Hartman, S. D. Hunter, G. Kanbach,
    D. A. Kniffen, A. Kraus, T. P. Krichbaum, Y. C. Lin, W. A. Mahoney,
    A. P. Marscher, H. A. Mayer-Hasselwander, P. F. Michelson,
    C. von Montigny, A. Mücke, P. L. Nolan, M. Pohl, O. Reimer,
    E. Schneid, P. Sreekumar, H. Teräsranta, D. J. Thompson, M. Tornikoski,
    E. Valtaoja, S. Wagner, and A. Witzel

Imaging Analysis of PKS0528+134 During Its Flare with A Direct Demodulation Technique... 1351
    S. Zhang, T. P. Li, M. Wu, and W. Yu

First Results of an All-Sky Search for MeV-Emission from Active Galaxies with COMPTEL... 1356
    J. G. Stacy, J. M. Ryan, W. Collmar, V. Schönfelder, H. Steinle,
    A. W. Strong, H. Bloemen, J. J. Blom, W. Hermsen, O. R. Williams,
    and M. Maisack

Variability Time Scales in the Gamma-ray Blazars Using Structure Function Analysis.. 1361
    G. Nandikotkur, P. Sreekumar, and D. A. Carter-Lewis

A Spectral Study of Gamma-ray Emitting AGN.......................... 1366
    M. Pohl, R. C. Hartman, P. Sreekumar, and B. B. Jones

EGRET Observations of PKS 2005-489................................. 1371
    Y. C. Lin, D. L. Bertsch, S. D. Bloom, B. L. Dingus, J. A. Esposito,
    S. D. Hunter, B. B. Jones, G. Kanbach, D. A. Kniffen,
    H. A. Mayer-Hasselwander, P. F. Michelson, C. von Montigny,
    R. Mukherjee, A. Mücke, P. L. Nolan, M. K. Pohl, O. L. Reimer,
    E. J. Schneid, P. Sreekumar, D. J. Thompson, and W. F. Tompkins

Whipple Observations of BL Lac Objects at E>300 GeV................ 1376
    M. Catanese, P. J. Boyle, J. H. Buckley, A. M. Burdett, J. Bussóns Gordo,
    D. A. Carter-Lewis, M. F. Cawley, D. J. Fegan, J. P. Finley, J. A. Gaidos,
    A. M. Hillas, F. Krennrich, R. C. Lamb, R. W. Lessard, C. Masterson,
    J. E. McEnery, G. Mohanty, J. Quinn, A. J. Rodgers, H. J. Rose,
    F. W. Samuelson, G. H. Sembroski, R. Srinivasan, T. C. Weekes,
    and J. Zweerink

Multiwavelength Observations of Markarian 421...................... 1381
    J. H. Buckley, P. Boyle, A. Burdett, J. Bussóns Gordo, D. A. Carter-Lewis,
    M. Catanese, M. F. Cawley, D. J. Fegan, J. P. Finley, J. A. Gaidos,
    A. M. Hillas, F. Krennrich, R. C. Lamb, R. W. Lessard, C. Masterson,
    J. McEnery, G. Mohanty, J. Quinn, A. Rodgers, H. J. Rose, F. Samuelson,
    G. H. Sembroski, R. Srinivasan, T. C. Weekes, and J. Zweerink

Observation of Strong Variability in the X-Ray Emission from Markarian 421 Correlated with the May 1996 TeV Flare............. 1386
    M. Schubnell

The Energy Spectrum of Mrk 421..................................... 1391
    F. Krennrich, for the Whipple Collaboration

Study of the Temporal and Spectral Characteristics of TeV Gamma Radiation from Mkn 501 During a State of High Activity by the HEGRA IACT Array 1397

F. Aharonian, A. Akhperjanian, J. Barrio, K. Bernlöhr, J. Beteta,
S. Bradbury, J. Contreras, J. Cortina, A. Daum, T. Deckers, E. Feigl,
J. Fernandez, V. Fonseca, A. Fraß, B. Funk, J. Gonzalez, V. Haustein,
G. Heinzelmann, M. Hemberger, G. Hermann, M. Heß, A. Heusler,
W. Hofmann, I. Holl, D. Horns, R. Kankanian, O. Kirstein, C. Köhler,
A. Konopelko, H. Kornmayer, D. Kranich, H. Krawczynski, H. Lampeitl,
A. Lindner, E. Lorenz, N. Magnussen, H. Meyer, R. Mirzoyan, H. Möller,
A. Moralejo, L. Padilla, M. Panter, D. Petry, R. Plaga, J. Prahl, C. Prosch,
G. Pühlhofer, G. Rauterberg, W. Rhode, R. Rivero, A. Röhring, V. Sahakian,
M. Samorski, J. Sanchez, D. Schmele, T. Schmidt, W. Stamm, M. Ulrich,
H. Völk, S. Westerhoff, B. Wiebel-Sooth, C. A. Wiedner, M. Willmer,
and H. Wirth (HEGRA collaboration)

Multiwavelength Observations of a Flare from Markarikan 501 1402

M. Catanese, S. M. Bradbury, A. C. Breslin, J. H. Buckley,
D. A. Carter-Lewis, M. F. Cawley, C. D. Dermer, D. J. Fegan,
J. P. Finley, J. A. Gaidos, A. M. Hillas, W. N. Johnson, F. Krennrich,
R. C. Lamb, R. W. Lessard, D. J. Macomb, J. E. McEnery, P. Moriarty,
J. Quinn, A. J. Rodgers, H. J. Rose, F. W. Samuelson, G. H. Sembroski,
R. Srinivasan, T. C. Weekes, and J. Zweerink

Recent Observations of γ-rays above 1. 5 TeV from Mkn 501 with the HEGRA 5 m Air Čerenkov Telescope 1407

D. Kranich, E. Lorenz, and D. Petry for the HEGRA Collaboration

BeppoSAX Monitoring of the BL Lac Mkn 501 1412

E. Pian, G. Vacanti, G. Tagliaferri, G. Ghisellini, L. Maraschi,
A. Treves, C. M. Urry, F. Fiore, P. Giommi, E. Palazzi, L. Chiappetti,
and R. M. Sambruna

Multiwavelength Observations of the February 1996 High-Energy Flare in the Blazar 3C 279 1417

A. E. Wehrle, E. Pian, C. M. Urry, L. Maraschi, G. Ghisellini,
R. C. Hartman, G. M. Madejski, F. Makino, A. P. Marscher,
I. M. McHardy, J. R. Webb, G. S. Aldering, M. F. Aller, H. D. Aller,
D. E. Backman, T. J. Balonek, P. Boltwood, J. Bonnell, J. Caplinger,
A. Celotti, W. Collmar, J. Dalton, A. Drucker, R. Falomo, C. E. Fichtel,
W. Freudling, W. K. Gear, N. Gonzalez-Perez, P. Hall, H. Inoue,
W. N. Johnson, M. R. Kidger, R. I. Kollgaard, Y. Kondo, J. Kurfess,
A. J. Lawson, B. McCollum, K. McNaron-Brown, D. Nair, S. Penton,
J. E. Pesce, M. Pohl, C. M. Raiteri, M. Renda, E. I. Robson,
R. M. Sambruna, A. F. Schirmer, C. Shrader, M. Sikora, A. Sillanpää,
P. S. Smith, J. A. Stevens, J. Stocke, L. O. Takalo, H. Teräsranta,
D. J. Thompson, R. Thompson, M. Tornikoski, G. Tosti, P. Turcotte,
A. Treves, S. C. Unwin, E. Valtaoja, M. Villata, S. J. Wagner, W. Xu,
and A. C. Zook

Radio to γ-Ray Observations of 3C 454.3: 1993–1995 1423
    M. F. Aller, A. P. Marscher, R. C. Hartman, H. D. Aller, M. C. Aller,
    T. J. Balonek, M. C. Begelman, M. Chiaberge, S. D. Clements,
    W. Collmar, G. De Francesco, W. K. Gear, M. Georganopoulos,
    G. Ghisellini, I. S. Glass, J. N. González-Pérez, P. Heinämäki,
    M. Herter, E. J. Hooper, P. A. Hughes, W. N. Johnson, S. Katajainen,
    M. R. Kidger, A. Kraus, L. Lanteri, G. F. Lawrence, G. G. Lichti,
    Y. C. Lin, G. M. Madejski, K. McNaron-Brown, E. M. Moore,
    R. Mukherjee, A. D. Nair, K. Nilsson, A. Peila, D. B. Pierkowski,
    M. Pohl, T. Pursimo, C. M. Raiteri, W. Reich, E. I. Robson, A. Sillanpää,
    M. Sikora, A. G. Smith, H. Steppe, J. Stevens, L. O. Takalo, H. Teräsranta,
    M. Tornikoski, E. Valtaoja, C. von Montigny, M. Villata, S. Wagner,
    R. Wichmann, and A. Witzel

Multi-Wavelength Radio Monitoring of EGRET Sources and Candidates. 1428
    P. G. Edwards, J. E. J. Lovell, R. C. Hartman, M. Tornikoski,
    M. Lainela, P. M. McCulloch, B. M. Gaensler, and R. W. Hunstead

VLBI Observations of Southern Hemisphere Gamma-Ray Loud and Quiet AGN. 1433
    S. J. Tingay, D. W. Murphy, P. G. Edwards, M. E. Costa,
    P. M. McCulloch, J. E. J. Lovell, D. L. Jauncey, J. E. Reynolds,
    A. K. Tzioumis, R. A. Preston, D. L. Meier, D. L. Jones,
    and G. D. Nicolson

VLBA Monitoring of Three Gamma-Ray Bright Blazars: AO 0235+164, 1633+382 (4C 38. 41) & 2230+114 (CTA 102) 1437
    W. Xu, A. E. Wehrle, and A. P. Marscher

Coordinated Millimeter-Wave Observations of Bright, Variable Gamma-ray Blazars with the Haystack Radio Telescope 1442
    J. G. Stacy, W. T. Vestrand, and R. B. Phillips

The Burst Activity of Millimeter Wavelengths Compared to Gamma-Activity of AGN 1447
    H. Teräsranta

Relationships Between Radio and Gamma-ray Properties in Active Galactic Nuclei. 1452
    A. Lähteenmäki, H. Teräsranta, K. Wiik, and E. Valtaoja

Fast Variations of Gamma-Ray Emission in Blazars 1457
    S. J. Wagner, C. von Montigny, and M. Herter

A $z = 2.1$ Quasar as the Optical Counterpart of the MeV Source GRO J1753+57 1462
    A. Carramiñana, V. Chavushyan, and J. Guichard

ASCA Observations of Blazars and Multiband Analysis 1467
    T. Takahashi, H. Kubo, G. Madejski, M. Tashiro, and F. Makino

Spectral Modelling of Gamma-ray Blazars 1473
    M. Böttcher, H. Mause, and R. Schlickeiser

Modelling the Rapid Variability of Blazar Emission 1478
    J. G. Kirk and A. Mastichiadis

OVERVIEWS, SURVEYS, AND MISCELLANEOUS

BeppoSAX Overview 1485
    L. Piro, on behalf of the BeppoSAX team
**Initial Results from the High Energy Experiment *PDS*
Aboard *BeppoSAX*** 1493
    F. Frontera, D. Dal Fiume, E. Costa, M. Feroci, M. Orlandini,
    L. Nicastro, E. Palazzi, G. Zavattini, and P. Giommi
**The CFA BATSE Image Search (CBIS) as Used for a Galactic
Plane Survey** 1498
    D. Barret, J. E. Grindlay, P. F. Bloser, G. P. Monnelly, B. A. Harmon,
    C. R. Robinson, and S. N. Zhang
**TeV Gamma Ray Emission from Southern Sky Objects
and CANGAROO Project** 1507
    T. Kifune, S. A. Dazeley, P. G. Edwards, T. Hara, Y. Hayami,
    S. Kamei, R. Kita, T. Konishi, A. Masaike, Y. Matsubara, Y. Matsuoka,
    Y. Mizumoto, M. Mori, H. Muraishi, Y. Muraki, T. Naito, K. Nishijima,
    S. Ogio, J. R. Patterson, M. D. Roberts, G. P. Rowell, T. Sako,
    K. Sakurazawa, R. Susukita, A. Suzuki, R. Suzuki, T. Tamura, T. Tanimori,
    G. J. Thornton, S. Yanagita, T. Yoshida, and T. Yoshikoshi
**Saturated Compton Scattering Models for the Soft Gamma-Ray
Repeater Bursts** 1512
    I. A. Smith, E. P. Liang, A. Crider, D. Lin, and M. Kusunose
The GRB 970111 Error Box 19-Hours After the High Energy Event 1516
    A. J. Castro-Tirado, J. Gorosabel, N. Masetti, C. Bartolini,
    A. Guarnieri, A. Piccioni, J. Heidt, T. Seitz, E. Thommes, C. Wolf,
    E. Costa, M. Feroci, F. Frontera, D. Dal Fiume, L. Nicastro, E. Palazzi,
    and N. Lund
**The Duration-Photon Energy Relation in Gamma-Ray Bursts
and its Interpretations** 1520
    D. Kazanas, L. G. Titarchuk, and X.-M. Hua

FUTURE MISSIONS AND INSTRUMENTATION

IBIS: The Imaging Gamma-Ray Telescope on Board INTEGRAL 1527
    P. Ubertini, on behalf of the IBIS Consortium
SPI: A High Resolution Imaging Spectrometer for INTEGRAL 1535
    B. J. Teegarden, J. Naya, H. Seifert, S. Sturner, G. Vedrenne,
    P. Mandrou, P. von Ballmoos, J.-P. Roques, P. Jean, F. Albernhe,
    V. Borrel, V. Schonfelder, G. G. Lichti, R. Diehl, R. Georgii, P. Durouchoux,
    B. Cordier, N. Diallo, J. Matteson, R. Lin, F. Sanchez, P. Caraveo,
    P. Leleux, G. K. Skinner, and P. Connell
**The Spectral Line Imaging Capabilities of the
SPI Germanium Spectrometer on INTEGRAL** 1544
    G. K. Skinner, P. H. Connell, J. Naya, H. Seifert, S. Sturner,
    B. J. Teegarden, and A. W. Strong

The IBIS View of the Galactic Centre: INTEGRAL's Imager
Observations Simulations 1549
    P. Goldoni, A. Goldwurm, P. Laurent, and F. Lebrun

Can the INTEGRAL-Spectrometer SPI detect γ-ray Lines
From Local Galaxies? 1554
    R. Georgii, R. Diehl, G. G. Lichti, and V. Schönfelder

Contribution of Passive Materials to the Background Lines
of the Spectrometer of *INTEGRAL* (SPI)................................ 1559
    N. Diallo, B. Cordier, M. Collin, and F. Albernhe

MGEANT—A Generic Multi-Purpose Monte-Carlo
Simulation Package for Gamma-Ray Experiments......................... 1567
    H. Seifert, J. E. Naya, S. J. Sturner, and B. J. Teegarden

A Small Scan Angle-Dependent Background Systematic
in Non-Standard OSSE Observations..................................... 1572
    J. D. Kurfess, K. McNaron-Brown, W. R. Purcell, R. L. Kinzer,
    and W. N. Johnson

A Time Dependent Model for the Activation of COMPTEL................. 1577
    M. Varendorff, U. Oberlack, G. Weidenspointner, R. Diehl, R. van Dijk,
    M. McConnell, and J. Ryan

Statistical Analysis of COMPTEL Maximum Likelihood-Ratio
Distributions: Evidence for a Signal from Previously Undetected AGN 1582
    O. R. Williams, K. Bennett, R. Much, V. Schönfelder, J. J. Blom,
    and J. Ryan

Improved COMPTEL 10-30 MeV Event Selections for
Point Sources from Inflight Data...................................... 1587
    W. Collmar, U. Wessolowski, V. Schönfelder, G. Weidenspointner,
    C. Kappadath, M. McConnell, and K. Bennett

Earth Occultation Technique with EGRET Calorimeter Data
Above 1 MeV 1592
    B. L. Dingus, D. L. Bertsch, and E. J. Schneid

Maximum-Entropy Analysis of EGRET Data 1596
    M. Pohl for the EGRET collaboration, and A. W. Strong

Non-Parametric Estimates of High Energy Gamma-ray Source
Distributions 1601
    D. D. Dixon, E. D. Kolaczyk, J. Samimi, and M. A. Saunders

Development of Gas Micro-Structure Detectors for
Gamma-Ray Astronomy 1606
    S. D. Hunter, S. V. Belolipetskiy, D. L. Bertsch, J. R. Catelli,
    H. Crawford, W. M. Daniels, P. Deines-Jones, J. A. Esposito, H. Fenker,
    B. Gossan, R. C. Hartman, J. B. Hutchins, J. F. Krizmanic, V. Lindenstruth,
    M. D. Martin, J. W. Mitchell, W. K. Pitts, J. H. Simrall, P. Sreekumar,
    R. E. Streitmatter, D. J. Thompson, G. Visser, and K. M. Walsh

The Design of a 17 m Air Cerenkov Telescope for VHE
Gamma Ray Astronomy above 20 GeV 1611
    E. Lorenz for the MAGIC Telescope Design Group

Monte Carlo Simulations of the Timing Structure of Cherenkov
Wavefronts of Sub-100 GeV Gamma Ray Air Showers...................... 1616
    D. R. Peaper, C. L. Gottbrath, M. P. Kertzman, and G. H. Sembroski

The University of Durham Mark 6 VHE Gamma Ray Telescope 1621
    P. M. Chadwick, M. R. Dickinson, N. A. Dipper, J. Holder,
    T. R. Kendall, T. J. L. McComb, K. J. Orford, S. M. Rayner,
    I. D. Roberts, S. E. Shaw, and K. E. Turver
Solar Tower Atmospheric Cherenkov Effect Experiment (STACEE)
for Ground Based Gamma Ray Astronomy 1626
    D. Bhattacharya, M. C. Chantell, P. Coppi, C. E. Covault, M. Dragovan,
    D. T. Gregorich, D. S. Hanna, R. Mukherjee, R. A. Ong, S. Oser,
    K. Ragan, O. T. Tümer, and D. A. Williams
On the Potential of the HEGRA IACT Array 1631
    F. A. Aharonian (HEGRA collaboration)
A Site for Čerenkov Astronomy in the White Mountains of California 1636
    J. R. Mattox and S. P. Ahlen
Simulation of HEAO 3 Background 1642
    B. L. Graham, B. F. Phlips, R. A. Kroeger, and J. D. Kurfess
Activation of Gamma Detectors by 1.2 GeV Protons 1647
    J. L. Ferrero, C. Roldán, I. Arocas, R. Blázquez, B. Cordier, J. P. Leray,
    F. Albernhe, and V. Borrel

Participant List 1653
Author Index 1665

Preface

Over 300 scientists met at *The Fourth Compton Symposium* in Williamsburg, Virginia on April 27-30, 1997 to discuss the latest developments in gamma-ray astronomy. This meeting was hosted by the Naval Research Laboratory and the Compton Gamma Ray Observatory Science Support Center, and is the fourth in a series of conferences devoted to *Compton Observatory* science. It followed *The Violent Universe Workshop*, held on April 25-27, 1997, which provided a forum to instruct and excite educators and the public about gamma-ray astronomy.

Six years after the launch of the *Gamma Ray Observatory* on April 5, 1991, new results and discoveries continue to pour forth. This document reviews the scientific achievements stemming from the *Compton Gamma Ray Observatory* and chronicles ongoing research in the field of gamma-ray astronomy as of mid-1997. Papers summarizing related developments from X-ray astronomy missions such as the *Rossi X-ray Timing Explorer, Beppo-SAX* and *ASCA*, from ground-based high-energy gamma-ray observatories such as *Whipple* and *HEGRA*, and from multiwavelength campaigns correlated with gamma-ray observations are also included.

The Scientific Organizing Committee for *The Fourth Compton Symposium* consisted of J. D. Kurfess (NRL, Chair), C. D. Dermer (NRL), C. E. Fichtel (GSFC), G. J. Fishman (MSFC), N. Gehrels (GSFC), I. A. Grenier (Saclay), J. E. Grindlay (CfA), K. C. Hurley (UCB SSL), C. Kouveliotou (USRA), R. C. Lamb (ISU and CIT), M. D. Leising (Clemson), P. F. Michelson (Stanford), R. E. Rothschild (UCSD), J. M. Ryan (UNH), V. Schönfelder (MPE), R. Sunyaev (IKI), P. von Ballmoos (CESR), and A. E. Wehrle (IPAC). The Local Organizing Committee consisted of M. Strickman (NRL, Co-Chair), C. R. Shrader (GROSSC, Co-Chair), S. Barnes (GROSSC), J. E. Grove (NRL), R. C. Hartman (GSFC), W. N. Johnson (NRL), J. P. Norris (GSFC), T. Obrebski (NRL), and E. Pentecost (USRA). We would like to thank everyone for their hard work and planning which contributed to the success of this meeting. Special thanks go to Tina Obrebski and Liz Pentecost for their excellent administrative support, and to Mali Friedman for help with the proceedings. We gratefully acknowledge TRW, Inc., Ball Aerospace and Technologies Corporation, and Universities Space Research Association for generously supporting this symposium.

The growth in the size of the gamma-ray astronomy community compelled us, as in the past two symposia, to exclude gamma-ray burst and solar contributions. Even so, we received more than 330 abstracts, which exceeded our most optimistic expectations. Our wish to avoid parallel sessions unfortunately meant that time for oral contributions and poster viewing was severely limited. Judging from the size of the proceedings, the lesson here seems to be that gamma-ray astronomy is too vast a subject and three days is too short a time to survey adequately the richness and variety of the high-energy universe.

The success of the *Fourth Compton Symposium* left little doubt about the necessity for a *Fifth Compton Gamma Ray Symposium*, and attendees at the symposium banquet enthusiastically endorsed another such meeting. The banquet was held on a beautiful spring evening at the Sherwood Forest Plantation, the residence of John Tyler after he left office as the tenth president of the United States. It also provided an opportunty for the *Compton* commmunity to pay well-deserved tribute to Frank McDonald and Don Kniffen, who were so instrumental in the genesis, development, and success of the *Compton Gamma Ray Observatory*.

Charles D. Dermer
Mark S. Strickman
James D. Kurfess

Prologue

Revolution is an overused word, but nothing short can describe the impact that the *Compton Gamma Ray Observatory* has had on the field of gamma-ray astronomy. It has been such a short time, relatively speaking, that this subject went from the speculative imaginings of Philip Morrison [1] to a discipline too great for any single individual to master. As a result of the *Compton Observatory*, gamma-ray astronomy has completed a process of "subfield-specialization" [2]: for example, traditional subfields such as gamma-ray burst astronomy and high-energy solar astronomy are now large enough to support separate communities; TeV astronomy has been energized by the *Compton Observatory* results; blazar physics has been reinvented; and gamma-ray astronomy plays an ever larger role in cosmic-ray studies.

The discoveries of the *Compton Observatory* have been hardly less important to the mainstream of astronomical research than to the field of high-energy astrophysics, and no well-versed research astronomer can afford to be ignorant of the knowledge that this mission has provided about the universe. These proceedings provide a record of gamma-ray astronomy six years after the launch of the *Compton Observatory*. The papers presented at the *Fourth Compton Symposium* are assembled in the second volume of the proceedings, entitled "Papers and Presentations." In addition to this volume, it was agreed by the Scientific Organizing Committee to organize a review of the scientific achievements of the *Compton* mission, authored by the central figures in the various topics that comprise gamma-ray astronomy.

The outcome of this endeavor is Volume One of the proceedings, entitled "The Compton Observatory in Review." The goal here is to provide a pithy summary of the new science provided specifically by the Compton Observatory, either alone or in consortium with other ground-based or space-based observatories. The motivation for producing such a volume arises from the frustrating experience of trying to derive an overall understanding of the results of past missions in the absence of an astronomical *Baedeker*. Volume One is therefore intended less for gamma-ray astronomers, who know where to find the primary literature sources, than for astronomers from other disciplines who would like a technically accurate and relatively concise summary of the *Compton* results. Try as we might, no single volume could still hope to be complete. Topics such as the remarkable gamma-ray flashes from terrestrial thunderstorms [3] and the elusive and poorly understood Cygnus X-3 [4] are not covered here.

The need for a review volume was also brought home by a number of less felicitous circumstances, including the downward trajectory of the *Compton Observatory* budget and, most worrisome of all, the departure of many of the mission scientists. In a large number of cases, scientists are moving from one project to another, as for European scientists moving to *INTEGRAL* and US scientists to *GLAST*. That is all well and good, and it is hoped that we are using this opportunity to distill their expertise before they set their minds on another challenge. Far more distressing is the departure of younger scientists who see, oftentimes more clearly than their tenured seniors, that the prospect of a US gamma-ray astronomy mission in the 100 keV - 30 MeV range is quite bleak. This is a simple matter of physics versus policy. It takes a space-based telescope of substantial area and cross section to stop a gamma ray, which means a massive and expensive payload, contrary to the central tenet of the "faster, cheaper, better" philosophy.

Yet in spite of this management policy, the *Compton Gamma Ray Observatory* and the *Hubble Space Telescope*, each hardly cheap or fast, have consistently scored high on the Senior Review science-per-dollar assessment. We should therefore take this opportunity to celebrate the great discoveries and new science emerging from the *Compton Gamma Ray Observatory*. The success of the *Compton* mission attests first and foremost to its outstanding design, to the skill of the technicians who built it, and to the ingenuity and hard work of the *Compton* scientists. Its success furthermore demonstrates the variety and unpredictability of the gamma-ray universe, the value of multiwavelength campaigns to monitor sources over the entire electromagnetic spectrum, and the need for long observing periods to detect weak gamma-ray fluxes.

Next year's great discovery may well be travelling at the speed of light toward Earth if we are only there to see it.

¡*Viva Compton!*

Charles D. Dermer
September, 1997

[1] Morrison, P., *Il Nuovo Cimento* **7**, 858 (1958).
[2] Ziman, J., *Knowing everthing about nothing: Specialization and change in scientific careers* (New York: Cambridge University Press) (1987).
[3] Fishman, G. J., *et al.*, *Science* **264**, 1313 (1994).
[4] Matz, S. M., these proceedings.

COMPTON STATUS AND FUTURE

Status and Future of the Compton Gamma Ray Observatory

Neil Gehrels and Chris Shrader

Laboratory for High-Energy Astrophysics, NASA Goddard Space Flight Center

Abstract. The Compton Gamma Ray Observatory and its four experiment packages continue to function in a nearly flawless manner now well into the sixth year of mission operations. We discuss the status of the mission as of mid-1997, as well as the prospects of an extended mission lasting into the next century.

INTRODUCTION

The 15900 kilogram Compton Gamma Ray Observatory (herein *Compton*), which was launched on the Space Shuttle Atlantis STS-37 on 5 April 1991, is the second in the fleet of NASA's Great Astronomical Observatories. Mission goals include performing broad-band gamma-ray observations with better angular resolution and an order of magnitude better sensitivity than previous missions, performing the first gamma-ray full-sky survey (completed on 17 November 1992), and compiling a database of gamma-ray burst measurements unprecedented in size and scope. The scientific theme of the mission is the study of physical processes taking place in the most dynamic sites in the Universe, including supernovae, novae, pulsars, black holes, active galaxies, and gamma-ray bursts. It additionally provides a capability to study the high-energy properties of solar flares. It is currently used by an international Guest Investigator community with observing time allocated on the basis of peer-reviewed proposals in annual periods or "Cycles" (during the first three years of the mission a certain amount of guaranteed time was preallocated for the four Instrument Development teams). A total mission lifetime of 10 years or more is currently anticipated. The four scientific instrument packages, described briefly below, combine to cover the hard x-ray and gamma-ray energy regimes from 15 keV to 30 GeV [1,2,14].

THE *COMPTON* INSTRUMENT PACKAGES

The Burst and Transient Source Experiment (BATSE) is optimized to measure brightness variations in gamma-ray bursts and solar flares on time scales down to microseconds over the energy range 30 keV to 1.9 MeV [3,4]. To accomplish this, 8 large-area NaI detectors oriented towards the 8 octants of the sky are used. A smaller spectroscopy detector, optimized for broad energy coverage (15 keV to 110 MeV) and enhanced energy resolution is associated with each of the 8 large-area detectors. BATSE also continuously monitors all transient sources and bright persistent sources in the gamma-ray sky.

The Oriented Scintillation Spectroscopy Experiment (OSSE) is designed to undertake comprehensive spectral observations of astrophysical sources in the 0.05 to 10 MeV range, with capability above 10 MeV for solar gamma-ray and neutron observations [5]. The detectors are NaI/CsI scintillators with a field-of-view determined by a passive tungsten collimator. Each of the 4 detectors has a single axis pointing system which enables a rapid OSSE response to targets of opportunity, such as transient X-ray sources, explosive objects, and solar flares.

The Imaging Compton Telescope (COMPTEL) detects gamma rays by the occurrence of two successive interactions: the first one is a Compton scattering collision in a detector of low-Z material (liquid scintillator) followed by a second interaction in a detector with high-Z material (NaI), in which, ideally, the scattered gamma ray is totally absorbed [6]. Source mapping is provided over a field of view of about 1 steradian. COMPTEL has performed the first sky survey in the energy range from 1 to 30 MeV.

The Energetic Gamma Ray Experiment Telescope (EGRET) is the highest energy instrument on *Compton* and covers the broadest energy range, from $\sim$ 20 MeV to $\sim$ 30 GeV [7]. It has a wide field of view ($\sim$ 0.5 sr), good angular resolution and very low background. Because it is designed for high-energy studies, the detector is optimized to detect gamma rays when they interact by the dominant high-energy pair-production process which forms an electron and a positron within the EGRET spark chamber.

PROSPECTS FOR AN EXTENDED *COMPTON* MISSION

As of the time of this submission, the prospects for an extended *Compton* mission look very promising. On technical grounds, the spacecraft is functioning in a nearly flawless manner and there is widespread optimism that this will continue to be the case for the foreseeable future. The spacecraft has recently been reboosted to an orbit with an altitude of 515 km. This should allow for an extension of the mission well into the next century – to 2005 or even 2008.

An earlier problem with the onboard tape recorders has been largely circumvented by the implementation of enhanced real-time telemetry coverage, usage of onboard memory buffers, and use of the low-gain omni antenna. The failure of one of the spacecraft batteries led to concerns that this problem might be repeated in the remaining 5 batteries – but this now does not seem to be the case, as the remaining 5 batteries are aging normally. In any case, spacecraft engineers have devised plausible scenarios, involving for example the cycling of spacecraft thermal-control subsystems, in which a 10-year mission can be completed even in the event of two additional battery failures.

The EGRET instrument uses an expendable neon/argon spark-chamber gas, of which the spacecraft initially carried 5 refills. The gas lifetime is directly related to the spark rate, and thus the number of photons detected. During the current observing cycle, EGRET is generally being configured to an operating mode in which its field of view is reduced by about a factor of 4, leading to a nominal two-fold increase in gas lifetime. In certain instances where there may be targets within the Z-axis field of view that are of interest for study with COMPTEL, but not with EGRET, the spark chamber is being turned off, again leading to significant gas savings. As of this submission, there is approximately 1–2 years of gas use remaining with EGRET used in the mode described above. Small amounts of gas will be reserved for special targets of opportunity or other milestone observations.

GUEST INVESTIGATOR PROGRAM

In all areas of *Compton* research described in these proceedings, Guest Investigators (GIs) have contributed significantly to the data analysis, correlative observations, or theoretical interpretations. A total of over 750 scientists from 128 institutions in 23 countries have been involved since launch. All observing time on *Compton* is awarded competitively through scientific peer review held during successive annual cycles. Cycle 7 of the mission begins in November, 1997.

Some recent policy changes will be implemented for Cycle 7, based on recommendations made by the primary advisory body for the mission, the *Compton* GRO User's Committee. The proprietary data rights period has been reduced from one year to six months. Unproposed target-of-opportunity observations and serendipitous discoveries made during routine processing will result in a prompt release of the data products.

GIs are provided extensive assistance by the instrument teams and by the Science Support Center (SSC) at NASA/Goddard Space Flight Center.
The Guest Investigator program is a strong and essential component of the *Compton* mission. For more information on the *Compton* Guest Investigator Program, the reader should consult the SSC homepage on the World Wide Web at the URL http://cossc.gsfc.nasa.gov.

Compton Public Science Data Archive

All data from *Compton* are available through the *Compton* Observatory Archive. This archive is managed by the SSC and is open to astronomers for use in archival research, proposal preparation or for any other scientific purposes. Data are placed in the archive when delivered to the SSC by the instrument teams following the expiration of the proprietary period. The data are easily accessible via computer networks through an online service provided by the SSC. Magnetic tape distribution is also available to accommodate requests for extremely large data sets. Usage of the archive by the community has been extensive, with approximately 75 individual download operations per month. In addition to the on-line data archive, a CD ROM containing the Phase-1 EGRET data and a BATSE CD ROM containing gamma-ray burst and solar flare data have been distributed widely within the USA and Europe. An updated BATSE CD ROM containing the most current catalog of gamma-ray bursts and a set of the relevant data products and software, is being prepared for release in the August 1997 timeframe, and a revised EGRET CD with refined calibration files and platform-independent "JAVA" based quick-look analysis tools is under preparation. Refer to the SSC World Wide Web home page (URL given above) for additional information on the availability of *Compton* data products.

Recent Science Highlights

We will not attempt to present a comprehensive summary of the scientific highlights of the mission – that is the purpose of this proceedings volume, and the quality and quantity of its contents attest to the continuing success of the mission. A summary of accomplishments and remaining goals is presented in Table 1. Here we will make brief mention of a few particularly exciting results and refer the reader to more detailed discussions within these volumes.

One of the remarkable results announced at this meeting is the detection and characterization of an extended "spur" of 511-keV gamma–ray emission in the Galactic Center region [8]. Maps (see Fig. 1) of this emission show a significant asymmetric enhancement which appears to extend up to ~ 10 degrees above the Galactic Center region. While the exact morphology of the enhancement is not yet known, several possible production and ejection mechanisms have been suggested. These include enhanced supernova activity in the galactic center region during the past 10^5-10^6 years, pair jets originating from an accreting black hole, or a gamma ray burst-like event in the galactic center region. The enhancement may not originate near the galactic center region, however, but rather represent a site of more local origin. The 511-keV mapping results are also important because they demonstrate a new and useful mapping capability for the OSSE instrument.

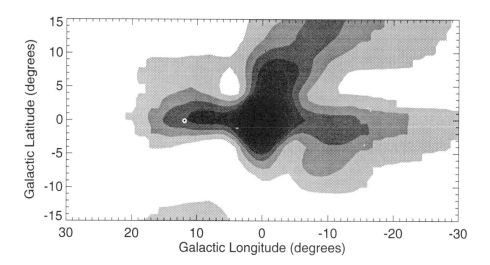

FIGURE 1. One of the highlights of this symposium was the presentation of this OSSE 511-keV map of the Galactic Center region [8]. The bright excess to the upper left of the bright central region is apparently associated with a cloud of positrons extending $\sim$ 1-2 kpc above the plane of the Galaxy. The origin of this cloud is unknown. Possibilities include a period when an enhanced rate of supernova explosions occurred in a concentrated region near the Galactic Center, or positron production through the heating of matter which accretes onto a black hole.

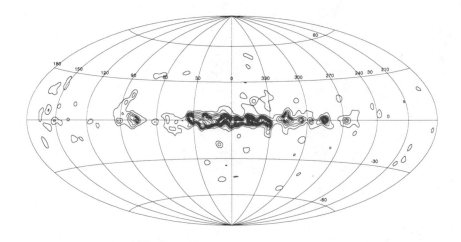

FIGURE 2. The COMPTEL team presented a map of the Milky Way in the light of the 1.809 MeV line from the nuclear decay of ^{26}Al, based on anlysis of 5 years of observation [11]. This emission traces sites of recent nucleosynthesis in the Galaxy.

A totally unexpected announcement, but one that must undergo further scrutiny, is the evidence for a residual signal in the timing solution for pulsed emission from the Geminga pulsar [9,10]. This small residual is consistent with what would be expected if a $1.7 M_\odot / \sin i$ mass planet were orbiting the pulsar at a distance of 3.3 AU. As such, it would represent the remarkable discovery of an extra-solar planet by gamma-ray astronomy alone! Caution must be advised however, as the result is only an exciting hint so far, and further data are needed for confirmation.

A map of the Galactic diffuse 1.809 MeV ^{26}Al line emission, based on the co-addition of over five years of data has been produced by the COMPTEL team, along with improvements in the understanding of instrumental backgrounds. The wide longitudinal extent and clumpiness of the emission is now seen much clearer, allowing one to discriminate between different source distribution models [11]. It was found that models of the galactic molecular gas distribution and sharply peaked nova distributions provide a relatively poor match to the data. Spiral structure in the ^{26}Al source distribution is now revealed.

A series of major announcements on gamma-ray burst counterparts at X-ray, optical and radio wavelengths were made at this conference [12], and some additional results have been announced subsequently [13]. Although these are not *Compton* results per se, they clearly represent a major breakthrough in gamma-ray burst research, for which in many regards *Compton* remains the premier facility. The database of over 1800 gamma-ray bursts (and climbing)

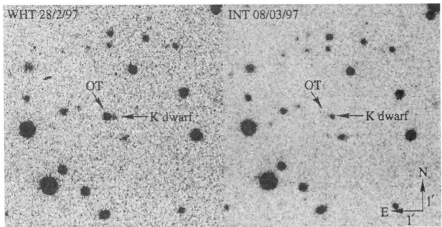

FIGURE 3. Although gamma-ray bursts were not intended to be a major topic of discussion at this meeting, the recent discovery of X-ray and optical counterparts prompted a special session. Discussion was lively. This figure depicts the optical counterpart to GRB 970228 about 1 and 9 days after the burst was recorded by the BeppoSAX satellite, as reviewed [12] in this volume.

will be central to any attempt to discern the underlying physics. The breakthroughs are credited largely to the Italian/Dutch satellite Beppo Sax. With its multiple capabilities for both wide-field monitoring and X-ray imaging, this observatory made direct detection of X-ray "afterglow" events associated with several bursts. Furthermore, the positional accuracy obtained was sufficient to allow for rapid followup studies including direct CCD imaging and radio mapping. The results thus far, notably two X-ray plus optical counterparts, seem to argue strongly for an extragalactic origin. In one case a "host" nebulosity with a spatial extent and surface brightness consistent with being a galaxy at $z \simeq 1$ seems to be present in deep images obtained with the HST and Keck telescopes. In the second case, an absorption line system identified as Fe II and Mg II lines commonly seen in our galaxy, but redshifted to $z \gtrsim 0.8$ has been reported.

Future Science

There are a number of exciting future studies that may be accomplished during the remainder of the mission. For example, continued study of the galactic center region, in particular obtaining improved maps of the newly discovered extended emission region, will be a major objective of OSSE. Deep observations at high galactic latitudes could detect faint quasars allowing a study of their spatial distribution, and thus constrain their contributions to the diffuse background. Contemporaneous operation with the Rossi X-Ray Timing Explorer, recently launched in December of 1995, will allow for unprecedented broad-band high-energy study of AGN and galactic binaries. With the sensitivity ultimately achievable by OSSE and COMPTEL over several years of surveying the Galactic plane, both diffuse ^{60}Fe and multiple ^{44}Ti supernova remnants should be detectable. These in principle yield the supernova rates over the past few million years and past few centuries, respectively. Despite the recent remarkable discoveries of gamma-ray burst X-ray, optical and radio counterparts, the phenomena still remains an enigma and the underlying physics is still far from understood. The database that is continuing to be compiled by BATSE will represent the most comprehensive set of information on the gamma-ray burst phenomena for the foreseeable future. Furthermore, a plan utilizing BATSE and RXTE for rapid detection of gamma-ray burst counterparts is currently being implemented, and has the potential for greatly expanding the available database of counterpart candidates. Also, an extended mission will provide unprecedented high-energy coverage of a complete solar cycle. Most significantly, other totally unanticipated discoveries may await us as well.

REFERENCES

1. Gehrels, N., Fichtel, C.E., Fishman, G.J., Kurfess, J.D., and Schönfelder, V., 1993, *Sci. Am.* **269**, 68.
2. Shrader, C.R., and Gehrels, N., 1995, *PASP*, **107**, 606.
3. Fishman, G.J., et al., 1992, in "The Compton Observatory Science Workshop," ed. C.R. Shrader, N. Gehrels, and B. Dennis, NASA CP-3137.
4. Band, D.L., et al., 1992, *Exp. Astr.*, **2**, 307.
5. Johnson, W.N., et al., 1993, *ApJS*, **97**, 21.
6. Schönfelder, V., et al., 1993, *ApJS*, **86**, 657.
7. Thompson, D.J., et al., 1993, *ApJS*, **86**, 269.
8. Smith, D. M., Purcell, W. R., and Leventhal, M., 1997, these proceedings.
9. Mattox, J.R., 1997, these proceedings.
10. Mattox, J.R., Halpern, J.P., and Caraveo, P., 1997, *ApJ*, in press.
11. Diehl, R., et al., 1997, these proceedings.
12. Meegan, C. et al., 1997, these proceedings.
13. Metzger, M.R., et al., 1997, *Nature*, **387**, 878.
14. Kurfess, J., Bertsch, D.L, Fishman., G.J., and Schönfelder, V., 1997, these proceedings.

Table 1. Summary of Objectives, Accomplishments, and Remaining Goals of the *Compton* Observatory

Source Type	Original Scientific Objectives
Gamma-Ray Bursts	Observations of gamma-ray bursts, their luminosity distribution, their spectral and temporal characteristics, and their spatial distribution.
Discrete Objects	A study of discrete objects such as black holes, neutron stars and objects emitting only at gamma-ray energies.
Nucleosynthesis	A search for sites of nucleosynthesis – the fundamental process for building the heavy elements in nature – and other gamma-ray emitting lines in astrophysical processes.
Diffuse Lines	Map the distribution of diffuse 0.511 MeV and ^{26}Al gamma-ray line emission and determine its origin.
Galactic Emission	Exploration of the Galaxy in gamma rays in order to study the origin and dynamic pressure effects of the cosmic-ray gas and the structural features revealed though the interaction of the cosmic rays with the interstellar medium.
Extragalactic	A study of the nature of other galaxies as seen at gamma-ray wavelengths, with special emphasis on radio galaxies, Seyfert galaxies and QSOs.
Cosmology	A search for cosmological effects through observations of the diffuse gamma radiation, and for possible primordial black hole emission.

Source Type	Accomplishments to Date
Gamma-Ray Bursts	1. Discovery that GRBs are isotropic and non-homogeneous 2. Discovery of high energy (> 1 GeV) gamma rays. 3. Discovery of delayed (hrs) GeV emission. 4. Evidence for time dilation and intensity-dependent spectral properties. 5. Establishment of multi-wavelength rapid counterpart-search networks.
Discrete Objects	1. Monitor 25 X-ray pulsars; orbital and spin-up/spin-down rates. 2. Discovered bursting X-ray pulsar. 3. QPOs and 110 keV line from A0535+26. 4. Discovery of 6 X-ray novae; first characterization of their gamma-ray properties; empirical link between disk and jets. 5. Discovery of 5 new rotation-powered gamma-ray pulsars. 6. Thermal and non-thermal classes of BH candidates.
Nucleosynthesis	1. ^{57}Co emission from SN 1987a. 2. Possible ^{44}Ti emission from Cas A. 3. Possible ^{56}Co emission from SN 1991t. 4. Limits on ^{22}Na emission from ONeMg novae. 5. Possible ^{26}Al from Vela SNR.
Diffuse Lines	1. OSSE mapping of spatially extended 511-keV emission in Galactic center region. 2. COMPTEL map of galactic ^{26}Al. 3. Establishment of a sensitive monitoring capability for transient line phenomena.
Galactic Emission	1. Detailed maps and improved spectra of high energy galactic diffuse gamma radiation, leading to dramatically improved physical models. 2. Low-energy continuum spectrum of galactic gamma radiation. 3. Detection of diffuse high-energy gamma rays from the LMC.
Extragalactic	1. Discovery of gamma ray blazars. 2. 100-keV spectral turnover of Seyfert galaxies. 3. Discovery of intra-day variability in both blazars and Seyferts. 4. Discovery of heavily-obscured Seyfert 2, NGC 4945.
Cosmology	1. Limits on the rate of Primordial BH events.

Source Type	Remaining Goals
Gamma Ray Bursts	1. Determine the distance scale for GRBs. 2. Obtain precise positions through BACODINE. 3. Determine nature of gamma ray emission. 4. Expanded searches for lensed events; begin to trace the large-scale distribution of matter (if GRBs are extragalactic).
Discrete Objects	1. Provide high-energy monitor support. 2. Support multi-wavelength campaigns; first simultaneous hard/soft monitoring of X-ray novae (with CGRO and RXTE). 3. Further understanding of jet-producing phenomena in BH accretion disks. 4. Expanded database of accretion-pulsar spin histories. 5. Additional unforeseen phenomena such as the Bursting Pulsar.
Nucleosynthesis	1. ^{56}Ni emission from Type 1a SN within 10 Mpc. 2. ^{22}Na emission from one or more novae within 2 kpc. 3. Line emissions from nearby Type 2 supernovae. 4. Determine nova rate from ^{44}Ti map. 5. Confirm ^{44}Ti emission from Cas A.
Diffuse Lines	1. Refine 0.511 MeV maps of the inner Galaxy. 2. Improve ^{26}Al maps with additional exposure. 3. Search for ^{60}Fe emission from inner Galaxy. 4. Investigate recent results from balloon experiments on new sources of 0.511 MeV emission and broadened ^{26}Al profiles.
Galactic Emission	1. Map 50-500 keV continuum flux in galactic plane. 2. Improved exposure of the Orion region to better constrain the nuclear interaction lines C and O; further test the line splitting model to explain the apparent broadening.
Extragalactic	1. Contribution of AGN to diffuse background. 2. Conduct extensive multi-wavelength campaigns to understand the nature of central engines in AGN. 3. ToOs of significant flares from previously γ-ray quiet AGN.
Cosmology	1. Continue to develop improved models for the diffuse extragalactic gamma-ray background and determine possible implications on large-scale structure.

SOLAR AND STELLAR GAMMA RAY ASTRONOMY

Solar and Stellar Gamma Ray Observations with COMPTON

Gerald H. Share and Ronald J. Murphy

E.O. Hulburt Center for Space Research, Code 7652, Naval Research Lab., Washington, D.C. 20375

James Ryan

University of New Hampshire, Durham, New Hampshire 03824

Abstract. Observations by the Compton GRO are providing fundamental insights into the high-energy physics of solar flares. Most of the analyses have been made of the flares emitted by active region number 6659 in 1991 June. Hard X-ray pulses in flares show energy dependent delays of up to $\sim$ 200 ms due to time of flight, consistent with the location of an acceleration region above the soft X-ray loop and an impact zone in the footpoints. Comparison of electron bremsstrahlung and 80 GHz synchroton emissions provides information on magnetic field strengths in the corona. Temporal variations as short as 0.1 s are observed at energies above 10 MeV constraining acceleration models. Protons are observed to be accelerated up to energies in excess of a few GeV and relativistic ions are still present in sufficient intensities up to eight hours after the impulsive phase of the 11 June flare to be detected in the γ rays they produce at energies above 50 MeV. Comparison of the energy dependence of time profiles suggests that particle acceleration is continuing hours after the impulsive phase and that trapped particles do not provide the dominant emission. Individual γ-ray lines are resolved and provide information on the accelerated ion spectra and the ambient elemental composition. The 1991 June 4 flare produced the highest line fluence observed to date and had an ambient abundance close to that observed in the corona. A rather high accelerated α/p ratio, $\sim$ 0.5, is inferred from the line measurements. The energy deposited in ions >1 MeV/AMU is at least an order of magnitude larger than that of electrons in this event. Temporal variation is observed in the accelerated electron/ion ratio over a broad energy range, in the ambient abundances, and in the spectral hardness in the June flares. Comparisons of the neutron capture and de-excitation lines provide measurements of the ^{3}He/H ratio, and information on the ^{4}He/H ratio in the photosphere. We also report on a search for hard X-rays from selected stars.

STELLAR FLARES

To provide a transition from the celestial emphasis of this meeting, we begin our review by discussing a search for hard X-ray emission from active stars [1]. A positive detection of hard X-rays from nearby stars would be significant because they are on the order of 10^7 times more distant than the Sun. Thus any measurable flux from a star 10 pc away corresponds to an enormous hard X-ray flux if the star were at 1 AU. Since solar and stellar flares are inherently plasma phenomena, it would illustrate the ability of plasma processes to accelerate an enormous number of energetic particles. Conversely, it would also represent an previously unknown capacity of stars to produce transient hard X-ray and possibly gamma-ray fluxes that, in principle, could be a source of cosmic gamma-ray bursts.

The searches are being conducted with the earth occultation techniques applied to BATSE data >20 keV. The preliminary work included 12 stars and a comparable number of background fields. A list of the fields observed and the excesses observed are given in White et al. in these proceedings [1]. The stars include flares stars, RS CVn and Algol binaries. Two of objects exhibited what appears to be significant hard X-ray emission on one day time scales. These two are AB Dor and AD Leo. The work is continuing with the objective of confirming that these events are truly from flares on these active stars.

These results heighten the importance of solar measurements, since solar flares may now allow us to study what may be small versions of truly cosmic phenomena. In other words, the physics of the Sun may be more applicable to the field of high-energy astrophysics than has been widely assumed. With these measurements as a backdrop we can now review in a new light the solar measurements performed with the CGRO and other spacecraft.

TEMPORAL STUDIES OF ELECTRON BREMSSTRAHLUNG

Location of Acceleration Region and Magnetic Trapping

Electron acceleration is a difficult phenomenon to study. Although energetic electrons are clearly present in all flares at some level, their origin remains controversial. The spectrum of the bremsstrahlung from these electrons is rather featureless, ranging from thermal-like exponential spectra to power-laws. Greater structure is found in the temporal nature of the X-ray flux. Transient spikes can be rapid and numerous and it is quite likely that the

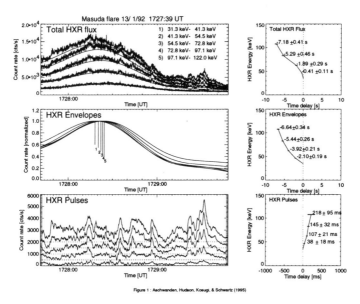

FIGURE 1. Time history of 1992 Jan. 13 flare separated into envelope and pulses and showing delays with energy. Reprinted courtesy of the Astrophysical Journal published by the Univ. of Chicago Press (c) 1996 (Amer. Ast. Soc.)

transport of the electrons is related to the processes that accelerated them. Therefore, studies of temporal structure in X-ray bursts can provide information about the prevailing conditions that govern the behavior of the electron transport and their acceleration.

Aschwanden et al. [2] have used high time resolution (64 ms) data from 42 flares observed by BATSE to separate individual pulse structures from the envelope of hard X-ray emission. A clear pattern can be discerned in these intensity-time profiles: higher energy pulses arrive earlier than lower energy pulses and broad envelopes decay faster at lower energies. These patterns are evident in Figure 1 [3]. The time delay shown in the abscissa is the time difference between a feature measured at 30 keV and a feature at energy E. Thus, the negative 6.64 s 'delay' in the envelope at 115 keV suggests that the lower energy photons are lost from the system sooner. Aschwanden et al. interpret this early loss as being due to trapping of electrons in magnetic loops on the order of about 10 s, where the lower energies scatter out of the loop at their mirror points earlier than do the higher energy electrons.

In contrast, the earlier peaking of high-energy pulses shown in the lower panel of Figure 1 is interpreted simply as being due to time of flight differences for electrons moving along the flare loop from the acceleration region to the foot points of the loop. This is depicted in Figure 2 which shows the *YOHKOH*

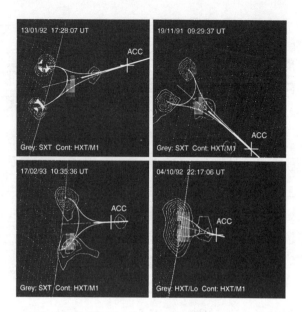

FIGURE 2. Four limb flares observed by Yohkoh depicting the soft X-Ray (grey scale) and hard X-ray (contours) and location of the acceleration region derived from the timing analysis.

soft and hard X-ray images of four limb flares. The hard X-ray emission comes from the two footpoints, and a source above the soft X-ray loop. A time-of-flight analysis of the hard X-ray data indicates an acceleration region that also sits above the flaring soft X-ray loop and perhaps just above the hard X-ray source. This geometry is representative of the larger sample of 42 events studied by Aschwanden et al. [2].

Electron Bremsstrahlung Temporal Studies at Higher Energy

The relationship between hard X-ray bremsstrahlung and microwave emission has long been controversial. The central question is whether the electrons responsible for the bremsstrahlung are the same as those responsible for the microwave emission. Here again, electron transport may play an important role since if the same electron population produces both forms of emission, the transport processes must allow those electrons to populate both emitting regions. The transport phenomenon reveals itself in the temporal behavior of both forms of emission. In Figure 3 we show a comparison of time profiles of the 1991 June 4 flare taken from Murphy et al. [4]. Separated electron bremsstrahlung rates above 1 MeV from OSSE are compared with renormal-

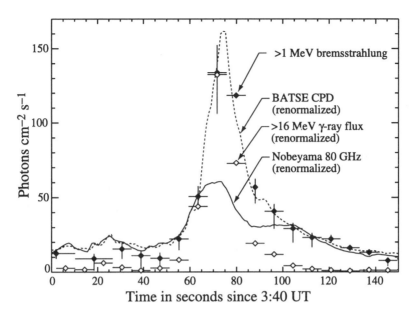

FIGURE 3. Time history near the peak of the 1991 June 4 flare. Reprinted courtesy of the Astrophysical Journal published by the Univ. of Chicago Press (c) 1997 (Amer. Ast. Soc.).

ized rates above 1 MeV from the BATSE charged particle detectors studied by Ramaty et al. [5], rates above 16 MeV from OSSE, and the renormalized 80 GHz radio emission from Nobeyama [6]. There is a lag in the peaking of the bremsstrahlung emission relative to peaking in the microwave. This early microwave emission is interpreted as being due to gyrosynchrotron emission of electrons trapped in the corona where the matter density is too low for significant bremsstrahlung. Using a trapping model developed by Ramaty et al. [5] coronal magnetic fields of a few hundred G are derived by comparing the ratios of the radio and bremsstrahlung emissions. This trapping model may be in conflict with the high energy data obtained by OSSE [4], however. The faster decay of the >16 MeV bremsstrahlung on the same 10 s time scale is opposite to what one would expect if trapping occurs in the corona.

Even faster temporal variations have been observed using the OSSE high-energy data for the 1991 June 4 flare. Plotted in Figure 4 is the early time history of the flare at a resolution of 16 ms [7]. There is clear evidence for temporal variations as short as a few tenths of seconds even at energies above 10 MeV. Spectral measurements suggest that this emission is from electrons and not from pions produced by high-energy protons [4]. This fast time structure is consistent with solar measurements made using the *GAMMA-1* experiment [8].

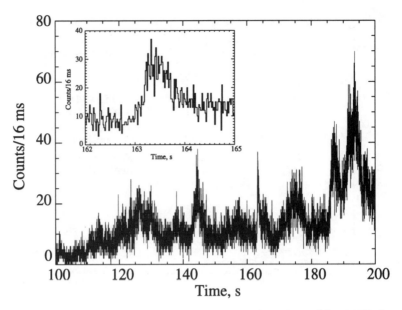

FIGURE 4. High time resolution plot of 1991 June 4 flare observed by OSSE above 10 MeV.

FLARE PARAMETERS DERIVED FROM NUCLEAR LINE SPECTROSCOPY

Broad Band Spectra

The complete coverage of a solar flare hard X-ray/gamma-ray spectrum is difficult to obtain. Different observing schedules and variable instrument modes conspire to make broad-band measurements of solar flares rare. However, this has been accomplished for several flares observed during the last solar maximum.

Broad-band spectra of flares covering the range from 1 to 200 MeV have been obtained with the EGRET TASC instrument. Spectra from the early stages of four of the intense 1991 June flares are plotted in Figure 5 [9]. The contribution from nuclear lines is evident below about 8 MeV. There is also a high energy component extending from 10 MeV to at least 200 MeV in all four flares. OSSE measurements [4] suggest that this high-energy emission in the June 4 flare is primarily from electron bremsstrahlung and not from production of charged and neutral pions. The line features at 2.223 MeV (neutron capture on H), 4.43 MeV and 6 - 7 MeV (de-excitation lines from C and O) are resolvable in the spectra, especially in the June 9 and June 11 flares when there was less scattering due to overlying material.

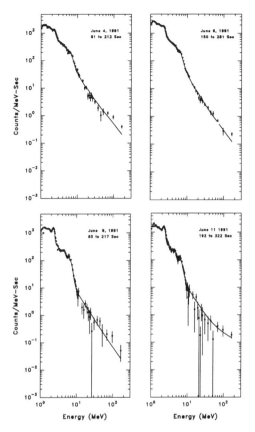

FIGURE 5. Time integrated spectra from four 1991 June flares observed by the EGRET TASC.

Rank et al. [10] produced a broad band spectrum of the decay phase of the 1991 June 15 solar flare from 25 keV to 4 GeV. This spectrum plotted in Figure 6 is a composite from measurements taken with the BATSE and COMPTEL instruments on CGRO and the GAMMA-1 instrument on the GAMMA spacecraft after corrections are made for the different observing times. One can see the transitions from hard X-ray bremsstrahlung to nuclear-line emission to the continuum produced by the decay of charged and neutral pions. Rank et al. infer a proton spectrum at the Sun with a power-law shape with index -3.3 softening above 300 MeV. The continuum emission above 8 MeV measured by both COMPTEL and GAMMA-1 is at least two orders of magnitude above the extrapolation of the BATSE hard X-ray spectrum in the decay phase of the flare. This suggests, contrary to the conclusions of Murphy et al. [4] for the June 4 flare, that the continuum derives from the radiation of the secondary electrons from pion decay.

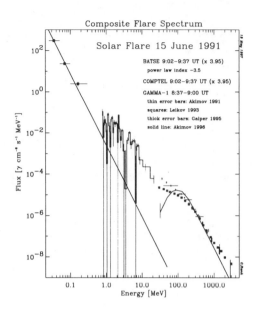

FIGURE 6. Composite spectrum for the extended phase of the 1991 June 15 flare. CGRO spectra have been scaled by a factor of 4 to compensate for different observation times.

Abundance Studies using γ-Ray Lines

The only in situ diagnostic of the composition of the accelerated ion population is nuclear-line spectroscopy. Whatever accelerates ions in a solar flare selects different species according to a process unknown to us. It is important to measure and, if possible, understand the composition of the accelerated ions. Abundance anomalies in the ambient solar material that is the target for the energetic ions can also be studied. One such anomaly is that of neon [11].

The best spectral resolution for nuclear-lines on CGRO is provided by the OSSE instrument [4]. An OSSE spectrum from one of its four detectors is plotted in Figure 7 along with the components used to fit the data: bremsstrahlung power law, annihilation line and 3 γ continuum, neutron-capture line, and narrow and broad de-excitation lines. (Narrow de-excitation lines are produced when accelerated protons and α-particles impact on ambient elements such as C, Mg, and Ne. Broad de-excitation lines are produced when accelerated heavy nuclei impact on ambient H; these lines are broadened by about ~ 1 MeV.) These model components fit the data quite well. Detailed fits revealing the presence of the various narrow de-excitation lines are shown in Figure 8 where the power law has been subtracted for clarity. The broad feature due to α-particle fusion with ambient helium is clearly visible below the annihilation line. Fits to both OSSE [4] and SMM [12] flare spectra have revealed that the

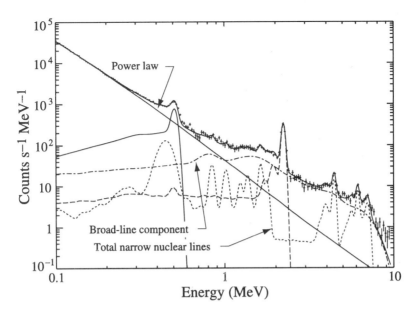

FIGURE 7. Time-integrated spectra from the 1991 June 4 flare observed by OSSE along with the best fitting components.

accelerated α-particle/proton ratio is closer to 0.5 than it is to the ambient abundance ratio of ~ 0.1 in the photosphere. Line profile studies of two strong disk flares with SMM have also demonstrated that the accelerated α particles do not follow a downward-beam geometry. Both a fan beam and a downward isotropic distribution provide acceptable fits to the data [12].

Nuclear de-excitation lines are clearly resolved in the OSSE spectra plotted in Figure 8. By comparing the relative line intensities one can infer information on the ambient elemental abundances [11]. It was found that flares show a pattern based on first ionization potential (FIP). Low FIP elements include Mg, Si, and Fe while high FIP elements include Ne, C, and O. The ratio of low FIP/high FIP lines appear to vary from flare-to-flare, but its average is consistent with a coronal composition [13]. The ratios for the 19 SMM flares are plotted in Figure 9 along with the ratios derived from the OSSE observations of the June 4 flare. The flare-averaged OSSE ratio is consistent with the average for the SMM observations [4]; however, there is evidence for an increase in the low FIP/high FIP ratio as the flare evolved. The increase is shown in greater detail in Figure 10. One explanation for such an increase is that the location where the particles mirror and interact may move higher in the solar atmosphere as the flare progresses.

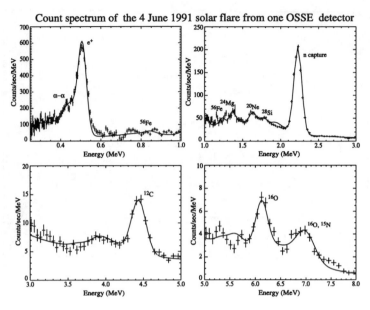

FIGURE 8. Time-integrated spectra from the 1991 June 4 flare observed by OSSE in four energy bands after removing the bremsstrahlung power law.

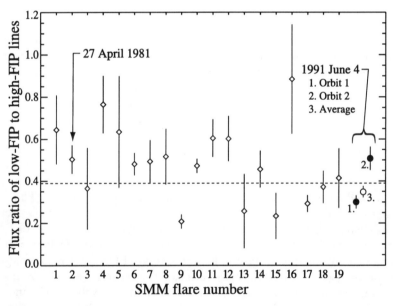

FIGURE 9. Low FIP/High FIP line ratios for 19 SMM flares and the OSSE 1991 June 4 flare.

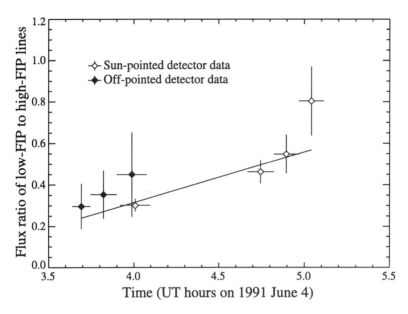

FIGURE 10. Variation in Low FIP/High FIP line ratios during the 1991 June 4 flare observed by OSSE.

Accelerated Particle Spectra and Energetics

Comparison of the intensities of different lines also provides information on the spectra of the accelerated particles [14]. The excitation of C to its first excited state occurs at lower proton energies than the production of neutrons that are then captured to form deuterium and its 2.223 MeV line. Thus the ratio of the 4.43 MeV/2.223 MeV lines provides one measure of the hardness of the particle spectrum. This technique has been used for many years. Unfortunately, there is roughly an average 100 s delay between the production of the neutrons and the consequent 2.223 MeV line. This delay is a function of the average neutron energy and the photospheric density in which it thermalizes. Thus the ratio is typically only computed after integrating over long periods of time or the entire flare. However if statistics are adequate, one might be able to detect temporal variation in the hardness of the accelerated particle spectrum and also determine the time constant for production of the capture line. The latter provides fundamental information on the abundance of ^{3}He in the photosphere because this isotope provides a radiationless competitor for capture on H. Quick capture of free neutrons can be due to a high hydrogen density (deep in photosphere) or by a large abundance of ^{3}He.

Examples of high statistical line data obtained from the four 1991 June flares observed by the EGRET TASC are plotted in Figure 11 [9]. The intensity-time

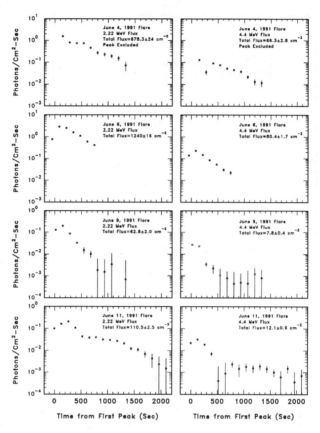

FIGURE 11. Time histories of the 2.22 and 4.4 MeV line fluxes for four flares observed by EGRET.

profiles of these two lines generally follow one another but there is evidence that the ratio may change in at least one of the flares, the June 11 flare. Data from the COMPTEL instrument suggest that the spectral index of the accelerated particle spectrum and perhaps the time constant for production of the delayed line vary over the course of this flare [15]. This variation is shown in Figure 12. For all but the first few intervals, the integration time is long compared to the nominal 100 s decay time of the 2.223 MeV line; thus a direct comparison can be made between the emission in the 4-7 MeV band and in the 2.223 MeV line. There is significant variation in the relative intensities of these indicators. The relative intensities of the 4-7 MeV and 2.223 MeV emission in the declining phase of the flare (after 8000 s UT) are nearly equal, indicating a harder spectrum than during the peak where the 4-7 MeV flux exceeds that of the of 2.223 MeV line by a factor of 2 to 3. The transition period from 7630 to 8000 s is characterized by an intense and sustained 2.223

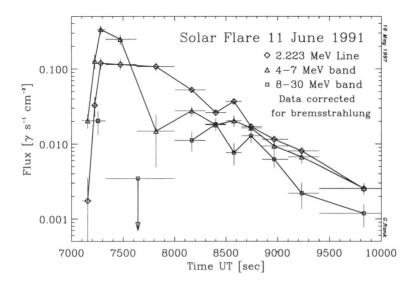

FIGURE 12. Time histories of the 2.223 MeV line, 4-7 MeV and 8-30 MeV bands from the 1991 June 11 flare observed by COMPTEL.

MeV flux and a low 4-7 MeV flux.

In contrast, there is no evidence for variability in either the spectral hardness of accelerated particles or in the time constant for production of the 2.223 MeV line observed in the June 4 flare by OSSE [4]. The time profile in the 2.223 MeV line, plotted in Figure 13, is compared with a model generated using an interaction rate based on the de-excitation lines from elements with high first ionization potential, a constant spectral index, and a single time constant. The fit is good and provides a flare-averaged accelerated proton power-law index of -4.1 ± 0.1 using the 4.44 MeV/2.223 MeV ratio. The derived time constant for the 2.223 MeV line yields an upper limit on the photospheric ^{3}He/H ratio of 2.5 $\times 10^{-5}$ using a relatively simple model [16]. Significant improvements in the accuracy of this isotopic abundance ratio derived for this and other flares are expected when the results of Monte Carlo simulations [17] are used, updated with new information on the accelerated particle composition and spectra. It has also recently been suggested that the γ-ray derived photospheric ^{3}He/H ratio can be multiplied by the observed solar wind ^{3}He/^{4}He ratio to determine the ambient photospheric ^{4}He/H ratio [18].

There is an alternative method for determining the spectral index of the accelerated protons in flares. This uses the 6.13 MeV/1.63 MeV de-excitation line ratio [11]; the 6.13 MeV line comes primarily from ^{16}O* and the 1.63 MeV line from ^{20}Ne*. Because both lines are prompt this ratio provides a direct

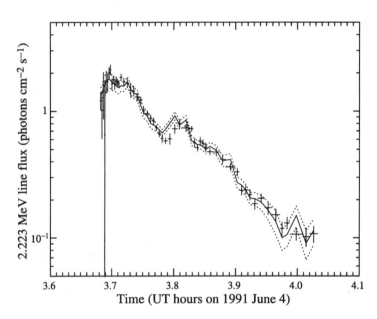

FIGURE 13. Time history of the 2.22 MeV line from the 1991 June 4 flare observed by OSSE; curves are fits to the data used to estimate the solar ^{3}He/H ratio. Reprinted courtesy of the Astrophysical Journal published by the Univ. of Chicago Press (c) 1997 (Amer. Ast. Soc.).

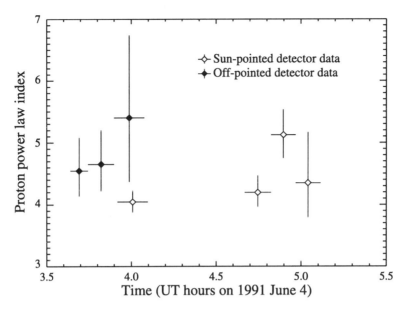

FIGURE 14. Variation of the power-law spectral index of accelerated protons during the 1991 June 4 flare from OSSE line observations.

measure of the spectral variability during the flare. The OSSE measurements for the June 4 flare suggest that the index is constant with a flare-averaged value of -4.27 ± 0.13 [4]; the constancy of the power-law spectral index is shown in Figure 14. The flare averaged index is consistent both with the index derived for the same flare using the 4.43 MeV/2.223 MeV line ratio (see above) and with the average ratio derived for 9 flares observed by SMM [13].

Where the energy content resides in accelerated particles has been a question for a long time. Protons and ions with energies below the first gamma-ray producing threshold are almost impossible to detect, much less measure, in solar flares. These protons (and ions) could contain considerable and, perhaps, the majority of energy of energetic particles. Electrons by virtue of their efficient radiation properties are measured down into the thermal distributions. Using the 6.13 MeV/1.63 MeV line ratio one can push the nuclear-line diagnostic capabilities for proton spectra down to about 1 MeV. Murphy et al. [4] integrated the accelerated proton spectrum above 1 MeV using the above power-law index and assumed a flat spectrum below 1 MeV to derive a total energy in protons of $\sim 3 \times 10^{32}$ ergs. If they assume impulsive-flare abundances for accelerated particles and an α/proton ratio of 0.5, the total energy in accelerated ions is $\sim 2 \times 10^{33}$ ergs. This compares with an upper limit on the energy in accelerated electrons >0.1 MeV of $\sim 1 \times 10^{32}$ ergs. Thus the energy in accelerated ions appears to be at least ten times that in accelerated

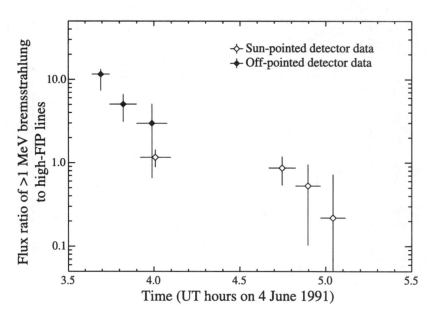

FIGURE 15. Variation of the ratio of bremsstrahlung and nuclear line radiation observed by OSSE during the 1991 June 4 flare.

electrons for this flare. This is consistent with what has been found in a study of SMM flares [13].

EXTENDED EMISSIONS FROM FLARES

The bremsstrahlung/nuclear ratio for the June 4 flare appears to vary, as is shown in Figure 15 [4]. The bremsstrahlung flux above 1 MeV falls much more rapidly than does the nuclear line emission with time. Evidence for a variation in this ratio has also been suggested based on γ-ray measurements by EGRET at higher energies during the June 11 flare [9]. Plotted in Figure 16 are four sets of spectra from 30 MeV to $\sim$ 2 GeV with EGRET; they have been fit by a power-law bremsstrahlung spectrum and a pion spectrum [19]. Bertsch et al. suggest that there is a decrease in the pion component over the first three time intervals, but the decrease is not of marked statistical significance. The fourth interval follows about 8 hours later and it clearly shows a small pion contribution; however, this emission may come from a 'M'-class flare which occurred at that time and may have nothing to do with the earlier 'X' class flare emission.

The long term emission from the June 11 flare is illustrated in Figure 17 which displays the >50 MeV flux observed by EGRET as a function of time. Extended emission from the X12 class flare was observed for at least 8 hours

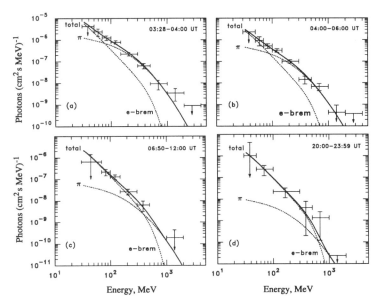

FIGURE 16. Fits to high energy spectra observed by EGRET at different times of the 1991 June 11 flare.

and appears to return to background after that time. The emission at 20 Hr UT appears to be consistent with an M5.3 flare. Both the γ-ray and X-ray fluxes were roughly a factor twenty lower during this smaller flare relative to near the peak of the X12 flare.

COMPTEL also made long term observations of the June 9, 11, and 15 flares [20] using its excellent sensitivity to the 2.223 MeV neutron capture line. The time histories of the line in the June 11 and 15 flares are plotted in Figure 18. The line was clearly observed in each of the three flares for up to at least 6 hours following the impulsive phase. Although not evident in the data plotted in this figure, Rank et al. [20] note there are small enhancements in the line fluxes roughly 2 to 3 hours after the onsets for each of the flares. They conclude that this can only be explained by newly accelerated particles and not by storage alone.

In Figure 12 we also see that the 2.223 MeV and high-energy emissions fall at roughly the same rate late in the flare. One would expect that particle trapping exhibits some fractionation property whereby particles of different species or the same particles at different energies would precipitate from the trapping region at different rates. This does not seem to be the case for the extended phase of the 11 June 1991 flare where the 2.223 MeV, the 8-30 MeV continuum (secondary electron bremsstrahlung) and the 4-7 McV nuclear-line region fluxes all show a similar decay. It is likely, therefore, that the emission

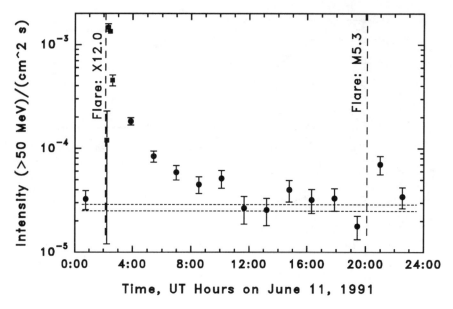

FIGURE 17. Intensities of >50 MeV photons observed by EGRET around the time of 1991 June 11 flare.

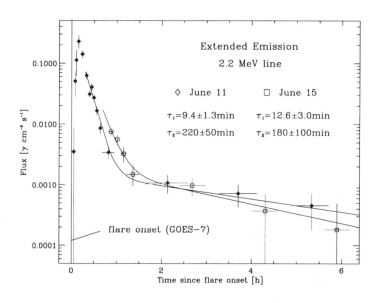

FIGURE 18. Intensities of 2.223 MeV photons observed by COMPTEL displaying the extended emission from the 1991 June 11 and 15 flares.

is from freshly accelerated particles and not from particles stored in the Sun's magnetic fields.

COMPTON OBSERVATIONS DURING CYCLE 23

Solar activity is expected to rise in 1998 as Solar Cycle 23 approaches its expected maximum in 2000. The Compton Observatory provides the most comprehensive and sensitive instrumentation in orbit to observe hard X-rays, nuclear-line γ-rays, pion-decay γ rays, and neutrons throughout this cycle. We anticipate that extensive use of Compton will be made to provide the necessary high-energy coverage required for comparison with flare observations conducted at other wavelengths by SOHO, TRACE, ACE, SAMPEX, Ulysses, and ground-based observatories.

Acknowledgment This work was supported under NASA DPR S-92556-F, S-10987-C, and S-57769F.

REFERENCES

1. White, S.M., Harmon, B.A., Lim, J., & Kundu, M.R., poster presentation at this workshop (1997).
2. Aschwanden, M.J., Kosugi, T., Hudson, H.S., Wills, M.J. & Schwartz, R.A., *Ap. J.* **470**, 1198 (1996).
3. Aschwanden, M.J., Hudson, H., Kosugi, T., & Schwartz, R. *Ap. J.* **464**, 985 (1996).
4. Murphy, R.J., et al., *Ap. J.* **490** in press (1997).
5. Ramaty, R., Schwartz, R., Enome, S., & Nakajima, H., *Ap. J.* **436**, 941 (1994).
6. Enome, S. private communication (1995).
7. Share, G.H., Grove, J.E., & Murphy, R.J., Priv. Comm. (1996).
8. Gal'per, A.M., et al., *JETP Letters* **59(3)**, 153 (1994)
9. D.L. Bertsch, et al., Submitted for publication in *Ap. J.* (1997).
10. Rank, G. et al., *Proc. 25th Int. Cosmic Ray Conf*, **Vol. 1:** Session SH 1 - 3, Durban, SA:Potchefstroomse Universiteit, p. 1 (1997).
11. Share, G.H., & Murphy, R.J., *Ap. J.* **452**, 933 (1995).
12. Share, G.H., & Murphy, R.J., *Ap. J.* **485** , 409 (1997).
13. Ramaty, R., et al., *Ap. J. Letters* **455**, L193 (1995).
14. Ramaty, R., et al., *Ap. J. Supp.* **40**, 487 (1979).
15. Rank, G., *Ph. D. Dissertation*, Technical University of Munich (1995)
16. Prince, T.A. et al., *18th Int'l. Cos. Ray Conf.* **4**, 79 (1983).
17. Hua, X.-M., & Lingenfelter, R.E., *Ap. J.* **319**, 555 (1987).
18. Share, G.H., & Murphy, R.J. *Ap. J. Letters* **484** L165 (1997).
19. Mandzhavidze, N., Ramaty, R., Bertsch, D.L., & Schneid, E.J., in *High Energy Solar Physics*, ed. R. Ramaty, N. Mandzhavidze, X-M Hua, AIP Conf. Proc. 374, New York:AIP, p. 225 (1996).

20. Rank, G. et al., *Proc. 25th Int. Cosmic Ray Conf*, **Vol. 1: Session SH 1 - 3**, Durban, SA:Potchefstroomse Universiteit, p. 5 (1997).

GALACTIC GAMMA RAY ASTRONOMY

Gamma-Ray Pulsars: the Compton Observatory Contribution to the Study of Isolated Neutron Stars

D.J. Thompson[1], A.K. Harding[1], W. Hermsen[2], M.P. Ulmer[3]

[1]*NASA/GSFC,* [2]*SRON/Utrecht,* [3]*Northwestern*

Abstract. The four instruments on the Compton Gamma Ray Observatory have identified at least seven isolated neutron stars by their pulsed gamma-ray emission. For all of these, the gamma radiation represents the largest observable fraction of the spin-down luminosity, making the gamma rays important diagnostics of particle acceleration and interaction in the neutron star magnetospheres. Several other "candidate" pulsars have tentative identifications based on pulsed radiation or positional association. The ensemble of CGRO pulsar detections, possible detections, and upper limits yields a number of patterns, most of which have exceptions. Some examples: (1) All except PSR B1509−58 have light curves consistent with double pulses; (2) All except the Crab have radio pulses out of phase with the gamma-ray pulses (the relative phase of the recently-reported radio pulse from Geminga is not known as of this writing); (3) Except possibly for PSR B1509−58 and PSR B0656+14, the gamma-ray energy spectra flatten with increasing pulsar age; (4) All except PSR B1951+32 show evidence of having high-energy spectral cutoffs in the CGRO energy band; (5) The pulsed flux from all the pulsars is relatively constant with time; (6) There is a trend for the pulsar luminosity to increase with increasing polar-cap current (or open field line voltage, or $\dot{E}$, all of which are similar). The diversity in the phenomenology of gamma-ray pulsars presents challenges to pulsar models. For example, double-pole models, often invoked by radio astronomers, now seem problematic. Most models have evolved significantly in response to the CGRO discoveries, and no comprehensive explanation for all the observations is yet in hand.

INTRODUCTION

Among the wide range of results from the four instruments on the Compton Gamma Ray Observatory (CGRO) are the identifications of at least seven gamma-ray pulsars: Crab, PSR B1509−58, Vela, PSR B1706−44, PSR B1951+32, Geminga, and PSR B1055−52. These results have stimulated considerable theoretical work on gamma-ray pulsars (e.g., [40], [12], [36], [16],

[35]). This article reviews the observational progress made with the Compton Observatory and the impact these observations have had on models.

GAMMA RAY PULSARS AND RADIO PULSARS

The quantity $\dot{E}/4\pi d^2$ (where d is the distance) is a useful measure of gamma-ray pulsar observability (see Figure 1). All six known pulsars with $\dot{E}/4\pi d^2 > 3\times 10^{-9}$ erg/cm²s [44] are gamma-ray pulsars. With the recent report of very weak radio pulsations from Geminga [26] all of these may also be radio pulsars, implying a strong overlap between the radio and gamma-ray beams. PSR B1055−52, the seventh gamma-ray pulsar, is the exception to the correlation with $\dot{E}/4\pi d^2$. At least a dozen pulsars (more if the millisecond

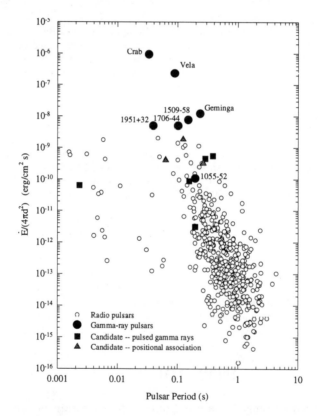

FIGURE 1. Radio and gamma-ray pulsars as a function of $\dot{E}/4\pi d^2$. The detection threshold for sources with the telescopes on the Compton Observatory lies a factor of a few below PSR B1055−52. Pulsars with significantly smaller values of $\dot{E}/4\pi d^2$ could not be seen in gamma rays unless the gamma-ray beam sweeps out significantly less than 1 sr.

pulsars are included) have higher values of $\dot{E}/4\pi d^2$ without showing strong evidence of gamma-ray pulsation. Whether PSR B1055−52 is unusually efficient in producing gamma rays, has a preferred beaming geometry, or possibly is much closer than implied by the radio dispersion measure is an open question.

There are also "candidate" gamma-ray pulsars, all of which have $\dot{E}/4\pi d^2$ close to or larger than the value for PSR B1055−52. At present, none of these are considered as definite detections. PSR B0144+59 [48], PSR B0355+54 [46], PSR B0656+14 [34], [21], PSR J0631+10 [52], PSR B1821−24 [2], and PSR J0218+4232 [50] all have gamma-ray light curves that are statistically improbable for a single trial, but they are part of a large sample of pulsars that have been examined for gamma-ray pulsation. Note that PSR J0218+4232 is a ms pulsar in a binary system, so it is not an isolated neutron star as the others are. PSR B1046−58, PSR J1105−6107 [23], and PSR B1853+01 are radio pulsars positionally consistent with gamma-ray sources (though the error boxes are large), also having $\dot{E}/4\pi d^2$ in the same range as the known gamma-ray pulsars. The gamma-ray energy spectra of these sources are flat power laws like those of known pulsars [13]. These gamma-ray sources are not time variable. Gamma-ray pulsars are steady; many other gamma-ray sources are highly variable. They are energetically consistent with a pulsar origin (the gamma-ray luminosity represents a small fraction of $\dot{E}$). None of the three, however, shows evidence of pulsation in gamma rays [33], [24]. Another group of "candidate" gamma-ray pulsars are the unidentified EGRET sources not positionally associated with radio pulsars but having similar gamma-ray characteristics, i.e. steady sources with flat spectra having a possible high-energy turnover. Examples are 2EG J2020+4026 [5] and 2EG J0008+7307 [6].

PULSAR LIGHT CURVES

Figures 2 (young pulsars. References: OSSE: Crab [47]; PSR B1509−58 [28]; Vela [39]; COMPTEL, Crab [30]; PSR B1509−58 [9]; Vela [25]; EGRET [22]) and 3 (older pulsars. References: [21]; [22]) show the light curves of the seven known gamma-ray pulsars as seen by the instruments on the Compton Observatory. PSR B0656+14, the strongest "candidate" pulsar, is separated from the others, because the detection is of much lower statistical significance. The agreement in phase between the highest point in the COMPTEL and EGRET light curves strengthens the case for this as a detection [21].

Among the features and implications of this compilation are:
1. Only the Crab has gamma-ray pulses which are in phase with the radio (although the radio precursor pulse is not seen in gamma rays).
2. The gamma-ray light curves tend to show two pulses with a bridge of emission between. Only PSR B1509−58 is inconsistent with such a picture,

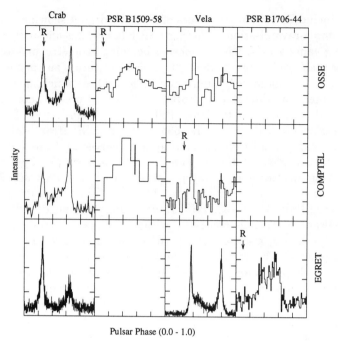

FIGURE 2. CGRO light curves of four pulsars with age less than 1×10^5 yr. The BATSE light curves are very similar to those of OSSE, because the energy ranges overlap. The "R" marks the phase of the radio pulse. A typical error bar is shown for some light curves.

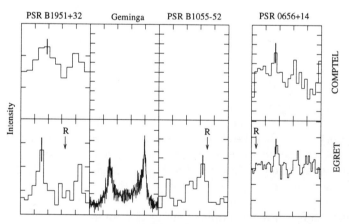

FIGURE 3. CGRO light curves of four pulsars with age greater than 1×10^5 yr. None of these pulsars are seen in the BATSE/OSSE energy range. The "R" marks the phase of the radio pulse. A typical error bar is shown for some light curves.

although the statistics for PSR B0656+14 and PSR B1055−52 are too limited to draw any firm conclusions. The gamma-ray light curve for PSR B1706−44 contains at least two pulses, with possible evidence for a third between the two.

3. Only Geminga has gamma-ray pulses with a 180° separation.

4. The commonality of double pulses, coupled with the absence of many 180° separations, argues against orthogonal rotator models in which both poles are sources of gamma rays.

5. The gamma-ray light curves are consistent with single-pole, hollow-cone models of emission [16].

6. The changing shapes of the light curve from low to high gamma-ray energies, seen especially in the Crab, imply that different phases of the radiation have different spectra.

MULTIWAVELENGTH SPECTRA OF GAMMA-RAY PULSARS

Figure 4 summarizes the $\nu F \nu$ phase-averaged spectra for the seven definitely-detected gamma-ray pulsars. This format shows the observed power per logarithmic energy interval. Pulsed emission is shown in all cases, with upper limits given for sources detected but without observed pulsation. This is an update of the figure from [45]. Updated data points are: PSR B1509−58, COMPTEL [20]; PSR B1951+32, COMPTEL, [21]; TeV, [38]; Geminga, radio, [26]; X-ray, [14].

COMMON FEATURES

1. In all cases, the maximum observed energy output is in the gamma-ray band. The peak ranges from photon energies of about 100 keV for the Crab to photon energies above 10 GeV for PSR B1951+32.

2. All these spectra have a high-energy cutoff or break. For PSR B1509−58, it occurs not far above 1 MeV photon energy, between the COMPTEL points and EGRET upper limits; for PSR B1951+32 it must lie somewhere above 10 GeV, between the highest-energy EGRET point and the TeV upper limit.

Figure 5 shows the location of the high-energy cutoff as a function of the calculated pulsar surface magnetic field. The high-field PSR B1509−58 stands out in having both the highest field and the lowest cutoff energy. For a discussion, see [19]. PSR B1951+32 is also notable in having the lowest field and the highest cutoff energy.

3. Over at least part of the gamma-ray band, the pulsar energy spectra can be approximated with power laws. Except for PSR B1509−58, the pulsars show a hardening of the spectrum with increasing characteristic age, visible in Figure 4.

INDIVIDUAL PULSARS

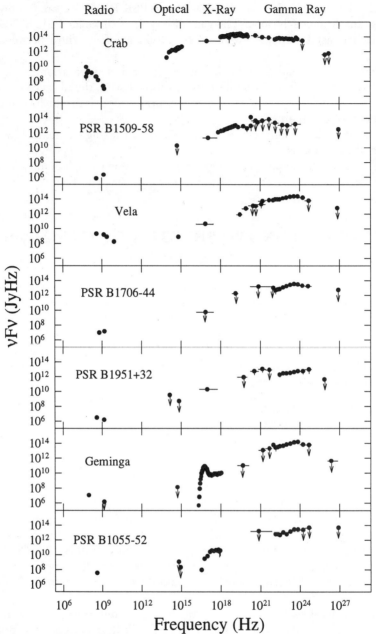

FIGURE 4. Multiwavelength energy spectra for the known gamma-ray pulsars, shown in order of increasing age.

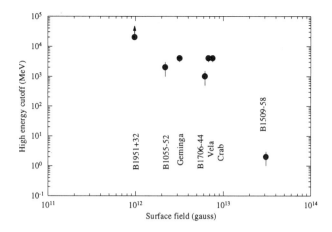

FIGURE 5. High-energy cutoff in pulsed emission as a function of calculated magnetic field. The cutoff for PSR B1951+32 lies somewhere between 10 GeV and the Whipple energy range.

1. Crab – the spectrum appears continuous from optical to high-energy gamma-ray energies.
2. PSR B1509−58 – all the points above 1 MeV (about 10^{21} Hz) are upper limits. The spectrum must have a break between 1 and 100 MeV.
3. Vela – the pulsed X-ray emission has a thermal component. The spectrum turns over sharply at a few GeV.
4. PSR B1706−44 – the gamma-ray spectrum is well described by two power laws, with a slope change at 1 GeV.
5. PSR B1951+32 – the high-energy gamma-ray spectrum can be represented as a single power law out to 30 GeV. It must turn over sharply above this energy. The COMPTEL data indicate a possible MeV excess.
6. Geminga – the gamma-ray spectrum has a sharp turnover in the few GeV energy range. The X-ray spectrum has two components, one of which is thermal. The non-thermal component does not appear to connect to the gamma-ray spectrum, although the uncertainties are large [14].
7. PSR B1055−52 – limited gamma-ray statistics preclude a detailed study of this energy spectrum. The X-ray spectrum has two components, one of which may extrapolate to gamma-ray energies.

Figures 6 and 7 show the phase-resolved gamma-ray energy spectra for the Crab and Vela pulsars, using multiple Compton instruments. For the Crab, the data are taken from [48]. For Vela, the compilation is from [25] and this meeting. The Vela pulse component definitions and the EGRET spectra above 30 MeV are from [13]: Leading Wing 1 (LW1), Peak 1 (P1), Trailing Wing 1 (TW1), Interpulse 1 (IP1), Interpulse 2 (IP2), Bridge, Leading Wing 2 (LW2), Peak 2 (P2), Trailing Wing 2 (TW2)) In both cases, the "bridge"

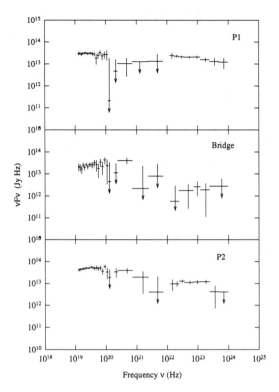

FIGURE 6. Phase-resolved spectra for the Crab pulsar.

region between the two pulses has the flattest spectrum. This is not the case for Geminga, where the flattest spectrum is associated with one of the two peaks [29], [13].

HIGH-ENERGY PULSAR LUMINOSITY

In terms of the observed energy flux F_E, the luminosity of a pulsar is
$$L = 4\pi f F_E d^2,$$
where d is the distance and f is the fraction of the sky into which the pulsar radiates. The pulsar beaming solid angle $4\pi f$ is highly uncertain and probably not the same for all pulsars, nor for the same pulsar at different wavelengths. It must lie between the neutron star polar cap solid angle and 4π. In the absence of a compelling argument for any particular value, we adopt a simple value for f, setting it equal to $1/4\pi$ (assuming that the pulsar radiates into 1 steradian).

The energy flux F_E is determined by integrating the broad-band energy

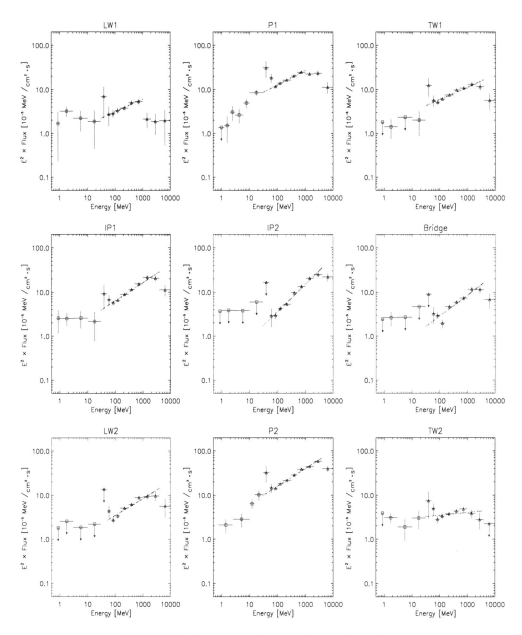

FIGURE 7. Phase-resolved spectra for the Vela pulsar.

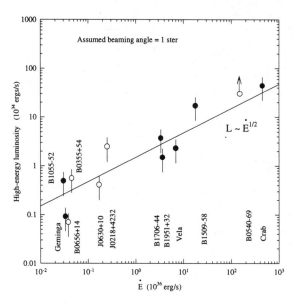

FIGURE 8. Pulsar luminosities, assuming radiation into one steradian, vs $\dot{E}$. Filled circles: high-confidence gamma-ray pulsars. Open circles: possible gamma-ray pulsars.

spectra, making realistic assumptions about regions where no measurements exist. In most cases, the observed gamma-ray spectrum is itself a good approximation, because most of the observed energy falls in this band. For the Crab and PSR B1509−58, the X-ray spectra must also be considered. PSR B0540−69 in the Large Magellanic Cloud is not seen in gamma rays, but its hard X-ray component suggests that its spectrum may extend to higher energies. The X-ray flux then represents a lower limit. PSR B0656+14 has pulsed X-ray emission with both a thermal and nonthermal component.

Figure 8 shows one possible pattern for pulsar luminosities. The high-energy luminosity increases with the spin-down luminosity $\dot{E}$. This is similar to a pattern seen in 0.1 - 2.4 keV X-rays by Becker and Trümper [4]. The slope of the line in Figure 8 is flatter than the one found by Becker and Trümper, because the integrated luminosity is dominated by the gamma rays, and the pulsars with smaller $\dot{E}$ are also the older pulsars that have flatter energy spectra.

A similar trend is seen in Figure 9 between the high-energy luminosity and the open field line voltage ($\sim B/P^2$), which is also proportional to the polar cap current [15]. This is expected, because the polar cap current is proportional to $\dot{E}$. A linear fit to the observed pulsars is a reasonable approximation, extending over several orders of magnitude.

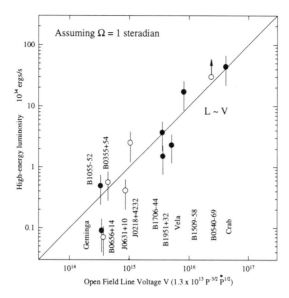

FIGURE 9. Pulsar luminosities, assuming radiation into one steradian, vs open field line voltage (proportional to polar cap current or B/P^2). Filled circles: high-confidence gamma-ray pulsars. Open circles: possible gamma-ray pulsars.

GAMMA-RAY PULSAR EMISSION MODELS

The new detections of γ-ray pulsars by CGRO and the vastly improved quality of the measurements of light curves and spectra have both constrained and accelerated model development. The patterns found in the properties of the pulsars detected by CRGO are revealing fundamentals of the physical processes involved in the emission. For example as seen in Figure 9, the high energy luminosity, L, is proportional to the quantity B/P^2 (or $P^{-3/2}\dot{P}^{1/2}$), which is itself proportional to both the rate of particle flow along the open field lines, $\dot{N}_{PC}$, and the voltage across the open field lines, V_{PC}. These quantities have the same dependence because they are both related to the electric field, $E \sim vB/c$, induced by the rotation. If $L \propto \dot{N}_p E_p$, where $\dot{N}_p$ is the rate of primary particle flow and E_p is the energy to which the primary particles are accelerated, then $L \propto B/P^2$ can be explained if either 1) a particle flow that varies from pulsar to pulsar, $\dot{N}_{PC}$, is accelerated to the same energy, $E_p \simeq 2 \times 10^{13}$ eV, or 2) a constant particle flow, $N_p \simeq 6 \times 10^{30}$ s^{-1}, is accelerated to varying energy, V_{PC}. Of course both N_p and E_p could vary, but their variation with the quantities P and B are constrained to the product B/P^2.

The generic nature of double-peaked pulses with bridge or interpeak emission has forced the models to change from a double-pole picture, where the high-energy emission is seen from both magnetic poles, to a single-pole picture, where the observer sees emission from one pole only. Figure 10 shows a schematic view of several proposed single-pole emission geometries (details of the models will be described below). In polar cap models, the γ-ray emission beam is a hollow cone centered on the magnetic pole. In outer gap models, the γ-ray emission is a wide, curved fan beam that is formed by the surface of the last open field line in the outer magnetophere. In both models, double pulses will occur most of the time and single pulses occur when the observer grazes the edge of the cone.

Polar Cap Models

The class of polar cap models for γ-ray pulsars are all based on the idea, dating from the earliest pulsar models of [43] and [37], of particle acceleration and radiation near the neutron star surface at the magnetic poles. Within this broad class, there is a large variation, with the primary division being whether or not there is free emission of particles from the neutron star surface. This question is still somewhat subject to debate due to our incomplete understanding of the neutron star surface composition and physics.

One subclass of polar cap models is based on free emission of particles of either sign, assuming that the surface temperature of the neutron star

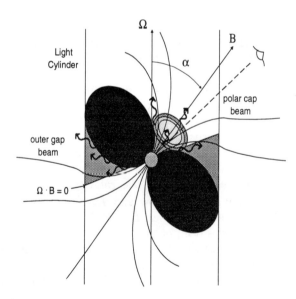

FIGURE 10. Schematic of emission geometry in single-pole polar cap and outer gap models. Ω and **B** define the spin and magnetic axes, with inclination angle α.

(many of which have now been measured in the range $T \sim 10^5 - 10^6$ K) exceeds the ion, T_i and electron, T_e, thermal emission temperatures. With the flow of particles limited only by space charge, an accelerating potential will develop due to the inability to supply the corotation charge (required to short out $E_\parallel$) all along each open field line. The models of Sturner & Dermer [40] and Sturner, Dermer and Michel [42] (DSM) assume that *only* the space-charge-limited flow generates the $E_\parallel$, which gives an accelerated particle energy $\gamma \sim 10^5 - 10^6$. In this case, the accelerated particles lose energy primarily to inverse-Compton radiation from resonant scattering of thermal X-ray photons from the hot polar cap. The scattered photons are emitted in a hollow cone around the pole and initiate a cascade by magnetic pair production and synchrotron radiation to form the observed γ-ray spectrum. The models of Daugherty and Harding [10], [11], [12] (DH) assume that the much higher $E_\parallel$ due to field line curvature [1] and general relativistic inertial frame dragging [31] will accelerate particles to $\gamma \sim 10^7$, in which case curvature radiation energy loss ($\propto \gamma^4/\rho_c$) is dominant. A magnetic pair production/synchrotron cascade is then initiated by the curvature photons, which are emitted in a hollow cone defined by the tangent to the last open field line, where radius of curvature, ρ_c, is smallest. Both of these polar cap models can produce double-peaked light curves with interpeak emission similar to that observed (see Figure 11). However, both suffer from the potential problem that a small tilt angle between the magnetic and spin axes ($\sim 5^0$ for surface emission with standard polar cap size) is required to produce the peak spacings as large as 0.4 - 0.5 seen in most of the γ-ray pulsars [41], [17]. This problem can be remedied more easily in the curvature-radiation initiated cascade models, where delayed acceleration one or two stellar radii above the surface is possible [12] and where flaring of the field lines can significantly increase the γ-ray beam opening angle to $20^0 - 30^0$. These models both predict cutoffs in the spectrum due to magnetic pair production attenuation around 1 - 5 GeV. The energy of these cutoffs should roughly scale inversely with magnetic field strength and could account for the trend seen in Figure 5.

Another subclass of polar cap models assumes that ions are bound in the surface layers of the star ($T < T_i$). In pulsars where $\mathbf{\Omega} \cdot \mathbf{B} < 0$, the corotation charge above the polar caps is positive, and a vacuum gap forms above the surface. Cascade breakdowns of the gap will occur when stray γ-rays entering the gap produce pairs that are then accelerated (positron upward, electrons downward). In the original version of this type of model (Ruderman & Sutherman [37] RS), the accelerated particles produced curvature radiation to initiate the cascades. A γ-ray pulsar model based on this picture was developed by Usov & Melrose [49], who included the effect of bound pair creation which delays gap breakdown and thus increases the acceleration potential in fields higher than $\sim 4 \times 10^{12}$ G. Zhang et al. [53], [54] consider an alternative variation in which the RS type gaps are controlled by inverse-Compton scattering of the particles instead of curvature radiation. At present, none of the polar vacuum gap models have developed light curves or spectra for comparison with the CGRO data.

Outer Gap Models

The outer gap models for γ-ray pulsars are based on the existence of a vacuum gap in the outer magnetosphere which may develop between the last open field line and the null charge surface (see Figure 10) in charge separated magnetospheres. Particles flowing along the open field lines are accelerated in the gaps, radiating curvature or inverse-Compton emission. The first outer

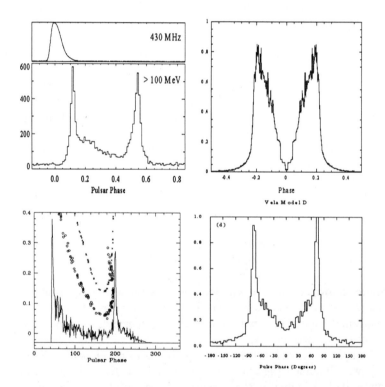

FIGURE 11. Observed (upper left) and computed Vela pulsar light curves for two polar cap models: DSM (upper right) and DH (lower right), and one outer gap model: Romani (1996) (lower left).

gap γ-ray pulsar models [8] assumed that emission is seen from gaps associated with both magnetic poles, but this picture, although successful in fitting the spectrum of the Crab and Vela pulsars, did not successfully reproduce the observed pulsar light curves. The pairs needed to control the gap are produced by photon-photon pair production with either non-thermal synchrotron X-rays (Crab-like pulsars) or with infra-red photons (older Vela-type pulsars) from pairs produced in the gap (introducing a high degree of non-linearity). However, the Vela-type model predicted large fluxes of TeV emission from inverse Compton scattering of the infra-red photons by primary electrons, which violates the observed upper limits [32] by several orders of magnitude. Cheng [7] revived the outer gap model for Vela-type pulsars by proposing another self-sustaining gap mechanism where thermal X-rays from the neutron star surface replace the infra-red radiation (which was never observed anyway). With the new data from CGRO showing the strong pattern of double-peaked light curves with interpeak emission, outer gap models, like polar cap models, were driven to single pole emission. Romani & Yadigaroglu [36] (RY) developed a "geometrical" single-pole outer gap model in which the gap width, acceleration and radiation were not modeled in detail, but the radiation was assumed to be emitted tangent to the open field line boundary (See Figure 10). This model can reproduce the observed light curves very well (See Figure 11) and has some success accounting for the radio/γ-ray phase offsets when the radio pulse is assumed to come from the opposite polar cap (see Figure 10 and the next section). In addition, the single-pole outer gap models can produce widely separated double pulses and interpeak emission with large inclination angles. Romani (1996) computed the gap width (the area of the open field region crossing the gap) assuming that this width is controlled by breakdown from pair production of curvature photons and thermal X-rays from the surface. The observed emission is curvature radiation from radiation-reaction limited primary electrons accelerated to different energies in various parts of the gap. The spectrum of the accelerated primary particles in this model, when convolved with the hard curvature spectrum (photon index $2/3$), reproduces the softer (photon index $\sim 1.5 - 2.0$) observed spectrum. By constrast, the spectrum of accelerated particles in polar cap models is monoenergetic, but the observed γ-ray spectrum, produced above the accelerating region while the particles are losing energy, is softened both through energy losses and cascading. Both outer gap [35] and polar cap models (DH) are capable of producing phase dependent spectral variations similar to those observed.

Comparison with Other Wavelengths

One notable difference in the computed light curves of the various γ-ray pulsar models is the predicted phase position ($0°$) of the closest approach to the magnetic pole (CAMP) relative to the γ-ray pulses, a critical issue in comparison with the light curves in other wavebands and a key to distinguishing between models. Figure 11 shows that the position of the CAMP in the polar cap models is midway between the two pulses, while in outer gap models the CAMP is outside the γ-ray pulses (although a distant CAMP for the other pole lies between the pulses). Traditionally, radio pulse core components originate near the magnetic axis and conal components are hollow cones centered on the magnetic axis. The standard model for the Vela single radio pulse is core emission, which is inconsistent with the polar cap model, but is consistent with the outer gap model as emission from the opposite pole (RY). However, a less standard model for the Vela pulse as the leading edge of a wide cone [27], is consistent with the polar cap if the radio conal emission is further out along the field lines than the γ-ray cone.

The thermal X-ray pulses are broad pulses centered on the CAMP. The phase positions of the observed ROSAT low-energy pulses of Vela and Geminga are centered between the two γ-ray pulses, which if they are indeed emission from a hot polar cap is consistent with the polar cap models [18], but not the outer gap models, which must interpret this emission as hard X-rays from the outer gap (RY). However, the hard X-ray pulses of Vela recently detected by RXTE [39] and Geminga by ASCA [14] do not look like the ROSAT pulses, but very much like the EGRET pulses.

Many more radio-quiet γ-ray pulsars are expected in the outer gap model [51] than in polar cap models (e.g. [41]), because the radio and γ-ray emission regions are in very different locations. In the polar cap models, γ-ray cascades produce the pairs which produce the radio emission, so the two are physically associated and one expects a high degree of coincidence.

Conclusion

Gamma-ray studies of isolated neutron stars are important to a multiwavelength picture and an understanding of both the geometry and energetics of these sources. Studies of pulsars by the Compton Observatory have added tremendously to our knowledge of pulsar behavior in the gamma-ray range, revealing tantalizing correlations. It appears that gamma-ray studies are strongly favoring a picture where emission is only seen from one-pole of a rotating magnetic field. This may be in some cases at odds with the two-pole picture adopted from radio studies, and it may have profound implications for neutron star evolution. Because there are no other simple patterns that hold for *all* gamma-ray pulsars, the results present some challenges for theoretical work. Both polar cap and outer cap models have their strengths and weaknesses, but both have versions that are consistent with the Compton results. The location of the high energy emission region is therefore still unresolved and must await further observations. In particular, results from X-ray telescopes such as ROSAT, ASCA and RXTE are showing that magnetospheric emission is important for many pulsars. These results in combination with those from CGRO may be able to finally differentiate between models. Extrapolating the trends seen among the presently detected gamma-ray pulsars, there should be a sizeable population of additional pulsars whose output is dominated by gamma radiation. The fraction of these that are only visible at high-energies is an important and still unresolved issue. They lie beyond the present threshold of CGRO. Given the success of CGRO, it will be crucial to follow-up these observations with the next generation of more sensitive gamma-ray telescopes.

ACKNOWLEDGEMENTS

We thank Lucien Kuiper, Mark Strickman, and Radhika Srinivasan for providing pulsar data in tabular form for this review paper.

REFERENCES

1. Arons, J., *ApJ*, **266**, 215 (1983)
2. Backer, D.C., private communication (1997)
3. Becker, W., Brazier, K.T.S., and Trümper, J., *A&A*, **298**, 528 (1995)
4. Becker, W., and Trümper, J., *A&A*, submitted (1997)
5. Brazier, K.T.S. et al., *MNRAS*, **281**, 1033 (1996)
6. Brazier, K.T.S. et al., these proceedings (1997)
7. Cheng, K. S., *Proc. Toward a Major Atmospheric Cherenkov Detector*, ed. T. Kifune (Tokyo: Universal Academy), 25 (1994)
8. Cheng, K. S., Ho, C. and Ruderman, M. A., *ApJ*, **300**, 500 (1986)
9. Carramiñana, A. et al., 1997, these pproceedings (1997)
10. Daugherty, J. K. and Harding, A. K., *ApJ*, **252**, 337 (1982)
11. Daugherty, J. K. and Harding, A. K., *ApJ*, **429**, 325 (1994)
12. Daugherty, J. K. and Harding, A. K., *ApJ*, **458**, 278 (1996)
13. Fierro, J.M., PhD Thesis, Stanford U. (1995)
14. Halpern, J.P. and Wang, *ApJ*, **477**, 905 (1997)
15. Harding, A.K., *ApJ*, **245**, 267 (1981)
16. Harding, A.K., in *Pulsars: Problems and Progress*, IAU Colloquium 160 (ed. M. Bailes, S. Johnston, M. Walker) (1996)
17. Harding, A.K. and Daugherty, J.K., *Space Sci. Rev.*, in press (1997)
18. Harding, A. K. & Muslimov, A., these proceedings.
19. Harding, A.K., Baring, M. and Gonthier, P.L., *ApJ*, **476**, 246 (1997)
20. Hermsen, W., private communication (1997)
21. Hermsen, W. et al., in *Proceedings 2nd INTEGRAL Workshop*, "The Transparent Universe,", ESA SP-382, p. 287 (1997)
22. Kanbach, G., NATO Institute, in press (1997)

23. Kaspi, V.M. et al., *AJ*, in press (1997a)
24. Kaspi, V.M. et al., these proceedings (1997b)
25. Kuiper, L. et al, 1997, in *Proceedings of NATO Advanced Study Institute,* "The Many Faces of Neutron Stars," Lipari, October, (1996)
26. Kuzmin, A.D. and Losovsky. B. Y., *IAU Circ.* 6559 (1997)
27. Manchester, R. N., *Proc. IAU Coll. 160*, ed. S. Johnson, M. A. Walker and M. Bailes, 193 (1997)
28. Matz, S.M. et al., *ApJ*, **434**, 288 (1994)
29. Mayer-Hasselwander, H. A., et al.,*ApJ*, **421**, 276 (1994)
30. Much, R. et al., these proceedings (1997)
31. Muslimov, A. G. and Tsygan, A. I., *MNRAS*, **255**, 61 (1992)
32. Nel, H. I. et al.,*ApJ*, **418**, 836 (1993)
33. Nel, H.I. et al., *ApJ*, **465**, 898 (1996)
34. Ramanamurthy, P.V. et al., *ApJ*, **458**, 755 (1996)
35. Romani, R.W., *ApJ*, **470**, 469 (1996)
36. Romani, R.W. and Yadigaroglu, I.-A., *ApJ*, **438**, 314 (1995)
37. Ruderman, M.A. and Sutherland, P. G., *ApJ*, **196**, 51 (1975)
38. Srinivasan, R. et al., *ApJ*, in press (1997)
39. Strickman, M.S. et al., it ApJ, **460**, 735 (1996)
40. Sturner, S. J. and Dermer, C. D., *ApJ*, **420**, L79 (1994)
41. Sturner, S. J. and Dermer, C. D., *A & AS*, **120**, 99 (1996)
42. Sturner, S. J., Dermer, C. D. and Michel, F. C., *ApJ*, **445**, 736 (1995)
43. Sturrock, P.A., *ApJ*, **164**, 529 (1971)
44. Taylor, J.H., Manchester, R.N. and Lyne, A.G. , *ApJS*, **88**, 529 (1993)
45. Thompson, D.J., in *Pulsars: Problems and Progress*, IAU Colloquium 160 (ed. M. Bailes, S. Johnston, M. Walker) (1996)
46. Thompson, D.J. et al., *ApJ*, **436**, 229 (1994)
47. Ulmer, M.P. et al., *ApJ*, **432**, 228 (1994)
48. Ulmer, M.P. et al., *ApJ*, **448**, 356 (1996)
49. Usov, V. V. and Melrose, D. B., *Aust. J. Phys.*, **48**, 571 (1995)
50. Verbunt, F. et al., *A&A*, **311**, L9 (1996)
51. Yadigaroglu, I.-A. and Romani, R. W., *ApJ*, **449**, 211 (1995)
52. Zepka, A., Cordes, J. M., Wasserman, I., and Lundgren, S.C. *ApJ*, **456**, 305 (1996)
53. Zhang, B. and Qiao, G. J, Λ & Λ, **310**, 135 (1996)
54. Zhang, B., Qiao, G. J., Lin, W. P. and Han, J. L., *ApJ*, **478** 313 (1997)

Recent Results from Observations of Accreting Pulsars

Mark H. Finger* and Thomas A. Prince[†]

*NASA Marshall Space Flight Center, Huntsville, AL
[†]California Institute of Technology, Pasadena, CA 91125

Abstract. Since its launch in 1991, the Compton Gamma Ray Observatory (CGRO) has provided a wealth of new information on binary systems containing a magnetized neutron star accreting from a stellar companion. In particular, the all-sky monitoring capability of the BATSE instrument has allowed the first long-term continuous measurements of the flux and torque histories of the brighter accreting pulsar systems. These have given new insights into the processes by which mass and angular momentum are exchanged between the accretion disk and the neutron star, and the role of radiation and X-ray driven winds in these systems. Observations of transient pulsars with BATSE and OSSE have provided new quantitative tests of standard theories of disk-magnetosphere interactions. Although many new insight have been gained, some of the observations present new puzzles and challenge current models of accretion processes. We review these observations, as well as a sampling of other recent results from CGRO and the ROSSI X-ray Timing Explorer (RXTE).

INTRODUCTION

Spinning magnetized neutron stars with stellar companions provide unique sites for studying mass and angular momentum transfer in binaries, as well as the winds and mass loss from evolved stars. Measurement of the x-ray and hard x-ray flux yields a measure of the mass accretion rate ($\dot{M}_{ns}$), while measurement of the time dependence of the spin frequency (ν_{spin}) can yield estimates of the torque on the neutron star (N), the mass of the companion (M_c), the orbital period (P_{orb}), and the eccentricity (e).

Since the first discovery of an accreting pulsar system, Cen X-3, over 25 years ago, observations have identified the basic classes of accreting pulsars, and have motivated simple "standard models" for the accretion torques and capture of material in these systems (see the reviews: [45,61,26,40,62]). Prior to the launch of CGRO, observations of accreting pulsars were carried out primarily with pointed X-ray telescopes. Since the launch of CGRO in 1991, the Large Area Detectors (LADs) of the Burst and Transient Source Experiment (BATSE) with their 4π steradian field of view have provided continuous monitoring of the brighter accreting pulsars, except for brief periods of earth-occultation. This has made it possible to establish the torque histories of these sources on time scales from days to years. Continuous monitoring has also allowed detailed study of the outburst characteristics of transient pulsars.

In the next section we present an overview of accreting pulsars and the CGRO observations of them. This is followed by a review of results on several selected topics of recent interest including, torque-flux measurements in transient systems, torque switching behavior in wind- and disk-fed systems, characteristics of transient outbursts, the population and orbits of transient systems, the recently detected very-high energy emission from Cen X-3, and some recent results from RXTE. Much of this review covers results obtained by BATSE since the launch of CGRO. We draw heavily from the extensive article by Bildsten et al. [5] (hereafter referred to as Paper I) which discusses the BATSE results in detail.

OVERVIEW

As of the time of this review, there are 44 known accreting pulsars, 26 of which are transients, with spin periods, P_{spin}, ranging from $0.069\,\text{s}$ through $1413\,\text{s}$ (see Table 1). Half of these pulsars have been observed by CGRO, with many of these observations being quite extensive. These sources are thought to be rotating, highly magnetized neutron stars, which radiate X-rays and γ-rays due to accretion of matter. For many of these pulsars a binary companion, from which the accreting matter originates, has been identified. The strong magnetic field ($B \sim 10^{11-13}$ G) controls the accretion flow close to the neutron star forcing captured material to flow along magnetic field lines to the magnetic poles. Misalignment of the rotation axis with the dipole field and asymmetric emission from the accreting polar cap leads to pulsed emission at ν_{spin} with a luminosity $L_{\text{acc}} \sim GM_{\text{ns}}\dot{M}_{\text{ns}}/r_{\text{ns}}$, where M_{ns} and r_{ns} are the mass and radius of the neutron star. The angular momentum of the accreting material and interactions between the accretion flow and the magnetic field produce torques on the neutron star, which can result in observable spin frequency changes over a few days.

Binary accreting pulsars can be classified according to the mass of the donor star as either *low-mass* ($M_c \lesssim 2.5\,M_\odot$) or *high-mass* ($M_c \gtrsim 6\,M_\odot$) systems. The high mass binaries may be divided into those with main-sequence Be star companions and those with evolved OB supergiant companions. The supergiant binaries may be further subdivided into those accreting via Roche-lobe overflow mediated by an accretion disk (disk accretors) and those accreting from the companion's stellar wind (wind accretors). As shown in Figure 1 these classes appear to occupy distinct regions in a P_{spin}-P_{orb} diagram [11,60]. Su-

Table 1: Known Accretion-Powered Pulsars (as of Feb. 1997)

	System[a]	l_{II}	b_{II}	P_{spin} (s)	P_{orb} (d)	Companion (MK Type)
	Low-mass binaries					
BO	GRO J1744−28	0.0	+0.3	0.467	11.8	
BO	Her X-1	58.2	+37.5	1.24	1.70	HZ Her (A9-B)
B	4U 1626−67	321.8	−13.1	7.66	0.0289	KZ TrA (low-mass dwarf)
BO	4U 1728−247 (GX 1+4)	1.9	+4.8	120		V2116 Oph (M6III)
	High-mass supergiant and giant systems					
	SMC X-1	300.4	−43.6	0.717	3.89	Sk160 (B0 I)
BE	Cen X-3	292.1	+0.3	4.82	2.09	V779 Cen (O6−8f)
	RX J0648.1−4419	253.7	−19.1	13.2	1.54	HD 49798 (O6p)
	LMC X-4	276.3	−32.5	13.5	1.41	Sk-Ph (O7 III-V)
B	OAO 1657−415	344.4	+0.3	37.7	10.4	(B0−6Iab)
BO	Vela X-1	263.1	+3.9	283	8.96	HD77581 (B0.5Ib)
	1E 1145−614	295.5	−0.0	297	5.65	V830 Cen (B2Iae)
	4U 1907+09	43.7	+0.5	438	8.38	(B I)
B	4U 1538−52	327.4	+2.1	530	3.73	QV Nor (B0Iab)
B	GX 301−2	300.1	−0.0	681	41.5	Wray 977 (B1.5Ia)
	Transient Be-binary systems					
	A 0538−67	276.9	−32.2	0.069	16.7	(B2 III-IVe)
BO	4U 0115+63	125.9	+1.0	3.61	24.3	V635 Cas (Be)
	V 0332+53	146.1	−2.2	4.37	34.2	BQ Cam (Be)
BO	2S 1417−624	313.0	−1.6	17.6	42.1	(OBe)
BO	EXO 2030+375	77.2	−1.3	41.7	46.0	(Be)
BO	GRO J1008−57	283.0	−1.8	93.5	≈ 248	(Be)
BO	A 0535+26	181.4	−2.6	105	110	HDE245770 (O9.7IIe)
	GX 304−1	302.1	+1.2	272	133 (?)	V 850 Cen (B2Vne)
B	4U 1145−619	295.6	−0.2	292	187	Hen 715 (B1Vne)
B	A 1118−616	292.5	−0.9	405		He3-640 (O9.5 III-Ve)
O	4U 0352+309	163.1	−17.1	835		X Per (O9 III-Ve)
	RX J0146.9+6121	129.9	−0.5	1413		LSI +61° 235 (B5 IIIe)
	Persistent systems with an undetermined companion					
	RX J1838.4−0301	28.8	+1.5	5.45		
	1E 1048−593	288.2	−0.5	6.44		
	1E 2259+586	109.1	−1.0	6.98		
	RX J0720.4-3125	244.2	−8.2	8.38		
	4U 0142+614	129.4	−0.4	8.69		
	Transient systems with an undetermined companion					
B	RX J0059.2−7138[c]	302.1	−45.5	2.76		
	RX J0502.9−6626	277.0	−35.5	4.06		
B	GRO J1750−27	2.4	+0.5	4.45	29.8	
	2E 0050.1−7247	302.9	−44.6	8.9		
	2S 1553−54	327.9	−0.9	9.26	30.6	
B	GS 0834−430	262.0	−1.5	12.3	106	
B	GRO J1948+32	64.9	1.8	18.7		
B	GS 1843+00	33.1	+1.7	29.5		
	GS 2137+57 (Cep X-4?)	99.0	+3.3	66.2		
B	GS 1843−024	30.2	−0.0	94.8		
	Sct X-1	24.5	−0.2	111		
B	GRO J2058+42	83.6	−2.7	198	∼110	
	GPS 1722−363	351.5	−0.6	414		

[a] Sources labeled by B,O, and E have been detected with BATSE, OSSE, or EGRET respectively.

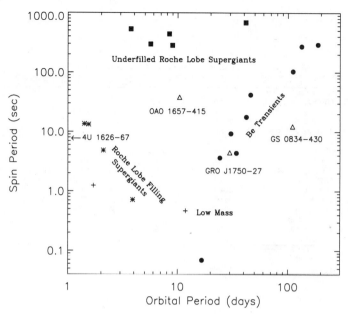

FIGURE 1. Spin period of accreting pulsars versus the binary period. Be transients are plotted as open circles, low-mass objects as plus signs, disk-fed OB supergiants with asterisks and wind-fed OB supergiants with squares. Triangles denote sources with unknown companions. For clarity the low mass system 4U 1626-67 has been left off the plot.

pergiant systems (asterisks) filling, or close to filling their Roche lobe have short spin ($P_{\rm spin} \lesssim 10$ s) and orbital periods ($P_{\rm orb} \lesssim 4$ d). They are persistent, quite luminous ($L \gtrsim 10^{37}$ erg s^{-1}), and tend to show long episodes ($\gtrsim P_{\rm orb}$) of relatively steady torques. Wind-fed supergiant binaries (squares) have longer orbital and spin periods. They are persistent, but less luminous ($L_{\rm x} \sim 10^{35}$–10^{37} erg s^{-1}), and display rapid fluctuations in accretion torque on timescales much shorter than their orbital period. The pulsars with Be star companions (circles) populate a third region of the diagram, displaying a correlation between their spin and orbital periods. These are are transients, with outbursts associated with periastron passage in their eccentric orbits. Low mass systems (pluses) vary widely in their properties. Three of these sources are persistent, while the newly discovered GRO J1744-28, which is covered in a separate review, is a transient.

An additional class, the so-called "6 second pulsars" has also been proposed [37]. These persistent sources with pulse periods near 6 s shown no evidence of binary companions. It has been suggested they are isolated pulsars which accrete from a circumstellar disk formed from the remnants of a previous evolutionary stage [54].

BATSE has continuously monitored the spin frequency and 20~70 keV pulsed flux of 8 persistent pulsars (see Table 1) on a daily basis, and monitored the spin frequency and pulsed flux during ~60 outbursts from 13 transients. These observations have uncovered 5 new accreting pulsars, have allowed the first measurement of 6 orbits and the refinement of 7 others, and provide the first long-term (6 yr), continuous, spatially uniform survey of the activity of transient accreting pulsars. Extensive references are given in Paper I.

OSSE has observed 3 persistent accreting pulsars and 7 transients (see Table 1). Pulse profiles for GX 1+4 are given by Maisack et al. [34]. The OSSE observations of the Be/X-ray transient A0535+262 revealed a cyclotron line in the phase averaged flux spectrum at 110 keV, implying a magnetic field near 10^{13} G [23]. The phase resolved spectra have been analyzed by Maisack et al. [33,35]. Negueruela et al. [41] compare the OSSE observations of 4U 0115+63 with BATSE, optical, and IR observations. A summary of Be/pulsar binary results is given by Grove [24]. OSSE observations of the transient low-mass accretion pulsar GRO J1744-28 revealed pulse phase lags within its bursts [51].

A detection by EGRET of an outburst of > 100 MeV photons from the persistent pulsar Cen X-3 has been reported by Vestrand et al. [56]. This discovery is discussed below.

RECENT RESULTS

Torques In Transient Pulsar Outbursts

Dramatic spin frequency increases are observed during the largest transient pulsar outbursts, implying the presence of accretion disks. Comparisons of the torques inferred from BATSE frequency observations during large transient outbursts with simultaneous flux and QPO measurements have provided new tests of accretion theory.

At the boundary between an accretion disk and a neutron star magnetosphere, the accretion flow changes from nearly Keplerian motion to an inflow constrained to move along the magnetic field lines, which rotate with the neutron star. This poorly understood boundary region is located at a radius [44,31]:

$$r_m = K\mu^{4/7}(GM_{ns})^{-1/7}\dot{M}_{ns}^{-2/7} \qquad (1)$$

where K is a constant thought to be between 0.5 and 1 and μ the neutron star magnetic moment. Equation 1 with $K = 0.91$ gives the Alfvén radius for spherical accretion. Assuming the gas deposits its angular momentum at the magnetospheric boundary and that field lines transport all of this angular momentum to the star, the accreting pulsar will experience a spin-up torque

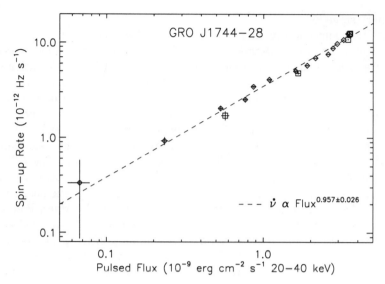

FIGURE 2. The spin-up rate versus flux relationship for GRO J1744-28 during the discovery outburst. The square symbols are on the rise of the outburst, the diamonds are on the decline. The line gives the best fit power-law, which has an index of 0.957 ± 0.026.

$$N = I 2\pi \dot{\nu}_{\text{spin}} \approx \dot{M}_{\text{ns}} (GM_{\text{ns}} r_{\text{m}})^{1/2} \qquad (2)$$

where I is the moment of inertia of the neutron star. $(GM_{\text{ns}} r_{\text{m}})^{1/2}$ is the specific angular momentum of matter in a Keplerian orbit at the inner disk radius r_{m}.

At lower mass accretion rates torques caused by the interaction of the magnetic field with the accretion disk [22] must be considered. The torques are predicted to be negative and dominate the direct accretion of angular momentum (eq. 2), causing spin-down, when the magnetosphere approaches the corotation radius. At the corotation radius, $r_{\text{co}} = (GM_{\text{ns}})^{1/3} (2\pi \nu_{\text{spin}})^{-2/3}$, the period of the Keplerian orbit is synchronous with the rotation of the neutron star. At yet lower mass accretion rates, the magnetospheric radius expands outside the corotation radius, and the accretion flow shuts-off by the centrifugal force acting on any material corotating with the magnetic field. This centrifugal inhibition of accretion is expected during the quiescent state of transient pulsars [50].

At high mass accretion rates the spin-up rate $\dot{\nu}_{\text{spin}}$ is predicted to depend on the flux F as $\dot{\nu}_{\text{spin}} \propto F^\alpha$ with $\alpha = 6/7$ (assuming that $F \propto \dot{M}_{\text{ns}}$). The first observational test of this prediction was with the EXOSAT observation of EXO 2030+375, which found $\dot{\nu}_{\text{spin}} \propto F^\alpha$ with $\alpha \sim 1.2$. The test was over a fairly small dynamic range in $\dot{\nu}_{\text{spin}}$, because of the difficulty of simultaneously determining the spin-up rate and the pulsar's orbital parameters.

The observations with BATSE of the bursting pulsar GRO J1744-28 [19] provide a test of the predicted relation between spin-up and flux over a larger dynamic range. This is shown in Figure 2, which compares the observed spin-up range with the 20-40 keV pulsed flux. The power-law behavior is clear, with points from the rise and fall of the outbursts matching nicely. On the other hand, minor systematic variations from the best fit power-law are evident. The best fit power-law index of $\alpha = 0.954 \pm 0.026$ is near, but formally inconsistent with the expected 6/7. This could be due either to a deviation from the torque predicted by equation 2, or a non-linear relationship between the hard X-ray pulsed flux and mass accretion-rate.

In general, a non-linear dependence of the flux on mass accretion-rate is expected. Pulsar beaming patterns change with luminosity, changing the spectra, pulse profile and the overall degree of beaming. Changes in the pulse profile (and hence the pulse fraction) will particularly effect the pulsed flux used in Figure 2. Empirical determination of the dependence of the magnetospheric radius on mass accretion rate (eq. 1) by torque-flux comparisons may therefore not be feasible.

Observations in 1994 of Quasi-Periodic Oscillations (QPO) in A 0535+262 [18] provided an alternative probe of the dependence of the magnetospheric radius on mass accretion rate. The relationships between the QPO frequency, the spin-up rate, and the flux observed in this outburst are shown in Figures 3 and 4. Also shown are the predictions of the beat frequency model [1] and the Keplerian frequency model [53]. In the beat frequency model the QPO

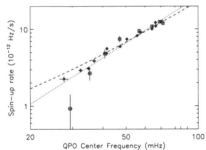 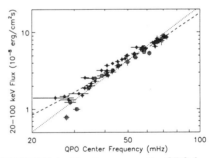

FIGURE 3. The spin-up rate versus the QPO frequency for A 0535+262. Squares are on outburst rise, diamonds on outburst decline. The lines give the predicted relationship for the beat frequency (dashed) and Keplerian frequency (dotted) models, with best fit normalizations.

FIGURE 4. The flux versus the QPO frequency for A 0535+262 (Finger, Wilson & Harmon 1996). Squares are on outburst rise, diamonds on outburst decline. The lines give the predicted relationship for the beat frequency (dashed) and Keplerian frequency (dotted) models, with best fit normalizations.

frequency is given by

$$\nu_{\text{QPO}} = \nu_k(r_m) - \nu_{\text{spin}} = (GM)^{1/3} r_m^{-2/3} - \nu_{\text{spin}} \tag{3}$$

where $\nu_k(r_m)$ is the Keplerian orbital frequency at the inner edge of the accretion disk and $\nu_{\text{spin}} = 9.67$ mHz is A 0535+262's spin frequency. In the Keplerian frequency model, $\nu_{\text{QPO}} = \nu_k(r_m)$. In either model the QPO provides a measure of the location of the accretion disk-magnetosphere boundary. Comparisons of the flux or torque with the QPO frequency therefore test equation 1 fairly directly.

The data follow the expected trends of $\dot{\nu}_{\text{spin}} \propto \nu_k^2$ and $F \propto \nu_k^{7/3}$, and the measured normalization constants in these relationships are consistent with reasonable neutron star parameters and the measured magnetic field of $B = 9.5 \times 10^{12}$ G (Grove et al. 1995) if we assume $K = 1$. The Ghosh & Lamb value $K = 0.5$ [22] is, however, ruled out. This suggests that the neutron star's magnetic field fully threads the disk [32].

Torque Reversals in Persistent Accreting Pulsars

Continuous monitoring by BATSE has revealed that torque switching between spin-up and spin-down is a common feature in the bright persistent pulsars, a result anticipated in earlier statistical analyses [6,14]. In the BATSE data the switching appears as abrupt transitions between torque states of comparable magnitude but opposite sign. Figure 5 shows the distinctive sawtooth quality of the spin-frequency histories of two pulsars, Cen X-3 and OAO 1657-415. In these systems, the torque reversals occur on timescales of weeks. In other systems such as GX 1+4 [8] and 4U 1626-67 [9] the reversals occur at intervals of years. Interestingly, in all four torque-reversal systems, the spin-up torque is somewhat larger in magnitude than the spin-down torque.

A further important clue is provided by the GX 1+4 system: during periods of prolonged spin-down, the magnitude of the spin-down torque is correlated with the observed 20-100 keV flux [8]. This is opposite in sign to the correlation expected in some models (see below).

Three classes of models have been suggested to explain the phenomenon of torque reversal:

Accretion from a prograde disk interacting with the pulsar magnetosphere. In this model, the pulsar accretes matter near equilibrium from a radius comparable to the corotation radius. In one variant [22] positive torques arise from the specific angular momentum of the material as well as by threading of the disk by the pulsar magnetic field inside r_{co}. Spin-down torques arise from fields which thread the disk outside r_{co}. The sum of these torques can be positive or negative depending on the radius from which mass is accreted, which is governed by $\dot{M}_{\text{ns}}$. Wang [58] has recently analyzed this type of model for the case of oblique rotators. In other variants of this class

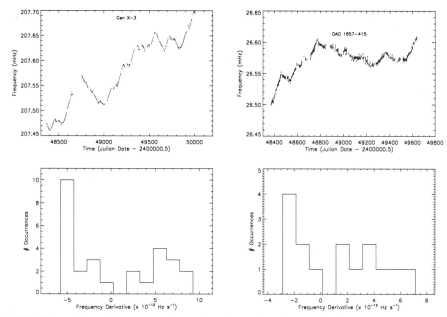

FIGURE 5. Torque histories and $\dot{\nu}_{spin}$ distributions for Cen X-3 (left) and OAO 1657-415 (right). The frequency derivatives ($\dot{\nu}_{spin}$) were obtained by performing a linear fit to the ν_{spin} measurements in each spin-down or spin-up period (see Koh 1997).

of model [4,3] the spin-down torque is due to loss of angular momentum in a magnetohydrodynamic outflow. The sign of the torque is again determined by the radius from which matter is being accreted, which in turn depends on $\dot{M}_{ns}$. To explain torque reversals, these models require abrupt changes in $\dot{M}_{ns}$, between values that would produce spin-up and spin-down torques of approximately equal magnitude. These models do not naturally explain the correlation between enhanced flux and spin-down torque observed in GX 1+4.

Transition from Keplerian to advection-dominated sub-Keplerian flow. Recently, Yi et al. [64] have proposed that torque reversals are due to transitions from the familiar Keplerian flow to an advection-dominated flow. This happens at a critical mass accretion rate, $\dot{M}_{crit}$ below which the disk becomes optically thin and has very low density. The specific angular momentum of material at the disk inner edge is lower in the advection-dominated flow, resulting in a lower inflow of angular momentum. An attraction of this model is the prediction of an $\dot{M}_{crit}$ at which a sharp torque transition is expected, like those observed in the four systems mentioned above. It remains to be seen whether allowable values of $\dot{M}_{crit}$ span the range of observed values in the torque-reversal systems.

Disk reversal: prograde-retrograde. Nelson et al. [42] have recently discussed how torque reversals might be the result of disk reversals, similar to the suggestion of Makishima et al. [36] for the torque reversal seen in GX 1+4. If matter is accreted via Roche lobe overflow, disk reversals are not expected because of the high specific angular momentum of the matter flowing from the companion. However, simulations [20,46] have shown that retrograde disks are expected to form episodically in wind-fed systems.

Koh [30] discusses the likelihood that all four of the systems for which torque reversals are observed have disks that are fed by wind accretion. For Cen X-3, Day & Stevens [13] have argued that X-ray heating of the companion will suppress any radiative wind and give rise to a X-ray driven thermal wind. OAO 1657-415 lies in the Corbet diagram between the wind accretors and the disk accretors [7] and might thus be expected to show signs of both. Koh shows that the X-ray and normal radiative wind components of OAO 1657-415 may be comparable. Chakrabarty & Roche [10] suggest that V2116 Oph, the companion to GX 1+4, may be an FGB star with a slow dense stellar wind. The system 4U 1626-67 is less well characterized. The companion is expected to be very low-mass but has not yet been identified. Nelson et al. [42] estimate that an X-ray driven wind from the companion could account for the inferred mass accretion rate if there is efficient capture of the wind, a reasonable assumption given that the neutron star dominates the gravitational potential of the system. Thus the torque switching observed by BATSE could be due to the presence of a wind component, in some cases driven by X-ray heating. However, these models do not naturally explain why the observed spin-down torque is always slightly smaller in magnitude than the spin-up torque.

Further observations of the torque reversals with RXTE will be helpful in discriminating between the various models.

Outbursts of Be/X-ray Pulsar Systems

The majority of the known accreting pulsars are transients, normally detected only during outbursts which last from days to months. The identified optical counterparts of transient accreting pulsars which have been identified are spectral class Be (or Oe) stars. With the exception of the LMXB transient pulsar GRO J1744-28, the transient pulsars with no identified counterpart are also likely to have Be star companions. BATSE has observed more the 60 transient pulsar outbursts in six years of observations.

Be stars are main-sequence stars of spectral type B that show (or have shown) Balmer emission lines (see [48] for a review). This line emission is thought to be associated with circumstellar material shed by the rapidly rotating star into its equatorial plane, forming a dense stellar wind. In Be star / neutron star binaries, capture of material from this dense wind by the neutron star causes the outbursts. These outbursts are often associated with the

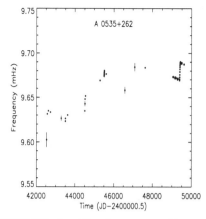

FIGURE 6. The long term history of A0535+262 frequency measurements (Bildsten et al. 1997).

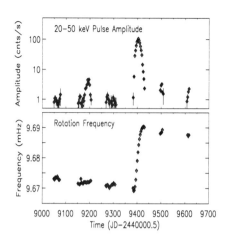

FIGURE 7. A0535+262 spin frequency and pulse amplitude measurements from BATSE observations (Finger, Wilson & Harmon 1996).

periastron passage of the pulsar's wide eccentric orbit. Direct wind accretion is often assumed to occur during the smaller outbursts, however it is now clear that an accretion disk is present in the largest outbursts. A review of these "Be/X-ray binary" systems is given by Apparao [2].

One of the best observed Be/X-ray pulsars is A0535+262 [21]. An orbital period of 111 days was inferred from the regular spacing of the x-ray outbursts [39] and confirmed by pulse timing measurements using BATSE [16]. The long-term history of frequency observations is shown in Figure 6. There is an overall trend of increasing spin frequency. Spin-up of the pulsar occurs almost entirely during the largest outbursts. Motch et al. [38] classify the outbursts as either "normal", with low flux and little or no spin-up, "giant", with high flux, rapid spin-up and a peak at a later orbital phase than normal outbursts, or "missing", with no outburst detected at an orbital phase where they might be expected. Prior to 1991 the most extensive observations were of the giant outbursts which were mainly observed by target-of-opportunity programs.

The BATSE observations are shown in Figure 7. These show a series of six outbursts, one per 111 day orbit, with the fourth (February-March 1994) a giant outburst, and the others normal outbursts. No outbursts were detected during the six orbits prior to or the eight orbits after those shown in Figure 7. Outside of the giant outburst the pulsar is spinning-down. The presence of series of normal outbursts, and the association of giant outbursts with them, is not evident in the pre-1991 data due to the selectivity of coverage.

Giant outbursts and series of normal outbursts are observed in many other sources. Both behaviors have been seen in V0332+53 [50]. The initial dis-

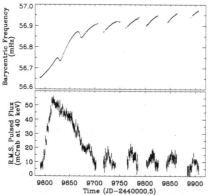

FIGURE 8. The barycentric pulse frequency (not orbitally corrected) and pulsed flux of 2S 1417-624 observed by BATSE (Finger, Wilson & Chakrabarty 1996).

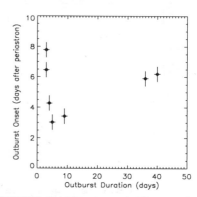

FIGURE 9. The time of 4U 0115+634 outburst onsets relative to periastron passage versus the outburst durations from BATSE observations. 4U 0115+634 has a 24.3 day orbit.

covery of EXO 2030+375 [43] was during a giant outburst, while BATSE observations show a series of 13 consecutive normal outbursts [52]. Figure 8 shows a series of outburst of 2S 1417-634 [17] which began with a giant outburst and ended with five normal outbursts. BATSE observations also show consecutive series of outbursts from GRO J1008-57, GRO J2058+42, 4U 0115+375, 4U 1145-619 and GS 0834-430 (Paper I). The persistent wind-fed source GX 301-2 also shows behavior that is similar in many aspects [29].

The peaks of the giant outbursts have been observed to be delayed in orbital phase from the peaks of the normal outbursts, with the delay correlated with the peak luminosity [63,38]. The BATSE observations suggest this effect is a result of the correlation of the peak luminosity and outburst duration. For a given source the outburst onsets observed by BATSE are found to be restricted in orbital phase, with these phases uncorrelated with peak flux or outburst duration. This is shown in Figure 9 for 4U 0115+634. This suggests that the giant outbursts begin in the same manner as normal outbursts. The standard explanation for the normal outbursts is direct wind accretion of the Be stellar disk material [59], while the giant outbursts are explained as an episode of enhanced outflow from the Be star. Based on the BATSE observations, it was proposed in Paper 1 that an accretion disk exists for extended periods, with tidal torques on the outside of the disk causing inflow near periastron that results in the normal outbursts. With a sufficient build-up of material in the outer disk, some of these normal outbursts trigger a thermal disk instability similar to that in dwarf nova [49] and soft x-ray transients [55,28], initiating a giant outburst.

The Population and Orbits of Accreting Be/Pulsar Systems

BATSE has detected 13 transient pulsars with high-mass companions, all known or suspected Be/X-ray pulsars (see Table 1). This constitutes a complete sample of 20-50 keV transient sources with pulsed flux in excess of $\sim 2 \times 10^{-10}$ erg cm^{-2} s^{-1}. Paper I and Koh [30] have used this sample to draw conclusions about the population of high-mass transient pulsars in the galaxy. The transients detected by BATSE are primarily Be/X-ray binaries.

The progenitor system of a Be/X-ray binary is likely a compact system of two B stars. The more massive star evolves more rapidly and transfers mass to the companion via Roche-lobe overflow, resulting in a helium star. The transfer of angular momentum to the companion is consistent with the observation that many Be stars exhibit very high rotational velocities. Next, the helium star undergoes a supernova explosion producing a neutron star in a wide eccentric orbit, if the system remains bound. Because the kick velocities are expected to be larger than typical orbital velocities of the pre-supernova system, most such systems are expected to be disrupted. In the limit of large kick velocities, Kalogera [27] has derived an expression for the distribution of eccentricities. Paper I compares the distribution of eccentricities for the BATSE sample of pulsars against this predicted distribution. The results are

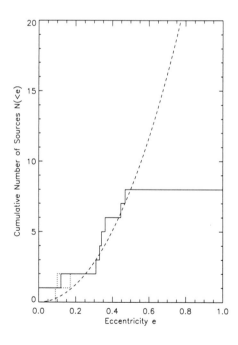

FIGURE 10. The cumulative distribution of Be/X-ray pulsar binary eccentricities. The predicted distribution (normalized to 8 sources with $e < 0.5$) is given by the dashed line. The dotted lines give the ranges of eccentricity for the cases of 2S 1553-54 and GS 0834-430 (see Paper I).

shown in Figure 10.

The predicted and observed distributions are consistent for $e < 0.5$, but diverge significantly above. Whereas approximately 30 sources would be expected for $e > 0.5$, none are observed. Paper I interprets this as a selection effect: neutron stars in high-eccentricity systems may undergo outbursts only near periastron, making such outbursts less frequent. The orbits of such systems will then be difficult to determine. A prediction is that almost all of the Be/X-ray pulsars for which no orbit has been determined (see table) have high eccentricity, with $e > 0.5$.

Cen X-3: Detection of GeV Radiation from an Accreting Pulsar

Recently, Vestrand et al. [56,57] reported a detection of the pulsar Cen X-3 by the EGRET instrument on CGRO. This is the first detection of an accreting pulsar with EGRET, which is sensitive to gamma-rays above 30 MeV. No detections have been reported using the Comptel instrument, which is sensitive to emission above 1 MeV (W. Collmar, private communication, 1997).

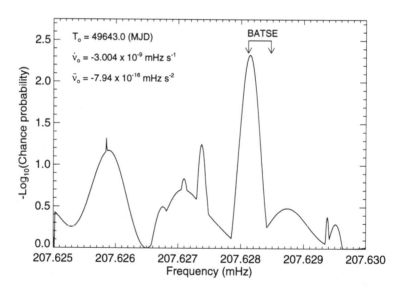

FIGURE 11. Chance probability versus the Cen X-3 spin frequency (at epoch TJD 9643) for the statistic used by Vestrand et al. (1997) to detect pulsations from Cen X-3 in the EGRET data. The 68% confidence range determined from the BATSE data is also shown.

The EGRET observations were made in October 1994 and revealed a $\sim 5\sigma$ significance source located at $l = 292.20°$ and $b = 0.48°$, $10.7'$ from Cen X-3 and consistent with its position. The source was not detected in earlier observations indicating that it is variable. The location in the galactic plane, and the strength of the source argue against it being an accidentally coincident blazar. Assuming a distance of 8 kpc, Vestrand et al. report a luminosity of $L_\gamma(0.1 - 10 \text{ GeV}) \sim 5 \times 10^{36}$ ergs^{-1}.

Using an ephemeris for the pulse frequency derived from BATSE data, Vestrand et al. tested for the presence of a periodic modulation of the flux from the detected EGRET source. The results are shown in Figure 11, taken from Vestrand et al. For a single harmonic, the chance probability for no modulation in the 0.1-10 GeV range at the BATSE-derived frequency is 0.47%. For two harmonics, the chance probability of no modulation is 0.16%.

Recent Observations by RXTE

The ROSSI X-ray Timing Explorer (RXTE) has observational capabilities that are an excellent complement to those of CGRO. The Proportional Counter Array (PCA) on RXTE with its 6250 cm^2 area and 1° field of view can make observations of individual systems at a much lower luminosity than BATSE. The All Sky Monitor (ASM) on RXTE does not have the timing capabilities of the BATSE detectors, but has sensitivity comparably to BATSE's (for an accreting pulsar) and can be used to determine the total X-ray flux. BATSE can make sensitive measurements of pulsed hard X-ray flux, and lower sensitivity measurements of total hard X-ray flux using Earth occultation. With both RXTE and CGRO observations, both the total and pulsed fluxes can be monitored continuously. The High-Energy X-ray Timing Experiment (HEXTE) also provides complementary capabilities, in particular good timing and spectroscopic capabilities at hard X-ray energies.

The PCA is performing accurate measurements of accreting pulsars at low luminosities, revealing some interesting new phenomena and testing ideas of about the dependence of the accretion and pulse properties on the accretion rate. The 8 day binary 4U1907+09 was observed to have dips in flux throughout its binary orbit from a luminosity of 10^{36} ergs s^{-1} to a few times 10^{35} ergs s^{-1}, during which the spectrum was substantially the same and not associated with increased column density [65]. GX 1+4 and GRO J1744-28 were observed when the pulses dropped below detection level, suggesting the possibility of propeller action of the magnetic dipole of the neutron star predicted for low accretion rates [12].

The observations of the pulse shape evolution in Her X-1 [47] confirm the complex pattern first observed in detail with Ginga and adds new details, such as transition to a quasi-sinusoidal profile at the end of the Main High state. The wealth of complex profile changes observed in Her X-1 are not associated with intrinsic luminosity variations as in transient pulsar outbursts and provide stringent constraints on pulse evolution models such as neutron star free precession or an inner disk occultation of the neutron star. The HEXTE instrument has observed Her X-1 on 3 occasions [25] and has measured the cyclotron line absorption energy of the pulsar. A power law fit with an exponential cutoff plus gaussian absorption yielded a cyclotron line energy of 39.0 keV in each observation, with statistical errors of 1% or better. There was no evidence for a harmonically spaced overtone. The line width was σ = 7.5 keV, large compared to the instrumental resolution of 3.0 keV at 39 keV. Significant residuals between 30 and 50 keV on the scale of the detector resolution indicate that the description of a Gaussian line on an exponentially cut-off continuum is now inadequate.

ACKNOWLEDGEMENTS

We thank in particular Jean Swank and Brian Vaughan for significant contributions to sections of this review.

REFERENCES

1. Alpar, M. A. & Shaham, J. 1985, Nature 316, 239
2. Apparao, K. M. V. 1994, Space Science Rev. 69, 255
3. Arons, J. , Burnard, D., Klein, R. I., McKee, C. F., Pudritz, R. E., & Lea S. M. 1984, in High Energy Transients in Astrophysics, ed. S. E. Woosley, AIP Press, 215
4. Anzer, U. & Borner, G. 1980, A&A, 83, 133
5. [Paper I] Bildsten, L., Chakrabarty, D., Chiu, J. Finger, M. H., Koh, D. T., Nelson, R. W., Prince, T. A., Rubin, B. C., Scott, D. M., Stollberg, M., Vaughan, B. A., Wilson, C. A., & Wilson, R. B. 1997, ApJ Supp. (accepted)
6. Baykal, A. & Ogelman, H. 1993, A&A, 267, 119
7. Chakrabarty, D. et al. 1993, ApJL, 403, L33
8. Chakrabarty, D. et al. 1997a, ApJ, 474, 414
9. Chakrabarty, D. et al. 1997b, ApJL, 481, L101
10. Chakrabarty, D. & Roche, P. 1997, ApJ (accepted)
11. Corbet, R. H. D. 1986, MNRAS, 220, 1047
12. Cui, 1997, ApJ, 482, L163
13. Day, C. S. R. & Stevens, I. R. 1993, 403, 322
14. de Kool, M. & Anzer, U. 1993, MNRAS, 262, 726
15. Finger, M. H., Wilson, R. B., & Fishman, G. J. 1994, in The Second Compton Symposium, Fichtel, C. E., Gehrels, N. & Norris, J. P. eds., (AIP, New York), 304

16. Finger, M. H., Cominsky, L. R., Wilson, R. B., Harmon, B. A., & Fishman, G. J. 1994, in AIP Proc. 308, The Evolution of X-rya Binaries, ed. S. Holt & C. S. Day (New Your:AIP), 459
17. Finger, M. H., Wilson, R. B., & Chakrabarty, D. 1996, A&A Supp. 120, 209
18. Finger, M. H., Wilson, R. B., & Harmon, B. A. 1996, ApJ 459, 288
19. Finger, M. H. et al. 1996, Nature, 381, 291
20. Fryxell, B. A. & Taam, R. E. 1988, ApJ 335, 862
21. Giovannelli, F. & Graziati, L. S. 1992, Space Sci. Rev. 59, 1
22. Ghosh, P. & Lamb, F. K. 1979, ApJ, 234, 296
23. Grove, E. et al. 1995, ApJ 438, L25
24. Grove , J.E. 1996, Mem. Soc Astron Ital, 67, 127-141
25. Gruber, D. E., Heindl, W. A., and Rothschild, R. E. 1997, Paper 19-14, this conference
26. Hayakawa, S. 1985, Physics Reports, 121 317
27. Kalogera, V. 1996, ApJ, 471, 352
28. King, A.R., Kolb, U. & Buderi, L. 1996, ApJ 464, L127
29. Koh, D.T. et al. 1997, ApJ 479, 933
30. Koh, D.T. 1997, PhD thesis, Caltech
31. Lamb, F. K., Pethick, C. J., & Pines, D. 1973, ApJ, 184, 271
32. Li, X.-D. 1997, ApJ 476, 278
33. Maisack, M. et al. 1996 A&AS 120 179
34. Maisack M. et al. 1996, Mem. Soc. Astron. Ital. 67, 373
35. Maisack, M. et al. 1997, submitted to ApJ
36. Makishima et al. 1988, Nature, 333, 746
37. Mereghetti, S. & Steela, L. 1995, ApJ 442, L17
38. Motch, C., Stella, L., Janot-Pachenco, E., & Mouchet, M. 1991, ApJ 369, 490
39. Nagase, F., et al. 1982, ApJ 263, 814
40. Nagase, F. 1989, PASJ, 41, 1
41. Negueruela, I. et al. 1997, MNRAS 284, 859
42. Nelson et al. 1997, submitted to ApJL
43. Parmar, A. N., White, N. E., Stella, L., Izzo, C. & Ferri, P. 1989, ApJ 338, 359
44. Pringle, J. E. & Rees, M. J. 1972, A&A 21, 1
45. Rapaport, S. & Joss, P. 1977, Nature 266, 123
46. Ruffert, M. 1997, A&A 317, 793
47. Scott, D. M., Leahy, D., Finger, M. H., & Wilson, R. B. 1997, Paper 19-15, this conference
48. Slettebak, A. 1988, PASP, 100, 770
49. Smak, J. 1983, ApJ 272, 234
50. Stella, L., White, N. E., & Rosner, R. 1986, ApJ, 308, 669
51. Strickman, M.S. et al. 1996 ApJ 464 L131
52. Stollberg, et al. 1994, in The Evolution of X-ray Binaries, ed. S. S. Holt and C. S. Day, AIP Press, 255
53. van der Klis, M., Jansen, F., van Paradijs, J.P., Lewin, W. H. G., Trumper, J., & Sztajno 1987, ApJ, 313, L19
54. van Paradijs, J., Taam, R. E., & van den Heuvel, E. P. J. 1995, A&A 229, L41

55. van Paradijs, J. 1996, ApJ 464, L139
56. Vestrand, W. T., Sreekumar, P., Mori, M. 1997, ApJ 483, L49
57. Vestrand, W. T., Sreekumar, P., Mori, M. 1997, Paper 19-22, this conference
58. Wang, Y. M. 1997, ApJL, 475, L135
59. Waters, L. B. F. M., de Martino, D., Habets, G. M. H. J., & Taylor, A. R. 1989, A&A 223, 207
60. Waters, L. B. F. M. & van Kerkwijk, M. H. 1989, A & A, 223, 196
61. White, N. E., Swank, J. H., & Holt, S. S. 1983, ApJ, 270, 711
62. White, N. E., Nagase, F., & Parmar, A. N. 1995, in X-Ray Binaries, ed. W.H.G. Lewin, J. Van Paradijs & E.P.J. van den Heuvel, Cambridge Univ. Press, 1
63. Whitlock, L., Roussel-Dupré, D., & Priedhorsky, W. 1989, ApJ 338, 381
64. Yi, I., Wheeler, C., Vishniac, E. T. 1977 (preprint: astro-ph9704269)
65. In't Zand, Strohmayer & Baykal 1997, ApJ, 479, L47

Low-Mass X-Ray Binaries and Radiopulsars in Binary Systems

Marco Tavani* and Didier Barret[†]

*Columbia University, New York (USA) and IFCTR-CNR, Milan (Italy)
[†]CESR-CNRS, Toulouse (France)

Abstract.
We review the study of high-energy emission from accreting neutron stars and radio pulsars in binary systems with particular emphasis on results from the Compton Observatory. BATSE detected emission in the energy range 20-200 keV from several low-mass X-ray binaries believed to contain weakly magnetized neutron stars. Twelve low-mass X-ray binaries (also including sources detected by SIGMA) have been significantly detected in the hard X-ray band. Before the SIGMA and BATSE discoveries only black hole candidates were known to emit hard X-ray emission. We review the existing observations in the hard X-ray energy band, address the issue of neutron star vs. black hole high-energy emission, and discuss theoretical implications.

Binaries containing energetic radiopulsars are also a potential source of high-energy emission. We discuss the important CGRO detection of the Be star/pulsar system PSR B1259-63 near periastron and its implications.

HARD X-RAY EMISSION FROM LOW-MASS X-RAY BINARIES

Hard X-ray emission (photon energy larger than $\sim$ 20 keV) from compact star binaries is a manifestation of energetic processes in accretion disks. Indeed, typical temperatures for thermal-like emission near the surface of neutron stars (NSs) or in optically thick regions of disks surrounding black hole candidates (BHCs) are below 5 keV (e.g., [78]). Hard X-ray emission is usually observed from X-ray pulsars (probably as a consequence of shock emission of material in accretion columns channelled by large magnetic fields with $B_s \sim 10^{12}$ G) [52], and from BHCs (most notably Cyg X-1 [79,57,58]). 'Thermal' disk models of BHC emission (e.g., [73,74,54]) may account for the hard X-ray emission in terms of unsaturated Comptonization of soft X-ray photons in an optically thin corona [77]. For BHCs, the corona is assumed to be heated by a mechanism related to the black hole nature of the compact object. Also in the case of weakly magnetized neutron stars (with $B_s \lesssim 10^{10} - 10^{11}$G) believed to be the accreting sources of X-ray bursters (XRBs), the inner part of the accretion disk is close to the compact object surface. However, contrary to the BHC case, a large additional flux of soft X-rays originating from the neutron star surface can be a major source of cooling photons. In thermal models of X-ray emission from XRBs, this additional soft component should suppress emission above ~ 10 keV [78]. The typical persistent X-ray emission of low-mass X-ray binaries in their high states confirms this expectation (e.g., [92]).

However, recent XRB observations by SIGMA on board of GRANAT and BATSE on board of CGRO drastically changed this idealized picture. In the next section we briefly review the hard X-ray detections of XRBs, summarizing both the temporal and spectral characteristics of the emission. We also discuss the similarities and differences between neutron star and black hole systems regarding their hard X-ray emission.

Hard X-ray detections of low-mass X-ray binaries

Historically, the first XRB detected at energies above 50 keV was the soft X-ray transient Cen X-4 during a major outburst in 1979 [9]. Since then, the detection capability improved with the operation of coded-aperture imaging detectors. The GRIP instrument produced the first detailed hard X-ray image (20-120 keV) of the Galactic center region and revealed two sources; the black hole candidate 1E1740-2942 and the X-ray burster GX354-0 (MXB1728-34) [15]. The SIGMA instrument pointed at the Galactic plane and center many times since 1990. Five more X-ray bursters were detected in the Galactic center regio during seven years of operation: 1E1724-3045 in the Globular cluster Terzan 2 [2], KS1731-260 [3], A1742-294 [12], SLX1732-304 in the globular cluster Terzan 1 [8], and SLX1735-269 [25] recently discovered to emit X-ray bursts [7]. In addition SIGMA detected GX354-0 several times [14] confirming the GRIP results. Four new X-ray bursters were later detected in the hard X-ray band by BATSE using the Earth occultation monitoring technique: 4U1608-522 [95], Aql X-1 [34], 4U0614+091 [17], 4U1705-44 [5] (also detected by SIGMA [13,69]), and the new X-ray burster GS1826-24 [88] (detected by both OSSE [75] and BATSE [36]). If we also include the detection of Cir X-1 by HEXE [60], the number of X-ray bursters currently detected above 50 keV is set to twelve (see Table 1). We consider these results a major breakthrough in the study of high energy emission from compact objects. As XRBs are intrinsically faint hard X-ray sources [6], this number is expected to grow as the sensitivity of current (RXTE, SAX) and future (INTEGRAL) experiments is improved. As BATSE monitoring progresses, several more XRBs might be added to the list of Table 1 (for a review, see ref. [5]. See also ref. [76] for a possible detection of Scorpius X-1 by OSSE).

TABLE 1. X-ray to hard X-ray observations of all the X-ray bursters detected above 50 keV. Exp. corresponds to the name of the experiment. Range defines the energy range of the hard X-ray observations. Models (parameters) are: PL=Power Law (α: photon index), TB=Thermal bremsstrahlung (kT: temperature), ST=Sunyaev & Titarchuk Comptonization model (kT_e: electron temperature and τ: half optical thickness of the disk), BPL=Broken Power Law (α_1, α_2: photon index and E_{Break}: energy of the break). CPL = Power-Law with Cutoff (α: photon index and E_{Cutoff} cutoff energy). F_{HX} is the hard X-ray flux in the energy range of the detection given in units of 10^{-9} ergs cm^{-2} s^{-1}.

	Date	Exp.	Range (keV)	Model	Parameters	F_{HX}	Ref.
4U0614+09	08-09/91	BATSE	20-100	PL	2.7	0.70	[17]
Cen X-4	05/79	SIGNE2	12-160	CPL	1.0, 45.0	58.5	[9]
Cir X-1	01-02/89	HEXE	20-80	ST	8.1, 10.6	3.3	[60]
4U1608-52	06/91-12/91	BATSE	20-200	BPL	1.75, 3.2, 65.0	2.9	[95]
				ST	23.0, 1.9		
4U1705-44	02-03/94	SIGMA	35-200	PL	2.6	0.94	[69]
	12/93-01/94	BATSE	20-200	BPL	1.5, 2.7, 70	1.3	[5]
1E1724-3045	03-04/90[a)]	SIGMA	35-200	PL	1.7	0.98	[2]
	03/90-10/93[b)]	SIGMA	35-150	PL	3.0	0.26	[23,24]
GX354-0	04/12/88	GRIP	35-120	PL	3.4	1.0	[15]
				TB	31.0		
GX354-0	02-04/92	SIGMA	30-200	BPL	1.4, 4.7, 58.0	0.7	[14]
				ST	12.0, $\gtrsim 3.8$		
				TB	38.0		
GX354-0	02/92-04/92	BATSE	20-100	TB	30.	0.3	[33]
				PL	2.7		
GX354-0	10/92-12/92	BATSE	20-100	TB	30.	0.6	[33]
				PL	3.0		
KS1731-260	03/14/91	SIGMA	35-150	TB	40.0	1.4	[3]
SLX1732-304	03/90-04/94	SIGMA	40-75	TB	33.0	0.035	[8]
				PL	3.2		
SLX1735-269	03/90-09/94	SIGMA	30-200	TB	53	0.21	[25]
				PL	2.9		
				ST	26, 1.2		
A1742-294	09-10/92	SIGMA	35-100	TB	41.0	0.35	[12]
				PL	3.1		
GS1826-34	10/94	OSSE	50-200	PL	3.1	0.3	[75]
				CPL	1.6, 57.0		
Aql X-1	08-09/91	BATSE	20-100	PL	2.6	1.7	[34]

[a)] Hardest and brightest state ever observed above 30 keV.
[b)] Hard X-ray spectrum averaged over the 4 years of SIGMA observations.

Temporal properties of the hard X-ray emission

SIGMA and BATSE observations have shown that XRBs could be divided in three classes according to the temporal behavior of their hard X-ray emission. The prototype of the first class is 1E1724-3045 in Terzan 2 showing persistent (though variable on a monthly timescale) hard X-ray emission [24,91]. Also SLX1732-301, SLX1735-269 and GS1826-24 probably belong to this class.

The second class contains sources showing *episodic* hard X-ray emission. In this class are the persistent and (variable) X-ray sources such as KS1731-260, A1742-294, GX354-0, 4U1705-44 and 4U0614+091. We cannot qualify the hard X-ray events from these sources as transient events as the maximum hard X-ray fluxes observed ($\sim 50 - 100$ mCrab) are only a factor of ten (or less) larger than the upper limits derived for their quiescent hard X-ray emission. A good example is given by GX354-0 whose long term light curve shows quiescent states ($\lesssim 5$ mCrab, 40-75 keV), flare-like events (~ 50 mCrab) together with low level of persistent emission (~ 10 mCrab) [91] (see also [33]). The recurrence time of these events is poorly known. However, for GX354-0 it cannot be very large as the source has been already detected several times since its first GRIP detection [15]. For most XRBs, the duration of the hard X-ray event is typically of the order of $\sim 10 - 20$ days [3,12,13,69,5,14,33,17].

The third class consists of *true* transient sources: Cen X-4, Aql X-1, and 4U1608-522. Transient hard X-ray emission was observed from Cen X-4 in coincidence with the onset and decay phase of a bright X-ray outburst [9]. For Aql X-1 and 4U1608-522, hard X-ray emission appears to be associated with low-level X-ray emission [95,70].

Several XRB hard X-ray outbursts have been monitored by BATSE with a time resolution of 1 to 5 days. This is illustrated in Fig. 1 where we show the remarkable long-term BATSE light curve of Aql X-1 [34] together with optical monitoring and X-ray data. It is interesting to note that the optical flux from Aql X-1 correlates well with the X-ray flux [70]. Fig. 4 shows a typical BATSE spectrum during a hard X-ray outburst from Aquila X-1 [34]. In many occasions hard X-ray emission from the same XRB was observed more than once.

Because of sensitivity limitations, the time parameters of the outbursts (recurrence time, duration, rise and decay times) are poorly constrained. Nevertheless, a wide variety of transient behaviors is evident. As shown in Fig. 1, Aql X-1 is characterized by long duration outbursts (50 days or longer) with slow rises and decays [34,70]. On the other hand, 4U1608-522 displayed a very long outburst (~ 6 months) with relatively short rise and decay times ($\sim 5 - 10$ days). Within the outburst the source remained on a plateau state varying between 75 and 150 mCrab (20-100 keV) [95].

Only a few simultaneous X-ray and hard X-ray observations of XRBs are currently available. The few existing data indicate that hard X-ray emission from XRBs is generally *anticorrelated* with X-ray emission below 10 keV (e.g.,

Cen X-4 [9]). This property agrees with the idea that XRBs emit hard X-rays in their low X-ray states, and that the hard X-ray emission is progressively quenched as the X-ray intensity increases. The source list of Table 1 is remarkable in its lack of 'Z-sources' [90]. On the contrary, several sources of Table 1 can be classified as 'atoll sources' (4U0614+09, 4U1608-52, 4U1705-44, 1E1724-3045, GX354-0) which are on the average less luminous than the 'Z-sources' [90]. This may reflect a fundamental property of the hard X-ray emission from low-mass X-ray binaries (e.g., [4,5]) since no selection effect discriminates against detection by BATSE monitoring of Z-sources vs. atoll sources. For these atoll sources, hard X-ray emission preferentially occurs during the so called 'island' state, according to the classification scheme of ref. [37]. The best example to date of an anticorrelated hard X-ray/soft emission is provided by 4U 0614+091 [17]. Simultaneous RXTE and BATSE monitoring data shown in Fig. 2 determined a clear anticorrelation between 3-12 keV and 30-100 keV data during the $\sim$ 10 day outburst of April 1996. Another example is given by the recent monitoring results of Aql X-1 by BATSE and RXTE [70]. Hard X-ray emission was detected during a quite faint X-ray outburst in June 1996 (of peak luminosity $\sim 6 \times 10^{35}$ erg s^{-1}, 2-10 keV), with no detectable emission above 20 keV detected during the more luminous outburst during February 1997 (of peak luminosity $\sim 10^{37}$ erg s^{-1}, 2-10 keV) [70]. Such

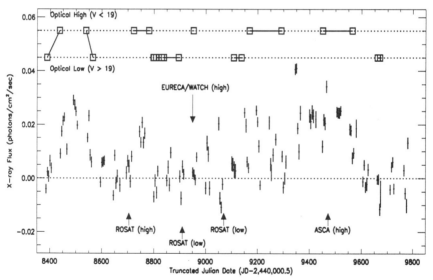

FIGURE 1. Hard X-ray lightcurve detected by BATSE from Aquila X-1 (from ref. [34]). Optical V-band monitoring data are marked at the top of the plot [39]. X-ray observations by ROSAT and ASCA are also indicated. A hard X-ray flux of 0.03 ph cm^{-2} s^{-1} corresponds to 100 mCrab in the 20-100 keV band.

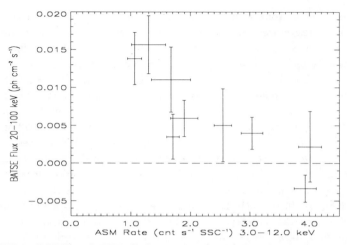

FIGURE 2. BATSE and XTE all-sky monitor (ASM) data for the April 1996 emission episode of 4U 0614+091 showing a clear anticorrelation between soft and hard X-ray emission (from ref. [17]).

an anticorrelation between hardness and intensity in the X-ray band below 20 keV is well known for low-mass X-ray binaries [92].

It is also interesting to note that high-frequency (kilo-Hz) quasi-periodic oscillations (QPOs) have been recently discovered by RXTE from seven of the sources listed in Table 1 (4U0614+091, 4U1608-522, 4U1705-522, 1E1724-3045, GX354-0, KS1731-260 and Aql X-1, see ref. [18]) Furthermore, the timing properties of several sources such as SLX1732-304, SLX1735-269, and GS1826-34 have not been investigated yet. There might be a connection between the occurrence of high-frequency QPOs and hard X-ray emission from XRBs, because of a common dependence on fundamental parameters of these systems such as the neutron star spin period, surface magnetic field, and mass accretion rate. More data are necessary to address this important issue. We also note that if the kilo-Hz QPOs of these systems reflect the underlying spin periods of neutron stars, the evolutionary connection between low-mass X-ray binaries and millisecond pulsars would be dramatically confirmed [1,68].

Neutron star hard X-ray spectra

For most sources, because of the weakness of the observed fluxes, the spectral analysis has been restricted to fitting their hard X-ray spectra with simple models such as a power law (PL) or a Thermal bremsstrahlung (TB). These models usually provide an equally good fit. This is the case for A1742-294 [12], SLX1732-304 [8], SLX1735-269 [25], 4U0614+091 [17], 1E1724-3045 [24].

Furthermore for all sources, the energy range of the detections is limited to ~ 200 keV. (For comparison, black hole detections extend well above 500 keV [28].) Table 1 summarizes the best fit spectral parameters obtained to date for the twelve XRBs detected in hard X-rays.

For the faintest sources, the PL photon index inferred is typically in the range $\sim 2.5 - 3.5$ indicating a rather steep hard X-ray spectrum. For a TB fit, this usually corresponds to a TB temperature around 30-50 keV. However, for 1E1724-3045 on one occasion its high energy spectrum was found to be significantly harder (photon index ~ 1.7, see [2] and [24]).

The first high quality hard X-ray spectrum of an XRB was provided by the SIGMA observation of the February 1992 hard X-ray event from GX354-0 [14]. The 35-200 keV time-averaged spectrum can be fit with a broken power law (BPL) with a break energy at ~ 60 keV. The PL and TB models did not fit the data; only the Sunyaev-Titarchuk (ST) [77] model could fit the data as well as the BPL.

For 4U1608-522, a good spectrum was obtained by BATSE during its long hard X-ray outburst [95]. Fortunately, in the middle of this outburst, GINGA observed 4U1608-522 while it was in its low X-ray state [94]. A joint fitting of the BATSE and GINGA data showed that the spectrum could be fit by either a BPL or ST model. For these models, the break energy/cutoff energy

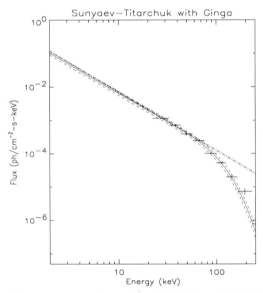

FIGURE 3. Hard X-ray spectrum detected by BATSE from 4U1608–52 [95] superimposed with the estrapolated GINGA spectrum during a quasi-simultaneous detection in the 2-10 keV band [94] for a ST spectral deconvolution (dashed curves) and PL fit (solid lines). From ref. [95].

is near 65 keV (see Fig.3). This was the first XRB broad-band X-ray spectrum showing a hard tail [95].

There are two more sources for which a cutoff at these energies has been observed: GS1826-24 and Cir X-1. The OSSE spectrum of GS1826-24 could be fitted with a power law of index 3.1 [75]. Combining the OSSE data, with non-simultaneous GINGA data suggests the existence of a cutoff near 57 keV. As for Cir X-1, the broad band (2-80 keV) TTM/HEXE spectrum shows that the hard X-ray part is well represented by a ST model with the electron temperature $\sim 6-8$ keV. Such a low temperature is equivalent to an energy cutoff in the spectrum near ~ 25 keV [60]. Note that for the weak sources for which a TB fit was performed, the TB temperature also sets the cutoff around 40-50 keV. Under the assumption that the hard X-ray emission is associated with low intensity states in X-rays, a simple comparison between the power law spectral index in X-rays (~ 2) and the one observed in hard X-rays (more like ~ 3) naturally implies the presence of such a spectral break (which could actually be a cutoff) (e.g., 4U1705-44, see [5]).

The Aql X-1 hard X-ray spectrum determined by BATSE is also of good quality. Contrary to GX354-0 and 4U1608-522, the hard X-ray spectrum of Aql X-1 is well fitted by a PL model. Although the source intensity precludes a precise determination of the spectrum, a power-law model (photon index $\sim 2.1-2.7$) gives a slightly better fit than the TB model with no evidence for any break or spectral cutoff below 100 keV as shown in Fig. 4 [34].

If the case of Aql X-1 is not unique, there may be two spectral classes of

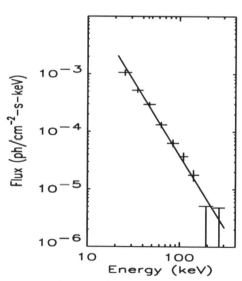

FIGURE 4. Hard X-ray spectrum detected by BATSE from Aquila X-1 (from ref. [34]).

XRBs in the hard X-ray band. A class characterized by a cutoff or a break near 50-70 keV or below, and a class with no detectable break up to ~ 100 keV. Whether the hard X-ray continuum is a power-law, a broken power-law or a Comptonized-like spectrum is a critical issue for theoretical interpretations (see below).

Neutron stars versus black holes

Several papers have appeared recently emphasizing the similarities/differences between XRBs and black hole candidates, and we refer the reader to these articles (see for instance refs. [89,96]). In the following, we focus on comparing their hard X-ray spectral properties up to ~ 200 keV. OSSE observations show that BH *transients* define two distinct spectral classes [28]. The first class is made of systems with hard Comptonized spectra and exponential cutoffs around 100 keV (e.g. GRS1716-249, GRO J0422+32 [51]). The second class contains systems showing soft power-law spectra with photon index $\sim 2.5 - 3.0$ (e.g. GRS1915+105, GRO J1655-40). A substantial fraction of the total luminosity $\gtrsim 10 - 20\%$ may be emitted at hard X-ray energies in BHC systems with Comptonized spectra. For the second BHC class, the fractional luminosity in the hard X-ray band is low ($\lesssim 10\%$) and similar to that of XRBs showing soft power-law spectra. Recent BATSE monitoring and OSSE observations of the prototypical BHC, Cyg X-1, remarkably found that this source shows both these spectral behaviors for different accretion states [97,66]. Fig. 5 shows the spectral states of Cyg X-1 detected by OSSE: the 'soft' state is indistiguishable from those of canonical X-ray bursters. BHCs usually produce spectra harder than NSs. However, BHC 'soft' states appear very similar to those of XRB 'hard' states, suggesting a common mechanism of emission. We also note that what is a rare soft state for Cyg X-1 appears to be the common state for the BHC superluminal transients GRO J1655-40 [98] and GRS 1915+105 [35].

Similarities and differences between BHCs and XRBs emerge. First, a steep power-law spectrum (with index $\gtrsim 2.5$) is observed from both BHCs and XRBs (when active), demonstrating that a steep hard X-ray spectrum is not a signature unique to XRBs. However, contrary to the BHC case, hard X-ray emission from XRBs is always associated with low X-ray intensity states. This is illustrated in Fig. 6 where the 1-20 keV and 20-200 keV luminosities are compared for BHCs and XRBs. BHCs include systems for which there is dynamical evidence that the mass of the compact object is larger than ~ 3 $M_\odot$ (see [6] for details). BHCs can emit hard X-ray tails over a range of luminosities wider than XRBs and their hard X-ray luminosities are also larger on the average.

Another difference is the value of the energy break observed in the hard X-ray spectrum. In the XRB case the cutoff is somewhere below 50 keV, whereas

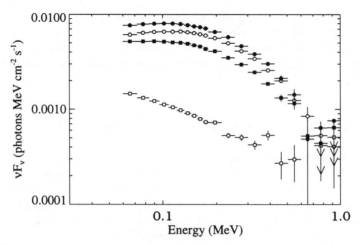

FIGURE 5. Logarithmic power per photon energy decade (νF_ν spectrum) from Cyg X-1 as observed by OSSE (from ref. [66]). The 'soft' state was detected during a target of opportunity OSSE observation of Cyg X-1 in response to the BATSE detection of a soft state transition during the period February-September 1996 [97].

for BHCs is generally around 100 keV. This obviously reflects the standard picture that on average BH spectra are harder than XRB spectra [91]. This difference could be explained by the additional cooling due to the neutron star surface acting as a thermostat capable of limiting the maximum electron temperature achievable in these systems (e.g., [77,48] and see discussion below). Fig. 5 showing the hard and soft spectral states of Cyg X-1 can be used to visualize the different emissions of BHCs and XRBs. Whereas a power-law spectrum of photon index $\gtrsim 2.5$ is commonly observed in both BHCs and XRBs, no weakly magnetized NS system was ever observed producing the equivalent of the Cyg X-1 spectral high state.

It is important to point out that no XRB was ever detected by CGRO above ~ 1 MeV. No COMPTEL or EGRET sources are clearly associated with accreting compact objects. Only Cyg X-1 has been probably detected by COMPTEL above ~ 1 MeV [62].

Theory

Results of hard X-ray observations of low-mass X-ray binaries are compelling. Weakly magnetized neutron stars are able to produce hard X-rays for relatively low values of the overall X-ray luminosity. Contrary to what expected in some idealized models of emission from BHCs and NSs [77] the presence of the NS solid surface reprocessing and emitting soft radiation does

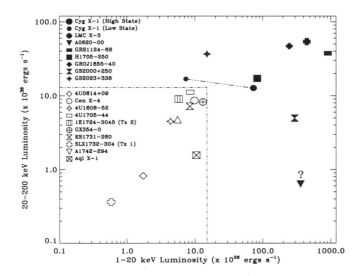

FIGURE 6. Soft L_x (1-20 keV) and hard L_{hx} (20-100 keV) typical luminosities of XRBs and BHCs (from ref. [6]). BHCs are binaries with dynamical evidence that the primary mass is larger than $3M_\odot$. For BHCs L_x and L_{hx} were computed from simultaneous X-ray to hard X-ray observations. For XRBs for which no simultaneous X-ray/hard X-ray observations exist, L_x was computed assuming the most likely spectral shape normalized on the observed hard X-ray fluxes (see ref. [6] for more details). Cyg X-1 is shown in its high/low X-ray states.

not completely quench the process of high energy emission. It is clear that simple idealizations based on strong Comptonization cooling in the inner region of accretion disks surrounding XRBs [77] are in disagreement with observations. What seemed a typical signature of black holes is now confirmed to be produced also by neutron stars. The theoretical models that survive the comparison with the observations give different weight to the two distinctive properties of NS systems with respect to BHCs: (1) the presence of a inner solid surface as a fixed boundary of the accretion flow, and (2) the existence of a magnetic field anchored on the NS surface and sheared by the inner part of the accretion disk. Models of the first type can be called *quasi-thermal* because are based on quasi-equilibrium properties of accretion flows and radiative transport. Comptonization of soft photons either in the accretion disk itself [87] or in hot coronae above the disk (e.g., [31,71]) can produce hard X-rays. Alternately, radiative processes involving boundary layers for neutron stars smaller than the marginally stable orbit of Schwarzschild geometry can in principle produce a hard spectrum [47]. Accretion flow solutions different from standard optically thin disks may also have some relevance for a particular choice of parameters. For a strong ion-electron thermal imbalance within the accretion flow, low mass accretion rate ($\dot{m}$) flow solutions can be advection-dominated with inefficient radiative cooling (e.g., [64]). In this case, hard X-ray emission is more likely from BHCs than neutron stars. Mixed Keplerian and sub-Keplerian accretion flows of low-viscosity can lead to inefficient radiative cooling and advection-dominated accretion [11]. In models characterized by inefficient radiative cooling, the high-energy emission of accretion disks is a non-trivial function of $\dot{m}$, and hard X-ray emission may be produced by a thermal mechanism in a hot two-temperature plasma [64] or by an accretion shock mechanism [11]. Bulk motion photon upscattering in spherical inflows onto compact objects was also recently shown to produce power-law emission in the hard X-ray regime [87]. In all quasi-thermal models except those based on bulk motion upscattering [87] the hard X-ray spectrum is of the Comptonized type, with parameters (temperature and optical depth) dependent on radiative equilibrium states of the disk and corona. Effective Comptonization requires seed soft photons and therefore in quasi-thermal models hard X-ray emission episodes of XRBs should be correlated with enhanced mass accretion rates (assuming that seed soft photons track values of $\dot{m}$). Whether or not this implies an anticorrelation between observable hard and soft X-ray emission depends on specific assumptions and model parameters. Strong Comptonization of soft photons producing hard X-rays is required from observations to be suppressed in NS systems in their high-luminosity states. This is just the opposite of what observed in several BHC transients (e.g., GRO J0422+32 [51]) that usually have their enhanced Comptonized spectral component during high-$\dot{m}$ states.

A different approach is the basis of *non-thermal* models of hard X-ray emission from NS systems [46]. In these models the NS magnetic field plays a crucial role in energizing the plasma of a hot corona being sheared and threaded by reconnecting magnetic loops formed in the inner part of the accretion disk. Within this theoretical framework, hard X-ray emission from weakly magnetized neutron star systems was predicted before the SIGMA and BATSE detections [46]. Particle acceleration occurring near the inner edge of weakly magnetized neutron stars can be influenced by magnetic field reconnection and pair cascade formation [46], moderately energetic magnetic field reconnection [82], or relativistic double layers [30]. The emergence of magnetic loops from the convective accretion disk surface, their turbulent state and reconnection instabilities are similar to those of current loops emerging from the sun's surface and producing solar flares (e.g., [67]). Emission from BHCs relies only on the self-interaction and reconnection of disk magnetic loops [16,19]. However, in NS systems a rotating magnetosphere anchored to the NS surface enhances magnetic field energy dissipation. As observed in solar flares, a fraction $f_r \gtrsim 10\%$ of the initial magnetic field energy of reconnection events can be released and used in particle acceleration (e.g., [29,56]). In non-thermal turbulent magnetic-driven models, particle acceleration is produced by transient electric fields generated in current sheets formed by magnetic field reconnection events. The number of simultaneously active acceleration sites in a magnetically threaded accretion disk near a NS can be $\sim 10^2 - 10^3$, and they can generate a total non-thermal hard X-ray luminosity near $10^{36} - 10^{37}$ erg s^{-1} only for low values of the soft X-ray luminosity at $\sim 0.1-0.01$ below Eddington [82]. For larger soft X-ray luminosities, the magnetic-driven acceleration process is quenched by strong radiative cooling, and a thermal-like Comptonized spectrum of relatively low effective temperature is expected. The synchrotron and inverse Compton (IC) spectrum of an XRB of low luminosity is typically a broken power-law with break energy $E_b \propto B_s^2 L_{x,38}^{-1}$, where B_s is the NS surface magnetic field and $L_{x,38}$ the soft X-ray luminosity in units of 10^{38} erg s^{-1} [46]. The observed spectral differences between BHCs and XRBs can then be understood in terms of additional radiative cooling applied to NS systems. In these models accreting NS spectra are typically softer than BHCs because of stronger inverse Compton cooling and additional synchrotron emission caused by the neutron star magnetic field reconnecting in the disk [82].

Current data provide interesting information. The soft/hard X-ray anticorrelation observed in many NS systems is naturally explained in non-thermal models and require fine-tuning of quasi-thermal models. More information can be extracted from high-quality soft/hard X-ray spectra of low-mass X-ray binaries. Emission models make different predictions about the spectra and intensity states of NS systems. Figs. 3–4 show two examples of different XRB spectra in the hard X-ray range: a spectrum compatible with a broken power-law or Comptonization for 4U1608-52, and a spectrum compatible with a single power-law for Aql X-1. Both quasi-thermal and non-thermal models of emission can in principle fit the current data. However, it is clear that a variety of spectral parameters apply to the XRB population, and that we are only at the beginning of a learning process towards a satisfactory understanding of XRB spectral states. Current data are encouraging, and future high-quality spectral information including soft and hard X-ray simultaneous detections of XRBs obtainable by Rossi-XTE, Beppo-SAX, Jet-X, and Integral will complement the information provided by BATSE. A deeper study of emission processes of accretion disks surrounding NSs and BHCs will be possible in the near future.

RADIOPULSARS IN BINARIES

Powerful radiopulsars lose a substantial fraction of their rotational energy as a relativistic particle wind [44]. As the Crab Nebula shows, unpulsed high-energy emission up to ~ 100 MeV can be produced by the interaction of a pulsar wind with a surrounding gaseous medium [65]. It is then natural to search for high-energy emission from binaries containing powerful radiopulsars. If a sufficiently dense gas is present in the binary, we expect a nebular interaction of the pulsar wind in a possibly time variable environment (e.g.,

[81]. The interacting gas can originate in the mass outflow from the companion star. Alternately, the interaction of a pulsar wind with a companion star can cause a self-induced outflow from the companion's surface [72]. X-ray and gamma-ray emission is expected from the shock interaction of the pulsar wind and its gaseous surroundings in compact binaries. We summarize here the CGRO observations of the intermediate age pulsar system PSR B1259-63 and of low-mass binaries with millisecond pulsars.

Massive systems with young/intermediate age pulsars: the case of PSR B1259-63

The PSR B1259-63 system provides the first binary where the interaction of a powerful pulsar with a gaseous wind from a companion star can be studied. It consists of a rapidly rotating radio pulsar (spin period $P = 47.76$ ms and spindown luminosity $\dot{E}_R \simeq 8 \times 10^{35}$ ergs s^{-1}) orbiting around the Be-star SS 2883, a 10th magnitude Be star of luminosity $L_* = 5.8 \times 10^4\ L_\odot$, estimated radius $R_* \sim (6-10)\ R_\odot$ and estimated distance from Earth between 1.5 and 3 kpc [41,42,61]. The pulsar orbit is highly eccentric and of period ~ 3.4 yrs. Fig. 7 summarizes the X-ray and gamma-ray observations of the PSR B1259-63 system. Of particular importance is the 3-week CGRO observation carried out in coincidence with the January 1994 periastron. During this observation, the pulsed radio signal from PSR B1259-63 was eclipsed by intervening gaseous material [42]. The source was detected by OSSE in the 30-200 keV range at the level of a few mCrab [$(2.8 \pm 0.6) \cdot 10^{-3}$ ph cm^{-2} s^{-1} MeV^{-1} at 100 keV], with a spectrum consistent with a power law of photon index ~ 1.8 [27]. For a 2 kpc distance, the inferred luminosity is $L_{x,hard} \simeq 3 \cdot 10^{34}$ erg s^{-1}, i.e., $\sim 4\%$ of the total pulsar spindown luminosity. The OSSE spectrum agrees with the time-averaged ASCA spectrum extrapolated in the hard X-ray range [27,83].

No other CGRO instrument detected the PSR B1259-63 system despite extensive searches in BATSE, COMPTEL and EGRET data for unpulsed and pulsed emission [83]. The EGRET upper limit to unpulsed emission is particularly relevant (the 95% confidence limit in the 100-300 MeV band is $4.8 \cdot 10^{-11}$ erg cm^{-2} s^{-1}) because it clearly implies a spectral cutoff in the MeV range [83]. Fig. 8 shows the ASCA and CGRO spectra obtained near periastron. The X-ray flux clearly changes both in intensity (by a factor of ~ 10, peak luminosity at periastron $\sim 10^{34}$ erg s^{-1}) and in spectral shape (power-law photon index 1.5–2.0) by comparing apastron and periastron observations [86]. Within the statistics of the OSSE detection, no variability in the 30-200 keV energy range is detected [27].

These results are of great importance for establishing the non-thermal nature of the high-energy source from the PSR B1259-63 system. We follow here the discussion of ref. [86]. A variety of mechanisms can in principle be

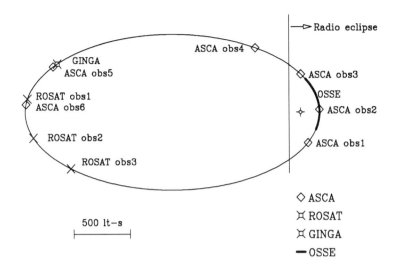

FIGURE 7. Orbit of the PSR B1259-63 system with OSSE (CGRO), GINGA, ROSAT and ASCA observations carried out during the period 1991-1994 (from refs. [38,86]). The thick part of the orbit near periastron indicates the period of the GRO observation of January 3-23 1994 [83,86]. Pulsed radio emission from PSR B1259-63 was eclipsed near periastron [42] as indicated in the figure.

invoked to interpret the observations: (1) accretion onto the surface of the neutron star, (2) propeller-driven emission near the magnetospheric radius [40,45,59,10,21], (3) pulsar magnetospheric emission, (4) Be star coronal emission, (5) bremsstrahlung emission of heated gas in the Be star outflow [50], (6) pulsar/outflow relativistic shock emission [80]. A series of detailed calculations exclude all except the last possibility to explain all the X-ray/gamma observations [86]. The unpulsed nature of the observed emission (both X-rays [43] and gamma-rays [27]) clearly excludes mechanism (3). The highly non-thermal spectrum up to ~ 200 keV excludes model (4), and model (5) is in disagreement with the observed X-ray flux and spectrum. Accretion onto the surface of the neutron star or gravitational 'trapping' of material near the magnetospheric boundary inducing a propeller regime require a very strong outflow from the Be star near periastron. Pulsar wind pressure can be overcome only for unreasonably high values of the mass outflow rate and/or outflow velocity [86]. The very small (constant along the orbit) value of the X-ray absorbing column density ($N_H \simeq 6 \cdot 10^{21}$ cm^{-2}), argues against the presence of dense material along the line of sight to the region of high-energy emission. A fourth ASCA observation of the PSR B1259-63 system carried out ~ 1 month after periastron as the radio pulsar became again visible [42] shows a non-thermal spectrum and intensity which are very similar to during periastron. For typical values of Be star outflow parameters ($\dot{M} \simeq 10^{-7}\, M_\odot\, \text{yr}^{-1}$, initial outflow speed $\sim 10^6\,\text{cm}\,\text{s}^{-1}$) the pulsar wind pressure of PSR B1259-63 is expected to break open the Be star equatorial wind [86]. We are left with model (6) that agrees with all radio and high-energy data and theoretical expectations [80].

Detailed modelling of the X-ray intensity and spectral variations of the PSR B1259-63 system was carried out in ref. [86]. A model of pulsar/outflow interaction dominated by synchrotron radiation of relativistically shocked particles provides an adequate description of the data. EGRET upper limits are of great importance since they support an upstream value of the Lorentz factor γ_* for the pulsar wind particles (electrons/positrons and possibly ions) not smaller than $\sim 10^6$, i.e., similar to the value for the Crab pulsar wind [44,20]. In the presence of a relatively strong soft photon background from the Be star surface near periastron, smaller values of γ_* would cause inverse Compton cooling to produce observable gamma-ray flux in the EGRET band. Gamma-ray emission above ~ 10 MeV might be observable in other pulsar systems with different parameters [81]. In the case of PSR B1259-63, a spectral cutoff is observed near $1-10$ MeV (see Fig. 8) as a consequence of strong cooling near periastron.

These results are of great importance for the theory of particle acceleration in relativistic shocks. We can summarize the results of the 1994 CGRO/ASCA campaign of observations of the PSR B1259-63 system near periastron as follows [86]:

(1) the PSR B1259-63 system behaves as a *binary plerion* showing characteristics of *diffuse* and *compact* plerions near the apastron and periastron, respectively. The pulsar radiation and MHD wind pressure is able to avoid surface accretion or magnetospheric propeller-like processes throughout the whole period of high energy observations from February 1992 through February 1994. The shock radius is established at relatively large distance from the pulsar, $r_s \sim 10^{12}$ cm (near periastron) corresponding to $\sim 10\%$ of the orbital distance. This radius corresponds to the distance of the 'apex' of the pulsar cavity in the plane of the pulsar orbit. The pulsar wind pressure is able to 'break open' both the polar and equatorial parts of the Be star outflow (even at periastron), and a relatively dilute gaseous environment is produced along the line of sight for all orbital phases. There is no evidence for a non-spherical pulsar wind to a distance of $\sim 10^{12}$ cm;

(2) the observed high-energy emission from the PSR B1259-63 system agrees with the predictions of a cooling model for the post-shock particle energy distribution function that assumes fast particle acceleration within a timescale smaller than the radiative timescales $\sim 10^2 - 10^3$ s. The index δ of the post-shock energy distribution function, $N(\gamma) \sim \gamma^{-\delta}$, (before radiative cooling) is constrained to be constant throughout the PSR B1259-63 orbit and of value $\delta \simeq 2$. The best cooling model which explains simultaneously the X-ray intensity and spectral properties of the PSR B1259-63 system near periastron

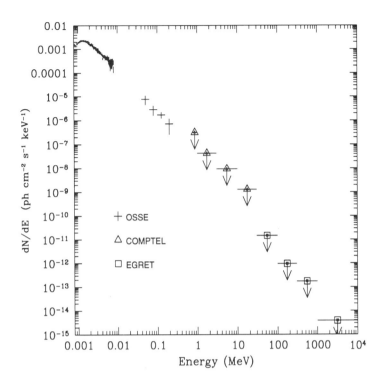

FIGURE 8. Combined ASCA and CGRO photon spectra of the PSR B1259-63 system during the January 1994 periastron (from refs. [83,86]).

is given by a combination of synchrotron and IC cooling [86]. IC cooling is most effective near periastron where the 'screening' by the Be star outflow is at its minimum. Strong screening of the optical/IR flux relevant to IC cooling within the flow timescale of the pulsar cavity can occur in the equatorial disk. High-energy radiation is produced more effectively for cooled (by a factor of ~ 10 compared to the original post-shock energy) $e^{\pm}$-pairs of the pulsar wind being advected away in the inner parts of the pulsar cavity. The observable radiation in the ASCA-OSSE energy band is synchrotron emission, with the IC contribution calculated to be radiated above ~ 10 GeV near periastron. The range of particle energies ($[\gamma_1, \gamma_m]$) contributing to the ASCA-OSSE energy spectrum between 1 and 200 keV can be deduced as $10 \lesssim \frac{\gamma_m}{\gamma_1} \lesssim 100$ from the absence of detectable emission in the EGRET energy range near periastron [83]. As the pulsar recedes from periastron, the quantity γ_m is expected to progressively increase for a synchrotron/IC power-law emission extending to energies $\gtrsim 10$ MeV;

(3) the pulsar orbit is most likely offset from the plane of the Be star equatorial outflow by an angle $\gtrsim 30°$. Models assuming a coplanar geometry for

the pulsar orbit and Be star equatorial outflow are not in agreement with the X-ray data. The double hump of the X-ray luminosity light curve [86] and the apastron/periastron contrast of intensity of about one order of magnitude cannot be reproduced by coplanar pulsar/outflow interaction models;

(4) time variable X-ray/soft γ-ray emission from the PSR B1259-63 system is established to originate from non-thermal particle acceleration processes most likely occurring at the pulsar wind termination shock. The efficiency of conversion of pulsar spindown power into visible radiation in the ASCA-OSSE energy range is $\sim 4-5\%$, with a possible comparable contribution calculated to be in the MeV energy range. This conversion efficiency is of the same order of magnitude as that one observed in the case of the Crab Nebula, suggesting a similar physical process of pulsar/nebula interaction and radiation. A series of constraints can be derived for both the pulsar wind parameters and for the details of the post-shock particle distribution function [86]. The acceleration mechanism timescale is constrained (for the first time for a relativistic object producing a MHD wind) to be less than $\sim 10^2 - 10^3$ seconds near periastron.

Other systems

The results obtained for the PSR B1259-63 system have general validity and can be applied to a variety of astrophysical objects. The PSR B1259-63 system shows that X-ray emission in a binary system (not related to intrinsic stellar emission) can be efficiently produced by a non-thermal mechanism which is drastically different from accretion-driven or magnetospheric-driven emissions. Binary pulsars are ideal astrophysical systems in this context, because their shocked relativistic pulsar wind provides the crucial ingredient to produce high-energy emission of moderate luminosity. For a generic binary pulsar we expect the high-energy emission to be of a power-law form (unless the resulting plerion turns out to be ultra-compact) and strongly dependent on synchrotron and IC cooling. The X/γ-ray emission is predicted to be unpulsed and characterized by a small intrinsic column density, since nebular outflows which avoid gravitational 'trapping' near the compact star are expected to be optically thin along the line of sight in a way similar to PSR B1259-63.

Several low/intermediate luminosity X-ray and γ-ray sources in our Galaxy might contain a pulsar similar to PSR B1259-63 orbiting around a low-mass (with millisecond pulsars) or high-mass companion (e.g, [81]). PSR B1259-63 is orbiting around its Be star companion in a very eccentric and long orbital period orbit, with its radio emission detectable during most of its revolutions, except for the 40-day eclipse near periastron. It is easy to imagine similar binary pulsars (with low-mass or high-mass companions) of smaller orbital periods and therefore likely to be eclipsed for most if not all of their orbits.

Candidate sources include a class of time variable unidentified EGRET sources near the Galactic plane such as GRO J1838-04 [85], 2CG 135+1

[49,84], dim X-ray sources in globular clusters, and OB-associations with anomalously large X-ray emission [63]. These systems show significant excess of their high-energy emission over the expected level of stellar emission. From the results obtained for the PSR B1259-63 system, temporal variations of the intensity and spectrum of Galactic and X/γ-ray unidentified sources can provide evidence for underlying non-thermal processes powered by 'hidden' pulsars.

We thank our collaborators for invaluable help and support in completing the projects described here. We are particularly grateful to J. Arons, E. Ford, J. Grindlay, E. Grove, B.A. Harmon, P. Kaaret, E. Liang, J. Mattox, M. Ruderman, M. Strickman, S.N. Zhang. The CGRO Science Support Center provided crucial support.

REFERENCES

1. Alpar, M.A., Cheng, A.F., Ruderman, M.A. & Shaham, J., 1982, *Nature*, **300**, 728.
2. Barret, D., et al., 1991, *ApJ*, **379**, L21.
3. Barret, D., et al., 1992, *ApJ*, **394**, 615.
4. Barret, D. and Vedrenne, G., 1994, *ApJS*, **92**, 505
5. Barret, D., et al., 1996, *A&AS*, **120**, 121.
6. Barret, D., McClintock, J. E., and Grindlay, J. E., 1996, *ApJ*, **473**, 963.
7. Bazzano, A., et al., 1997, IAU Circular no. 6668.
8. Borrel, V., et al., 1996, *ApJ*, **462**, 754.
9. Bouchacourt, P., et al., 1984, *ApJ*, **285**, L67.
10. Campana, S., Stella, L., Mereghetti, S. & Colpi, M., 1995, *A&A*, **297**, 385.
11. Chakrabarti, S. & Titarchuk, L., 1995, *ApJ*, **455**, 623.
12. Churazov, E., et al., 1995, *ApJ*, **443**, 341.
13. Churazov, E., et al., 1997, *Adv. Space Res.*, **19**, 55.
14. Claret, A., et al., 1994, *ApJ*, **423**, 436.
15. Cook, W.R., et al., 1991, *ApJ*, **372**, L75.
16. Eardley, D.M., and Lightman, A.P., 1975, *ApJ*, **200**, 187.
17. Ford, E., et al., 1996, *ApJ*, **469**, L37.
18. Ford, E., et al., 1997, in Proceedings of the 4th Compton Symposium, in press.
19. Galeev, A.A., Rosner, R. & Vaiana, G.S., 1979, *ApJ*, **229**, 318.
20. Gallant, Y.A. & Arons, J. 1994, *ApJ*, **435**, 230.
21. Ghosh, P, 1995, *ApJ*, **453**, 411.
22. Goldwurm, A., et al., 1993, in AIP Proc. no. 280, p. 366.
23. Goldwurm, A., et al., 1994, *Nature*, **371**, 589.
24. Goldwurm, A., et al., 1993, in AIP Conf. Proc. no. 304, p. 421.
25. Goldwurm, A., et al., 1996, *A&A*, **310**, 857.
26. Grebenev, S., et al., 1993, *A&ASuppl.Ser.*, **97**, 281.

27. Grove, J.E., Tavani, M., Purcell, W.R., Johnson, W.N., Kurfess, J.D., Strickman, M.S. & Arons, J., 1995, *ApJ*, **447**, L112.
28. Grove, J., Kroeger, R. & Strickman, M., 1996, in *Proceedings of the 2nd INTEGRAL Workshop*, ESA SP-382, p. 197.
29. Haisch, B.M. et al., 1991, *ARA&A*, **29**, 275.
30. Hamilton, R.J., Lamb, F.K. & Miller, M.C., 1993, in AIP Conf. Proc. no. 280, p. 433.
31. Hanawa, T., 1990, *ApJ*, **355**, 585.
32. Harmon, B.A., et al., 1992, in AIP Conf. Proc. no. 280, p. 314.
33. Harmon, B.A., et al., 1993, in AIP Conf. Proc. no. 304, p. 456.
34. Harmon, B.A., et al., 1996, *A&AS*, **120**, 197.
35. Harmon, B.A., et al., 1997, *ApJ*, **477**, L85.
36. Harmon, B. A., 1997, Abstract submitted for the HEAD Meeting, Estes Park.
37. Hasinger, G. & van der Klis, M., 1989, *A&A*, **225**, 79.
38. Hirayama, M., et al., 1996, *PASJ*, **48**, 833.
39. Ilovaisky, S.A. & Chevalier, C., 1991-1993, IAU Circulars no. 5281, 5297, 5551, 5665, 5831.
40. Illarionov A.F. & Sunyaev, R.A., 1975, *A&A*, **39**, 185.
41. Johnston S., et al., 1992, *ApJ*, **387**, L37.
42. Johnston, S., et al., 1996, *MNRAS*, **279**, 1026.
43. Kaspi, V., Tavani, M., Nagase, F., Hoshino, M., Aoki, T., Kawai, N. & Arons, J., 1995, *ApJ*, **453**, 424.
44. Kennel, C.F., and Coroniti, F.V., 1984, *ApJ*, **283**, 694.
45. King A., & Cominksy, L., 1994, *ApJ*, **435**, 411.
46. Kluźniak, W., Ruderman, M., Shaham, J., and Tavani, M., 1988, *Nature*, **336**, 558.
47. Kluźniak, W. & Wilson, R.J., 1991, *ApJ*, **372**, L87.
48. Kluźniak, W., 1993, *A&AS*, **97**, 265.
49. Kniffen, D. et al., 1997, *ApJ*, in press.
50. Kochanek C., 1993, *ApJ*, **406**, 638.
51. Kroeger, R.A., et al., 1996, *A&AS*, **120**, 117.
52. Lewin, W.H.G. and Joss, P.C., 1983, in *Accretion-driven Stellar X-ray Sources*, eds. W.H.G. Lewin and E.P.J. van den Heuvel (Cambridge, Cambridge University Press), p. 41.
53. Liang, E.P. & Nolan, P., 1984, *Space Sci. Rev.*, **38,**, 353.
54. Liang, E.P., and Dermer, C.D., 1988, *ApJ*, **325**, L39.
55. Liang, E., 1992, in AIP Conf. Proc. no. 304, p. 396.
56. Longcope, D.W. & Strauss, H.R., 1994, *ApJ*, **426**, 742.
57. Ling J.C., et al., 1987, *ApJ*, **321**, L117.
58. Ling, J.C., et al., 1993, in AIP Conf. Proc. no 304, p. 220.
59. Lipunov, V.M., Nazin, S.N., Osminkin, E.Yu. & Prokhorov, M.E., 1994, *A&A*, **282**, 61.
60. Maisack, M., et al., 1995, *Adv. Space Res.*, **16**, 91.
61. Manchester, R.N., et al., 1995, *ApJ*, **445**, L137.
62. McConnell, M., et al., 1993, in AIP Conf. Proc. no. 304, p. 230.

63. Motch, C. & Pakull, M., 1996, IAU Circular no. 6285.
64. Narayan, R. & Yi, I., 1995, *ApJ*, **452**, 710.
65. Nolan, P.L., et al., 1993, *ApJ*, **409**, 697.
66. Phlips, B.F., et al., 1996, *ApJ*, **465**, 907.
67. Priest, E.R., 1987, *Solar Magnetohydrodynamics* (Dordrecht: Reidel).
68. Radhakrishnan, V. & Srinivasan, G., 1984, *Curr. Sci.*, **51**, 1096.
69. Revnivtsev, M., et al., 1996, in the Proceedings of the 2nd INTEGRAL Workshop, *The Transparent Universe*, ESA SP-382, p. 277.
70. Rubin, B. C., et al., 1997, in Proc. of the 4th Compton Symposium, in press.
71. Rudak, B. & Meszaros, P., 1991, *ApJ*, **383**, 269.
72. Ruderman, M., Shaham, J. & Tavani, M., 1989, *ApJ*, **336**, 507.
73. Shakura, N. & Sunyaev, R.A., 1973, *A&A*, **24**, 337.
74. Shapiro, S., et al., 1976, *ApJ*, **204**, 187.
75. Strickman, M., et al., 1996, *A&AS*, **120**, 217.
76. Strickman, M. et al., 1996, paper presented at the HEAD meeting, San Diego.
77. Sunyaev, R. & Titarchuk, L., 1980, *A&A*, **86**, 121.
78. Sunyaev, R. & Titarchuk, L., 1989, ESA SP-296, p. 627.
79. Tanaka, Y., 1989, ESA SP-296, p. 3.
80. Tavani, M., Arons, J. & Kaspi, V., 1994, *ApJ*, **433**, L37.
81. Tavani, M., 1995, in *The Gamma-Ray Sky with COMPTON GRO and SIGMA*, eds. M. Signore, P. Salati & G. Vedrenne (Dordrecht: Kluwer Academics), p. 181.
82. Tavani, M., and Liang, E., 1996, *A&AS*, **120**, 133.
83. Tavani, M., Grove, et al., 1996, *A&AS*, **120**, 221.
84. Tavani, M., et al., 1997, in preparation.
85. Tavani, M., et al., 1997, *ApJ*, **479**, L109.
86. Tavani, M. & Arons, J., 1997, *ApJ*, **477**, 439.
87. Titarchuk, L., Mastichiadis, A. & Kylafis, N.D., 1996, *A&AS*, **120**, 171.
88. Ubertini, P., et al., 1997, IAU Circular no. 6611.
89. van der Klis, M., 1994, *ApJS*, **92**, 511.
90. van der Klis, M., 1995, in *X-Ray Binaries*, eds. W. H. G. Lewin, J. Van Paradijs, and E. P. J. van den Heuvel, Cambridge University Press, p.225.
91. Vargas, M., et al., 1996, in *Proceedings of the 2nd INTEGRAL Workshop*, ESA SP-382, p. 129.
92. White N., et al., 1988, *ApJ*, **324**, 363.
93. White, N., Nagase, F. & Parmar, A., 1995, in *X-Ray Binaries*, eds. W. H. G. Lewin, J. Van Paradijs, and E. P. J. van den Heuvel, Cambridge University Press, p. 1.
94. Yoshida, K., et al., 1993, *PASJ*, **45**, 605.
95. Zhang, S.N. et al., 1996, *A&AS*, **120**, 279.
96. Zhang, S.N., 1996, in *Accretion Phenomena and Related Outflows*, eds. D.T. Wickramasinghe, G.V. Bicknell & L. Ferrario, ASP Conf. Proc. Series, Vol. 121.
97. Zhang, S.N. et al., 1997, *ApJ*, **477**, L95.
98. Zhang, S.N. et al., 1997, *ApJ*, **479**, 381.

GRO J1744−28: the Bursting Pulsar

Chryssa Kouveliotou* and Jan van Paradijs[†]

*Universities Space Research Association
ES-84, NASA/MSFC, Huntsville, Alabama 35812
[†] University of Amsterdam and
University of Alabama in Huntsville.

Abstract. During more than 25 years of studies of X-ray binaries several dozens of X-ray pulsars and X-ray bursters were discovered; not a single one of these showed both pulsations and bursts. Here we describe the properties of the Bursting Pulsar, a unique X-ray source, discovered in December 1995 with the Burst And Transient Source Experiment (BATSE), which seems to be the first to defy the dictum 'pulsars don't burst, and bursters don't pulse'.

DISCOVERY-IDENTIFICATION

An unusually rapid sequence of a series of rather weak outbursts, each lasting between 10–15 s, was first noticed on December 4, 1995, in the background data of one of the 8 Large Area Detectors (LADs) that comprise the Burst And Transient Source Experiment (BATSE) on board the Compton Gamma-Ray Observatory (CGRO) [16,25]. The events had actually appeared during the second half of December 2, 1995; on this day, the intervals between the bursts were variable but for a period of three hours they clustered around 170±15 s. Their rate settled after December 2nd to an average of ∼20–25 events per day, until about April 27, 1996. It declined rapidly therafter and burst activity stopped altogether on May 3, 1996 [26].

The 8°(radius) error box of the burst source location, as determined from the relative strength of the bursts in the different BATSE detectors alone, included the Galactic center. The large uncertainty of the location was due to the low intensity and the soft spectrum of the events. They were visible only between 20–70 keV, which made them softer than Gamma Ray Bursts (GRBs) and Soft Gamma Repeaters (SGRs), but much harder than Type-I bursts emitted from X-ray bursters.

In mid-December, a new transient source of hard X-ray emission, GRO J1744−28, was detected in the BATSE occultation data [33], from the same general area near the Galactic Center also associated with the new burster. The brightness of this source increased gradually during the next several weeks and reached a peak in mid-February 1996; it declined slowly afterward and reached the background level at the end of April. From the initial analysis of GRO J1744−28 data, it was soon discovered that the new source emitted coherent X-ray pulsations, at a frequency of 2.1 Hz. Doppler shifts showed that the pulsar was a member of a binary system with orbital period $P_{\rm orb} = 11.8$ days, and a very small mass function, indicating that the compact star had a low mass companion [14,15].

During the months following their detection the peak intensity of the bursts from the Galactic center region also increased steadily. Finally, it became possible to detect pulsations within the bursts. The pulsations arrived at the same frequency of 2.1 Hz and were present at a much enhanced amplitude but at the same fractional rms variation as in the persistent source [27]. This confirmed the suspicion that the pulsar and the burster were one and the same source, which soon became known as the 'Bursting Pulsar'. This result was confirmed by subsequent observations of GRO J1744−28 with the Rossi X-ray Timing Explorer (RXTE), which also provided the first ∼1 arcmin radius error box for the source [38].

The Bursting Pulsar is a transient; both the (pulsed) persistent and the burst emission rose and decayed in a remarkably similar fashion. The peak

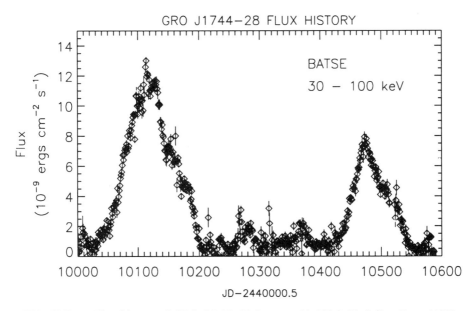

FIGURE 1. Flux history of GRO J1744−28 between 30−100 keV. Julian Date 10053 corresponds to December 2, 1995. The plot extends to the beginning of June 1997. Each point represents a daily average for the source flux.

strength of both was reached in early February 1996, after which their decay started. The last burst detected with BATSE occurred very close to the cessation of the persistent and pulsed emission, on 1996 May 3 and ∼7, respectively.

Like many transient LMXBs, GRO J1744−28 shows repetitive outbursts: on December 2, 1996, exactly one year after the onset of the first outburst, the Bursting Pulsar became active again as a source of both persistent and burst emission. The second outburst was almost a carbon copy of the first: during the first day the bursts came at intervals of a few minutes, but a day later the burst rate decreased to ∼ 20 per day, at which level it remained until the end of the outburst, which lasted until the end of April 1997. The second outburst, however, was dimmer by a factor of 2 (see Figure 1).

In this paper we review the properties of the Bursting Pulsar, as derived mainly from observations made with BATSE and RXTE. With BATSE we have a virtually uninterrupted view of the source throughout the entire duration of both outbursts (Figure 1). At least half of the over 5500 bursts detected during both active episodes led to an onboard trigger and these were, therefore, observed with a variety of high time and spectral resolution modes. BATSE, however, has limited sensitivity between 20–100 keV as compared to the RXTE Proportional Counter Array (PCA), which has an energy range between 2–60 keV. In addition, the RXTE/PCA is collimated, which results in a very low background, thus lending to high signal to noise ratio observations. The adverse side of this is that very intense bursts from GRO J1744−28 saturate the PCA and cannot be easily analyzed. Combined data from BATSE and PCA are nicely complimentary.

SOURCE LOCATION

The initial BATSE error circle of GRO J1744−28 included the direction to the Galactic center. Scanning [38] of this error box with RXTE on January 18-19, 1996, just two weeks after its launch, produced an ∼ 1′ error box, and confirmed that the bursts did not come from the Galactic center. Subsequent observations with SIGMA/Granat [5], Mir-Kvant-TTM [6] and ASCA [13] produced error boxes with radii between ∼ 1′ and ∼ 3′, which confirmed this result (see Figure 2). Finally, a ROSAT observation [28] made on March 14, 1996 (earlier observations being impossible because of the small angle between the source and the Sun) led to a source position, at R.A. = $17^h44^m33^s.1$, dec = $28°44'29''$ (equinox 2000) with an accuracy of ∼ 10″.

In addition, the ASCA observations on February 27, 1996 also provided evidence for heavy absorption in the source direction ($N_H = 5 \times 10^{22} \text{cm}^{-2}$), which indicates that the source is at a distance consistent with that of the Galactic Center [12].

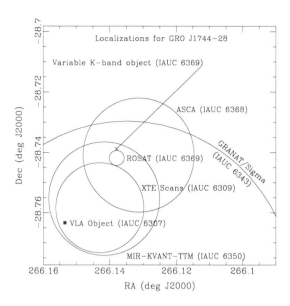

FIGURE 2. Some of the arcmin localizations and error circles as reported in the IAU Circulars indicated in parenthesis. The uncertainties are of different significances (see references). The figure is courtesy of Dr. R. Rutledge.

PULSATIONS

Pulsations were detected from the source during the entire period of its bursting activity the first and the second transient outburst. The earliest pulsed flux was detected on December 1, 1995, approximately one day before the source started its series of rapid bursts. The results of the analysis of the pulsar data have been given by Finger et al. (1996b). In their paper, they note that the pulse profile is nearly sinusoidal comprising two harmonics with amplitudes of 6.2 and 1.4 % of the fundamental, respectively. The first harmonic maximum occurs 77 ms before the second, which results in a steeper pulse rise than decay (see also their Figure 1). There is no dependence of the pulse profile on energy, *i.e.*, it remains the same from 20 to 70 keV.

Measurements of the Doppler delay of the pulse arrival times, caused by orbital variations, provided an orbital period for the system of 11.8 days. The orbit is almost circular, with eccentricity $e < 1.1 \times 10^{-3}$ (at the 90% confidence level). The mass function of the system is surprisingly small: $f_x(M) = 1.36 \times 10^{-4}$ $M_\odot$ (see Finger et al. 1996b for a detailed description of the orbital parameters). This indicates that the secondary is likely a low-mass star ($\sim 0.25 M_\odot$) [10,42] and that the inclination of the orbit is relatively small ($\sim 10°$). According to the above, the Bursting Pulsar system is a low-mass X-ray binary (LMXB).

Superposed on the orbital modulation of the pulse arrival times is a long-term pulsar spin up, whose rate, $\dot{\nu}$, is strongly correlated with the observed flux (see Figure 3), according to

$$\dot{\nu} \propto Flux^{0.957\pm0.026} \qquad (1)$$

This relation, which is followed both during the rise of the outburst and its (slower) decay, is in approximate agreement with simple versions of models of accretion torques exerted on magnetic neutron stars [34] and provides constraints to the pulsar magnetic field. Assuming a purely dipole field, the polar field strength will be $<6\times10^{11}$ G (see Finger et al. 1996b), which places the system between typical X-ray pulsars (with $B\sim10^{12}$ G) and X-ray bursters ($B\sim10^{8-9}$ G). This intermediate magnetic field value may in part explain the peculiar properties of this source, not seen yet in any X-ray pulsar or burster.

The spin-up rate versus pulsed flux relation is following the same law during the second outburst of the source between December 1996–April 1997. In fact, the plots for both outbursts are identical in shape, but during the second outburst the pulsed flux intensity and maximum spin-up rate reached only ~70% of their values during December 1995–May 1996 (e.g. spin-up rate values for the two outbursts are $\sim 1.2 \times 10^{-11}$ Hz/sec and $\sim 7 \times 10^{-12}$ Hz/sec, respectively). This reduction factor agrees well with most of the source proper-

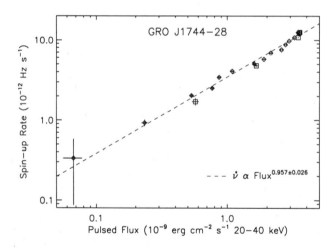

FIGURE 3. Observed relationship between the 20–40 keV r.m.s. pulsed flux and pulsar spin-up rate $\dot{\nu}$ of GRO J1744−28 during its first outburst (December 1995–May 1996). The square symbols are from the outburst rise and the diamond symbols are from the outburst decay. The dashed line is a best fit power law. The figure is courtesy of Dr. M. Finger.

ties during both outbursts, (such as total duration, burst intensity, persistent source intensity), as we will also see in the following sections.

X-RAY SPECTRA

The energy range covered by the BATSE LADs is 20–2000 keV. GRO J1744−28 is detected between 20 to ∼70 keV. We have fitted the spectra of each burst with an optically thin thermal bremsstrahlung function. The evolution of their temperature (kT) is shown in Figure 4 for the entire first outburst of the source. We note that the temperature remains remarkably constant throughout the outburst at $kT \sim 10.5$. The average burst temperature throughout the second outburst seems to rise from 6 to 10 keV at the peak of the outburst and it remains constant at ∼ 9 keV thereafter [43].

The RXTE/PCA burst spectra cover the range 2–60 keV. Most of the bursts from GRO J1744−28 , however, suffer from severe deadtime effects. Fox et al. (1998) are currently developing software to correct burst rates for deadtime effects; they have also reported on their first results using hardness ratios instead of performing detailed spectral fits. They find that the burst and persistent X-ray spectra are quite similar, with the bursts approximately 5–10% harder than the persistent emission; in cases where a burst is accompanied by a 'shoulder', the latter appears to be harder. Such 'shoulders' have occa-

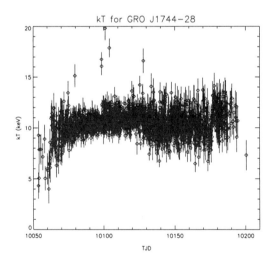

FIGURE 4. Daily temperatures (kT) for the bursts from GRO J1744−28 during its first outburst (December 1995–May 1996). The spectra are fitted with a thermal bremsstrahlung function over 20–100 keV. The first data points with higher error bars, represent the rising part of the outburst when the events were rather weak [7].

sionally other interesting properties, as we will describe in the section on fast variability.

This hardness difference between the two components of a burst is the only indication of spectral variability from the source. From an analysis of the BATSE data Briggs et al. (1998) find that the bursts show no spectral evolution throughout the entire outburst. The same result is obtained by Woods et al. (1998), for the second outburst of the source [43].

The persistent emission of GRO J1744−28 has a typical X-ray pulsar spectrum; RXTE and ASCA have both reported [38,12] power-law best fits with an exponential cut off at ∼10 keV. In the BATSE energy range, the spectrum can be fitted well with an optically thin thermal bremsstrahlung model. The temperature remains at $kT \simeq 11$ keV throughout the entire outburst [11] (except perhaps the first ∼10 days); similarly, during the second outburst the kT rises from 5 to 10 keV during the first 30 days, and remains constant thereafter.

The same picture emerges from the RXTE/PCA observations, which show [19] that during the first outburst (after mid-January 1996) the persistent spectrum did not show substantial variations. Like the burst spectra, it is well described by an exponentially cut off power law, with $\alpha = 1.2$, $kT = 15$ keV, and an absorbing column density, $N_H = 3 - 4 \times 10^{22}$ cm^{-2} (∼1.5 times less than the ASCA value given above).

In summary, BATSE and RXTE/PCA show remarkably little variation in the spectral properties of GRO J1744−28 as a function of time during the two outbursts, both for the persistent emission and the bursts. The persistent emission spectrum is similar to that observed during bursts and is approximately the same in both outbursts.

VARIATION OF THE BURST AND PERSISTENT EMISSION DURING THE OUTBURSTS

Figure 1 shows the variation of the flux of the persistent emission from GRO J1744−28 as determined from the daily averaged BATSE occultation data. An independent measure of the brightness of GRO J1744−28 is provided by the pulsed flux intensity [15]. The two curves track each other well; their similarity indicates that throughout the two outbursts the relative amplitude of the pulsations did not change much.

The daily averaged burst peak fluxes and fluences during the two outbursts also track each other well, and are both quite similar to the persistent emission light curve. From all the above similarities we can infer the following:

• The effective burst duration $\tau = E_b/F_{\max}$ does not vary systematically during the outbursts. Indeed, for the first outburst the average burst duration (after the first day of emission) is found [23] to be constant at ∼4 seconds. The analysis is still in progress for the second outburst.

- Since (except for the first day) the burst rate remained constant during each of the two outbursts, the similarity between the variations of the persistent flux and the burst fluences implies that also the total amount of energy emitted in bursts during, say, a day, tracks the amount of energy in the persistent emission during that day. The ratio of the average persistent flux to time-averaged burst flux is customarily indicated by α. The variation of α during the first outburst is shown in Figure 5, which shows that except for the first day, α is mainly in the range 20 to 100. During the first day α is much smaller: $\alpha = 3.0 \pm 0.7$ on December 2, 1995. During the second outburst α ranged between 10 and 80 with a minimum of $\alpha = 1.5 \pm 1.3$ on December 2, 1996.

The very low value of α near the onset of each outburst indicates that the bursts on that day cannot be caused by thermonuclear flashes. For such events (type I bursts) mass conservation requires [30,40] $\alpha > 20$ when averaged over a sufficiently large number of bursts. This implies that the first-day bursts from GRO J1744−28 are caused by an accretion instability. The strong similarity between the bursts emitted during the first day and those emitted later on in the outbursts extends this conclusion to all bursts. Note that this conclusion could not have been drawn on the basis of the α values if only the later bursts would have been detected with BATSE.

The above results are based on average burst properties obtained from the BATSE observations, covering the high-energy part of the persistent and burst spectra. However, the RXTE/PCA observations during the first outburst

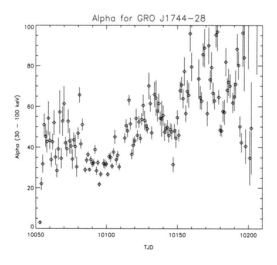

FIGURE 5. Daily averages for α, the ratio between the persistent over the burst emission flux, from GRO J1744−28 during its first outburst (December 1995−May 1996). The flux is integrated between 30−100 keV [7].

confirm these results in detail [17].

FAST VARIABILITY

From their analysis of the RXTE/PCA observations of GRO J1744−28, which cover the first outburst, Zhang et al. (1996) find that its power density spectrum shows a strong and broad Quasi-Periodic Oscillation (QPO) peak, at a frequency near 35 Hz, and much weaker peaks near 20 and 60 Hz. The centroid frequency of the strong QPO peak ranged between 33 and 41 Hz; it is not correlated with the persistent X-ray flux over an order of magnitude range in counting rate. The strength of the ∼35 Hz QPO, as measured by its r.m.s. fractional variation, is correlated with the X-ray flux, and varies between ∼ 1.5% and ∼ 6%.

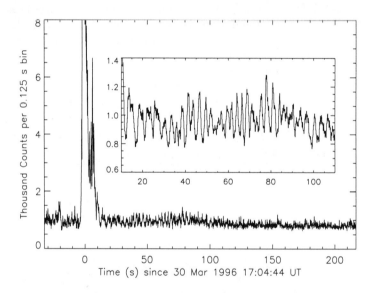

FIGURE 6. One of the 10 RXTE/PCA bursts from the 94 detected during the first outburst of GRO J1744−28 with PCA exhibiting a 'shoulder', where 0.4 Hz QPO are evident. The insert is a blow-up of the shoulder portion demonstrating the strong oscillations observed during this part of the burst [22].

Zhang et al. (1996) consider several possible models for the QPO, and conclude that none of them provide an adequate quantitative explanation. These include the magnetospheric beat frequency model, models in which a near Eddington limit X-ray luminosity affects the optical depth of the accretion flow, neutron star oscillations, and photon bubbles.

QPO with a frequency near 0.4 Hz has been detected from GRO J1744−28 following 10 out of the 94 X-ray bursts detected with the RXTE/PCA during its first outburst [22]. Figure 6 exhibits one of these events; note that the QPO appears during the 'shoulder' part of the event mentioned above. The period of these oscillations decreased over their 30–80 s lifetime, and when they occur the spectrum of the emission is relatively hard (see also previous section). These oscillations are strikingly similar to the ~ 40 mHz oscillations in the persistent emission of the Rapid Burster, following ~ 10 % of the Type II bursts from this source [20]. As in the Bursting Pulsar, the period of the oscillations decreased during their ~ 100 s lifetime, and when they occured, the persistent X-ray spectrum was relatively hard. This strengthens the similarity between the Rapid Burster and the Bursting Pulsar (see also the section below with their detailed comparison).

PULSE PHASE DELAYS DURING BURSTS

The bursts and pulsations of GRO J1744−28 show a strongly coupled behaviour which was discovered by Strickman et al. (1996), Jung et al. (1996) and Stark et al. (1996) using CGRO/OSSE and RXTE/PCA data, respectively: associated with each burst is a pulse phase delay. This delay starts increasing at the beginning of the burst, to a maximum of ~ 0.15 cycles (~ 75 ms) near the peak of the burst; afterwards the delay decreases, but retains a value of ~ 20 ms at the end of the burst. This residual post-burst delay decays more slowly, on a time scale of order 10^3 s.

A statistical analysis of BATSE data on triggered bursts shows that the average variation of the pulse delay during bursts change very little (Figure 7) during the two outbursts; this implies that the delay history during the burst does not depend strongly on the burst peak flux (or fluence), over a range of a factor ~ 5 in these quantities.

The phase delay likely reflects a change in the angular distribution of the X-ray emission, caused by the temporary increase of the mass inflow rate during a burst. A specific model for such a change was proposed by Miller (1996). In his model the increased mass inflow corresponds to an increase in the azimuthal deformation of the magneric-field lines near the magnetospheric boundary. In the case of a magnetic oblique rotator this results in an azimuthal change of the location of preferred 'pick up' of matter by the neutron star, and a corresponding azimuthal shift in the 'footprint' of the accretion stream on the neutron star surface (this footprint is a segment of a projected accretion ring, analogous to the auroral ring structure).

It is not clear how in this model the phase shift would be independent of the strength of the burst. Perhaps both the burst and the phase shift are caused by an instability involving a change in the magnetosphere which is independent of the increase in mass flow rate onto the neutron star.

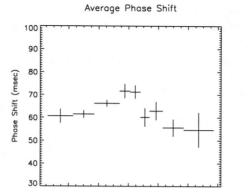

FIGURE 7. Phase-shift versus time in days of GRO J1744−28 during its first outburst. Notice that the shift changes very little during the entire outburst [24].

COUNTERPARTS AT OTHER WAVELENGTHS

For the interstellar column density, $N_H \sim 5 \times 10^{22}$ cm^{-2}, inferred from the low-energy cut-off in the X-ray spectrum [12], one expects an interstellar extinction in the V band of some 23 magnitudes. It therefore appears that identification of an optical counterpart is impossible, and that the infrared and radio regions, where the interstellar extinction drops sharply, are the spectral regions of choice for counterpart searches. Observations with the VLA [18] during the first outburst did not reveal [28] a radio counterpart to GRO J1744−28 . From a comparison of K band (2.2 μm) images obtained with the ESO NTT on February 8, 1996 and March 28, 1996, Augusteijn et al. (1996a) found a possible counterpart in the first image, which is absent in the second one. However, detailed scrutiny of the images [2] revealed that this object was detected only in three of the ten consecutive exposures of which the February 28 image is composed. Whether this reflects rapid infrared variability, or an instrumental artifact is a debated issue [9,3].

STRUCTURE AND EVOLUTION OF THE BINARY SYSTEM

The very low mass function of GRO J1744−28 strongly suggests that the secondary star in this system is a low-mass star; this has been accepted in discussions made so far of the evolutionary state of this system [10,36,42].

The long orbital period indicates that GRO J1744−28 is a late stage in the evolution of a binary in which the mass transfer is driven predominantly by the secular expansion of the secondary. For such systems the orbital period is determined by the mass of the degenerate helium core of the secondary, which for $P_{\rm orb} = 11.8$ days equals about $0.25\,{\rm M}_\odot$.

According to the analysis of the properties of transient LMXBs by Van Paradijs (1996) and King et al. (1996) the transient nature of GRO J1744−28 requires its *average* mass transfer rate to be below a critical value which, for an orbital period of 11.8 days, is $\sim 5 \times 10^{-10}\,{\rm M}_\odot\,{\rm yr}^{-1}$. For secondaries expanding because of their nuclear evolution, such a low mass transfer rate implies that the system is at the end of its phase of mass transfer; the secondary mass is then not much different from its core mass of $\sim 0.25\,{\rm M}_\odot$. Since the initial mass of the (evolved) secondary must have been at least $0.9\,{\rm M}_\odot$, this star must have transferred at least $0.5\,{\rm M}_\odot$. There is substantial evidence [39] that neutron stars, which have accreted such a large amount of matter, have magnetic fields that are much weaker than the value ($\sim 10^{11}$ G) inferred for GRO J1744−28 from its spin up rate during the first outburst. This led Van Paradijs et al. (1997) to suggest that the neutron star was formed by the accretion-induced collapse of a white dwarf during an earlier stage in the same phase of mass transfer in which we currently observe the system. However, as pointed out to us by Dr. R. Wijers, this conclusion depends on the assumption that no other factors besides the amount of accreted matter have a significant effect on the magnetic field of the neutron star. We concur.

COMPARISON OF GRO J1744−28 WITH THE RAPID BURSTER (RB)

The bursts from GRO J1744−28 do not show the characteristics of the type I bursts, which are caused by thermonuclear flashes; in particular they do not show the softening of the spectrum during the decay of the burst. From a phenomenological point of view they are therefore type II bursts, similar to those observed from the Rapid Burster [29,30].

A detailed comparison of GRO J1744−28 with the Rapid Burster was made by Lewin et al. (1996). In addition to the lack of spectral evolution during burst decay they find that the two sources have in common: (i) very low ratios of the burst parameter α (at least during periods that the RB emits type II bursts); (ii) the presence of dips following some of the bursts (in addition, many of the bursts from the RB are also preceded by a dip, which is not the case for the GRO J1744−28).

The bursts from the RB are characterized by a strong correlation between the burst fluence and the time interval to the next burst. This relaxation oscillator behaviour [29] is not obvious in the GRO J1744−28 bursts, whose intervals on any given day do not show a large dynamic range. However, according to the analysis of Kommers et al. (1997) also for GRO J1744−28 the burst fluences (normalized to the persistent X-ray flux) are significantly correlated with the intervals following the burst, but not with the intervals preceding it.

QPO have been detected following about 10% of the bursts, with centroid frequencies of ~ 0.4 Hz during their 30 to 80 s lifetime, and they occur during spectrally hard 'shoulders' which follow the burst [22]. This is qualitatively similar to the 0.04 Hz QPO found by Lubin et al. (1992) following about 10% of the type II bursts from the RB, which likewise showed a frequency increase during their $\sim 10^2$ s lifetime, and occurred during spectrally hard humps in the persistent emission.

These detailed similarities between the RB and GRO J1744−28 suggest that a similar accretion instability produces the type II bursts in both sources [22]. If this is correct, it would imply that the instability is independent of the presence of a strong neutron star magnetic field, and allow one to reject magnetospheric instability models for Type II bursts [30]. Cannizzo (1996) proposed that the bursts in GRO J1744−28 are caused by the Lightman-Eardley instability. If the same model applies to the RB as well, which during the periods of its burst activity has an accretion rate at least an order of magnitude below that of GRO J1744−28 , the question arises why the rapidly repetitive type II bursts do not occur in many other LMXBs. We are thus caught between Scylla and Charybdis: either we have to allow for different mechanisms for the Type II bursts in the RB and GRO J1744−28 , or admit that these bursts, in spite of their rich phenomenology, pose a still unsolved problem.

ACKNOWLEDGEMENTS

The authors would like to thank M.S. Briggs, J.M. Kommers, T.M. Koshut and P. Woods for providing unpublished results from their analysis of GRO J1744−28 for this review.

REFERENCES

1. Augusteijn, T., et al. *IAU Circular 6369*, (1996a).
2. Augusteijn, T., Lidman, C., and Blanco, P. *IAU Circular 6484*, (1996b).
3. Augusteijn, T., et al. ApJ **486**, 1013 (1997).
4. Blanco,
5. Bouchet, L., et al. *IAU Circular 6343*, (1996).
6. Borozdin, K., et al. *IAU Circular 6350*, (1996).
7. Briggs, M.S., et al. *ApJ in preparation*, (1998).
8. Cannizzo, J.K. *ApJ Letters* **466**, L31 (1996).
9. Cole, D.M., et al. *ApJ* **480**, 377 (1997).
10. Daumerie, P., Kalogera, V., Lamb, F.K., and Psaltis, D. *Nature* **382**, 141 (1996).

11. Deal, K. *private communication* (1996).
12. Dotani, T., et al. *IAU Circular 6337*, (1996a).
13. Dotani, T., et al. *IAU Circular 6368*, (1996b).
14. Finger, H.M., et al. *IAU Circular 6286*, (1996a).
15. Finger, H.M., et al. *Nature* **381**, 291 (1996b).
16. Fishman, G.J., et al. *IAU Circular 6272*, (1995).
17. Fox, D., et al. *ApJ, in preparation*, (1998).
18. Frail, D., et al. *IAU Circular 6307*, (1996).
19. Giles, A.B., et al. *ApJ* **469**, L25 (1996).
20. Jung, G.V., et al. *IAU Circular 6321*, (1996).
21. King, A.R., Kolb, U., and Burderi, L. *ApJ Letters* **464**, L127 (1996).
22. Kommers, J.M. et al. *ApJ Letters* **482**, L53 (1997).
23. Kommers, J.M., et al. *ApJ, in preparation*, (1998).
24. Koshut, T., et al. *ApJ Letters, in preparation*, (1998).
25. Kouveliotou, C., et al. *Nature* **379**, 799 (1996a).
26. Kouveliotou, C., et al. *IAU Circular 6395*, (1996b).
27. Kouveliotou, C., et al. *IAU Circular 6286*, (1996c).
28. Kouveliotou, C., et al. *IAU Circular 6369*, (1996d).
29. Lewin, W.H.G., et al. *ApJ Letters* **207**, L95 (1976).
30. Lewin, W.H.G., van Paradijs, J., and Taam, R.E. *X-ray Binaries*, eds. W.H.G. Lewin, J. van Paradijs, and E.P.J. van den Heuvel, Cambridge University Press, 1995, pp. 175.
31. Lewin, W.H.G., et al. *ApJ Letters* **462**, L39 (1996).
32. Lubin, L.M. et al. *MNRAS* **258**, 789 (1992).
33. Miller, G.S. *ApJ Letters* **468**, L29 (1996).
34. Paciesas, W.S., et al. *IAU Circular 6284*, (1996).
35. Rappaport, S., and Joss, P. *Nature* **266**, 683 (1977).
36. Stark, M., et al. *apJ Letters* **470**, L109 1996).
37. Sturner, S.J., and Dermer, C.D. *ApJ Letters* **465**, L31 (1996).
38. Strickman, M.S., et al. *ApJ Letters* **464**, L131 (1996).
39. Swank, J.H., et al. *IAU Circular 6291*, (1996).
40. van den Heuvel, E.P.J., and Bitzaraki, O. *A&A Letters* **297**, L41 (1995).
41. van Paradijs, J., Penninx, W., and Lewin, W.H.G. *MNRAS* **233**, 437 (1988).
42. van Paradijs, J. *ApJ Letters* **464**, L139 (1996).
43. van Paradijs, J., et al. *A&A Letters* **317**, L9 (1997).
44. Woods, P., et al. *ApJ in preparation*, (1998).
45. Zhang, W. et al. *ApJ* **469**, L29 (1996).

The Soft Gamma-Ray Repeaters

I.A. Smith

*Department of Space Physics and Astronomy, Rice University, MS-108,
6100 South Main, Houston, TX 77005-1892*

Abstract. The Soft Gamma-Ray Repeaters (SGR) are sources of brief intense outbursts of low energy gamma rays. Most likely they are a new manifestation of neutron stars. Three sources were known prior to the launch of CGRO, and bursts from two of these have been seen by BATSE. However, no new sources have been discovered, which means that they are either a very rare type of object, or the bursting stage is a very short phase of the life of many neutron stars.

The BATSE and RXTE observations have shown that SGR 1806−20 remains the most prolific of the three. An important breakthrough was that a BATSE burst was also seen by ASCA, allowing a good location to be determined. A quiescent X-ray source was found at this location, as well as a compact and an extended radio source. These suggested that this is a pulsar-powered supernova remnant. However, also present at this location was a heavily reddened star that is consistent with being a rare Luminous Blue Variable. In addition, there appear to be compact and extended dust sources present. The relationship of these different components to the emitter of the gamma-ray bursts remains unclear.

BATSE also detected bursts from SGR 1900+14. Using the network synthesis technique, relatively small error boxes were determined, and a candidate ROSAT X-ray counterpart was found. A search for a radio counterpart there did not reveal any sources. However, a pair of heavily reddened stars that appear to be variable M5 supergiants were consistent with this location: their spectra are remarkably similar. As for SGR 1806−20, compact and extended dust sources were also discovered. But again, it is not currently understood what, if any, relevance these objects have to the source of the gamma-ray bursts.

To date, BATSE has not seen any bursts from SGR 0525−66. The error box for this source is consistent with the N49 supernova remnant in the LMC, and a possible soft X-ray counterpart has been found. However, in spite of detailed searches, no point-like counterpart has been detected at any other wavelength. Given the bizarre counterparts to the other two SGR, this lack has only increased the confusion regarding these enigmatic sources.

Assuming the sources are correctly identified as being in our Galaxy and in the LMC, it is not easy to model the super-Eddington luminosities during the bursts. The most popular models currently span a very large range from ultra-strong magnetic field scenarios ("magnetars") to self-absorbed synchrotron emission from cooling relativistic electrons in a moderate neutron star magnetic field to pulsar glitches to accretion of planets or comets onto neutron stars.

PRE-CGRO REVIEW

The Soft Gamma-Ray Repeaters (SGR) are a small but highly unusual class of sources of brief intense outbursts of low energy gamma rays. It appears that they are new manifestations of neutron stars. However, accounting for the wide variety of multiwavelength observations as well as the bursts themselves has produced no clear picture for these sources.

There are three known SGR. Two of them (SGR 1900+14 and SGR 1806-20) are in our Galaxy, while the third (SGR 0525-66) is generally believed to be in the LMC. There are several features of the bursts that distinguish them as a separate class of objects [41,26,48]:

(1) The burst spectra are harder than in X-ray bursters, but softer than in gamma-ray bursters. While an optically thin thermal bremsstrahlung can usually fit the spectra above ~ 20 keV, there is a definite turn-over below this in the SGR 1806-20 bursts seen by ICE [13]. The burst spectra are usually similar from burst to burst. The apparent exception was the well known 1979 March 5 event from SGR 0525-66, whose initial spectrum may be consistent with those of gamma-ray bursters [14].

(2) Unlike the gamma-ray bursters, where it remains uncertain whether there is any evidence for repetition, the SGR definitely repeat. Well over 100 bursts have been detected from SGR 1806-20, 7 for SGR 1900+14, and 16 for SGR 0525-66.

(3) The bursts are short with simple time profiles. Most of their durations are $\lesssim 1$ second [21]. The rise and fall times can be as fast as 1-5 ms, indicating that a compact object is involved [12,26].

(4) The recurrence time can range from seconds to years, with no correlation between the intensity and the time between the bursts [34]. The burst time intervals for the ICE SGR 1806-20 bursts appeared to satisfy a lognormal distribution [23]. These behaviors are similar to the glitches in pulsars, and also to earthquakes [4]. A wide range of burst fluences have been observed for each of the sources. They are therefore not standard candles.

This review will focus on the burst and counterpart observations since the launch of CGRO. Earlier satellite observations of the bursts are discussed in the following papers (and references therein): KONUS [40-43,17], PVO [12,14], SIGNE 2MP (Prognoz 9) [3,1], SMM [26], ICE [34,62,13,14], Phobos [19], Ulysses [20].

CGRO OBSERVATIONS

Three sources were known prior to the launch of CGRO, and bursts from two of these have been seen by BATSE [27–30]. BATSE has not seen any bursts from SGR 0525–66. Also, no new sources have been discovered, which means that they are either a very rare type of object, or the bursting stage is a very short phase of the life of many neutron stars [28].

SGR 1900+14

The first SGR observed by BATSE was SGR 1900+14 [27]. Four bursts were detected in the space of two months, with fluences ~ 5 times fainter than the weakest burst seen by KONUS. BATSE was able to resolve fine time structures in the bursts, but their time series remained relatively simple. Their spectra were not unusual, and no indication of spectral evolution was found.

The BATSE bursts alone are consistent with the error box of the previous KONUS bursts [27]. Using the network synthesis technique with the BATSE and Ulysses observations [20,35] two error boxes were determined outside SNR G42.8+0.6. If the compact object that produces the SGR bursts was formed in this supernova explosion, it would have to have a transverse velocity $\gtrsim 1000$ kms^{-1} [20].

A quiescent, steady, ROSAT X-ray point source was found consistent with the primary error box [22]. Very bizarre multiwavelength counterparts were also found at this ROSAT location, as detailed later. However, it should be cautioned that until more bursts are detected by multiple satellites — particularly with a pointed X-ray instrument — the location of SGR 1900+14 remains uncertain.

SGR 1806–20

Although SGR 1806–20 was the second SGR seen by BATSE, its recurring activity has confirmed that it remains the most prolific of the three [28–30]. As before, the pulse profiles are relatively simple, with no pulsations on millisecond time scales. There has been nothing unusual in their spectra.

The error boxes of the BATSE bursts are consistent with the source locations from previous satellites. But a key breakthrough was that a BATSE burst on 1993 October 9 was also seen by ASCA, allowing a good location to be determined [47]. A steady quiescent X-ray point source AX 1805.7–2025 was found at this location [7,47,57], and it was found to be inside a compact radio nebula in SNR G10.0–0.3 [31]. This spurred a multiwavelength observing campaign that has produced fascinating but contradictory information, as detailed in the next section.

The 1996 October BATSE bursts triggered a Target of Opportunity using RXTE. In addition to the large bursts seen by BATSE, numerous small bursts were seen to come in groups with burst rates as high as one per minute, and with peak fluxes about 2 orders of magnitude fainter [30].

THE BIZARRE QUIESCENT COUNTERPARTS

The discovery of possible quiescent counterparts to the SGR has been a significant breakthrough that has spurred detailed multiwavelength observations of these objects. Rather than clarifying the situation, the fact that the three counterparts differ markedly has deepened the confusion. See Hurley [21] for the most recent review.

SGR 1806-20

The current status of the radio through near infrared observations of the quiescent counterparts to SGR 1806-20 is shown in Figure 1 [53].

The discovery of an extended radio nebula suggested that this source is a plerion [32]. This was bolstered by the detection of a central radio core, with an evolving jet-like extension [66,15].

This picture was completely confused when a heavily reddened star was also found at the X-ray location. It is consistent with being a very rare Luminous Blue Variable [33,64]. From ^{12}CO(1-0) and ^{13}CO(1-0) millimeter observations, several molecular clouds were found in the direction of the source, and a careful analysis indicated that the distance to SGR 1806-20 is 14.5 kpc [8]. At this distance, the stellar counterpart could be one of the most luminous stars in our Galaxy.

The photospheric emission from this star dominates in the near infrared. K-band spectra showed that the most prominent Brγ emission line has a blueshifted absorption component (i.e. a P Cygni profile), indicating the presence of an outflow with a terminal velocity of $\sim$ 400 kms^{-1} [64,55]. The spectra also reveal the presence of both low- and high-excitation lines. This surprising result could indicate that there are two physically distinct circumstellar regions, e.g. a disk in addition to the wind.

A detailed study of the IRAS observations of SGR 1806-20 revealed even more components [65]. The emission at 12 and 25 μm is point-like and in excess of the photospheric emission, while the emission is definitely extended at 60 μm. However, careful studies in the sub-millimeter and millimeter regions have not found any point sources [52-54]. Monoenergetic synchrotron radiation and black body spectra are too broad to be consistent with both these infrared and millimeter point source observations [54]. However, simple dust models such as those of Dwek & Werner [10] can explain the combined observations. It therefore appears as if the star may have shed massive dusty shells, and there is also an unknown point source of dust. The lack of millimeter emission indicates that the 'core' radio emission is not due to free-free emission from a fully ionized stellar wind [53].

The relationship of these multiple components to the source of the bursts

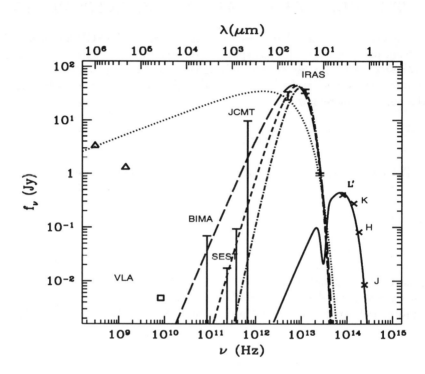

FIGURE 1. SGR 1806-20 [53]. Open triangles: VLA radio emission for the entire 9×6 arcmin^2 nebula [32]. Open square at 3.6 cm is for the central 'core' of the radio emission [66]. 3σ point source upper limit at 85 GHz (3.5 mm) from the Berkeley-Illinois-Maryland Array (BIMA) [52]. 3σ upper limit at 235 GHz (1.3 mm) from the Swedish-ESO Submillimeter Telescope (SEST) [53]. 3σ upper limits at 800 μm and 450 μm from the UKT14 on the James Clerk Maxwell Telescope (JCMT) [54]. IRAS flux densities are shown with 1σ error bars; the source is point-like at 12 μm and 25 μm, and extended at 60 μm [65]. Crosses: L', K, H, and J-band flux densities for star A in Kulkarni et al. [33]. Dotted line is a monoenergetic synchrotron spectrum with $\nu_c = 5 \times 10^{12}$ Hz [16]. Long dashed line is a 120 K black body spectrum. Short dashed line is a Population I dust model ($Q_\nu \propto \nu$) with $T_{gr} = 100$ K [10]. Short dashed-dotted line is a Population II dust model ($Q_\nu \propto \nu^2$) with $T_{gr} = 92$ K [10]. Solid curve is a 30,000 K black body attenuated by the interstellar extinction law of Rieke & Lebofsky [50] with $A_V = 30$ [64]. Note that care should be used in interpreting this figure, since it mixes both point and extended source components.

of gamma rays remains an enigma that only future observations can resolve.

SGR 1900+14

The current status of the radio through near infrared observations of the quiescent counterparts to SGR 1900+14 is shown in Figure 2 [53].

Unlike SGR 1806–20, no radio source was detected at the location of the X-ray counterpart to SGR 1900+14 [22]. Nor has one been seen in the millimeter or sub-millimeter [52,54].

However, a quite remarkable pair of infrared stellar counterparts has been found. They are heavily reddened, and appear to be variable M5 supergiants separated by 3.37" at a distance of $\sim 12 - 15$ kpc [67]. Their most striking feature is that the ratio of their flux densities is mostly constant across their entire optical and infrared spectra [67,55]. Particularly in the most recent higher resolution infrared observations [55], although a large number of atomic and molecular lines are present, the significant difference between the two stars is in the first overtone CO absorption band starting at 2.3 μm: the absorption is deeper in the fainter star B. However, for the second overtone CO bands at 1.6 μm, the ratio of the two stars is virtually constant.

An image of SGR 1900+14 at 10 μm showed that the emission is still dominated by the well-resolved stars, with a faint extended component [65]. A detailed study of the IRAS observations revealed other components [65]. The emission at 25 μm is point-like and in excess of the photospheric emission, and the emission is even brighter but also definitely extended at 60 μm. Monoenergetic synchrotron radiation and black body spectra are too broad to be consistent with both these infrared and the millimeter point source observations [54]. However, simple dust models can explain the combined observations. It therefore appears as if the star(s) may have shed massive dusty shells, and there is also an unknown point source of dust, making it similar to SGR 1806–20 in this feature.

Once again, the relationship of these multiple components to the source of the bursts of gamma rays remains a mystery.

SGR 0525-66

SGR 0525–66 has long been known to have a position consistent with the N49 supernova remnant in the LMC [40,18,5], and a quiescent soft X-ray counterpart has been found [51,39]. If the compact object that produces the SGR bursts was formed in this supernova explosion, it would have to have a transverse velocity $\gtrsim 1200$ kms^{-1}.

Figure 3 shows the current status of the other multiwavelength observations of SGR 0525–66 [53]. For comparison, the Crab continuum spectrum and Crab IRAS detections are included.

No point source has been detected at the location of SGR 0525–66 in the radio [9], millimeter [53], optical or near infrared [9]. The other two SGR counterparts are brightest in the mid infrared, making this the most promising place to look for one for SGR 0525–66. A detailed study of the IRAS data

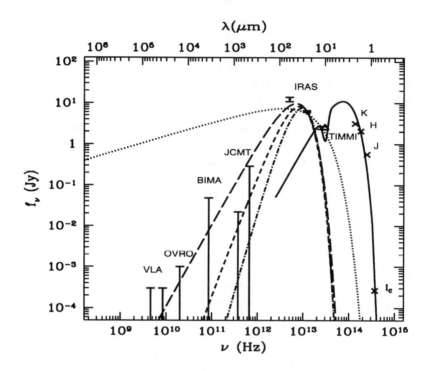

FIGURE 2. SGR 1900+14 [53]. 3σ upper limits at 4.56 and 8.44 GHz at the location of the X-ray peak using the VLA [22]. 3σ upper limit at 20 GHz using the Owens Valley Radio Observatory (D. Frail, private communication). 3σ point source upper limit at 85 GHz (3.5 mm) from BIMA [52]. 3σ upper limits at 800 μm and 450 μm from the UKT14 on the JCMT [54]. IRAS flux densities are shown with 1σ error bars; the source is point-like at 12 μm and 25 μm, and extended at 60 μm [65]. Triangle: N-band flux density for stars A plus B seen by TIMMI on the 3.6-m at ESO [65]. Crosses: K, H, J, and I_C-band flux densities for stars A plus B seen by the USNO [67]. Dotted line is a monoenergetic synchrotron spectrum with $\nu_c = 1.3 \times 10^{13}$ Hz [16]. Long dashed line is a 120 K black body spectrum. Short dashed line is a Population I dust model with $T_{gr} = 100$ K [10]. Short dashed-dotted line is a Population II dust model with $T_{gr} = 92$ K [10]. Solid curve is a 2900 K black body attenuated by the interstellar extinction law of Rieke & Lebofsky [50] with $A_V = 19.2$ [67]. Note that care should be used in interpreting this figure, since it mixes both point and extended source components.

indeed found a source [65]. However, it is extended at 12, 25, and 60 μm, and the IRAS colors are typical of other supernova remnants. This infrared emission likely originates from heated dust in N49. Images were made in the N-band to cover most of N49, but no point sources were found [53].

The lack of detection of a point source at other wavelengths has led to questions about the association with N49 [9]. For example, scaling the 12 μm IRAS flux densities of SGR 1806-20 or SGR 1900+14 to the distance of the LMC would give $\sim$ 100 mJy, which would have been easily detected in the

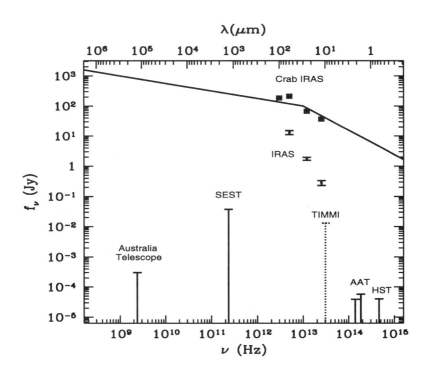

FIGURE 3. SGR 0525-66 [53]. Upper solid curve: Crab continuum spectrum [38]. Solid squares: Crab IRAS [58]. All other points are SGR 0525-66. 3σ point source upper limit at 12.6 cm from the Australia Telescope [9]. 3σ upper limit at 235 GHz (1.3 mm) from SEST [53]. IRAS flux densities are shown with 1σ error bars; the source is extended at all three wavelengths [65]. 3σ N-band ($\lambda_{\text{eff}} = 9.862$ μm) point source upper limit from the Thermal Imaging Multi-Mode Instrument (TIMMI) on the 3.6-m telescope at ESO [53]: dotted limit indicates result is preliminary. 3σ upper limit at 2.16 μm and 1.64 μm from IRIS on the 3.9-m Anglo Australian Telescope [9]. 3σ upper limit at 656.3 nm from the Hubble Space Telescope [9]. Note that care should be used in interpreting this figure, since it mixes both point and extended source components.

TIMMI N-band observations. One might also expect that if the X-ray source is in N49, its radiation would produce an effect at another wavelength. Only further observations will clarify the situation.

BURST THEORIES

Given the very complicated observational situation, it is difficult to determine what information is relevant for modeling these sources. Are these isolated neutron stars? If so, do they have high velocities? Or other unique features, such as very strong magnetic fields, large glitches, or planetary disks? Do the possible companion stars play an important role?

Assuming the sources are correctly identified as being in our Galaxy and in the LMC implies that the bursts have super-Eddington luminosities by factors up to $\sim 10^6$. This is a challenge for the models, though many inventive ones have been proposed; see Liang [36] for a recent review. Research has progressed recently on two avenues.

Spectral Fitting

Many possible physical mechanisms have been directly fitted to the burst data to determine whether they are allowed. The most important data set has proven to be the ICE bursts from SGR 1806−20, where there is a definite roll-over below the spectral peak [13].

Fenimore et al. found that a black body could not explain this data. While an optically thin thermal bremsstrahlung model can explain the spectral shape above the peak, some form of self-absorption is required to explain the roll-over, with only the size of the emitting surface varying from event to event. The ICE burst data is consistent with photoelectric absorption by neutral material, but the required 10^{24} N_H cm^{-2} is very large. It is also consistent with emission from the lowest harmonics from a 1.3×10^{12} G field [13].

Synchrotron radiation was initially suggested as the mechanism for the 1979 March 5 burst from SGR 0525−66 [49]. To explain the roll-over in the ICE bursts, it was found that a model using self-absorbed synchrotron emission from a cooling distribution of mildly relativistic leptons works, provided the particles are injected with monoenergetic or truncated power-law forms [37]. Moderate magnetic field strengths $\sim 10^{11}$ are implied.

Compton scattering models do not generally fit the ICE bursts, though a Sunyaev & Titarchuk [59] model with thermal particles and a black body injection was acceptable [13]. A problem with this model is that the spectrum drops too rapidly above the peak, for example when comparing it to the BATSE SGR 1900+14 spectrum [27], where there are better statistics at higher energies than in the ICE bursts. This problem was overcome by including a small but significant fraction of non-thermal particles [56].

Energy Source

Several different energy sources for the bursts have been studied in detail recently. These include:

(1) **Magnetic Field Energy.** For several reasons, most importantly explaining the required luminosity, this model requires an ultra-strong magnetic field $\gtrsim 10^{14}$ G, and the stars have been dubbed "magnetars" [60,61,46,2,63]. There is no direct observational evidence for or against the existence of this type of object.

(2) **Accretion.** Steady accretion onto the compact object does not work, because the intensity of the bursts is not dependent on the time between them. However, the irregular accretion of planets or comets does not suffer from this problem [6,24,25]. A general prediction of this class of models is that the burst duration should scale as the 1/3 or 4/9 power of the burst energy.

(3) **Radio Pulsar Glitches.** In this model, crustal disturbances in the surface of a neutron star with magnetic field $\sim 10^{12}$ G excite a spectrum of outwardly propagating sheared Alfvén waves [44,45,11]. A mildly relativistic flow is driven from the star. The resulting burst spectrum approximates that of a black body. Since Fenimore et al. found that a black body could not explain all the ICE observations, it will be important to perform more detailed fitting of the model results to these bursts.

FUTURE OBSERVATIONS

It is apparent that further observations will be needed to guide the theory further.

Many detailed studies of the quiescent counterparts are underway, including using ASCA, the Infrared Space Observatory, Hubble Space Telescope, the new sub-millimeter camera SCUBA on the JCMT, millimeter line studies at SEST, and high resolution infrared spectra at Kitt Peak to name just a few. With these, it is hoped that it will be possible to untangle the multiple multiwavelength components. Studies of the recent set of RXTE bursts from SGR 1806-20 will be very interesting, since they probe luminosities much lower than before.

The all-sky coverage of BATSE is a key sentinel for detecting the recurrence of burst activity so that coordinated multiwavelength observations can be performed. Unfortunately, the most recent series of bursts from SGR 1806-20 took place just when the source was very difficult to view because of Sun constraints. Future observations during an active phase should determine whether any of the other multiwavelength components also vary. Reactivation of SGR 1900+14 or SGR 0525-66 could potentially resolve the location uncertainties. Thus it is certainly hoped that BATSE will continue to provide this essential information for many years to come.

ACKNOWLEDGEMENTS

This work was supported by NASA grants NAG 5-1547, NAG 5-2772, and NAG 5-3824 at Rice University.

REFERENCES

1. Atteia, J.-L., et al. 1987, ApJ, 320, L105
2. Baring, M. G. 1995, ApJ, 440, L69
3. Boer, M., et al. 1986, Adv. Space Res., 6, 97
4. Cheng, B., Epstein, R. I., Guyer, R. A., & Young, A. C., 1995, Nature, 382, 518
5. Cline, T. L., et al. 1982, ApJ, 255, L45
6. Colgate, S. A., & Petschek, A. G. 1981, ApJ, 248, 771
7. Cooke, B. A. 1993, Nature, 366, 413
8. Corbel, S., et al. 1997, ApJ, 478, 624
9. Dickel, J. R., et al. 1995, ApJ, 448, 623
10. Dwek, E., & Werner, M. W. 1981, ApJ, 248, 138
11. Fatuzzo, M., & Melia, F. 1996, ApJ, 464, 316
12. Fenimore, E. E., Evans, W. D., Klebesadel, R. W., Laros, J. G., & Terrell, J. 1981, Nature, 289, 42
13. Fenimore, E. E., Laros, J. G., & Ulmer, A. 1994, ApJ, 432, 742
14. Fenimore, E. E., Klebesadel, R. W., & Laros, J. G. 1996, ApJ, 460, 964
15. Frail, D. A., Vasisht, G., & Kulkarni, S. R. 1997, ApJ, 480, L129
16. Ginzburg, V. L., & Syrovatskii, S. I. 1965, ARA&A, 3, 297
17. Golenetskii, S. V., Ilyinskii, V. N., & Mazets, E. P. 1984, Nature, 307, 41
18. Helfand, D. J., & Long, K. S. 1979, Nature, 282, 589
19. Hurley, K., et al. 1994, ApJ, 423, 709
20. Hurley, K., et al. 1994, ApJ, 431, L31
21. Hurley, K. 1996, in Huntsville Gamma-Ray Burst Symposium, ed. C. Kouveliotou, M. S. Briggs, & G. J. Fishman (New York: AIP), 889
22. Hurley, K., et al. 1996, ApJ, 463, L13
23. Hurley, K. J., McBreen, B., Rabbette, M., & Steel, S. 1994, A&A, 288, L49
24. Katz, J. I., Toole, H. A., & Unruh, S. H. 1994, ApJ, 437, 727
25. Katz, J. I. 1996, ApJ, 463, 305
26. Kouveliotou, C., et al. 1987, ApJ, 322, L21
27. Kouveliotou, C., et al. 1993, Nature, 362, 728
28. Kouveliotou, C., et al. 1994, Nature, 368, 125
29. Kouveliotou, C., et al. 1996, IAUC, 6501
30. Kouveliotou, C., et al. 1996, IAUC, 6503
31. Kulkarni, S. R., & Frail, D. A. 1993, Nature, 365, 33
32. Kulkarni, S. R., et al. 1994, Nature, 368, 129
33. Kulkarni, S. R., et al. 1995, ApJ, 440, L61
34. Laros, J. G., et al. 1987, ApJ, 320, L111

35. Li, P., et al. 1997, ApJ, in press
36. Liang, E. P. 1995, Ap&SS, 231, 69
37. Liang, E. P., & Fenimore, E. 1995, ApJ, 451, L57
38. Marsden, P. L., et al. 1984, ApJ, 278, L29
39. Marsden, D., Rothschild, R. E., Lingenfelter, R. E., & Puetter, R. C. 1996, ApJ, 470, 513
40. Mazets, E. P., Golenetskii, S. V., Il'inskii, V. N., Aptekar', R. L., & Guryan, Yu. A. 1979, Nature, 282, 587
41. Mazets, E., & Golenetskii, S. V. 1981, Ap&SS, 75, 47
42. Mazets, E., et al. 1981, Ap&SS, 80, 3
43. Mazets, E. P., Golenetskii, S. V., Guryan, Yu. A., & Ilyinskii, V. N., 1982, Ap&SS, 84, 173
44. Melia, F., & Fatuzzo, M. 1993, ApJ, 408, L9
45. Melia, F., & Fatuzzo, M. 1995, ApJ, 438, 904
46. Miller, M. C. 1995, ApJ, 448, L29
47. Murakami, T., et al. 1994, Nature, 368, 127
48. Norris, J. P., Hertz, P., Wood, K. S., & Kouveliotou, C. 1991, ApJ, 366, 240
49. Ramaty, R., Lingenfelter, R. E., & Bussard, R. W. 1981, Ap&SS, 75, 193
50. Rieke, G. H., & Lebofsky, M. J. 1985, ApJ, 288, 618
51. Rothschild, R. E., Kulkarni, S. R., & Lingenfelter, R. E. 1994, Nature, 368, 432
52. Smith, I. A., Chernin, L. M., & Hurley, K. 1996, A&A, 307, L1
53. Smith, I. A., et al. 1997, Adv. Space Res., in press
54. Smith, I. A., Schultz, A. S. B., Hurley, K., Van Paradijs, J., & Waters, L. B. F. M. 1997, A&A, in press
55. Smith, I. A., et al. 1997, ESA SP-382, ed. C. Winkler, T. J.-L. Courvoisier, & Ph. Durouchoux (Noordwijk: ESA Publications), in press
56. Smith, I. A., Liang, E. P., Crider, A., Lin, D., & Kusunose, M. 1997, these proceedings
57. Sonobe, T., Murakami, T., Kulkarni, S. R., Aoki, T., & Yoshida, A. 1994, ApJ, 436, L23
58. Strom, R. G., & Greidanus, H. 1992, Nature, 358, 654
59. Sunyaev, R. A., & Titarchuk, L. G. 1980, A&A, 86, 121
60. Thompson, C., & Duncan, R. C. 1995, MNRAS, 275, 255
61. Thompson, C., & Duncan, R. C. 1996, ApJ, 473, 322
62. Ulmer, A., Fenimore, E. E., Epstein, R. I., Ho, C., Klebesadel, R. W., Laros, J. G., & Delgado, F. 1993, ApJ, 418, 395
63. Usov, V. V. 1997, A&A, 317, L87
64. Van Kerkwijk, M. H., Kulkarni, S. R., Matthews, K., & Neugebauer, G. 1995, ApJ, 444, L33
65. Van Paradijs, J., et al. 1996, A&A, 314, 146
66. Vasisht, G., Frail, D. A., & Kulkarni, S. R. 1995, ApJ, 440, L65
67. Vrba, F. J., et al. 1996, ApJ, 468, 225

Galactic Black Hole Binaries: High-Energy Radiation

J.E. Grove[1], J.E. Grindlay[2], B.A. Harmon[3], X.-M. Hua[4], D. Kazanas[4], and M. McConnell[5]

[1] *E.O. Hulburt Center for Space Research, Naval Research Lab, Washington, DC 20375*
[2] *Harvard-Smithsonian Center for Astrophysics, Cambridge, MA 02138*
[3] *NASA/Marshall Space Flight Center, Huntsville, AL 35812*
[4] *NASA/Goddard Space Flight Center, Greenbelt, MD 20771*
[5] *University of New Hampshire, Durham, NH*

Abstract. Observations of galactic black hole candidates made by the instruments aboard the Compton GRO in the hard X-ray and γ-ray bands have significantly enhanced our knowledge of the phenomenology of the emission from these objects. Understanding these observations presents a formidable challenge to theoretical models of the accretion flow onto the compact object and of the physical mechanisms that generate high-energy radiation. Here we summarize the current state of observations and theoretical interpretation of the emission from black hole candidates above 20 keV.

The all-sky monitoring capability of BATSE allows, for the first time, nearly continuous studies of the high-energy emission from more than a dozen black hole candidates. These long-term datasets are particularly well-suited to multiwavelength comparison studies, from the radio upward in frequency (Zhang et al. 1997a, these proceedings). Energy spectral evolution and/or spectral state transitions have been observed from many of the black hole candidates. Moderately deep searches of the galactic plane suggest a deficit of weak γ-ray transients. Such population studies have implications for the origin of black hole binaries and the nature of accretion events.

Observations above 50 keV from OSSE demonstrate that in the γ-ray band there exist two spectral states that appear to be the extensions of the X-ray low (hard) and high (soft), or perhaps very high, states. The former state cuts off with e-folding energy $\sim$100 keV and has its peak luminosity near this energy; thus substantial corrections need to be made to historical estimates of the bolometric luminosity of black holes in the "low" state. In contrast, in the X-ray high (soft) state, the luminosity peaks in the soft X-rays and the spectrum extends with an unbroken power law, even up to energies above 500 keV in some cases. COMPTEL has detected emission above 750 keV from Cyg X-1 and the transient GRO J0422+32. In both cases the data suggest that an additional weak, hard spectral component is required beyond that observed by OSSE at lower energies, although the precise spectral form is yet to be determined.

The breaking γ-ray spectrum can be well modeled by Comptonization of soft photons from the accretion disk in a hot thermal plasma. However, recent studies of the combined X-ray and γ-ray spectrum of Cyg X-1 and GX339-4 cast severe doubts on the simple geometry of a hot corona overlying a thermal accretion disk. Furthermore, timing studies of the former source are inconsistent with spectral formation by Compton scattering in a uniform, compact hot cloud, suggesting instead a radial decline in the electron density. The power-law γ-ray spectral state creates more significant theoretical challenges, particularly in explaining the lack of a break at energies exceeding the electron rest mass. It has been suggested that in the X-ray high (soft) state, the high-energy emission arises from bulk-motion Comptonization in the convergent accretion flow from the inner edge of the accretion disk. Such a process can conceivably generate the γ ray spectrum extending without a cutoff, if the accretion rate approaches that of Eddington.

I INTRODUCTION

The instruments of the Compton GRO have made extensive observations in the hard X-ray and γ-ray bands of galactic black hole candidates (BHCs). With its all-sky capability, BATSE has monitored emission on a nearly continuous basis from at least three persistent sources (Cyg X-1, 1E1740.7−2942, GRS 1758−258) and eight transients (GRO J0422+32, GX339-4, N Mus 1991, GRS 1716−249, GRS 1009−45, 4U 1543−47, GRO J1655−40, and GRS 1915+105). OSSE has made higher-sensitivity, pointed observations of all of these sources. COMPTEL has detected emission above 750 keV from Cyg X-1 and GRO J0422+32. To date, there have been no reported detections of galactic BHCs by EGRET.

The most reliable evidence for the presence of a black hole in a binary system comes from determination of a mass function through optical measurements of the radial velocity of the companion star. If the resulting lower limit on the mass of the compact object exceeds $3\,M_\odot$, the upper limit for the mass of a stable neutron star based on current theory, then one can reasonably assume that the compact object is a black hole. There are at least nine X-ray binary systems with minimum mass estimates exceeding $3\,M_\odot$, of which three (Cyg X-1, GRO J0422+32, and GRO J1655−40) have been clearly detected by GRO instruments. Other objects are identified as BHCs based on the similarity of their high-energy spectra and rapid time variability to those of Cyg X-1. Such classification is, of course, somewhat tenuous. Before neutron stars and black holes can be reliably distinguished based on their X-ray and γ-ray spectra, the full range of spectral forms from both classes must be observed and characterized. Extensive knowledge of the X-ray emission of these objects has accumulated in the literature, but the broad nature of the γ-ray emission is only now coming to light, with the high sensitivity of current-generation instruments.

The French coded-aperture telescope Sigma on the Russian Granat spacecraft has imaged at least a dozen BHCs, including most of those in the list above, but with the addition of TrA X-1, GRS 1730–312, and GRS 1739–278. The latter two objects were weak transients discovered during a multi-year survey of the galactic center region and have been classified as BHCs by their outburst lightcurves and the hardness of their spectra (Vargas et al. 1997). In this survey, Sigma regularly detected the persistent, variable sources 1E1740.7–2942 and GRS 1758–258, both of which are classified as BHCs on spectral grounds. The most striking result from Sigma observations of BHCs is the detection of broad spectral features below 500 keV from 1E1740.7–2942 (Sunyaev et al. 1991, Bouchet et al. 1991, Churazov et al. 1993, Cordier et al. 1993) and N Mus 1991 (Goldwurm et al. 1992, Sunyaev et al. 1992). These features have been interpreted as thermally broadened and red-shifted annihilation radiation from the vicinity of the compact object.

II BATSE SURVEY FOR BLACK HOLE BINARIES IN THE GALACTIC PLANE

BATSE has proven to be a remarkably effective tool for the study of persistent and transient hard X-ray sources using the occultation technique. Not only can known sources be studied, with sensitivity of $\sim$100 mCrab for a 1-day integration (Harmon et al. 1992), but the powerful occultation imaging technique (Zhang et al. 1993) has opened the way for the study of relatively crowded fields and previously unknown sources. Grindlay and coworkers have begun a survey of the galactic plane with the objective of measuring or constraining the black hole X-ray binary (BHXB) population in the Galaxy. BHXBs, both those with low mass companions (e.g. the X-ray novae such as N Muscae 1991) and the high mass systems such as Cyg X-1, are distinguished by having relatively luminous hard X-ray (20–100 keV) emission as compared to the systems containing neutron stars (Barret et al. 1996a). Thus a survey for persistent or transient sources in the hard X-ray band is optimally suited for the detection and study of BHXB systems. Furthermore, since most BHXBs are now known to be transient—and indeed the X-ray novae have allowed the most convincing dynamical mass function measurements of the optical counterparts when they are in quiescence (cf. van Paradijs and McClintock 1995)—the transients are the systems most likely to be BHXBs (and see van Paradijs 1996 for a likely explanation).

Given the recurrence times of the prototypical soft X-ray transients (SXTs) to be $\gtrsim$10-50 years (e.g. $\gtrsim$10 years for Nova Muscae; cf. Barranco and Grindlay 1997), and the relatively nearby (1–4 kpc) optical distances for the bright SXTs identified at a rate of $\sim$1 per year by Ginga, WATCH and BATSE, it is straightforward to extrapolate that 3-4 per year should be detectable by BATSE in these deeper searches (i.e. with peak fluxes at or below 100 mCrab) if the BHXBs are distributed uniformly in a galactic disk with radius 12 kpc. Thus over the 6 years of archival BATSE data now available, a full analysis of the galactic plane should yield a significant number ($\gtrsim 20$) of new BHXBs.

The CfA BATSE Image Search (CBIS) system (Barret et al. 1997, Grindlay et al. 1997) has been run on 900d of BATSE data. Known sources, though perhaps not previously detected by BATSE (e.g. Cir X-1; Grindlay et al. 1997) are found in the survey. Both short-outburst BHXBs (e.g. 4U1543–47) and longer duration outburst from neutron star LMXBs (e.g. 4U1608–52) have been found in the data "automatically" at the times and approximate fluxes seen with direct occultation light curve analysis (Harmon et al. 1992). More importantly, the search has yielded at most 5 candidate new (i.e. uncatalogued) sources. All of these candidate sources are below $\sim$50–100 mCrab in peak flux; no new transients at $\sim$100–200 mCrab have been found in this initial survey. The preliminary results thus suggest, but do not yet prove, a lower rate of candidate new BHXB transients than the simple scalings above would suggest: a rate of $\sim$1–2 per year rather than the 3–4 possibly expected.

If the true number is in fact much less than expected, then important questions arise for the BHXB population and/or transient outburst models:

1. *What is the total number of BHXBs?* Many authors (e.g. Tanaka and Lewin 1995 and references therein) estimate on the basis of simple arguments (such as the SXT detection rate) that the total population of low-mass BHXBs in the galaxy is $\gtrsim$300–1000. If BATSE cannot find the predicted fainter systems, have the optical distances been systematically under-estimated or are the peak luminosities typically fainter?

2. *What is the formation rate of BHXBs?* Clearly, a measurement of, or constraints on, the total number of BHXBs is needed to determine if models for the formation of BHXBs vs. neutron star binaries (e.g. Romani 1992) that predict large numbers of low-mass BHXBs are correct.

3. *What are the characteristic spectra and lightcurves of SXTs?* A search for faint BHXBs with BATSE (and future more sensitive surveys) is crucial for comparison with ASM or WFC searches (RXTE and future ASMs) which will typically operate in the 2–10 keV band. Recent WFC detections of $\sim$10 mCrab transients in the galactic bulge region by SAX (e.g. in 't Zand and Heise 1997) may suggest a much higher rate of either fainter or softer transients.

4. *What are the characteristic recurrence times for SXTs?* If the BATSE detection rate is much lower than predicted, is the recurrence time much longer than the small-N statistics would now indicate? This of course directly affects questions 1 and 2 above as well as the outburst models. It is of particular interest to search for "mini"-outbursts of the nearby bright systems (e.g. A0620–00), which might be expected to be more detectable in the hard X-ray band than the soft band in the low (hard) state (cf. Fig. 1).

III TWO GAMMA-RAY SPECTRAL STATES

The historical record of X-ray (i.e. <30 keV) observations of galactic BHCs reveals at least four spectral states (see, e.g., Tanaka 1989 & 1997, Grebenev et al. 1993, and van der Klis 1994 & 1995). In the "*X-ray very high*" and "*X-ray high (soft)*" states, the X-ray spectrum is dominated by an "ultrasoft" thermal or multi-color blackbody component with kT $\sim$ 1 keV. A weak power-law tail, with photon number index $\Gamma \sim$ 2-3, is frequently present and dominant above $\sim$10 keV. The states differ in X-ray luminosity—the former being close to the Eddington limit, L_E, and the latter typically $\sim$3-30 times less than L_E—and in the character of their rapid time variability—the very high state usually has 3-10Hz quasi-periodic oscillations (QPOs) and stronger broadband noise. The "*X-ray low (hard)*" state exhibits a single power-law spectrum with $\Gamma \sim$ 1.5-2, and a typical X-ray luminosity of <1% of Eddington. Recent measurements by OSSE indicate that, in this state, the γ-ray luminosity of a typical BHC exceeds the X-ray by a factor $\sim$5 (Grove et al. 1997b). This state is characterized by strong, rapid, intensity variability, with rms variations of order a few tens of percent of the total emission. The "*X-ray off*" or "*quiescent*" state exhibits very low level emission with uncertain spectral shape at a luminosity $L_X < 10^{-4} L_E$.

OSSE observations of a number of transient BHCs (Grove et al. 1997a, 1997b) indicate that there are at least two distinct γ-ray spectral states, the "*breaking*" state, which corresponds to the X-ray low (hard) state, and the "*power law*" state, which corresponds to the X-ray high (soft) state or very high state. The identification of the breaking state with the low (hard) state is quite firm but, because of the paucity of simultaneous X-ray and γ-ray observations, the identification of the power law state with which of the two high states is less certain. It is clear, though, that the presence of a strong soft excess at least some of the time during an outburst is a requirement for the γ-ray power law state to be observed. Fig. 1 shows shows photon number spectra from OSSE for seven transient BHCs, along with the best-fit analytic model extrapolated down to 10 keV and contemporaneous X-ray data if they are available. The two spectral states are readily apparent.

Sources in the X-ray low (hard) [i.e. breaking γ-ray] state typically have an X-ray index $\Gamma \sim$ 1.5, begin breaking from the power law at $E_b \sim$ 50 keV, and cut off with exponential folding energy of $E_f \sim$ 100 keV. The bulk of the luminosity in this state is emitted near 100 keV. Indeed the spectra of GRO J0422+32 and GRS 1716-249 indicate that the luminosity above 50 keV (or 10 keV) exceeds that in the 0.5-10 keV band by a factor of $\simeq$4 (or $\simeq$6).

Sources in the γ-ray power-law state have a strong ultrasoft excess above a single power-law spectrum, with $\Gamma \sim$ 2.5 − 3.0 and no evidence for a high-energy break, even at energies exceeding $m_e c^2$. No spectral features (e.g. narrow or broad lines) are apparent near or above 511 keV, as would be

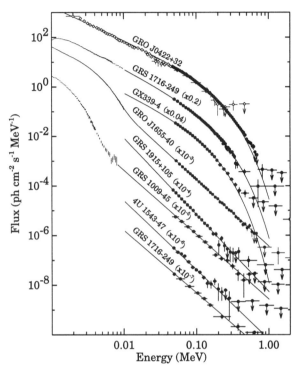

FIGURE 1. Photon number spectra from OSSE for seven transient BHCs. Spectra are averaged over all observing days for which there was detectable emission and, for clarity of the figure, have been scaled by arbitrary factors as indicated. Two spectral states are apparent. Contemporaneous TTM and HEXE data (open diamonds) and ASCA data (crosses) are shown for GRO J0422+32 and GRS 1716-249, respectively. Non-contemporaneous ASCA data (crosses) are shown for GRS 1009-45 and GRO J1655-40. ASCA data for GRS 1716-249 and GRS 1009-45 are from Moss (1997). Figure is from Grove et al. (1997b).

expected from standard nonthermal Comptonization models (Blumenthal & Gould 1970, Lightman & Zdziarski 1987). In this state, L_γ is only a fraction of the X-ray luminosity. The strong ultrasoft component is the signature of either the X-ray very high or high (soft) state, although the generally weak or absent *rapid* time-variability (<1 sec) suggests that the association is with the high (soft) state (Grove et al. 1997b). As a caveat, we note that there indeed can be a degree of independence of the ultrasoft and power-law spectral components exhibited in the long-term temporal behavior of the broadband emission: for example, while recent RXTE observations (Greiner, Morgan, & Remillard 1996) of the galactic superluminal source GRS 1915+105 demonstrate that the X-ray intensity is dramatically variable on timescales of tens to thousands of seconds, showing several repeating temporal structures, the γ-ray emission is steady and only slowly evolving (Grove et al. 1997b).

Recent observations of Cyg X-1 reveal a bimodal spectral behavior in the γ-ray band as well as the X-ray band (Phlips et al. 1996), equivalent to the above, and confirming the identification of X-ray and γ-ray states (Gierlinski et al. 1997; Phlips et al. 1997). These observations are discussed in greater detail below.

The transient GRS 1716–249 is shown twice in Fig. 1, apparently having undergone a spectral state change late in its second outburst (Moss 1997, Grove et al. 1997b). As is the case for Cyg X-1 (Phlips et al. 1996), the power-law state has lower L_γ than the breaking state. How the *bolometric* luminosity changes is less certain and will require further simultaneous broadband observations.

BATSE hard X-ray (20–100 keV) lightcurves for the same seven transient BHCs plus N Mus 1991 are shown in Fig. 2. Harmon et al. (1994) identified two types of transients based on such lightcurves. The first type has a relatively fast rise followed by an exponential decay, with a secondary maximum some weeks or months into the decline. The rise and decay times vary over a broad range, but are of the order of a few days and few tens of days, respectively. The second type has a longer rise time (of order weeks) and multiple, recurrent outbursts of highly variable duration and without a strong periodicity. There is no correlation between lightcurve type and spectral state; thus the mechanism that gates the accretion flow, and therefore regulates the production and evolution of outbursts, does not determine the physical process responsible for the γ-ray emission.

Elsewhere in this volume, Zhang et al. (1997a) report the detection of both γ-ray spectral states from 1E1740.7–2942, the persistent BHC near the galactic center. The breaking γ-ray state has higher L_γ than the power law state. They also report the detection of two luminosity states in GRS 1758–258, but because of the relative faintness of the source, do not present any spectral analysis. In Fig. 3 we have plotted the spectrum of 1E1740.7–2942 in the low-luminosity, power-law state from BATSE (adapted from Zhang et al.) and the high-luminosity, breaking state from OSSE. Because neither

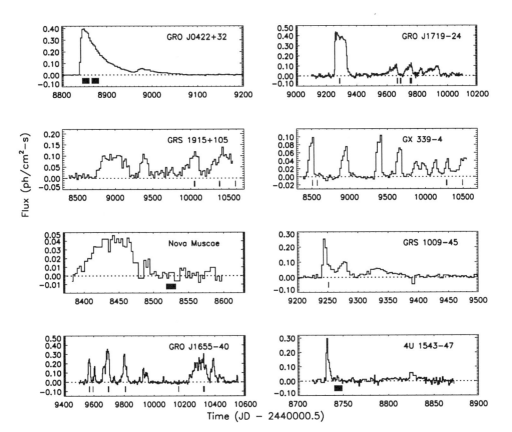

FIGURE 2. Lightcurves of transient black hole binary systems in the 20–100 keV band detected with BATSE. OSSE observing times are shown as bold solid lines along the horizontal (time) axis. All light curves include the primary and secondary outbursts until mid-1997, except for N Mus 1991. Its primary outburst occurred prior to launch of CGRO.

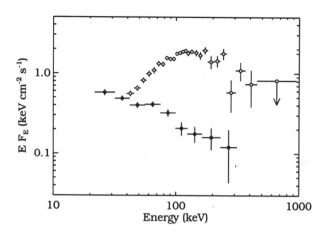

FIGURE 3. Spectra of 1E1740.7-2942 in the γ-ray breaking state (open circles; OSSE) and power-law state (solid circles; BATSE). The OSSE data have been corrected for nearby point sources and diffuse emission.

BATSE nor OSSE is an imaging instrument, both measurements are subject to contamination from the bright diffuse galactic continuum emission and nearby point sources. To minimize this potential source of error, the BATSE data were collected when the limb of the earth was at large angles to the galactic plane, so that the galactic ridge component would not produce any occultation modulation (S.N. Zhang, private communication). The OSSE spectrum shown is the result of a simultaneous fit to several known point sources, which were reasonably well separated by scanning the OSSE detectors along the ecliptic, and we have subtracted an estimate of the diffuse emission derived from an extensive series of galactic-center region mapping observations (G.V. Jung, private communication).

IV BROADBAND OBSERVATIONS AND SPECTRAL MODELING

In recent years, substantial progress has been made in our understanding of the environment of accreting black holes and the physical processes that drive the high-energy emission. Indeed the breaking γ-ray state can be relatively well modeled by Comptonization of soft photons from the accretion disk in a hot thermal plasma, with the plasma temperature determined by the cutoff energy. In some cases, additional spectral components are required which can be understood as effects of the hard radiation scattering off relatively cold electrons (for example as a reflection off a cold accretion disk or the

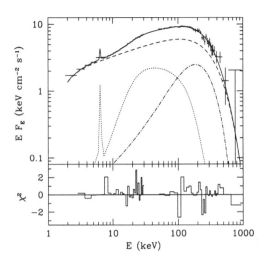

FIGURE 4. Broadband spectrum of Cyg X-1 observed simultaneously by ASCA & OSSE. Two-temperature thermal Comptonization model (dashed and dot-dashed curves), along with Compton reflection (dotted curve). Bottom panel gives contribution to total χ^2. From Gierlinski et al. (1997).

transmission of the hard radiation through a cold medium surrounding the compact source). The nature and geometry of the Comptonizing plasma is, however, highly debatable, and recent studies of the combined X-ray and γ-ray spectrum and temporal properties of Cyg X-1 indicate that a simple, slab geometry for the corona is implausible.

Simultaneous, broadband spectral measurements are especially powerful in elucidating the physical processes that drive the high-energy emission of BHCs. Two recent broadband studies, one from Ginga and OSSE observations of Cyg X-1 (Gierlinski et al. 1997) and the other from ASCA and OSSE observations of GX339-4 (Zdziarski et al. 1997), are the most detailed investigations to date on BHCs in the X-ray low (hard) state.

In the low (hard) state, the X-ray spectrum of Cyg X-1 consists typically of a power law with photon number index $\Gamma \sim 1.6-1.7$ and a Compton-reflection component that significantly affects the continuum above 10 keV and includes near 7 keV both an Fe K edge and an Fe Kα fluorescence line (Ebisawa et al. 1996 and references therein). The low-energy gamma-ray spectrum is steeply cut off above $\sim$150 keV and, while L_γ varies by a factor of at least several in this state, the spectral parameters vary only weakly (Phlips et al. 1996). The X-ray (Ueda et al. 1994) and γ-ray (Grabelsky et al. 1995) spectra of

GX339–4 in the low (hard) state are generally quite similar in form to those of Cyg X-1.

For both Cyg X-1 and GX339–4, the combined X-ray and γ-ray spectrum was modeled by thermal Comptonization in a spherical cloud with an isotropic source of soft seed photons from a blackbody distribution, as described in Zdziarski, Johnson, & Magdziarz (1996). The parameters of the hot plasma are the temperature, kT, and the X-ray photon number index, Γ, which is related to the Thomson optical depth, τ, of the Comptonizing plasma. The model allowed for Compton reflection with reflector solid angle Ω, and an Fe Kα line. An additional soft X-ray blackbody component was required below 3 keV from both sources; this component is likely to originate in a cold accretion disk in the vicinity of the hot Comptonizing plasma. This same cold disk would then also be responsible for the Compton reflection component. In the case of Cyg X-1, the spectrum cut off above 100 keV too sharply for an isotropic, single-temperature Comptonization model, so Gierlinski et al. (1997) added a Wien-like component from an optically-thick plasma at $kT \sim 50$ keV, speculating that it is the signature of a transition region between the hot and cold media in the accretion flow.

As shown in Fig. 4 and 5, these complicated, multi-parameter models give an excellent description of the X-ray and OSSE data in the entire energy range, from $\sim$2 keV to $\sim$1 MeV. The time-averaged spectrum of Cyg X-1 has been measured by COMPTEL out to several MeV (McConnell et al. 1997). A comparison with contemporaneous data from both OSSE and BATSE has so far yielded only limited insights because of discrepancies among the spectra from the three instruments. While the nature of these discrepancies is still being investigated, they may result merely from a difference in absolute flux normalization. In any case, it is becoming clear that the emission above 1 MeV implies more than a simple Comptonization process. A similar conclusion can be drawn from COMPTEL results for GRO J0422+32 (van Dijk et al. 1995).

For both Cyg X-1 and GX339–4, the description of Gierlinski et al. (1997) and Zdziarski et al. (1997) rules out a geometry with a corona above the surface of an optically-thick disk (Haardt & Maraschi 1993; Haardt et al. 1993). The solid angle Ω of the reflector is only 30–40% of the 2π expected for a corona above a flat disk. Furthermore, the intrinsic spectrum is so hard that the flux of the incident seed photons is about an order of magnitude less than the Comptonized flux (i.e. the source is "photon starved"), whereas the seed photon Comptonized photon fluxes are almost the same in the disk-corona model (Haardt & Maraschi 1993). In addition for Cyg X-1, the narrowness of the Kα line observed by ASCA (Ebisawa et al. 1996) implies that the reflecting medium is cold and therefore likely far from the central black hole. On the other hand, the fits *can* correspond to a geometry with a hot inner disk surrounded by a colder outer disk. The hot disk is geometrically thick and irradiates the outer cold disk. The irradiation gives rise to the Compton-reflection component with small Ω, and accounts for the modest blackbody

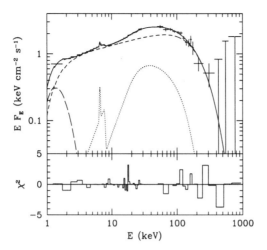

FIGURE 5. Broadband spectrum of GX339-4 observed simultaneously by Ginga & OSSE on 1991 Sep 11. Model consists of blackbody radiation (long dashes) as seed for thermal Componization in hot plasma (short dashes), which is Compton reflected from the disk (dotted). Bottom panel gives contribution to total χ^2. From Zdziarski et al. (1997).

emission of the disk observed at low energies.

There are currently no high-sensitivity, *simultaneous* X-ray and γ-ray studies in the literature for the X-ray high (soft) state, although several simultaneous datasets from RXTE and GRO exist for the superluminal sources GRO J1655-40 and GRS 1915+105. A simultaneous ASCA and BATSE observation of GRO J1655-40 *has* been published (Zhang et al. 1997b), which shows a strong ultrasoft excess and a soft power-law tail ($\Gamma \simeq 2.4$) extending beyond 100 keV.

The physical processes driving the hard emission in the X-ray high (soft) state are less well determined. It is generally well agreed upon that the ultrasoft component, which can be described by a multi-color disk blackbody spectrum (Mitsuda et al. 1984), is thermal emission from an optically thick and geometrically thin accretion disk roughly in the region $10^7 < r < 10^9$ cm from the black hole. The power-law tail seen by Ginga from a number of BHCs in the high (soft) state (e.g. GX339-4: Makishima et al. 1986) was ascribed to thermal Comptonization in the hot ($kT \sim 60$ keV) inner disk region. This interpretation is ruled out by Sigma observation of GX339-4 in this state with a spectrum that extends unbroken to at least 100 keV (Grebenev et al. 1993), well beyond the break for a Comptonized spectrum of this temperature. The unbroken power laws observed by OSSE from a number of other BHCs require temperatures of at least several hundred keV.

Ebisawa, Titarchuk, & Chakrabarti (1996) propose that the high-energy emission in the high (soft) state arises from bulk-motion Comptonization of soft photons in the convergent accretion flow from the inner edge of the disk. In the high (soft) state, the copious soft photons from the disk cool the electrons efficiently in the inner, advection-dominated region, and Comptonization due to bulk motion dominates over that due to thermal motion. In contrast, in the low (hard) state, there are fewer soft photons, hence higher temperatures in the Comptonization region (i.e. $\sim$50–100 keV, rather than $\sim$1 keV), and thermal Comptonization dominates. Calculations by Titarchuk, Mastichiadis, & Kylafis (1996) indicate that the bulk-Comptonization spectrum can continue unbroken well beyond $m_e c^2$, as is observed at least in the case of GRO J1655–40.

The detection of Compton reflection in the high (soft) state may prove to be a complication (A. Zdziarski, private communication). Compton reflection is clearly seen in the low (hard) spectrum of a number of sources, and is detected in both *spectral states* in N Mus 1991 (Ebisawa et al. 1994) and Cyg X-1 (Gierlinski et al. 1997). In the latter objects, the reflection parameter Ω approaches 2π in the high (soft) state, indicating that the source of the hard photons completely covers the reflector, e.g. the cool disk. The bulk-Comptonization model postulates an advecting, Thomson-thick medium above the disk, and it is in this advecting medium that the Comptonization takes place. However, because it is optically thick, the reflected component is trapped in the advecting flow and does not escape to the observer. Thus detection of reflection in the high (soft) state would cast doubt on the application of bulk Comptonization in these objects.

V THE SEARCH FOR LINE EMISSION

Transient, broad emission lines have been reported in the literature from Cyg X-1 (the "MeV bump"), 1E1740.7–2942, and N Mus 1991. In the latter two cases, the lines have been interpreted as red-shifted and split (due to disk rotation) annihilation features. The very high sensitivity of the GRO instruments relative to previous gamma-ray telescopes makes deep searches for such lines possible on many timescales. Long-term observations of Cyg X-1, by both COMPTEL and OSSE, at a range of hard X-ray intensities set strict upper limits on the magnitude and duty cycle of a possible MeV bump. The upper limits are more than an order of magnitude below the historical reports (Phlips et al. 1996, McConnell et al. 1994). Similarly, searches by Grove et al. (1997b) through hundreds of days of BHXB observations by OSSE have revealed no evidence for transient red-shifted annihilation lines at levels approaching those in the literature. For example, a broadened 480 keV line at 6×10^{-3} ph cm^{-2} s^{-1}, the intensity reported from N Mus 1991 by Sigma (Goldwurm et al. 1992, Sunyaev et al. 1992), would have been detected by OSSE at $\sim 40\sigma$ in an average 24-hour period.

In general these transient events are reported by a single instrument and cannot be confirmed or refuted because no other instrument is observing at the same time. However, the broad-line excess from 1E1740.7-2942 in 1992 Sep reported by Sigma (Cordier et al. 1993) is *not* confirmed by OSSE (Jung et al. 1995), which by chance was viewing the galactic center region before, during, and after the event. Confirmation of such transients would provide strong support for pair plasma models of black hole radiation, lending credence to the suggestion (e.g. Ramaty et al. 1994) that sources such as 1E1740.7-2942 might be significant sources of positrons in the central region of our galaxy.

VI RAPID VARIABILITY AND PHASE LAGS

Strong, rapid, aperiodic variability (i.e. on timescales of tens of seconds or less) is frequently reported in X-rays from BHCs in the X-ray low (hard) state, with the occasional appearance of peaked noise or quasi-periodic oscillations (QPO). The rms variability is typically of order tens of percent of the average intensity. For recent reviews, see van der Klis (1994 & 1995). The GRO instruments find similarly strong, rapid variability in gamma-ray emission from sources in the X-ray low (hard) state; see e.g. van der Hooft et al. (1996) for GRS 1716-249 and Grove et al. (1994) for GRO J0422+32.

Recent results from RXTE indicate that the superluminal sources GRO J1655-40 and GRS 1915+105 in the X-ray high (soft) state both show weaker rapid X-ray variability (of order several percent rms) and QPOs on many timescales, at frequencies up to 67 Hz in the latter case (Morgan, Remillard, & Greiner 1997), while such variability is undetected in the gamma-ray band, where the statistical limits are $\sim 5\%$ (Crary et al. 1996, Kroeger et al. 1996).

Using the BATSE instrument, van der Hooft et al. (1996) studied the evolution of the rapid time variability of GRS 1716-249 (GRO J1719-24) throughout an entire ~ 80-day outburst. The power density spectrum showed a strong QPO with a centroid frequency that increased from ~ 0.04 Hz at the onset of the outburst to ~ 0.3 Hz at the end. Interestingly, they reported that the power spectrum could be described with a single characteristic profile, the frequency scale of which stretched proportionally during the outburst, and that the total power, integrated over a scaled frequency interval, was constant throughout the outburst.

The upper panel of Fig. 6, adapted from Grove et al. (1994), shows the normalized power density spectra (PDS) in the 35-60 keV and 75-175 keV bands for the OSSE observation of GRO J0422+32. The shape of the power spectrum is essentially identical in the two energy bands. It shows breaks at a

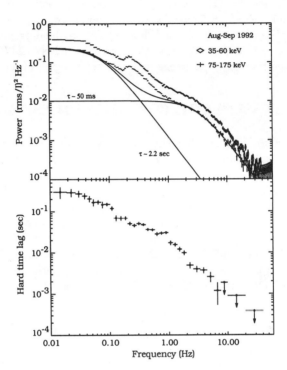

FIGURE 6. Upper panel: OSSE power density spectra of GRO J0422+32 in ~35–60 keV and 75–175 keV energy bands. From Grove et al. (1994). Lower panel: Time lag spectrum between same energy bands.

few 10^{-2} Hz and a few Hz, and a strong peaked noise component at 0.23 Hz, with FWHM ∼ 0.2 Hz. The peaked noise profile is broad and asymmetric, with a sharp low-frequency edge and a high-frequency tail; thus the physical process responsible for the peaked noise appears to have a well-defined maximum timescale. We note that, in contrast to GRS 1716–249, the characteristic frequencies of the shoulders and the peak are independent of source intensity. (In passing we also note the striking similarity of the power spectrum of GRO J0422+32 to that of the X-ray burster 1E1724–3045 derived from RXTE data, including both a low-frequency and a high-frequency break and a strong peaked noise component. See Olive et al. in these proceedings for details).

The lower panel shows the time lag spectrum between these two energy bands: the hard emission (75–175 keV) lags the soft emission (35–60 keV) at all Fourier frequencies, falling crudely as $1/f$, up to about 10 Hz, where there is no statistically significant lag or lead between the two bands. In the 10–30 Hz frequency range, time lags as small as 1 ms would be detectable at >99% confidence. At frequencies ∼0.01 Hz, hard lags as large as 300 ms are observed. There is no significant change in the lag at the frequencies dominated by the strong peaked noise component at 0.23 Hz.

The lag of GRO J0422+32 as a function of Fourier frequency is quite similar to the lags reported from Cyg X-1 (Miyamoto et al. 1988; Cui et al. 1997; Wilms et al. 1997), N Mus 1991 and GX339–4 (Miyamoto et al. 1993), in rather different energy ranges. The OSSE data from GRO J0422+32 provide more evidence indicating that the frequency-dependent time lag is a common phenomenon shared by many, if not all, accreting objects in binaries.

This form of PDS and the $1/f$ dependence of time lag on the Fourier frequency, first observed in Cyg X-1 by Ginga (Miyamoto et al. 1988), are very different from that expected in accretion models which presumably produce most of the X-ray and γ-ray emission from a region whose size is comparable to that of the last stable orbit around a black hole of mass a few $M_\odot$: The characteristic time scale associated with the dynamics of accretion in such an object is of order 10^{-3} sec, and consequently one would expect most of the associated power at this frequency range. By contrast there is a remarkable *lack* of power at this range.

The shape of the PDS of accreting BHC has been the source of much discussion, but there exists no widely accepted, compelling theory which provides a "reasonable" account of it. While most of the attention to date has been focused on the $1/f$-like noise in the intermediate frequency regime (∼ 0.01 − 10 Hz), the breaks at both the low and high frequencies deserve equal attention. It is puzzling both that there is a lack of high-frequency power and that most of the variability power appears concentrated at the low frequency break, i.e. 4–5 orders of magnitude lower than the frequencies associated with the dynamics involved with the production of X-rays.

Generally, the models of BHC variability attempt to reproduce the observed shapes of the PDS (in fact, mainly the slope of their power-law section) as an ensemble of *exponential* shots with a range of decay times and/or amplitudes. The ensemble of shots is usually derived by simulating the accretion onto the BH in terms of avalanche-type models with Self Organized Criticality (Bak 1988; Negoro et al 1995).

This approach gives very little physical insight into the breaks at low and high frequencies, and it may well be erroneous at the outset. The reason is shown in Fig. 6b, the time-lag spectrum. The roughly-$1/f$ dependence of time lag on the Fourier frequency is very different from that expected if the 35–175 keV radiation results from Comptonization of soft photons by hot electrons the vicinity of a black hole; under these conditions the time lags, which are indicative of the photon scattering time in the hot electron cloud, should be independent of the Fourier frequency and of order 10^{-3} sec, the photon scattering time in this region, which is roughly similar to the dynamical time scale.

The observed lags (Fig. 6b) are generally much longer and Fourier-frequency dependent, a fact noticed first by Miyamoto et al. (1988). The very long observed lags ($\simeq 0.3$ sec) and in particular their dependence on the Fourier frequency exclude from the outset models in which the variability is produced by mechanisms which effect a modulation of the accretion rate. These models, while they are constructed to produce the observed PDS, since they produce the hard radiation from Comptonization in vicinity of the black hole, would yield very short, frequency independent lags.

The discrepancy prompted Kazanas, Hua & Titarchuk (1997) to propose that the density of the hot electron cloud responsible for the formation of the high energy spectra, through the Comptonization process, is not uniform. Specifically they found that the $1/f$ dependence of the hard X-ray lags on Fourier frequency could be accounted for if the density profile of the hot scattering medium, $n(r)$, is of the form $n(r) \propto 1/r$ for radial distance r ranging from $\sim 10^6$ to 10^{10} cm. Because this model keeps Comptonization as the mechanism for the production of high energy photons, it can at the same time explain both energy spectra and the time variablity (Hua, Kazanas & Cui 1997). According to this interpretation, the aperiodic variability of these sources, to which little attention has been paid so far, can provide diagnostics to the density structure of the accreting gas around the black holes (Hua, Kazanas & Titarchuk 1997), an important information for understanding the accretion precess (e.g. Narayan & Yi 1994).

REFERENCES

1. Barranco, J. and Grindlay, J. 1997, A&A, in preparation
2. Barret, D., et al., 1996a, ApJ, 473, 963
3. Barret, D., et al., 1996b, A&As, 120C, 121

4. Barret, D. et al. 1997, Proc. 4th Comp. Symp., in press
5. Bouchet, L. et al. 1991, ApJ, 383, L45
6. Churazov, E. et al. 1993, ApJ, 407, 752
7. Crary, D. J. et al. 1996, ApJ, 463, L79
8. Cordier, B. et al. 1993, A&A, 275, L1
9. Cui, W. et al. 1997, ApJ, 484, in press
10. Ebisawa, K. et al. 1994, PASJ, 46, 37
11. Ebisawa, K. et al. 1996, ApJ, 467, 419
12. Ebisawa, K., Titarchuk, L., & Chakrabarti, S.K. 1996, PASJ, 48, 59
13. Gierlinkski, M. et al. 1997, MNRAS, in press
14. Goldwurm, A. et al. 1992, ApJ, 389, L79
15. Grabelsky, D.A. et al. 1995, ApJ, 441, 800
16. Grebenev, S. et al. 1993, A&AS, 97, 281
17. Greiner, J., Morgan, E. H., & Remillard, R. A. 1996, 473, L107
18. Grindlay, J. et al. 1997, Proc. INTEGRAL Wkshp., St. Malo, ESA SP-282, 551.
19. Grove, J.E. et al. 1994, AIP Conference Proc. 304, 192
20. Grove, J.E. et al. 1997a, in Proc. 2nd INTEGRAL Workshop (St. Malo), ESA SP-382, p. 197
21. Grove, J.E. et al. 1997b, submitted to ApJ
22. Haardt, F. 1993, ApJ, 413, 680
23. Haardt, F. & Maraschi, L. 1993, ApJ, 413, 507
24. Harmon, B.A., et al. 1992, AIP 280, p. 314
25. in 't Zand, J. and Heise, J. 1997, IAUC 6618
26. Hua, X.-M., Kazanas, D. & Titarchuk, L. 1997 ApJ, 482, L57 and these preceedings
27. Hua, X.-M., Kazanas, D. & Cui, W. 1997 ApJ, submitted
28. Jung, G.V. et al. 1995, A&A, 295, L23
29. Kazanas, D., Hua, X.-M. & Titarchuk, L. 1997 ApJ, 480, 735 and these preceedings
30. Kroeger, R. A., et al. 1996, A&AS, 120, 117
31. Makishima, et al. 1986, ApJ, 308, 635
32. Mitsuda, K. 1984, PASJ, 36, 741
33. Miyamoto, S. et al., 1988, Nature, 336, 450
34. Miyamoto, S. et al., 1993, ApJ, 403, L39
35. Morgan, E. H., Remillard, R. A., & Greiner, J. 1997, ApJ, 482, 993
36. Moss, M. J. 1997, Ph.D. Thesis, Rice University
37. Phlips, B.F. et al. 1996, ApJ, 465, 907
38. Phlips, B.F. et al. 1997, ApJ, in preparation
39. Ramaty, R. et al. 1994, ApJS, 92, 393
40. Romani, R. 1992, ApJ, 399, 621
41. Sunyaev, R. et al. 1991, ApJ, 383, L49
42. Sunyaev, R. et al. 1992, A&A, 280, L1
43. Tanaka, Y. 1989, in Proc. 23rd ESLAB Symp. on Two Topics in X-Ray As-

tronomy (Bologna), ESA SP-296, p. 3
44. Tanaka, Y. and Lewin, W. 1995, in *X-ray Binaries* (Lewin, Paradijs and Heuvel, eds.), Cambridge Univ. Press, p. 126
45. Tanaka, Y., 1997, in Proc. 2nd INTEGRAL Workshop (St. Malo), ESA SP-382, p. 145
46. Titarchuk, L., Mastichiadis, A., & Kylafis, N.D., 1996, A&AS, 120, 171
47. Ueda, Y. et al. 1994, PASJ, 46, 107
48. van der Hooft, F. et al. 1996, ApJ, 458, L75
49. van der Klis, M. 1994, ApJS, 92, 511
50. van der Klis, M. 1995, in *X-ray Binaries* (W. Lewin, J. v. Paradijs and E. v.d. Heuvel, eds.), Cambridge Press, p. 252
51. van Paradijs, J.. and McClintock, J. 1995, in *X-ray Binaries* (W. Lewin, J. v. Paradijs and E. v.d. Heuvel, eds.), Cambridge Press, 58
52. Vargas, M. et al. 1997, in Proc. 2nd INTEGRAL Workshop (St. Malo), ESA SP-382, p. 129
53. Wilms, J. et al., 1997, these preceedings
54. Zdziarski, A. A., Gierlinski, M., Gondek, D., & Magdziarz, P. 1997, A&A, 120, 553
55. Zdziarski A. A., Johnson N. W., & Magdziarz, P. 1996, MNRAS, 283, 193
56. Zhang, S.N. et al. 1993, *Nature*, 366, 245-247
57. Zhang, S.N. et al. 1997a, these proceedings
58. Zhang, S.N. et al. 1997b, ApJ, 479, 381

Galactic Black Hole Binaries: Multifrequency Connections[1]

S. N. Zhang**, I. F. Mirabel[†], B. A. Harmon*, R. A. Kroeger[‡],
L. F. Rodriguez[♯], R. M. Hjellming[b] and M. P. Rupen[b]

ES-84, Marshall Space Flight Center, Huntsville, AL 35812
*Universities Space Research Association
[†]*CEA-CEN Saclay, Service D'Astrophysique, 91191 Gif-Sur-Yvette, Cedex, France*
[‡]*Naval Research Lab., MS 4151, 4555 Overlook Ave., Washington, DC 20375*
[♯]*Instituto de Astronomia UNAM, Apdo Postal 70-264, DF04510 Mexico City, Mexico*
[b]*National Radio Astron. Obs., P.O. Box O, 1003 Lopezville Rd., Socorro, NM 87801*

Abstract. We review the recent multifrequency studies of galactic black hole binaries, aiming at revealing the underlying emission processes and physical properties in these systems. The optical and infrared observations are important for determining their system parameters, such as the companion star type, orbital period and separation, inclination angle and the black hole mass. The radio observations are useful for studying high energy electron acceleration process, jet formation and transport. X-ray observations can be used to probe the inner accretion disk region in order to understand the fundamental physics of the accretion disk in the strongest gravitational field and the properties of the black hole. Future higher sensitivity and better resolution instrumentation will be needed to answer the many fundamental questions that have arisen.

INTRODUCTION

Significant progress has been made in the study of the galactic black hole binaries since the launch of the Compton Gamma Ray Observatory in April 1991. It is now widely believed that some X-ray binary systems harbor a stellar black hole at the center of the accretion disk in each system. A variety of high energy spectra and light curves have been observed from many of them. Some of them also exhibited highly relativistic jets, which are found to be in correlation with the high energy radiation. Since the first mass determination of the assumed black hole in Cyg X-1 about 25 years ago [10,108], the second property, i.e., the spins, of black holes have also been inferred recently [116],

---

[1]) Invited Review Article for the 4th Compton Symposium, Williamsburg, VA, April, 1997

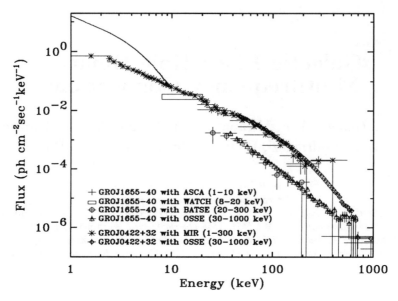

FIGURE 1. Broad band high energy spectra of GROJ1655-40 [115] (Zhang *et al.* 1997) and GROJ0422+32 [56] (Kroeger *et al.* 1996).

and are considered to be the missing link in the proposed unification scheme of all types of black hole binaries [116].

Several review articles on galactic black hole binaries now exist, including those in the same proceedings. In this review article, we try to avoid any significant overlapping with other review articles, by focusing on (a) summarizing these new and important multiwavelength observations, and (b) exploring the physical connections between observations made at different wavelengths to obtain more complete pictures of these systems.

BASIC HIGH ENERGY CHARACTERISTICS

All of these galactic black hole binaries were originally discovered in either X-ray (1-10 keV) or hard X-ray (20-300 keV) bands. Their X-ray or hard X-ray luminosity dominates their total electromagnetic radiation energy output. In this section we discuss briefly their basic high energy emission characteristics, in order to explore the multiwavelength aspects of these systems.

High Energy Continuum Spectra

The high energy spectra of galactic (candidate) black hole binaries are made of primarily only two components – a blackbody-like soft component and a power-law-like hard component (see [111] for a review of high energy continuum spectra observed in black hole and neutron star X-ray binaries), as shown in figure 1 for the superluminal jet and black hole source GRO J1655-40 [115]. In some sources, the soft component is absent above 1 keV *and* the hard power-law component becomes harder with a break between 50-200 keV, as shown in the same figure for GROJ0422+32. The soft component is believed to be emitted from the inner accretion disk region and can be well described by the so-called multi-color disk blackbody spectral model [60,70], in which the total emitted spectrum is the integration of the local blackbody emission from each annulus with a temperature $kT_{bb} \propto r^{-3/4}$. To obtain the physical parameters of the inner accretion disk, corrections for the relativity and spectral hardening effects must be taken into account [116]. The observed peak blackbody temperatures of the multi-color disk blackbody spectra vary between 0.1 to 2.5 keV (see figure 5 and the corresponding section for details).

The origin of the power-law-like hard component is still not well understood, although inverse Compton up-scattering of low energy photons by high energy electrons are usually believed to be involved. The nature and origins of both the low energy photons and the high energy electrons are still not identified unambiguously yet, despite the existence of many models. There usually exists an anti-correlation between the soft component luminosity and the hardness or flatness of the power-law whose photon spectral index varies between -1.5 to -3.5. Usually the power-law component is observed to fall into one of the two states: a hard state corresponding to a minimum luminosity ($< 5 \times 10^{36}$ ergs/s) [114] or the absence of soft X-ray component [22], with a photon spectral index between -1.5 to -2.0 and spectral break above 50-200 keV; or a soft state corresponding to a higher soft X-ray component luminosity [22,35,115,114], with a photon spectral index between -2.5 to -3.5 and no detectable spectral break up to 300-600 keV [35]. Figure 1 illustrates a typical spectrum of each type (soft or high state for GROJ1655-40 and hard or low state for GROJ0422+32). However, a source may remain in a different state for a time, and the transition or evolution between the states is usually continuous, for example in Cyg X-1 [114].

Both the soft and the hard components have been used as signatures or indicators of black hole binary systems. Similar components have, however, also been detected from neutron star binary systems, such as the type I X-ray burster 4U1608-52 [69,112], although the absolute hard X-ray luminosity is found to be systematically higher in black hole systems than in neutron star systems [5].

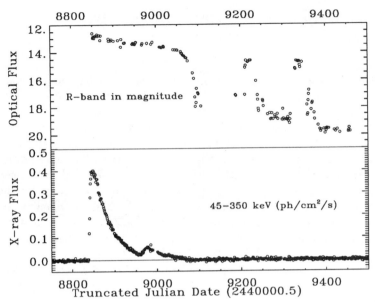

FIGURE 2. Optical and hard X-ray light curves of a type I or nova-like source GROJ0422+32 [11] (Callanan *et al.* 1996).

High Energy Long Term Light Curves

A variety of high energy light curves have been observed from galactic black hole binaries [18]. These light curves can be placed conveniently into three classes: transients with nova-like light curves, which form the majority of the black hole binaries, such as A0620-00, GS2000+25, GS1124-68, GS2023+23, H1705-250, GROJ0422+32, GROJ1719-24 = GRS1716-249 and GRS1739-278, with a fast rise (a few days) and slow exponential decay (a few tens of days); transients with multiple outbursts, such as the superluminal jet sources GRS1915+105, GROJ1655-40 and a possible jet source GX339-4; persistent sources such as Cyg X-1, LMC X-1, LMC X-3. We shall call them type I or nova-like, type II or multiple-outbursts and type III or persistent *high energy* light curves in the following discussion. Samples of some BATSE hard X-ray light curves, together with some optical, radio and soft X-ray light curves are shown in figures 2-4. Please refer to the references where some of the original light curves were published for more details. (Note that some of the sources listed above, i.e., GROJ1719-24, GRS1739-278 and GX339-4, are not yet dynamically established black hole binaries with the compact object masses in excess of 3 $M_\odot$). These different types of light curves cannot be understood with high energy observations alone. In the following, we will try to understand some of their high energy behaviors using multifrequency data.

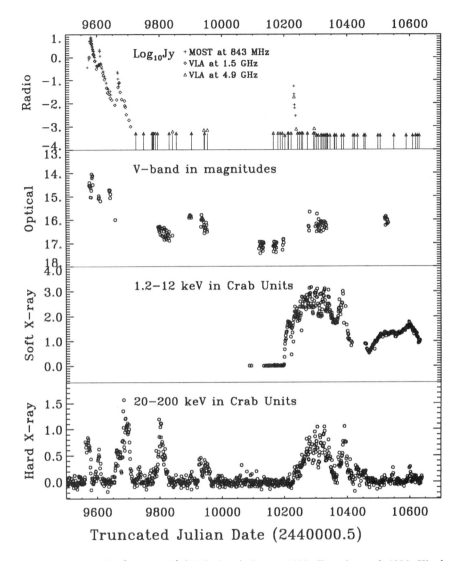

FIGURE 3. Radio [45,100,109] (Hjellming & Rupen 1995, Tavani *et al.* 1996, Wu & Hunstead 1996), Optical [2,3,76,78,80] (Bailyn *et al.* 1995a, 1995b, Orosz 1996, Orosz & Bailyn 1997, Orosz *et al.* 1997a, 1997b), soft and hard X-ray [90] (Robinson *et al.* 1997) light curves of a type II or multiple-outbursts source GROJ1655-40.

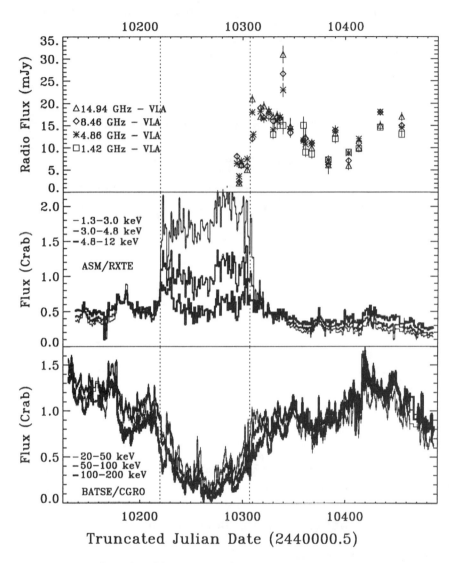

FIGURE 4. Radio, soft and hard X-ray light curves of the type III or persistent source Cygnus X-1 [114,117] (Zhang *et al.* 1997).

TABLE 1. Dynamically Established Black Hole Binary Systems.

Source Name (1)	Alternate Name (2)	f_M ($M_\odot$) (3)	i (deg) (4)	M_{BH} ($M_\odot$) (5)	Star Sp. (6)	M_c ($M_\odot$) (7)	X-ray Sp. (8)	Type l.c. (9)	Radio (10)	Ref. (11)
0538-641	LMC X-3	0.14	50°–60°	7–14	B3 Ve	4–8	S	III		19,23
0540-697	LMC X-1	2.3±0.3	40°–63°	4–10	O7-9 III	20–25	S	III	D	23,48,97
GRO J0422+32	XN Per 1992	1.21±0.06	48°±3°	3.57±0.34	M2 V	0.4	H	I	T	15,31,37,77
A 0620-00	XN Mon 1975	2.91±0.08	66°±4°	4.9–10	K4 V	0.4–0.7	S	I	T	43,46,62,81,94
GRS 1124-683	XN Mus 1991	3.01±0.15	54°–65°	5.0–7.5	K2 V	0.8	S, H	I	T	50,79
GRO J1655-40	XN Sco 1994	3.16±0.15	69°.5±0°.08	7.02±0.22	F3-F6 IV	1.2–1.5	S	II	J, M	3,45,78,101,103,115
H 1705-250	XN Oph 1977	4.0±0.8	70°±10°	4.9±1.3	K3 V	0.7	S	I	T	61,87,88
1956+350	Cyg X-1	0.24±0.01	10°–40°	7–20	O9.7 Iab	20–30	H, S	III	P	10,33,34,53,82,85,108,114,117
GS 2000+25	XN Vul 1988	4.97±0.10	65°±9°	8.5±1.5	K2-K7 V	0.4–0.9	S	I	T	12,46
GS 2023+338	V404 Cyg	6.08±0.06	56°±2°	12.3±0.3	K0 IV	0.9	H	I	T	13,14,38,53,93,95,106,107

Note: (3) mass function; (4) system inclination angle; (5) black hole mass; (6) optical/infrared spectral type; (7) companion mass; (8) high energy spectral type ('S': soft or high state, 'H': hard or low state); (9) high energy light curve type (I: nova-like transient, II: multiple outbursts, III: persistent); (10) radio properties ('P': persistent, 'T': transient, 'M': multiple-episodes, 'D': positive detection, but long term behavior unknown, 'J': superluminal jets.); (11) references: the numbers are referred to their order as appeared in the reference section of this paper. (Part of this table is adapted from the table 1 of Chen, Shrader and Livio 1997 [18].)

OPTICAL AND INFRARED STUDIES

In this section we discuss briefly the application of optical and infrared observations to the understanding of the properties of inner accretion disk regions and the black holes.

Determining the Black Hole Masses

Ellipsoidal optical and infrared light curve modulations of the companions in these systems can be used to determine their orbital periods. Spectroscopic measurements of the Doppler shifts of some spectral line features can be used to obtain their mass functions, i.e., their mass lower limits. Detailed modelling of their light curves allows the determination of the system inclination angles. The spectral types of the companion provide information on the nature of the companion stars. Therefore in principle the complete system parameters, such as the orbital periods, orbital separations, masses of both the companions and the compact objects (the black hole for the systems we are discussing in this article) can be determined from optical and infrared observations [83]. Perhaps the best example and success of this technique is the precise black hole mass (7 $M_\odot$) and other system parameters determination in GROJ1655-40 [103,78].

Up to now, reliable dynamical mass determinations from optical and infrared observations have resulted in ten systems in which the compact object masses are most likely greater than 3.2 $M_\odot$ (see table 1, which lists also their radio and X-ray properties discussed in the following sections), the observational and theoretical upper limits of a neutron star. In several systems, the lower limits to the compact object masses are already above 3.2 $M_\odot$. Arguably these constitute so far the strongest evidence of the existence of black holes in binary systems. Although probing only the surface of the companion star and the outer disk region, these optical and infrared studies can also be used for understanding the high energy emission (from the inner-most region of the disk) behaviors of these systems, especially the three types of light curves we mentioned in the previous section.

Understanding Their High Energy Light Curves

The companion stars in the systems producing the Type III or persistent light curves are all high mass (10-30 $M_\odot$) O or B stars. The type III light curves are apparently due to the high mass transfer rates from their high mass companions, either through Roche-lobe overflow or stellar wind accretion. Apparently they are all accreting below the Eddington rate. There are, however, some significant differences between the three sources listed above. Cyg X-1 spends most of its time in a hard state, in which the total luminosity is dominated by the hard power-law. Occasionally the source enters into its soft state, in which a prominent soft component (with a higher temperature than that in the hard state) appears, nearly simultaneous with the softening of the power-law [114]. LMC X-1 and LMC X-3, on the other hand, always remain in the soft state with prominent soft X-ray components and steeper power-law components, similar to the soft state of Cyg X-1. Even between LMC X-1 and LMC X-3, there exists also some major differences. For example, in LMC X-1 the soft X-ray component remains rather stable, while the power-law component can vary substantially [23]. While in LMC X-3, the two components may vary substantially and independently, and the intensity and spectral hardness below 13 keV show a positive correlation [25]. It is also interesting to note that LMC X-3 and Cyg X-1 are the only black hole binaries exhibiting long term periodic variations in their optical and X-ray light curves. In particular, orbital modulations in both the soft and hard X-ray bands have been seen from Cyg X- 1 [53,82,85,113]. Although the soft X-ray flux modulation may be interpreted as the modulation of the column density (ionized or neutral) due to stellar wind [53], the hard X-ray modulation is inconsistent with this interpretation [82,86]. Since $\sim 30\%$ of the total hard X-ray component is estimated to be made of the reflection of an input power-law component [21], one possible origin of the observed orbital modulation is that its companion star also contributes to the total reflection component. Then the contribution from the companion would produce orbital modulations, similar to the optical and infrared ellipsoidal modulations.

The companion stars in all these systems producing the Type I or nova-like light curves are low mass (<1 $M_\odot$) main sequence stars, and no such low mass companion black hole binaries are known to produce other types of light curves. The transient nature of such systems has been explained as due to their sufficiently low mass transfer rate and that the mass accretor is a black hole [51,105]. The outburst trigger mechanism is likely due to the thermal instability in the outer part of the disk [57]. In the quiescence state, the material is steadily transferred at a very low rate from the companion to the accretion disk via Roche-lobe overflow and a significant portion of the material is accumulated in the disk. Thus the quiescence state disk is quasi-steady. When an instability is developed in the disk, an outburst occurs, during which the accumulated material is transferred to the black hole with a much higher rate. In this scenario the rise time would correspond to the sound propagation time from the outer disk to the inner disk [65,92]: $t_{rise} = R(\alpha C_s)^{-1}$ (where R is the outer disk radius, α the dimensionless viscosity parameter and C_s the sound speed) which is on the order of a day for $\alpha \sim 0.1$. The decay time would correspond to the viscous time scale of the hot state disk, $t_{vis} = (R/H)^2(\alpha\Omega)^{-1}$ (where H is the vertical height of the disk and Ω the angular speed of disk at radius R), which is on the order of 100 days for $\alpha \sim 0.1$.

There are still some fundamental questions concerning the quiescent disk structure. The standard thin disk extending to the last stable orbit of the black hole is problematic, because the inner part of the disk cannot stay in the 'cold' branch of the 'S' curve (inferred from the observed quiescence X-ray emission), which is essential for the disk instability model [57]. The most recent version of the advection dominated disk model [73,42] with the cooling dominated inner part truncated is much more successful than the standard thin disk model, but also has some problems. First, location of the inner disk boundary, critical in the model, is not determined self-consistently, although the thermal instability condition can be applied to produce a constrain. Second, it cannot explain the delayed hard X-ray outburst in GROJ1655-40 [90], with respect to the soft X-ray outburst after the initial optical on-set [80]. Therefore a more consistent quiescent state disk model is still not available yet.

The companion stars in the Type II or multiple-outbursts systems may be intrinsically different from that in the Type I systems (the companion star in GX339-4 has not been identified yet). Although also called a low mass system, GROJ1655-40 has an evolved companion star of 2.3 $M_\odot$ [3]. Usually such donor stars would have sufficiently high mass transfer rate, preventing the system from having a transient nature. This discrepancy has been reconciled by assuming that GROJ1655-40 is in a short-lived evolutionary stage where the mass transfer rate is sufficiently low to allow instabilities to occur [55]. This is a possible way to explain the rarity of this type of systems. Its multiple outbursts are not periodic and so apparently not related to its orbital motion, but most likely due to another instability after the initial outburst is triggered. Perhaps the X-ray heating of the companion star induces the so-called mass transfer instability [20,41] or 'echoes' [1]. This may be due to its peculiar type (much massive and larger than other companion stars) of this companion star. Comparing the the K($2.2\mu m$) spectra of GRS1915+105 observed at different epochs [16,30,68], one finds the characteristic HeI and Brγ emission lines of O-Be stars [68] with time variable intensity. It has been shown that the absolute infrared magnitude, time variability, and spectral shape of GRS1915+105 is comparable to that of SS 433 and other high mass X-ray binaries [17]. Furthermore, the infrared afterglow of an X-ray/radio outburst observed on August 1995 revealed that at that time GRS1915+105 was enshrouded in a dusty nebula of $\sim 10^{16}$ cm radius [67]. In summary, the infrared observations suggest that the donor star belongs to the class of massive stars with transitional spectral classification due to the dynamically unstable stellar atmosphere or wind [72]. If this is confirmed, GRS 1915+105 would be a black hole with a peculiar massive companion which is losing mass at rates of 10^{-6}-10^{-5} $M_\odot$ yr^{-1}. Further more definitive observations are still needed to clarify its companion star type, which may be critical for understanding the nature of the system. Despite that GROJ1655-40 is the only one of the Type II sources whose companion is unambiguously identified to be a peculiar star, it is worthwhile to explore the relationship between the properties of the companions and the high energy behaviors of the systems.

RADIO STUDIES

Radio fluxes from these systems originate in the incoherent synchrotron radiation of high energy electrons. The radio emission region is usually comparable or larger than the size of the accretion disk or the whole binary system. For a typical source of 0.1 Jy at 3 kpc, since the surface brightness temperature cannot exceed 10^{12} K, the minimum size of the radio emission region is about 10^{12} cm. The size of a typical black hole binary with an orbital period of order one day is also about 10^{12} cm. For those spatially resolved systems, i.e., those jet sources, their sizes are usually between 10^{14} to 10^{16} cm. Therefore the radio emission region can be significantly larger than the whole binary system. At a distance far away from the accretion disk, the strength of the magnetic field is most likely of order 10^{-3} Gauss, therefore the typical electron energy is of order 10^9 eV for a characteristic radiation frequency of 10^{10} Hz. Therefore the radio emission from black hole binary systems is usually produced by high energy electrons far away from the central region of the accretion disk.

Corresponding to the three types of X-ray light curves, there are also three types of radio light curves, with similar morphologies. The systems producing the type I or nova-like X-ray light curves produce usually also type I radio light curves with fast rises (although their radio flux on-sets are rarely caught) and slow exponential decays, in good correlation with the X-ray light curves (see ref. [44] for a review). The radio emission regions in this class of systems are never spatially resolved and have been modelled as synchrotron bubble events, i.e., expanding spherical bubbles of relativistic plasma. The origin and acceleration mechanism of the high energy electrons is still not identified, but is believed to be related to the enhanced mass accretion rate in the inner accretion disk region, inferred from the correlation between the radio and X-ray light curves. It is worthwhile to note that such mass accretion and radio emission correlation cannot be firmly established in those active galactic nuclei (AGN), since the dynamical time scales in AGN are typically larger by a factor of 10^{6-9} than the black hole binary systems.

The radio light curve of Cyg X-1 is very similar to its X-ray light curve, i.e., persistent, variable and with two distinct states. Its high radio flux state corresponds to its hard X-ray high and soft X-ray low flux state, i.e., the so-called hard/low state [114,117]. Although the hard/low state mass accretion rate may be slightly lower than that in its soft/high state, any changes in mass accretion rate through the inner accretion disk region may not play a strong role [114]. Therefore the observed radio, soft and hard X-ray correlation suggests that the radio flux increase is not simply related to the mass accretion rate increase. Instead, the radio flux seems to respond to the hard X-ray flux positively. It is therefore possible that the much higher energy (10^9 eV) electrons responsible for the radio flux production are somehow related to the much lower energy (10^5 eV) electrons, despite that the former population of electrons are distributed far away from the central region of the disk, where the hard X-ray photons are produced. Perhaps the much higher energy electrons originate from the lower energy electrons and they share the same initial acceleration mechanism. Exactly how these electrons are accelerated and why the acceleration process can be maintained stably still remain poorly understood. Another type-III X-ray light curve source, LMC X-3, has never been observed at any radio frequencies, with an upper limit of < 0.3 mJy at 3 and 6 cm [30]. This would not be surprising if its radio luminosity is similar to that from Cyg X-1, since it is located at about a factor of 20 farther away than Cyg X-1 and the radio flux from Cyg X-1 is only between a few and tens of mJy. The detection of a strong radio flux has been reported from the third type III source LMC X-1 at 81 mJy at 6 cm [97]. Assuming isotropy, its absolute luminosity is about a factor of 3000 *stronger* than Cyg X-1, about a factor of 2 *brighter* than the peak luminosity of the superluminal jet sources GROJ1655-40, and *comparable* to the peak luminosity of the other superluminal jet source GRS1915+105. We caution the readers that a later observation of LMC X-1 with the same instrument did not detect the source at an upper limit of $\sim$ 1 mJy [30]. Therefore more radio observations of the source are still needed to clarify the situation.

The radio light curves of the two superluminal jet sources, i.e., GROJ1655-40 [45,101] and GRS1915+105 [66], exhibit close, but complicated relationships with their type II X-ray light curves. Their peak radio luminosity is also much higher than most of the type I or III systems discussed above. In GRS1915+105, there seems to exist an almost one to one correspondence between the radio flares and the *hard* X-ray outbursts [32]. However, there seems to be three different kinds of radio flare states [32]. The relationship between the particular type of the radio flare and the hard X-ray emission properties is not clear. The correlation between radio intensity and the hard X-ray intensity is positive below about 100 mJy; above this level, the correlation becomes negative [39]. In GROJ1655-40, a dynamically confirmed black hole binary, the initial hard X- ray outburst [110] was accompanied with a bright radio flare resulting from superluminal jet ejection [45,101]. All first three hard X-ray outbursts were well correlated with radio flares (superluminal ejection events) [40], although the overall level of the radio emission followed a slow exponential decay. Two subsequent hard X-ray outbursts, however, were not accompanied with any detectable radio emission above 5 mJy between 1.5-15 GHz [100]. Therefore, this seems to indicate that significant hard X-ray production is a necessary, but not sufficient condition for jet ejection. Therefore models [58,63] involving hard X-ray production from the relativistic jets seem unlikely [115]. Nevertheless, what we have learned from the radio and hard X-ray correlations are still important for understanding some problems.

The positive X-ray and radio flux correlation indicates that jet ejection events are related to increased mass accretion. It is widely accepted that a magnetic field should be an essential ingredient in the jet formation according to the Blandford and Payne model [6,7]. Therefore it is very likely that rather strong magnetic fields exist in these systems. If the hard X-ray component is indeed produced by the long suspected Comptonization in an optically thin (inferred from the rather steep power law) corona, then the correlation also supports the jet production model in which the jets are produced from the interaction between the corona and the magnetic field [64].

The black hole (candidate) binary system GX339-4 is also a variable radio source [96]. Intense ($\sim$ Jy) radio flares similar to those from GROJ1655-40 and GRS1915+105 have never been observed. We, however, cannot conclude that the source has never produced radio flares because of the rather sporadic observations from the southern hemisphere. Because of this, the relationship between its radio and X-ray fluxes is not completely clear. The recent observation of a possible radio jet in GX339-4 [29], during an X-ray active state, may provide evidence that its radio behavior is similar to the two superluminal jet sources. We may eventually be able to conclude that all the type II or multiple-outbursts systems share some common radio emission properties, related to their high energy, especially their hard X-ray, behavior.

An essential question to be answered concerns the physical connections between the types of the companion stars, orbital separations, optical and IR light curves, X-ray and hard X-ray light curves, and radio light curves. A possible scenario follows: their companion stars and orbital separations determine the mass transfer property from the companion stars to the outer regions of the disk, this in turn controls their optical properties, at least during the initial outburst after a quiescence period. The mass flow rates through the inner disk boundary onto the black hole modulate their soft and hard X-ray light curves. The soft X-ray photons are emitted from the optically thick inner accretion disk region, and the hard X-ray photons are believed to be produced by inverse Compton up-scattering of low energy photons (probably the soft photons from the disk) by higher energy electrons, although the origin of the electrons is still not unambiguously identified. Somehow at least a portion of these electrons are accelerated to higher energies and move far away from the central disk region to produce radio fluxes via incoherent synchrotron radiation. This global picture agrees qualitatively to the three types of X-ray and radio light curves, through our investigations of the multifrequency connections in these systems. The correlation between the three types of companion stars and their corresponding light curves suggest that in the type II or multiple-outbursts systems another instability, possibly the mass overflow instability in the companion star caused by the X-ray heating of the central soft and hard X-ray sources, may play a major role in the subsequent outbursts since the initial one. In the type III or persistent systems, perhaps the feedback of the X-ray heating to the high mass companion stars reaches a steady state so that these systems become persistent sources.

The above picture does not explain why some systems produce powerful jets, while others apparently do not do so. It is possible that in GROJ1655-40 the inferred mass transfer rate from the X-ray observations is only a portion of the total mass transfer rate from its peculiar companion star. The rest of the material is not transferred through the disk, but instead forms a disk 'wind', available for producing the powerful jets. In other systems the disk 'wind' may be very weak. In GRS1915+105, although the exact type of its companion is unknown as we discussed above, there indeed appears to be some evidence for a significant wind from the system [68]. This scenario, however, still does not explain why the outflows in GROJ1655-40 and GRS1915+105 are highly collimated. This may be partially explained by the properties of the black holes, as will be discussed below.

X-RAY PROBING OF THE INNER DISK REGION

X-ray observations at keV energies can be used for probing the accretion disk very close to the black hole horizon, thus providing important tools for testing accretion disk models and some general relativistic effects. This is because the peak emission energy of the blackbody spectra from the inner disk region is between one-tenth and several keV [116], and that these X-ray photons are much more penetrating than EUV photons emitting from the inner disk regions of AGNs. Because of the interactions between the black holes and the accretion disks, X-ray observations can also be used for studying the properties of the black holes [116]. In this section we will discuss three aspects of probing the accretion inner disk regions with X-ray observations.

Continuum X-ray Spectra

As we mentioned before, the continuum X-ray spectrum from the inner accretion disk region is well described by the multi-color disk blackbody model [60,70]. The spectrum depends only upon the location of the inner accretion boundary, mass of the black hole and the mass flow rate through the inner disk region. For a steady disk in a high mass accretion rate state, the inner disk boundary is assumed to be the last stable orbit of the black hole, beyond which the material falls into the black hole with near radial trajectories. For the purpose of the spectral fitting, the free parameters are the peak blackbody temperature, absorption column density, and the normalization at a given energy. The inner disk radius and the disk inclination angle are coupled in the normalization parameter, so they cannot be determined independently. In some cases the system inclination angle and/or the black hole mass can be well constrained through the optical light curve modelling and spectroscopic measurements (to determine the companion star type). Therefore the size of the inner disk boundary can also be constrained. The values of the black hole mass, spin and the rotation direction of the material moving around it are all coupled into the size of the inner disk boundary, since the location of the last stable orbit of a black hole depends upon all of them. If the mass

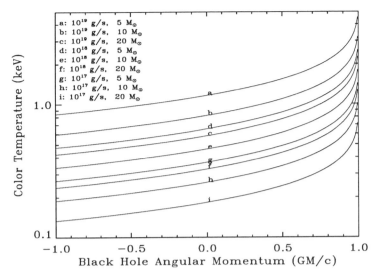

FIGURE 5. The peak color temperature (kT_{col}) of the accretion disk emission, for different black hole masses, accretion rates and angular momenta [116] (Zhang, Cui & Chen 1997).

of the black hole is also known from optical observations, we then have the possibility to constrain the spin of the black hole. It should be noted that various correction factors to account for spectral hardening (due to electron scattering) and relativity effects have to be included in this process. This has been done recently in a number of black hole systems, whose system parameters have been well constrained from optical observations [116]. In figure 5, the peak blackbody temperature is shown as the functions of the black hole mass, spin and mass accretion rate.

It has been found that the two superluminal jet sources GROJ1655-40 and GRS1915+105 (by combining the interpretation of its 67 Hz QPOs as the g-mode oscillations [75]) contain most likely a black hole spinning at near the maximally allowed rate, while several other black holes with the observed soft components, but no relativistic jets, contains slowly or non-spinning black holes. It is also proposed that several 'hard' X-ray transients, i.e., no soft component has ever been observed from them, may contain rapidly spinning black holes with retrograde disks [116]. Therefore all types of observed black hole binary systems are naturally unified within one simple scheme. The state transitions in Cyg X-1 are proposed to be due to the disk rotation direction reversal, caused by the so-called 'flip-flop' type instability in wind accreting systems [116]. Future higher sensitivity observations, especially below 1 keV, are needed to test this scheme.

Iron Line Diagnostics

The first spectral line feature from a black hole binary system was detected with the *EXOSAT GSPC* in Cyg X-1 [4]. Since then the Cyg X-1 spectral line feature has been observed with *Tenma* [54], *Ginga* [24,99] and more recently with *ASCA* [26]. From the high resolution ASCA observations [26], the overall feature can be modelled by a narrow (equivalent width 10-30 eV) iron K emission line at 6.4 keV and an iron K edge at 7 keV representing a reflection component, from the outer part of the accretion disk. The disk is inferred to be ionized, with a covering angle $\Omega/2\pi \sim 0.2 - 0.4$. Due to the limited statistics, minor contribution due to the reflections from the companion star and/or fluorescence emission from the inner accretion disk region cannot be excluded. The small or null contribution from the inner disk region may be due to the much higher degree of ionization of the inner disk region. However, it should be noted that these observations were made during the regular hard/low state of Cyg X-1. In the hard/low state, its inner disk radius is probably larger than $3R_s$ [114]. Therefore the inner disk contribution to the observed line profile is expected to be less important in the hard/low state. Moreover, the broadening and skewness of the expected line profile from the inner disk region is less significant for a small inclination angle (only 10°-40° in Cyg X-1). Future higher sensitivity and better resolution instruments are required for unambiguously identifying these different components in its iron line profile.

Recently, iron line features have also been observed from the two superluminal jet sources GROJ1655-40 and GRS1915+105 with ASCA [27], as shown in figure 6. Except during a light curve 'dip' in GROJ1655-40, all other observed line profiles show strong absorption features at around 7 keV. A possible interpretation is that the continuum spectrum passes through a slab-like ionized absorption medium (accretion disk corona) at a distance about 10^{10} cm away from the original X-ray source [102]. The degree of ionization of the corona increases for a higher continuum luminosity, indicating that the ionization is also caused by the X-ray illumination. This is perhaps the first evidence of the existence of a hot corona in a black hole binary system. It should be noted that these observed absorption features are significantly different from the apparent emission feature observed in Cyg X-1. It is not clear if this is due to the differences in their nature (for example, jet vs. non-jet systems), or the inclination angle difference (70° vs 10°-40°).

On the other hand, the apparent emission feature in the 'dip' spectrum of GROJ1655-40 should not be ignored completely. This feature is very similar to the broadened and skewed line profiles observed from many Seyfert I galaxies (e.g. ref. [28,49,98]). The only difference seems to be the much higher peak energy at ~ 7.0 keV, possibly blue-shifted, if the original peak is also the 6.4 iron K_α line. This is actually expected due to the much higher inclination angle of 70° in GROJ1655-40. It is also unlikely that the whole complex line features of other spectra (non-'dip') of the sources can be explained completely

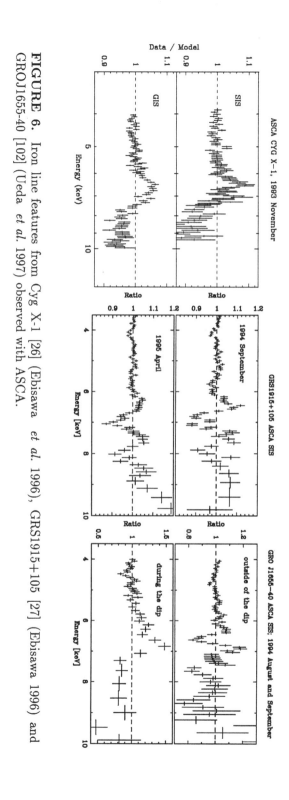

FIGURE 6. Iron line features from Cyg X-1 [26] (Ebisawa et al. 1996), GRS1915+105 [27] (Ebisawa 1996) and GROJ1655-40 [102] (Ueda et al. 1997) observed with ASCA.

with absorptions alone. It is possible that the non-'dip' line profiles consist of both the absorption by the disk-corona and the iron fluorescence emission (relativistically broadened and skewed) from the inner disk region, which is believed to be around 1 R_s [115,116]. One possible scenario is that during the 'dip' period the emitted X-ray luminosity from the central region of the disk becomes lower, due to perhaps the reduced mass transfer rate. The lower X-ray flux therefore changes the properties of the disk-corona (geometry and ionization state), so that the resonance absorption of X-ray photons by iron ions becomes negligible.

It is clear from the above discussion that rich information may be derived from iron line diagnostics of black hole binary systems. Many of the current interpretations are, however, quite uncertain, due primarily to the limited number of detections and the poor statistical quality of past observations.

Timing Diagnostics

Rapid variability studies of their X-ray light curves provide essential information about many dynamic processes in the central region of the accretion disk and in the corona. Many of the earlier results were reviewed by van der Klis [104]. Some recent results are reviewed in a companion review article in the same proceedings [36]. Here we only review briefly the high frequency QPOs detected from the two superluminal jet sources GROJ1655-40 at ∼300 Hz [89] and GRS1915+105 at ∼67 Hz [71]. An important feature in these QPOs is their *stable* frequency, compared to the majority of other QPOs observed in black hole and neutron star binaries, with the exception of the stable 34 Hz QPO in the bursting pulsar GROJ1744-28 [118].

It is thus natural to relate these QPOs to the last stable orbit of the black hole. Assuming that the QPO frequencies are actually the Keplerian frequencies of the last stable orbit, the masses of the assumed non-spinning black holes are around 33 $M_\odot$ and 7 $M_\odot$ (consistent with its dynamical mass estimate), in GRS1915+105 and GROJ1655-40, respectively. However, since the spin of the black hole in GROJ1655-40 is inferred to be near its maximal rate from X-ray spectroscopic measurements as we discussed above [116], the non-spinning black hole assumption is not likely to be valid and thus neither is this Keplerian frequency interpretation. The other interpretation is to attribute these QPOs to the g-mode oscillations due to the general relativity effects [74,75,84]. When applied to GROJ1655-40, the inferred black hole spin rate is remarkably consistent with that obtained from independent spectroscopic measurements [116]. Taking the same approach, the implied black hole spin in GRS1915+105 would also be near its maximal rate and the black hole mass is around 30 $M_\odot$. In this model, however, it is not clear why the QPOs seem to be associated with the hard X-ray component [71,89].

SUMMARY

We have reviewed the recent progress in the investigations of the multifrequency properties and connections of galactic black hole binaries. There is now sufficient evidence for the existence of stellar mass black holes in at least 10 systems. Their high energy and radio light curves are well correlated and can be divided into three classes, which may be related to their respective types of companion stars. Their X-ray spectral properties may be related closely to the spins of the black holes, in a newly proposed unification scheme of all types of black hole binaries. The correlation between the hard X-ray and radio flares suggests that the radio emission and therefore the associated outflows are related to the corona in the inner disk region. Iron line studies in the two superluminal jet sources provide possible evidence of a hot disk-corona. Only the black holes in the jet systems are found to be likely spinning at nearly the maximal rates, suggesting that the highly relativistic jets are related to the spin of the black hole. This supports strongly the previously suggestions that the highly relativistic jets in AGNs might be related to the rapid black hole spins [8,9,59]. Therefore the eventual unification of all types of black hole systems (galactic and extra-galactic) becomes possible using the black hole spin as one of the key parameters.

However, many of the current interpretations are still quite uncertain, due to both of our limited theoretical understanding of the detailed physics involved and the limitations of the current observations. Therefore more theoretical investigations and future multifrequency observations are required. Nevertheless, such recent progress suggests that the current observational and theoretical studies of the galactic black hole binary systems have begun to show the premise of using these systems as the laboratories for testing some physical laws in the strongest gravitational field. Compared to AGNs in which similar studies can also be carried out and in fact were started much earlier, the galactic black hole binary systems are more advantageous because a) the dynamical time scales in them are much shorter; b) their inner disk regions are directly observable in the X-ray frequency range; c) the fine details of their jets are observable because they are much closer to us; and d) the system parameters can be measured through optical/infrared observations.

We thank Wan Chen, Ken Ebisawa, Jerry Orosz, Craig Robinson and Kinwah Wu for kindly providing data and figures to be included in this review paper. Part of the V-band light curve of GROJ1655-40, provided by Jerry Orosz and obtained by Charles Bailyn and various Yale observers and by Jeff McClintock, has not been published. SNZ also appreciates those stimulating discussions with many of his colleagues, including Didier Barret, Sandip Chakrabarti, Wei Cui, Wan Chen, Ken Ebisawa, Rob Fender, Jerry Fishman, Eric Grove, Jean-Pierre Lasota, Hui Li, Edison Liang, Mario Livio, Jeff McClintock, Ramesh Narayan, Bill Paciesas, Ron Remillard, Craig Robinson, Marco Tavani, Lev Titarchuk, Jan van Paradijs and Kinwah Wu, etc.

REFERENCES

1. Augusteijn, T., Kuulkers, E. & Shahams, J. 1993, *A&A*, **279**, L13
2. Bailyn, C.D. *et al.* 1995, *Nature*, **374**, 701
3. Bailyn, C.D. *et al.* 1995, *Nature*, **378**, 157
4. Barr, P., White, N.E. & Page, C.G. 1985, *MNRAS*, **216**, 65
5. Barret, D., McClintock, J.E. & Grindlay, J.E. 1997, *ApJ*, **473**, 963
6. Blandford, R. 1976, *MNRAS*, **176**, 465
7. Blandford, R. & Payne, D. 1982, *MNRAS*, **199**, 883
8. Blandford, R. & Znajek, R.L. 1977, *MNRAS*, **179**, 433
9. Blandford, R. & Levinson, A. 1995, *ApJ*, **441**, 79
10. Bolton, C.T. 1972, *Nature*, **258**, 307
11. Callanan, P.J. *et al.* 1996, *ApJ*, **461**, 351
12. Callanan, P.J. *et al.* 1996, *ApJ*, **470**, L57
13. Casares, J., Charles, P. A., & Naylor, T. 1992, *Nature*, **355**, 614
14. Casares, J., Charles, & P. A. 1992, *MNRAS*, **255**, 7
15. Casares, J. *et al.* 1995, *MNRAS*, **276**, L35
16. Castro-Tirado, A.J., Geballe, T.R. & Lund, N. 1996, *ApJ* **461**, L99
17. Chaty, S. *et al.* 1996 *A&A*, **310**, 825
18. Chen, W., Shrader, C. & Livio, M. 1997, *ApJ*, in press.
19. Cowley, A.P. *et al.* 1983, *ApJ*, **272**, 118
20. Chen, W., Livio, M. & Gehrels, N. 1993, *ApJ*, **408**, L5
21. Done, C. *et al.* 1992, *ApJ*, **395**, 275
22. Ebisawa, K., Titarchuk, L. & Chakrabarti, S.K. 1996, *PASJ*, **48**, 59
23. Ebisawa, K., Mitsuda, K. & Inoue, H. 1989, *PASJ*, **41**, 519
24. Ebisawa, K. *et al.* 1992, *Frontiers of X-ray Astronomy*, ed. Y. Tanaka & K. Koyama (Tokyo: University Academy Press), p.301
25. Ebisawa, K. *et al.* 1992, in *ApJ*, **367**, 213
26. Ebisawa, K. *et al.* 1996, *ApJ*, **467**, 419
27. Ebisawa, K. 1996, in *X-ray Imaging and Spectroscopy of Cosmic Hot Plasmas*, ed. F. Makino & K. Mitsuda ((Tokyo: University Academy Press), p.427
28. Fabian, A.C. *et al.* 1995, *MNRAS*, **277**, L11
29. Fender, R.P. *et al.* 1997, *MNRAS*, **286**, L29
30. Fender, R.P. 1997, private communication
31. Filippenko, A.V. Matheson, T. & Ho, L.C. 1995, *ApJ*, **455**, 614
32. Foster, R.S. *et al.* 1996, *ApJ*, **476**, L81
33. Gies, D.R. & Bolton, C.T. 1982, *ApJ*, **260**, 240
34. Gies, D.R. & Bolton, C.T. 1986, *ApJ*, **304**, 371
35. Grove, J.E., Kroeger, R.A. & Strickman, M.S. 1996, in *Second Integral Workshop, The Transparent Universe*, St. Malo, France, ESA SP-382
36. Grove, J.E. *et al.* 1997, in *The 4th Compton Symposium*, these proceedings
37. Han, X. & Hjellming, R.M. 1992, *IAUC*, 5593
38. Han, X. & Hjellming, R.M. 1997, *ApJ*, in press (submitted in 1994)
39. Harmon, B.A. *et al.* 1997, *ApJ*, **477**, L85
40. Harmon, B.A. *et al.* 1995, *Nature*, **374**, 703

41. Hameury, J.-M., King, A.R., & Lasota, J.-P. 1986, *A&A*, **161**, 71
42. Hameury, J.-M. *et al.* 1997, *ApJ*, in press.
43. Haswell, C.A. *et al.* 1994, *ApJ*, **411**, 802
44. Hjellming, R.M. & Han, X. in *X-ray Binaries*, Eds. Walter H.G. Lewin, Jan van Paradijs & P.J. van den Heuvel, Cambridge University Press, 1995, ch. 7
45. Hjellming, R.M. & Rupen, M.P. *et al.* 1995, *Nature*, **375**, 464
46. Hjellming, R.M. *et al.* 1988, *ApJ*, **355**, L75
47. Hjellming, R.M. *et al.* 1997, in preparation
48. Hutchings, J.B. *et al.* 1987, *AJ*, **94**, 340
49. Iwasawa, K. *et al.* 1997, *MNRAS*, **279**, 837
50. Kesteven, M.J. & Turtle, A.J. 1991, *IAUC*, 5181
51. King, A.R., Kolb, U. & Szuszkiewicz, E. 1997, *ApJ*, **488**, in press
52. Kitamoto, S. *et al.* 1990, in *Accretion Powered Compact Binaries*, ed. C.W. Mauche (Cambridge University Press), p.21
53. Kitamoto, S. *et al.* 1990, *PASJ*, **42**, 85
54. Kitamoto, S. *et al.* 1997, submitted to *ApJ*
55. Kolb, U. *et al.* 1997, *ApJ*, **485**, L33
56. Kroeger, R.A. *et al.* 1996, *A&AS*, **120**, C117
57. Lasota, J.P. 1997, in *Accretion Phenomena and Related Outflows*, ed. D.T. Wickramasinghe, G.V. Bicknell & L. Ferrario, ASP Conf. Series Vol. 121
58. Levinson, A. & Blandford, R. *et al.* 1996, *ApJ*, **456**, L29
59. Livio, M. 1997, in *Accretion Phenomena and Related Outflows*, ed. D.T. Wickramasinghe, G.V. Bicknell & L. Ferrario, ASP Conf. Series Vol. 121
60. Makishima, K. *et al.* 1986, *ApJ*, **308**, 635
61. Martin, A.C. *et al.* 1995, *MNRAS*, **274**, L46
62. McClintock, J.E. & Remillard, R.A. 1986, *ApJ*, **308**, 110
63. Meier, D. 1996, *ApJ*, **459**, 185
64. Meier, D. *et al.* 1997, *Nature*, **388**, 350
65. Mineshige, S. 1996, *PASJ*, **48**, 93
66. Mirabel, I.F. & Rodriguez, L.F. 1994, *Nature*, **371**, 46
67. Mirabel, I.F. *et al.* 1996, *ApJ*, **472**, L111
68. Mirabel, I.F. *et al.* 1997, *ApJ*, **477**, L45
69. Mitsuda, K. *et al.* 1989, *PASJ*, **41**, 97
70. Mitsuda, K. *et al.* 1984, *PASJ*, **36**, 741
71. Morgan, E.H., Remillard, R.A. & Greiner, J. 1997, *ApJ*, **482**, 993
72. Morris, P.W. 1997, *ApJ*, in press
73. Narayan, R. 1997, in *Accretion Phenomena and Related Outflows*, ed. D.T. Wickramasinghe, G.V. Bicknell & L. Ferrario, ASP Conf. Series Vol. 121
74. Nowak, M.A. & Wagoner, R.V. 1993, *ApJ*, **418**, 187
75. Nowak, M.A. *et al.* 1996, *ApJ*, **477**, L91
76. Orosz, J.A. 1996, PhD Thesis, Yale University
77. Orosz, J.A., & Bailyn, C.D. 1995, *ApJ*, **446**, L59
78. Orosz, J.A. & Bailyn, C.D. 1997, *ApJ*, **477**, 876
79. Orosz, J.A. *et al.* 1996, *ApJ*, **468**, 380
80. Orosz, J.A. *et al.* 1997, *ApJ*, **478**, L83

81. Owen, F.N. et al. 1976, ApJ, **203**, L15
82. Paciesas, W.S. et al. 1997, in *The 4th Compton Symposium*
83. van Paradijs, J. & McClintock, J. E.., in *X-ray Binaries*, Eds. W.H.G. Lewin, J. van Paradijs & P.J. van den Heuvel, Cambridge University Press, 1995
84. Perez, C.A. et al. 1995, ApJ, **476**, 589
85. Priedhorsky, W.C., et al. 1983,ApJ **270**, 233.
86. Priedhorsky, W.C. et al. 1995, A&A, **300**, 415
87. Remillard, R.A. et al. 1994, BAAS, **26**, 1483
88. Remillard, R.A. et al. 1996, ApJ, **459**, 226
89. Remillard, R. 1996, in *Texas in Chicago, Texas Symposium*, Chicago
90. Robinson, C.R.. et al. 1997, in *The 4th Compton Symposium*
91. Rodriguez, L.F. & Mirabel, I.F. 1997, ApJ, **474**, L123
92. Rubin, B.C. et al. 1997, ApJ, in press
93. Sanwal, D., et al. 1996, ApJ, **460**, 437
94. Shahbaz, T., Naylor, T., & Charles, P.A. 1994, MNRAS, **268**, 756
95. Shahbaz, T. et al. 1994, MNRAS, 271, L10
96. Sood, R. & Campbell-Wilson, D. 1994, IAUC 6006
97. Spencer, R.E. et al. 1997, *Vistas in Astronomy*, 41(1), 37
98. Tanaka, Y. et al. 1995, Nature, **375**, 659
99. Tanaka, Y. 1991, in *Lecture Notes in Physics 385, Iron Line Diagnostics in X-ray Sources*, ed. A. Treves (Berlin: Springer), 98.
100. Tavani, M. et al. 1996, ApJ, **473**, 103
101. Tingay, S.J. et al. 1995, Nature, **374**, 141
102. Ueda, Y., et al. 1997, ApJ, in press
103. van der Hooft, F. et al. 1997, MNRAS, **286**, 43
104. van der Klis, M., in *X-ray Binaries*, Eds. Walter H.G. Lewin, Jan van Paradijs & P.J. van den Heuvel, Cambridge University Press, 1995, ch. 6, pp. 252-307.
105. van Paradijs, J. 1996, ApJ, **464**, L39
106. Wagner, R.M. et al. 1990, in *Accretion Powered Compact Binaries*, ed. C.W. Mauche (Cambridge University Press), p.
107. Wagner, R.M. et al. 1992, ApJ, **401**, L97
108. Webster, B.L. & Murdin, P. 1972, Nature, **235**, 37
109. Wu, K. & Hunstead, R. W. 1997, in *Accretion Phenomena and Related Outflows*, ed. D.T. Wickramasinghe, G.V. Bicknell & L. Ferrario, ASP Conf. Series Vol. 121
110. Zhang, S.N., et al. 1994, IAUC 6046
111. Zhang, S.N. 1997, in *Accretion Phenomena and Related Outflows*, ed. D.T. Wickramasinghe, G.V. Bicknell & L. Ferrario, ASP Conf. Series Vol. 121
112. Zhang, S.N., et al. 1996, A&A **120**, C279
113. Zhang, S.N., Robinson, C.R. & Cui, W. 1996, IAUC 6510
114. Zhang, S.N., et al. 1997, ApJ **477**, L95
115. Zhang, S.N., et al. 1997, ApJ **479**, 381
116. Zhang, S.N., Cui, W. & Chen, W. 1997, **482**, L155
117. Zhang, S.N., et al. 1997, ApJ, to be submitted.
118. Zhang, W., et al. 1995, ApJ **449**, 930

CGRO Studies of Supernovae and Classical Novae

Mark D. Leising

Department of Physics & Astronomy, Clemson University, Clemson, SC 29634

Abstract. We summarize CGRO observations of supernovae, including SN 1987A, SN 1991T, and SN 1993J, and classical novae, and their impact on current understanding of those phenomena. Many important fundamental questions about these objects remain, some of which might still be addressed by future CGRO observations or by data already in hand. We briefly sketch some of these prospects.

INTRODUCTION

In prelaunch GRO literature and propaganda, one finds the stated objectives of detection of gamma-ray lines from supernovae and novae at or near the top of every list. At the Fourth Compton Symposium, roughly six years after launch, these topics were not deemed worthy of even a single invited talk (a decision endorsed by this member of the scientific organizing committee). Have we, or CGRO, somehow failed to achieve major objectives upon which the mission was sold? Hardly. We have had several good opportunities to study rare events and these have contributed significantly to our understanding; in a few cases CGRO observations have probably helped to avoid substantial misunderstanding. That our line detections came early in, and even before, the CGRO mission perhaps made them seem commonplace. In retrospect we probably should have been more committed to our stated goals. More observations of SN 1987A and longer – and especially earlier – measurements of SN 1991T should have been acquired, but those were very hard to justify at the time. Perhaps we had forgotten that we study these targets of opportunity at the whim of Poisson – only quite rare events would be clearly detectable given our instruments' sensitivities. Here we discuss those we have got and those we might yet get. This is not intended to be an exhaustive review. Even the best of those tend to be boring to experts and, given the size and proliferation of tomes like this one, seldom read by others. We will simply summarize the CGRO observations and outline the essential remaining questions and the hopes for their resolution. The reader is directed to other useful summaries [1–3], which remain quite relevant. We will not discuss the important studies of ^{44}Ti in the Cas A supernova, as they are extensively reviewed elsewhere in this volume.

CORE COLLAPSE SUPERNOVAE

We have been afforded the chances to directly probe the very innermost region of one core collapse, SN 1987A, and the circumstellar environment of another, SN 1993J. SN 1987A came just slightly too early for optimal CGRO study, but of course provided a great target for many other observers. It was the first supernova detected in gamma-ray lines [4] and their scattered continuum photons [5]. Ultimately, at least four lines of ^{56}Co were detected [6] and two lines were resolved and shown to be unexpectedly broad and redshifted [7], as was also indicated by infrared lines [8]. The high-energy light curves, especially their early rise, required some radioactive ^{56}Co to be at low optical depths very early in the expansion [6]. Generally these observations confirmed wonderfully theoretical ideas of core collapse supernova nucleosynthesis, but still remain to be explained in detail.

CGRO/OSSE began to observe SN 1987A only after 1600 days post-explosion. It was still able to detect the 122 keV line of ^{57}Co [9] – another direct isotopic measurement of freshly synthesized radioactivity – dramatically confirming expectations of the nuclear processing in the innermost ejected layers of the supernova. The measured flux of the ^{57}Co line and scattered photons, $9.0\pm2.2\times10^{-5}$ cm^{-2} s^{-1}, suggested an initial ^{57}Ni mass of 2.7×10^{-3} M$_\odot$, subject to some uncertainty in the actual optical thickness of the ejecta at that epoch. This mass, about that expected from nucleosynthesis considerations [10], was a factor of three less than was suggested to be powering the late UVOIR light output [11]. So rather than forcing investigations of extreme nucleosynthesis scenarios, this gamma-ray observation led to consideration of other possible power sources [12], including the delayed recombination emission from ionization due to earlier radioactive decay. This effect was later shown in detail to be operating in SN 1987A [13]. Even though such a latecomer, CGRO's contributions to understanding this most-studied of all supernovae were indeed substantial. It is conceivable that there will be more, in the form of studies of the young compact remnant, but only if we decide to look.

SN 1993J was also a relatively rare event, occurring in M81 at a distance of 3.6 Mpc. As for almost any event so scrutinized, it was found to be quite unusual in many respects. It turned out to the defining member of supernovae of "Type IIb": "II" because hydrogen lines were seen in its spectrum, "b" because in essentially all other respects its evolution tracked that of the well known Type Ib supernovae. It clarified the nature of those events as envelope-stripped massive stars, with SN 1993J simply being one that exploded just before the last of its hydrogen envelope blew away. Early detections of X-ray emission [14,15] were naturally interpreted as thermal emission from reverse-shocked supernova ejecta at $\sim 10^8 K$ [16,17]. It came as something of a surprise that CGRO/OSSE began detecting hard X-rays (50 - 100 keV) from SN 1993J nine days after the explosion [18]. Interpreted as thermal emission, a temperature of 10^9 K was implied. This temperature was not unexpected for the forward-shocked circumstellar matter, but the implied luminosity - 5×10^{40} ergs s^{-1} - far exceeded expectations from that region because of the supposed low density, and therefore low emissivity, of the expected stellar wind. Radioactivity could be excluded as the source of the hard photons, because the mass of ^{56}Co and the relatively optically thick ejecta established from the optical light curve would provide too little flux, the spectrum was too steep, and the flux declined too fast with time. The best direct search for gamma-ray line emission, by CGRO/COMPTEL [19] provided upper limits consistent with all detailed models of the event.

It was shown that the thermal emission spectrum implied by OSSE accounted for essentially all of the observed soft X-ray emission [18], leaving no room for the supposedly much brighter soft flux from the shocked ejecta. In the end a nice picture of the situation consistent with all X-ray and radio observations was developed [20,21]. The essential features included a presupernova wind with a density profile shallower than r^{-2}, due to non-steady or non-spherical mass loss, or possibly a clumpy circumstellar medium. All the early hard emission is thought to come from the adiabatic forward shock, with a dense radiative shell behind the reverse shock hiding the X-rays originating in the supernova ejecta at early epochs.

THERMONUCLEAR SUPERNOVAE

Radioactivity so dominates these objects that it has been thought that they would be the subject to which gamma-ray astronomy contributes so much. This will very likely happen yet. With an expected sensitivity to gamma-ray lines from SN Ia within 10 Mpc, we had reasonable hope to get one or two during the life of CGRO. We have not quite had this chance yet. Actually we had one chance, but we did not look.

SN 1991T occurred just outside this range, at just about the time of the CGRO launch. With the sky survey to complete and uncertain instrument and spacecraft lifetimes, it was hard to argue that we should dedicate substantial observing time, or even the very first observation, to SN 1991T, but in retrospect we probably should have. A somewhat overluminous SN Ia, SN 1991T was observed three times by CGRO, around 70, 140, and 180 days post-explosion (in the second interval it was in only the OSSE field of view.) Analyses of those observations suggested no hint of any line emission, but somewhat constraining upper limits [22,23]. A number of published SN Ia models could be ruled out for distances of 10 Mpc, a few for the more favored distance of 13 Mpc. Accounting for the apparent brightness of SN 1991T and allowing for extinction suggested that SN 1991T must have been quite gamma-ray faint relative to its optical brightness, although a few models could still accommodate the observations [23].

Subsequent analysis [24] of the same COMPTEL data suggested a hint of ^{56}Co emission, with an average flux of $5.3\pm2.0\times10^{-5}$ cm^{-2} s^{-1} in the 847 keV line in the 70 and 180 day intervals. This analysis used the statistics of two lines in both intervals. The quoted flux is marginally consistent with the OSSE measurement, $0.5\pm1.6\times10^{-5}$ cm^{-2} s^{-1} for the 847 keV line at 73 days, given the uncertainties in both. Still this is a large average flux, as the later flux should be only $\simeq 0.55$ of the earlier one. For reasonable ejecta transparencies (0.6 and 0.95 847 keV escape fractions) in the two intervals, 5.3×10^{-5} cm^{-2} s^{-1} corresponds to 1.1 M$_\odot$ of initial ^{56}Ni at a distance of 10 Mpc, and 1.9 M$_\odot$ at the more likely distance of 13 Mpc. This is hard to reconcile with current models. Further analyses of COMPTEL data [25] suggest that the quoted statistical uncertainties should be taken seriously. They quote a higher detection significance of 3.3σ, but do not give fluxes. It would seem that both instruments, and models for SN 1991T such as W7DT [26] at a distance of 12–13 Mpc could all be consistent with a flux of about 3×10^{-5} cm^{-2} s^{-1} in the 847 keV line at 70 days. It is possible that further analysis of COMPTEL data simultaneously using both spectral and imaging information can reduce the uncertainty in the flux.

Since the launch of CGRO a particular SN Ia scenario, the He-triggered detonation of a sub-M_{Ch} whited dwarf [27,28] has come back into vogue, in part because the circumstances leading up to it should be much more common than the M_{Ch} white dwarf in a binary system, and because they would be expected to yield a range of observed characteristics, as is now apparent for SN Ia [29]. Such objects might even be required to provide the solar ^{44}Ca [30,31]. Ground-based SN Ia observations do not definitively rule out or confirm any theoretical SN Ia scenarios, except for excluding pure M_{Ch} detonations (which is also excluded for SN 1991T by the above gamma-ray measurements.) He-triggered SN Ia have, however, definitive gamma-ray signatures [32] that can be tested as soon as we get one close enough. Because most of the triggering He cap is burned to the iron peak and expands rapidly, significant fluxes of the lines of short-lived ^{56}Ni ($\tau=8.8$ days) should be detected. We simply looked too late for SN 1991T, but premaximum observations of another SN Ia even close to that distance should clarify whether that object is indeed a He-triggered white dwarf explosion.

There is also the hope that we will get a very nearby SN Ia, like SN 1972E or SN 1986G, during the life of CGRO. These are far out on the tail of plots of number in the catalogs versus distance, so extrapolating to their frequencies is not very meaningful. Still we know they happen, maybe once per decade, and we will get to observe one if NASA elects to operate CGRO long enough.

We barely missed an excellent chance to study a SN Ia in great detail. The Type Ia SN 1993af occurred in NGC 1808, whose distance is estimated as low as 6 Mpc. Unfortunately, it was discovered at least 3 months past maximum because of its proximity to the Sun. We still might have detected it had we looked immediately upon discovery, but we could not preempt an important multiwavelength campaign underway at the time. We will not get many more chances even this good.

CLASSICAL NOVAE

CGRO has not definitively detected any emission from a classical nova. Some part of the observed 1.809 MeV emission from ^{26}Al might come from classical novae, but there is no convincing evidence to that effect. In fact the apparently irregular angular distribution [33] suggests suggests rather younger sources. The theoretical expectations have not changed much, but we always talked in terms of novae closer than 1 kpc and we simply have not had any nearly so close. This is something of a puzzle in itself, but still we have some very interesting limits on line emission from novae.

The prompt ejection of material by classical novae is facilitated by the local deposition of large amounts of energy in the outer layers of the accreting white dwarf by positron emitters. Clayton & Hoyle [34] pointed out the potential for detecting the annihilation line from these positrons early in the outburst. Later calculations [35] of the spectrum and time profile showed that detection from novae within 1 kpc was possible, but that the emission is extremely dependent on details of the basic nova models. This is of course the signature of a useful diagnostic – if we actually detect this emission. Recent calculations [36] confirm the earlier expectations. CGRO/BATSE data have been searched for such emission around the times of the outbursts of N Her 1991 and N Cyg 1992, but no significant excesses have been found [37]. This is not entirely unexpected, although N Her 1991 - one of the fastest classical novae on record - should have been a good candidate, because of their relatively large distances. That these gamma rays do not readily escape insures that the less penetrating positrons do not themselves escape (see also [35]) and contribute to the diffuse 511 keV line emission, although one often finds claims to the contrary.

A by-product of nova nucleosynthesis, ^{7}Be provides another potentially interesting diagnostic of novae [38]. OSSE searches of spectra of nearby novae have not detected it [39]. This is also not in serious conflict with current calculations [40].

Prior to CGRO, limits [41] on 1.28 MeV emission from ^{22}Na synthesized in classical novae [34], including individual Ne-rich ones and the average of the collection expected in the central Galaxy, were already beginning to constrain the models. Thus it is quite surprising that this line still has not been detected [39,42,43].

TABLE 1. ^{22}Na Limits for Recent Ne-rich Novae

	Distance (kpc)	Ejected Mass ($10^{-4} M_\odot$)	^{22}Na Mass Limit ($10^{-7} M_\odot$)	X(^{22}Na) Limit (10^{-3} by Mass)
N Her 1991	3.4	0.64	1.2	1.9
N Cyg 1992	3.2	0.7–5	0.7	0.14–1.0
N Vul 1984 #2	2.0	15	7.0	0.5

The most interesting CGRO limits to date come from COMPTEL observations of individual Ne-rich novae [42]. They set upper limits to the 1.28 MeV line flux ranging from $2-6 \times 10^{-5}$ cm^{-2} s^{-1} for five recent Ne-rich novae and similar limits for six other "normal" novae. The two most constraining limits, along with one from SMM [41] are shown in the table. The larger distance for N Cyg 1992 [44] is used, as are other parameters from Shara (1994) [45].

The flux limit for N Her 1991 is below the predicted flux for that event [46]. This prediction was arrived at by multiplying the ^{22}Na mass fraction of the most massive ONeMg white dwarf model by the observed total ejected mass. Typical calculations of ^{22}Na production indicate mass fractions of 10^{-3} or greater in white dwarfs of 1.25 $M_\odot$ or larger [46,47], and substantially exceed the measured mass-fraction limits for 1.35 $M_\odot$ calculations. However, the calculated explosions on massive white dwarfs eject substantially less matter (typically $\leq 10^{-5}$ $M_\odot$, if any) than the observed Ne-rich events ($10^{-4} - 10^{-3}$ $M_\odot$).

It is not clear that we understand at all what these Ne-rich objects are. Models have produced the observed Ne abundances only by assuming them to begin with. Typically this has been explained as core material of ONeMg white dwarfs mixed into the envelope [48], requiring high-mass white dwarfs, which necessarily eject small masses. An alternative hypothesis, that lower mass CO white dwarfs experience He shell flashes that enrich their envelopes in O and Ne, might avoid the ejected mass problem [45,49]. However, these models have not ejected enough Ne to match the observations, and they might have an even worse conflict with the gamma-ray limits. At high temperatures ^{22}Na is fragile, so substantial amounts can be made if either peak nova temperatures are low or rapid convection and cooling removes it from high temperature regions [50]. The latter effect is shown for such metal-rich mixtures that the peak temperatures are always high [47] but only the most massive explosions eject the ^{22}Na before it burns. Dwarfs with low enough masses that they can eject up to 10^{-3} $M_\odot$ might have low enough temperatures, for realistic enrichment, that they are copious ^{22}Na producers. Such models have not yet been calculated. There might yet be a place in the parameter space of the current paradigm where ejected masses, Ne abundances, and ^{22}Na limits can all be accommodated, but the allowed region is shrinking. Gamma-ray detections would restrict that space severely.

This work was supported by NASA grant DPR S-10987C to NRL via subcontract to Clemson University and by NASA grant NAG5-3699 to Clemson University. A part of this work was undertaken while the author held a fellowship from the Alexander von Humboldt-Stiftung and enjoyed the hospitality of the Max-Planck-Institut für extraterrestrische Physik in Garching, Germany.

REFERENCES

1. D. D. Clayton, in *Essays in Nuclear Astrophysics*, edited by C. A. Barnes, D. D. Clayton, and D. N. Schramm (Cambridge University Press, Cambridge, 1982), p. 401.
2. M. D. Leising, in *Gamma-ray line astrophysics; Proceedings of the International Symposium, Paris, France, Dec. 10-13, 1990* (New York, American Institute of Physics) 1991, p. 173-182.
3. N. Prantzos, *Astronomy and Astrophysics Supplement* **120**, 303 (1996).
4. S. M. Matz et al., *Nature* **331**, 416 (1988).
5. R. Sunyaev et al., *Nature* **330**, 227 (1987).
6. M. D. Leising and G. H. Share, *Astrophys. J.* **357**, 638 (1990).
7. J. Tueller et al., *Astrophys. J.* **351**, L41 (1990).
8. J. Spyromilio, W. P. Meikle, and D. A. Allen, *Mon. Not. R. Astron. Soc.* **242**, 669 (1990).
9. J. D. Kurfess et al., *Astrophys. J. Letters* **399**, L137 (1992).
10. S. E. Woosley and R. D. Hoffman, *Astrophys. J.* **368**, L31 (1991).
11. E. Dwek et al., *Astrophys. J.* **389**, L21 (1992).
12. D. D. Clayton et al., *Astrophys. J. Letters* **399**, L141 (1992).
13. C. Fransson and C. Kozma, *Astrophys. J. Letters* **408**, L25 (1993).
14. Y. Kohmura et al., *Publ. Astron. Soc. Japan* **46**, L157 (1994).
15. H.-U. Zimmermann et al., *Nature* **367**, (1994).
16. T. Suzuki et al., *Astrophys. J. Letters* **419**, L73 (1993).
17. R. A. Chevalier and C. Fransson, *Astrophys. J.* **420**, 268 (1994).
18. M. D. Leising et al., *Astrophys. J. Letters* **431**, L95 (1994).
19. G. G. Lichti et al., *Astronomy and Astrophysics Supplement* **120**, 353–356 (1996).
20. T. Suzuki and K. Nomoto, *Astrophys. J.* **455**, 658 (1995).
21. C. Fransson, P. Lundqvist, and R. A. Chevalier, *Astrophys. J.* **461**, 993 (1996).
22. G. G. Lichti et al., *Astron. Astrophys.* **292**, 569 (1994).
23. M. D. Leising et al., *Astrophys. J.* **450**, 805 (1995).
24. D. J. Morris et al., in *Seventeenth Texas Symposium on Relativistic Astrophysics*, edited by H. Bohringer et al. (New York Academy of Sciences, New York, 1995), p. 397.
25. D. J. Morris et al., this volume (1997).
26. H. Yamaoka, K. Nomoto, T. Shigeyama, and F. K. Thielemann, *Astrophys. J.* **393**, L55 (1992).
27. K. Nomoto, in *IAU Symposium No. 93*, edited by D. Sugimoto, D. Lamb, and D. Schramm (Reidel, Dordrecht, 1980), p. 295.

28. S. E. Woosley, T. A. Weaver, and R. E. Taam, in *Type I Supernovae*, edited by J. C. Wheeler (University of Texas, Austin, 1980), p. 96.
29. M. M. Phillips, *Astrophys. J.* **413**, L105 (1993).
30. M. D. Leising and G. H. Share, *Astrophys. J.* **424**, 200 (1994).
31. S. E. Woosley, R. E. Taam, and T. A. Weaver, *Astrophys. J.* **301**, 601 (1986).
32. P. Höflich, J. C. Wheeler, and A. Khokhlov, *Astrophys. J.*, in press (1997).
33. U. Oberlack *et al.*, in preparation (1997).
34. D. D. Clayton and F. Hoyle, *Astrophys. J. Letters* **187**, L101 (1974).
35. M. D. Leising and D. D. Clayton, *Astrophys. J.* **323**, 159 (1987).
36. M. Hernanz *et al.*, this volume (1997).
37. J. Fishman, Private communication (1993).
38. D. D. Clayton, *Astrophys. J. Letters* **294**, L97 (1981).
39. M. D. Leising, *Astronomy and Astrophysics Supplement Series (ISSN 0365-0138)* **97**, 299 (1993).
40. M. Hernanz, J. Jose, A. Coc, and J. Isern, *Astrophys. J. Letters* **465**, L27 (1996).
41. M. D. Leising, G. H. Share, E. L. Chupp, and G. Kanbach, *Astrophys. J.* **328**, 755 (1988).
42. A. F. Iyudin *et al.*, *Astron. Astrophys.* **300**, 422 (1995).
43. M. J. Harris *et al.*, *Astronomy and Astrophysics Supplement* **120**, 343 (1996).
44. F. Paresce, *IAU Circ. No. 5814* (1993).
45. M. M. Shara, *Astron. J.* **107**, 1546 (1994).
46. S. Starrfield *et al.*, *Astrophys. J. Letters* **391**, L71 (1992).
47. M. Politano *et al.*, *Astrophys. J.* **448**, 807 (1995).
48. J. W. Truran and M. Livio, *Astrophys. J.* **308**, 721 (1986).
49. M. M. Shara and D. Prialnik, *Astron. J.* **109**, 1735 (1995).
50. M. Wiescher, J. Gorres, F. K. Thielemann, and H. Ritter, *Astron. Astrophys.* **160**, 56 (1986).

Supernova Remnants and Plerions in the Compton Gamma-Ray Observatory Era

Ocker C. de Jager* and Matthew G. Baring,[†][1]

*Space Research Unit, PU vir CHO, Potchefstroom 2520, South Africa
†Laboratory for High Energy Astrophysics, Code 661,
NASA Goddard Space Flight Center, Greenbelt, MD 20771, USA

Abstract. Due to observations made by the Compton Gamma-Ray Observatory over the last six years, it appears that a number of galactic supernova remnants may be candidates for sources of cosmic gamma-rays. These include shell-type remnants such as IC443 and γ Cygni, which have no known parent pulsars, but have significant associations with unidentified EGRET sources, and others that appear to be composite, where a pulsar is embedded in a shell (e.g. W44 and Vela), or are purely pulsar-driven, such as the Crab Nebula. This review discusses our present understanding of gamma-ray production in plerionic and non-plerionic supernova remnants, and explores the relationship between such emission and that in other wavebands. Focuses include models of the Crab and Vela nebulae, the composite nature of W44, the relationship of shell-type remnants to cosmic ray production, the relative importance of shock-accelerated protons and electrons, constraints on models placed by TeV, X-ray and radio observations, and the role of electrons injected directly into the remnants by parent pulsars. It appears as if *relic electrons* may be very important in the Vela and Crab remnants. The recent observation of the TeV hot spot in the Vela remnant, which is offset from the current pulsar position, is attributed to relic electrons that were left at the birthplace of the pulsar, the offset being due to the proper motion of the Vela pulsar during its 11,000 year lifetime. We also discuss the role of *freshly-injected* electrons in the remnants around the Crab, Vela and PSR B1706-44 pulsars. These electrons can acquire energies that tap up to at least 10 percent of the full pulsar polar potential, and can produce prominent synchrotron and inverse Compton radiation signatures. The various recent models for predicting gamma-ray emission in shell-type remnants are summarized. The constraining upper limits to TeV emission from such remnants obtained by the Whipple Observatory indicate that either the emission due to particles accelerated at remnant shocks is too faint to be detected by EGRET, or that conditions near their shells (e.g. high density, low magnetic field) limit the acceleration of particles to below a few TeV.

[1] Compton Fellow, Universities Space Research Association

PULSAR-DRIVEN NEBULAE (PLERIONS).

A number of plerions have been discovered in radio, optical, and X-rays [71], with the Crab as the youngest and most energetic source. Plerionic nature is usually indicated by a center-filled morphology, resulting from the continuous injection of pulsar electrons into the nebula, with the additional constraint that the spectra must be non-thermal (power-law) resulting from a statistical acceleration processes. The reason for the latter constraint is that some center-filled X-ray remnants show evidence for thermal emission such as W44 [48], resulting from the presence of hot gas from the central regions.

The radio, optical and X-ray emission observed in plerions is believed to be due to synchrotron emission. Observations map the product of the field strength (to some power) and the number of energetic particles via the synchrotron brightness, however the spectral components from different wavebands cannot be separated unless another emission mechanism is observed from the same particle population. An example of this is inverse Compton scattering, where the magnetic field is replaced with a photon density as a "target" for the high energy particles, and observations of both processes lead to a determination of the field strength. Inverse Compton emission is usually seen at higher energies compared to synchrotron emission, and a multiwavelength study of a source allows us to simultaneously probe the synchrotron and inverse Compton processes, thereby permitting measurement of both the magnetic field strength and the total energy budget of the electrons. In fact, Compton Gamma-Ray Observatory (CGRO) observations of the Crab Nebula provided the first detection of the transition from synchrotron to inverse Compton emission [19], whereas CGRO observations of the Vela remnant are limited to the OSSE instrument (see Fig. 2), and therefore allow us to probe only the synchrotron component of the Vela spectrum.

Constraints on the development of a plerion

The first condition for the development of a synchrotron nebula around a pulsar is that some fraction of the Poynting flux from the pulsar should be transferred to electrons. This means that the ratio σ of Poynting to particle energy fluxes should not be too large. The second condition is that either a shock (at an angular distance r_s), or an instability in the wind (at an angular distance r_o) should remove the electrons from the comoving frame of the pulsar wind, so that the electrons can "see" an effective perpendicular field component for synchrotron radiation [51]. For example, high resolution HST images of the Crab pulsar/nebula shows that the first knots (disturbances in the relativistic unperturbed flow) are seen at a distance of $r_o < 1$" on its polar axis [43], whereas the pulsar wind shock is expected at $r_s \sim 8" - 10"$ [51]. The combination of these two effects (energy transfer and the "visibility" of a field for synchrotron radiation) results in plerionic synchrotron emission outside a minimum distance from the pulsar. The development of a shock also allows the pulsar wind to be slowed down from relativistic speeds to the speeds of typical expanding supernova ejecta.

The Crab Nebula

The Crab Nebula is the prototype cosmic source of synchrotron radiation and inverse Compton (IC) scattering [39,82,18]. Since synchrotron photons are inverse Compton scattered by their parent relativistic electrons, the Crab Nebula is considered to be a synchrotron-self-Compton source.

The Structure of the Nebula

ROSAT and HST observations of synchrotron emission from the Crab Nebula led Hester et al. [43] to make the following fundamental observation about the structure of this nebula: almost all observations of the system at all scale sizes show a well-defined axis of cylindrical symmetry running from the southeast (SE) to the northwest (NW) through the center of the nebula, at an angle tilted by $20° - 30°$ with respect to the plane of the sky. This axis corresponds to the direction of elongation of the nebula as a whole, the axis of the X-ray and optical jets, the X-ray torus (first identified by [3]) and the alignment of the optical "wisps."

Hester et al. [43] also summarized the main properties of the pulsar wind which is responsible for the unpulsed emission over several decades in energy. The symmetry axis is probably associated with the pulsar spin axis, and the DC component of the rotating magnetic field results in a helical polar wind centered on the spin axis. The elongated optical synchrotron nebula appears to be associated with this high latitude structure. Equipartition between particle and field energy is probably reached in the optical nebula, with $B_{\text{optical}} = 3 \times 10^{-4}$ G [43]. This optical nebula is also expected to be the source of IC TeV γ-rays [39,18]. In fact, TeV observations of the Crab Nebula did confirm a field strength of $\sim 3 \times 10^{-4}$ G for the optical/TeV nebula [18]. Closer to the pulsar, [43] identified relatively small optical knots in the polar axis with equipartition field strengths as high as 2 mG, and [19] discussed the possibility that the variable structures near the pulsar may be associated with the variable γ-ray emission seen by EGRET (see also below).

The equatorial zone (identified as the X-ray torus by [3]) extends $\sim \pm 10°$ from the spin equator of the system, with a relatively low field (azimuthally wound up) strength distribution of $B_{\text{torus}} \sim 7 \times 10^{-5} (r/8")^{0.5}$ G for $r > 8"$, as derived by [11] from EINSTEIN HRI images. The X-ray emission up to at least 50 keV is associated with this torus [66], but above this energy we have no imaging capabilities to identify the site of gamma-ray emission, and we have to rely on spectral and temporal characteristics to infer constraints on the properties of the gamma-ray emission site (see [19] for a more detailed discussion of the gamma-ray properties).

The high energy synchrotron tail

CGRO probed the energy range above 50 keV where no imaging is possible, and a comparison between BATSE, OSSE, COMPTEL and EGRET observations of the Crab total emission was made by [62]. Whereas BATSE overestimated the the low energy gamma-ray flux, OSSE and COMPTEL produced consistent results below 1 MeV. The spectral steepening above ~ 100 to 200 keV reported by [49] and [7] is confirmed by OSSE observations. This indicates that the toroidal component terminates gradually above ~ 200 keV.

Conflicting results are however produced above ~ 1 MeV: whereas COMPTEL observed a consistent spectral hardening above 1 MeV, OSSE observed this hardening only during Observation 221 [62]. The flux in the 1–30 MeV range also appears to be variable with time [61], and if this hardening is real, it would be indicative of the presence of another γ-ray emitting site. The rapid variability associated with this component was interpreted by [19] as an emission site near the pulsar where $B \sim 0.1\mu G$. Hester et al. [43] identified the optical knots near the pulsar as shock-like sites with B of the same order as required by De Jager et al. [19].

A comprehensive analysis of EGRET observations of the Crab Nebula was reported by [19] and it was shown that the hard COMPTEL component should cut off above ~ 25 MeV to meet the steep EGRET spectrum between 70 MeV and 150 MeV. De Jager et al. [19] interpreted the steep spectrum between 30 MeV and 150 MeV as the exponential tail of the synchrotron cutoff in the Crab Nebula. The e-folding energy at 25 MeV is consistent with the interpretation that electron acceleration in a relativistic shock occurs at a rate equal to the gyrofrequency. This acceleration is constrained by synchrotron losses, and it was shown that the synchrotron characteristic energy associated with the highest electron energy is independent of the magnetic field strength, and depends only on fundamental constants and a factor $\epsilon \sim 1$ which depends on the Doppler factor and the average electron pitch angle. Thus,

$$h\nu_{\max} = \epsilon \Big(\frac{3}{4\pi}\Big)^2 \frac{hc}{r_e} = \epsilon \frac{9}{8\pi} \frac{m_e c^2}{\alpha_f} \approx 25\epsilon \text{ MeV}, \qquad (1)$$

where r_e is the classical electron radius and α_f is the fine structure constant.

Furthermore, this EGRET component was also found to be variable on a timescale similar to the COMPTEL variability timescale. Thus, whereas the stable emission below 1 MeV is known to be associated with the torus, the variable hard component associated with the synchrotron cutoff leads to the

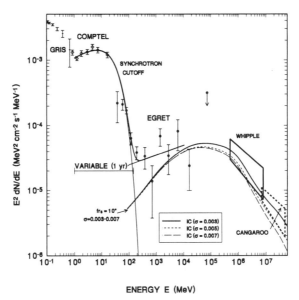

FIGURE 1. The Crab nebular unpulsed γ-ray spectrum ($E^2 dN/dE$) in the energy range 0.1 MeV to 20 TeV. The references are: GRIS [7], COMPTEL [61], & EGRET [19]. References to TeV points are given by [19]. The Whipple error box (including systematic and statistical errors) is from [16], and the CANGAROO error box is from [70]. A two-component fit (1-150 MeV) resulted in a power law with an exponential cutoff at 25 MeV, and an inflection point at 150 MeV. The inverse Compton model of [18] was used to generate spectra for $r_s = 10"$ and three values of σ (as labelled), as shown for energies between 100 MeV and 20 TeV. The model uncertainty is a factor 2 at any energy.

conclusion of an association with a high-B region close to the pulsar, which is removed from the torus. The polar region where high-B optical knots and variable "anvil" is found, is therefore a candidate region for this variable hard component. However, future γ-ray observations above 1 MeV with improved calibration is required to confirm these findings.

The Inverse Compton component

Above ~ 150 MeV, [19] found a steady hard component, which steepens gradually to meet the very high energy spectral points above 100 GeV (see Fig. 1). This steady hard component was predicted to be the inverse Compton component. The *relic* radio emitting electrons also scatter the optical background photons into the ~ 100 MeV to GeV energy range, whereas the younger optical and X-ray emitting electrons are expected to be responsible for the TeV γ-rays. The lifetimes of the radio to X-ray emitting electrons are much longer than the timescale of gamma-ray observations, which implies that we should not expect to see a variable inverse Compton γ-ray component. The observed γ-ray flux above 150 MeV was found to be larger than predicted by the synchrotron-self-Compton model [19]. The EGRET γ-rays above 150 MeV may be associated with inverse Compton γ-ray emission if the field strength in the larger radio nebula is $\sim 1.3 \times 10^{-4}$ G, which is smaller than the field strength in the smaller optical/TeV nebula. This would suggest a departure from the nebular field distribution derived by [51].

The Vela Supernova remnant

The Vela Supernova remnant appears to be associated with the Vela pulsar PSR B0833-45, which has a spindown age of about 11,000 years. A large-scale *ROSAT* image of thermal soft X-rays ($kT = 0.12$ keV) from the Vela SNR shows that the diameter of the remnant is 7.3°, with the pulsar at its center. In hard X-rays we do not see this thermal shell. Rather, the synchrotron nebula is resolved into an elongated (NE-SW) hard X-ray (2.5 - 10 keV) structure as shown by [83], and a compact nebula surrounding the pulsar. The energy-dependent geometry of this remnant was illustrated in Fig. 1 of [21].

The OSSE detection of Vela

Fig. 2 shows the non-thermal X-ray to low energy gamma-ray spectrum of the 1 arcmin compact synchrotron nebula around the Vela pulsar as seen by ROSAT, EINSTEIN, Birmingham and possibly OSSE as reported by [21] and [74]. Even though the OSSE instrument does not have imaging capabilities for a clear association, the spectra of other X-ray sources in the field-of-view cut off below the OSSE range, and only the Vela SNR remains as candidate. Furthermore, since the OSSE spectrum connects smoothly with the lower energy X-ray (imaging) spectra, the association is quite likely. The authors have also shown that the energetics of electrons (given the field strength scaled from the pulsar) is sufficient to produce γ-rays into the low energy part of the EGRET range (Fig. 2). The unbroken spectrum and the EGRET upper limit constrain the maximum electron energy to values smaller than those expected from the 3×10^{15} volt polar cap potential drop. More sensitive observations above 400 keV are needed to measure the cutoff energy, which should lead to a measurement of the maximum electron energy in the nebula.

An inverse Compton component from Vela X?

The relatively hard spectrum associated with the bright Vela X radio nebula (about 1° south of the pulsar) represents synchrotron emission from relic

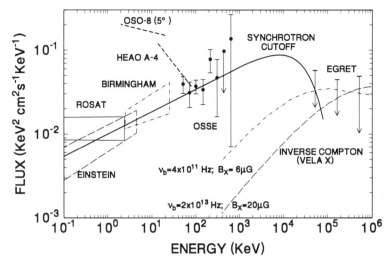

FIGURE 2. The EINSTEIN, ROSAT, & Birmingham X-ray spectra of the 1' Vela compact nebula (see [21] for references). See also ref. [21] for the corresponding OSSE spectrum of the Vela 1' radius compact nebula. The 30-100 MeV EGRET upper limit constrains the extent of this spectrum to an e-folding cutoff energy < 40 MeV [20]. EGRET upper limits above 100 MeV constrain the inverse Compton contribution from Vela X, with the break frequency ν_b and field strength B_X as free parameters. Two such Vela X spectra constrained by EGRET are shown by the dot-dash and long dashed lines (see ref. [20] for details).

electrons in the nebula, while the inverse Compton emission resulting from the scattering of the 2.7K cosmic microwave background produces MeV to GeV γ-rays. This component would be observable if the electron concentration in the Vela X remnant were sufficient. However, we only know the synchrotron brightness of Vela X, which represents a combination of the electron concentration and and the magnetic field distribution. A search for emission in the EGRET data base revealed marginal evidence for excess emission from this direction. De Jager et al. [20] have shown that we should not expect to see MeV to GeV inverse Compton γ-rays from the Vela X SNR if the particles and fields are in equipartition with $B \sim 30\mu G$. This is consistent with the EGRET upper limits, which lie above the flux expected for equipartition.

Fig.2 shows some non-equilibrium inverse Compton spectra which are just below the EGRET upper limits, and it is clear that we must know the maximum electron energy (which is seen as a cutoff at a frequency ν_b in the radio spectrum) before we can give a meaningful lower limit on B. Unfortunately the detection of ν_b in the radio to far-infrared region is hampered by the presence of dust emission in the far IR. Again, a next generation gamma-ray observatory may lead to a detection of Vela X, which should allow a more detailed study of the parameters involved.

Relic electrons at the birthplace of the Vela pulsar?

The CANGAROO telescope has reported unpulsed emission above 2.5 TeV from the Vela pulsar. The peak of this emission is located 8' southeast of the pulsar, which is the projected birthplace of the pulsar, given the 11,000 year spindown age of the pulsar, and the observed proper motion vector [85]. A hint of radio emission at this birthplace is also seen (Frail, D.A. 1996, private communication). Using archival ASCA data, Harding et al. [41] have shown that the X-ray and TeV γ-ray observations indicate the presence of relic electrons left in the trail of the moving pulsar if the magnetic field strength at this site is between 1.3 and 3 μG. Such a low field would have allowed electrons to have survived since the birth of the pulsar. While the X-ray emission is the result of synchrotron emission from relic TeV electrons in a weak field, the TeV γ-ray emission results from the inverse Compton scattering of the same electrons on the cosmic microwave background.

The Unpulsed TeV Source: PSR B1706-44.

Frail, Goss & Whiteoak [35] identified a 4' halo ($\sim$ 60 mJy flux density at 1.5 GHz) of extended plerionic radio emission around the pulsar PSR1706-44. The 20' trailing emission, which, if attributed to proper motion effects, implies that the pulsar and SNR G343.1-2.3 are not associated. Becker, Brazier, & Trümper [8] identified unpulsed power law ($\alpha_x = 1.4 \pm 0.6$ energy index) X-ray emission from this pulsar, with a spatial extent which is consistent with the $\sim$ 0.5' spatial resolution of the ROSAT PSPC. No soft X-ray emission has been detected from the 4' radio halo, and Becker, Brazier, & Trümper [8] made the point that the unresolved X-ray source associated with this pulsar is a compact nebula, similar to the compact nebula of the Vela pulsar.

This pulsar was also seen to be emitting unpulsed TeV γ-rays [54], similar to the Vela TeV detection, which strengthens the conclusion of the similarity between the Vela and PSR1706-44 compact nebulae. However, no unpulsed γ-ray emission in the CGRO range was seen from the direction of PSR1706-44, and the pulsed flux from the pulsar is consistent with the total flux in the EGRET range [80]. The non-detection of γ-rays in the CGRO range is not surprising: any OSSE type emission is expected to be $\sim$ 10 times weaker than the corresponding flux seen from Vela, given the larger distance to PSR1706-44 and similar source parameters. The absence of a radio synchrotron nebula due to relic electrons would also rule out the possibility of an inverse Compton component in the 100 MeV to 1 GeV range.

W44: A COMPOSITE (SHELL + PLERIONIC) SNR

The previous discussion was concerned with the detection of gamma-rays from pulsar-injected electrons (plerions). Some remnants are composite, exhibiting both pulsar/plerionic and shell structures, and we may expect an interesting interaction of electrons injected by the pulsar into the shell, where further electron acceleration may take place, resulting in a bright radio shell. If this remnant is also interacting with a molecular cloud, we may expect relativistic bremsstrahlung to produce gamma-rays in addition to inverse Compton scattering. Furthermore, a γ-ray component from cosmic ray proton acceleration in the shell may also be expected. In fact, the EGRET instrument on the Compton Gamma Ray Observatory detected high energy γ-rays from the vicinity of radio-bright shell-type supernova remnants γ Cygni, IC443, W44 and W28 [34]. These remnants are all associated with molecular clouds, which provide the natural target material for the production of γ-rays via either relativistic bremsstrahlung, or the spallation products of proton-gas interactions (discussed below).

Morphological properties of W44

Fig. 4b (see next section) gives us an indication of the morphology of W44: The EGRET γ-ray source 2EG J1857+0118 [80,34] is located on the eastern side of this remnant, and from Fig. 1 of [22] it is clear that this remnant is also interacting with a molecular cloud on the eastern side, as inferred from ^{13}CO and other molecular line observations [84,24]. The presence of ^{13}CO is indicative of gas densities in excess of 10^3 cm^{-3} in localized interstellar "baseballs," whereas the inter-clump densities are of the order of ~ 1 cm^{-3} or less [69]. De Jager & Mastichiadis [22] found an average hydrogen density of ~ 50 cm^{-3} in the shell of the remnant, which is large enough for the production of γ-rays via various processes.

Whereas the radio is shell-like, the X-ray emission [69] is centrally-peaked, with the radio pulsar PSR B1853+01 offset from the center of the remnant. The association of the pulsar with the remnant is strengthened by the detection of a weak cometary tail (the pulsar wind nebula) pointing towards the center of the remnant [36]. The transverse speed of the pulsar is comparable to the expansion speed of the radio shell [36]. Harrus, Hughes, & Helfand [42] also discovered a smaller X-ray PWN, but the luminosity of the PWN is however negligible compared to the total X-ray luminosity of W44. Arendt [2] identified infrared emission from W44, showing a good spatial correlation with W44. This emission is probably a result of swept-up dust during the SNR expansion, and the infrared energy density is larger than the other galactic radiation fields [22]. This radiation field was included in the inverse Compton calculations of de Jager & Mastichiadis [22].

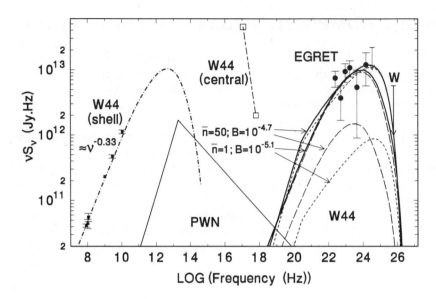

FIGURE 3. The observed multiwavelength spectrum of W44. See ref. [22] for references to data/model fits at all wavelengths. The dot-dash line indicates a model fit to the radio data, which includes a cutoff at $\nu_b = 3 \times 10^{12}$ Hz; open boxes — thermal X-ray energy fluxes of W44 (the thermal turnover at $\nu < kT/h$ is not included); solid circles with error bars — EGRET energy fluxes; upper limit "W" — Whipple upper limit at $\nu = 9 \times 10^{25}$ Hz. *Observed synchrotron spectrum of the pulsar wind nebula*: solid triangle with base between 10^{11} Hz and 10^{20} Hz, with intersection at $\nu_p = 2 \times 10^{13}$ Hz. *Fits to γ-ray data* (for two choices of $\bar{n}$ and B as indicated): short dashed lines — relativistic bremsstrahlung; long dashed lines — inverse Compton; thick solid lines — bremsstrahlung + inverse Compton.

Spectral properties of W44

Fig. 3 shows the spectral properties of W44 as summarized by [22]. The emission from the shell dominates the pulsar emission (indicated by "PWN" in Fig. 3). Whereas the weak radio/X-ray plerionic component is non-thermal, only the radio component of the bright shell is non-thermal. The absence of non-thermal X-ray emission associated with the shell indicates that the radio spectrum must terminate at frequencies $\ll 10^{16}$ Hz. The dot-dash line in Fig. 3 is a model fit through the radio spectrum, and the turnover at 3×10^{12} Hz (constrained by TeV γ-ray observations) assumes an exponential cutoff in the electron spectrum as discussed by [22].

The bremsstrahlung/inverse Compton origin of 2EG J1857+0118

De Jager & Mastichiadis [22] have shown that the same electrons that are responsible for the radio emission can also explain the observed EGRET γ-ray spectrum between 70 MeV and 10 GeV. *The required cutoff in the high frequency radio spectrum accounts for the non-detection of W44 in non-thermal X-rays [69] and TeV γ-rays [58]*. This emission is expected to be concentrated towards the eastern shell where the molecular densities are highest. The dust emission will also contribute to a weaker inverse Compton component [22].

The maximum electron energy in the shell, as derived from multiwavelength observations, is (in terms of the characteristic synchrotron frequency ν_{12} corresponding to the cutoff frequency ν_b)

$$E_{\max} = 0.14\,(B_{-5})^{-1/2}\nu_{12}^{1/2} \text{ ergs.} \tag{2}$$

This maximum energy is orders of magnitude below the maximum obtained for SN1006 - another shell remnant which shows evidence for electron acceleration [59]. The uncertainty in the exact hydrogen density makes it difficult to infer the magnetic field strength from coupled synchrotron and bremsstrahlung equations, but values around 10μG are expected given the approximate hydrogen densities. De Jager & Mastichiadis [22] also found that inverse Compton scattering dominates relativistic bremsstrahlung for molecular densities around 1 particle cm^{-3}, whereas bremsstrahlung dominates for $\bar{n} \sim 50$ cm^{-3}, the density expected for this remnant.

The implications of γ-rays from W44

By integrating the electron energy spectrum up to the maximum electron energy, [22] obtained a total electron energy content of $E_{\rm el} = 5.8 \times 10^{49} B_{-5}^{-1.33}$ ergs, which is about 6 times larger than the value found for SN1006 by [59]. The conversion efficiency ($\eta_{\rm el} = E_{\rm el}/E_{\rm SN}$) of SN explosion energy $E_{\rm SN} = 6.7 \times 10^{50}$ ergs to electrons is then

$$\eta_{\rm el} = 0.087 \left[\frac{6.7 \times 10^{50} \text{ ergs}}{E_{\rm SN}}\right] B_{-5}^{-1.33}, \tag{3}$$

which is relatively high for primary electron acceleration in SNR shocks [32,60]. De Jager & Mastichiadis [22] addressed this problem by investigating the total output from the pulsar PSRB1853+01 since the birth of this $\sim 20,000$ year old SNR. They found that the shell electrons which we are now seeing in radio and γ-ray emission, may have been the result of relic electrons which have been injected into the SNR since birth, provided that the initial spindown power of the pulsar exceeded the present spindown power of the Crab pulsar.

Further support for this interpretation comes from the fact that the radio (and hence electron) spectrum of the shell of W44 is hard and Crab-like, rather than reminiscent of the softer shell-type spectra. Thus, the radio/γ-ray spectrum of the shell of W44 may be the result of relic Crab-like electrons injected into the shell of W44 during the past 20,000 years.

Proton – gas interactions may result in the production of pions from which γ-rays are expected (see below). This process is believed to account for the γ-ray emission from shell remnants interacting with molecular clouds as observed by EGRET. De Jager & Mastichiadis [22] have however shown that the average gas density in W44 is not sufficient to account for a dominant proton contribution to the EGRET γ-rays, and the contribution from this hadronic component is probably not more than 20%. Extrapolation of this relatively weak component to the TeV range may also account for the unobservability of W44 at TeV energies, regardless of whether or not wave damping due to neutrals [26] truncates the proton spectrum below the TeV range.

SHELL-TYPE SUPERNOVA REMNANTS

It has been widely-perceived that supernova remnants (SNRs) are a principal, if not the predominant, source of galactic cosmic rays (e.g. see [57]) up to energies of around $\sim 10^{15}$ eV, where the so-called *knee* in the spectrum marks its deviation from almost pure power-law behaviour. Such cosmic rays are presumed to be produced by diffusive shock (Fermi) acceleration in the environs of supernova shocks. Remnants are a convenient origin for cosmic rays below the knee because their ages (between 100 and 10^5 years) and sizes permit the diffusive process to accelerate up to such high energies, they have the necessary power to amply satisfy cosmic ray energetics requirements, and current estimates of supernova rates in our galaxy can adequately supply the observed cosmic ray density (e.g. see [10]).

The evidence for cosmic ray acceleration in remnants is, of course, circumstantial. Nevertheless, the ubiquity of polarized, non-thermal radio emission in remnants (e.g. see references in the SNR compendium of Green [40]) argues convincingly for efficient acceleration of electrons if the synchrotron mechanism is assumed responsible for the emission. X-rays also abound in remnants, and are usually attributed to thermal emission from shock-heated electrons (because of the appearance of spectral lines, e.g. see [12] for Cas A). The striking spatial coincidence of radio and X-ray images of shell-type remnants (e.g Tycho [10] and SN1006; see [52] for a radio/X-ray correlation analysis for Cas A) suggests that the same mechanism is responsible for emission in both wavebands. This contention has recently received a major boost with the discovery of non-thermal X-ray emission in SN1006 [56], which implies [68] the presence of non-thermal electrons at super TeV energies. In addition, very recent ASCA spectra (Keohane et al. [53]) for the remnant IC 443 and

RXTE observations of Cas A (Allen et al. [1]) exhibit non-thermal X-ray contributions, adding to the collection of super-TeV electron-accelerators. A nice review of radio and X-ray properties of SNRs is given in [33].

A product of cosmic ray acceleration in SNRs is that such energetic particles can generate gamma-rays via interactions with the ambient interstellar medium (ISM), just as in models of the diffuse gamma-ray background [45,46]. Despite theorists' early expectations [44,17] that remnants will be gamma-ray bright, no definitive detection of such emission from a supernova remnant has been reported. Prior to the launch of CGRO, associations of remnants with gamma-ray sources has been limited to two unidentified COS-B sources [67] (for γ-Cygni and W28), however CGRO has played an important role in advancing this field. Our emphasis in the discussions below will be on shell-type remnants such as γ-Cygni, IC 443 and W28.

Gamma-Ray Observations in the CGRO Era

The potential importance of the Compton Observatory's contribution to the study of supernova remnants was identified on a theoretical level by Drury, Aharonian and Völk [27], and in an observational context by Sturner and Dermer [75] in relation to unidentified (UID) EGRET sources. While [75] indicated that the latitudinal distribution of the UID sources may suggest a supernova origin (discussed more extensively in [64]), it was also pointed out that the chance probability of coincidental association for a handful of unidentified EGRET sources with known radio SNR counterparts was small. The best candidates for such associations are presented in the work of Esposito et al. [34], who focused on unidentified EGRET sources (with approximately E^{-2} spectra at above 100 MeV) in or near the galactic plane and proximate to relatively young radio/optical/X-ray-emitting remnants. Such associations, which are at first glance very enticing, suffer from the large uncertainty [34] in exact directional location of the (assumed point) sources for the EGRET detections, of the order of 0.5–1 degrees, i.e. the size of typical nearby remnants (see the images depicted in Figure 4). Hence a definitive connection between *any* of these gamma-ray sources and the young SNRs is not yet possible.

The situation is complicated by the presence of a pulsar (PSR B1853+01) in the field [34,22] of the 95% confidence contour of the EGRET source 2EG J1857+0118, whose association with the remnant, W44, is discussed above. Such a pulsar, or its wind nebula, could easily generate the observed gamma-ray emission, although no evidence of pulsation exists in the EGRET data [79]. There is also the recent suggestion [14] of a pulsar counterpart to the CTA 1 remnant's EGRET source 2EG J0008+7307. In addition, the improvement of the localization of 2EG J2020+4026 by the consideration of only super-GeV photons leads to the conjecture [13] that this source is not associated with the shell of γ Cygni, but rather with a distinct ROSAT source that may also be a

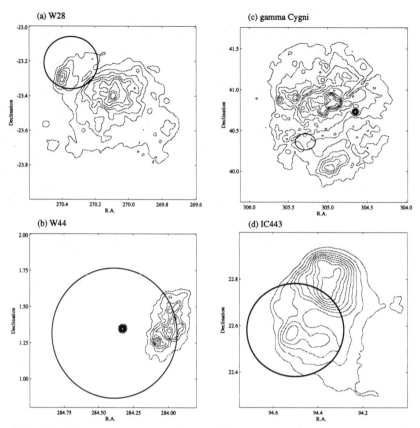

FIGURE 4. The X-ray/gamma-ray images of the supernova remnants W28, W44, γ-Cygni and IC443, as presented in Esposito et al. [34]. These consist of X-ray contours from the ROSAT telescope's HRI, and the 95% (elliptical) confidence contours of emission above 100 MeV in the associated EGRET unidentified sources.

pulsar. The possibility that pulsars could be responsible for most unidentified EGRET sources near the galactic plane [50] (see also [64]) currently precludes any assertions stronger than just suggestions of a remnant/EGRET source connection. Notwithstanding, it is quite possible that such remnants could plausibly emit gamma-rays at levels below EGRET's sensitivity.

The absence of TeV emission associated with the remnants surveyed by Esposito et al. [34], as reported by the Whipple collaboration [58,15], is as important for this field as the detections embodied in the EGRET unidentified sources. Upper limits obtained by Whipple to a number of remnants can severely constrain models, dramatically impacting the hypothesized contributions of hadronic processes, bremsstrahlung and inverse Compton scattering. Very recently, the CANGAROO experiment detected [78,55] the barrel-shaped remnant SN1006 above 3 TeV, a source for which the existence of highly super-TeV electrons has already been established [56,68] from the X-rays. Future observations in the TeV band will provide powerful model diagnostics.

Modelling of Gamma-Ray Production in SNRs

The modelling of γ-ray emission from supernova remnants was limited to a few very preliminary analyses (e.g. [44,17,9]) prior to the launch of CGRO. This field of research began in earnest with the seminal paper of Drury, Aharonian and Völk [27], who computed (as did ref. [65]) the photon spectra expected from the decay of neutral pions generated in collisions between shock-accelerated ions and cold ions in the ISM. Since then, there has been a small flurry of activity, with different groups using alternative approaches, extending the considerations to include bremsstrahlung and inverse Compton emission. Here we review the handful of SNR gamma-ray emission models developed over the last four years that invoke shock acceleration.

Drury et al. [27], determined the photon spectra and fluxes expected from the decay $\pi^0 \to \gamma\gamma$ of neutral pions generated in hadronic collisions ($pp \to p\pi^0 X$ etc.) between power-law shock-accelerated ions and cold ions in the ISM; they neglected the other (electromagnetic) processes mentioned just above. Due to the isotropy of decay in the pion rest frame, the decay kinematics yield [72] a photon spectrum that is symmetric about $m_\pi/2 \approx 67\,\text{MeV}$, an unmistakable signature of the production of pions in astrophysical systems (see [45,46] for its role in determining the gamma-ray background spectrum). Supernova remnant models of pion production and decay normally use some variant of a hybrid approach (e.g. see Dermer [23]), where low energy pion creation (for shock-accelerated proton momenta p_p below around 3 GeV/c) is mediated by an isobaric state $\Delta(1232)$ (Stecker [72]), or a collection of different states, and the complexities of pion creation at high energies (for $p_p \gtrsim 10$ GeV/c) is described by some adaptation [73,77] of Feynman scaling. Drury et al. used the *two-fluid* approach [28,25,4] to explore shock acceleration hydrodynamics, treating the cosmic rays and thermal ions as separate entities (electrons go along for the ride). This technique, which is extremely useful for time-dependent problems such as SNR expansions, obtains solutions that conserve particle number, momentum and energy fluxes, thereby describing some of the non-linear effects [47,4] of diffusive shock acceleration. Drury et al. determined that the luminosity peaked in the Sedov phase, in accord with maximal shock dissipation arising when the supernova ejecta is being compressed and significantly decelerated by the ISM.

In order to match the EGRET flux, Drury et al.'s pion decay model requires a *high target density* ($> 100\,\text{cm}^{-3}$). Gamma-ray bright remnants might therefore be expected to border or impinge upon dense regions of the ISM, perhaps giant molecular clouds, in accord with the earlier suggestions of [9].

Drury et al. [27] predicted that such remnants should become limb-brightened with age, an effect that arises because, as the shock weakens with time, the dominant γ-ray flux is always "tied" somewhat to a region near the shock. While such a limb-brightening is seen in radio and X-ray images of remnants (e.g. Tycho and SN1006), higher angular resolution observations are needed in gamma-ray telescopes before its existence, or otherwise, can be probed at high energies. Such a definitive connection of the γ-ray emitting regions to a remnant's shell (which may be ruled out for γ Cygni according to [13]) would argue strongly for a shock-acceleration origin of the energetic particles responsible for emission in the gamma-ray and other wavebands. Note that Drury et al. did not incorporate physical (spatial and temporal) limits imposed [57,59,68,4,6] on the shock acceleration mechanism by the supernova shell, so that they permitted particles to be accelerated to at least 100 TeV. This omission promoted observational investigations by the Whipple collaboration that produced upper limits in the TeV energy range [58,15] that contradicted the Drury et al. predictions. While this conflict has been proposed as a failure for shock acceleration models of SNRs, realistic choices [22,76,6] of the maximum energy E_{max} of particle acceleration actually yield model spectra that are quite compatible with Whipple's observational constraints to γ Cygni and IC 443.

A number of substantial model developments has ensued since Drury et al.'s enunciative work. Among these was the work of Gaisser, Protheroe and Stancv [37]. They computed emission fluxes and luminosities for the decay of π^0s produced in hadronic collisions, bremsstrahlung and inverse Compton scattering, however they omitted consideration of non-linear shock dynamics in any form, did not treat time-dependence, and assumed test-particle power-law distributions of protons and electrons. In their model, the inverse Compton scattering used both the microwave background and an infrared/optical background field local to the SNRs as seed soft photons. Their bremsstrahlung component was due to cosmic ray electrons colliding with ISM protons. Their model has difficulty with the TeV upper limits obtained by Whipple, unless sufficiently steep particle distributions are assumed. Gaisser et al. imposed a high matter density ($> 300\,\mathrm{cm}^{-3}$) to enhance the bremsstrahlung and π^0 decay to inverse Compton flux ratio, thereby generating steeper spectra for the sources associated with γ Cygni and IC443. Note that for all models discussed here, the $\pi^\pm \to e^\pm$ secondaries are always unimportant for the SNR problem since the ion cooling time in pion production is much longer than typical remnant ages.

Recently Sturner et al. [76] have developed a time-dependent model, where they solve for electron and proton distributions subject to cooling by inverse Compton scattering, bremsstrahlung, π^0 decay and synchrotron radiation. Like Gaisser et al., the work of Sturner et al. assumes canonical power-laws but does not include any treatment of non-linear shock acceleration effects. One feature of their model is the dominance of inverse Compton emission, which intrinsically has a flatter spectrum than either bremsstrahlung or pion decay radiation. This arises because they generally opt to have the same energy density in non-thermal electrons and protons, so that the shock-accelerated electrons are more populous than their proton counterparts. Sturner et al.'s work introduced cutoffs in the distributions of the accelerated particles (first done by [59]), which are defined by the limits on the achievable energies E_{max} in Fermi acceleration. Hence, given suitable model parameters, Sturner et al. can accommodate the constraints imposed by Whipple's upper limits [58] to γ Cygni and IC 443.

The most recent development among gamma-ray production models has been the work of Baring et al. [5,6] on the application of non-linear shock acceleration theory to the SNR problem, an appropriate step given that remnant shocks are strong enough that the generated cosmic rays are endowed with a significant fraction of the total particle pressure. This work utilizes the steady-state Monte Carlo simulational approach (described in the reviews of [47,4]), a kinematic technique that can self-consistently model the feedback of the accelerated particles on the spatial profile of the flow velocity, which in turn determines the shape of the particle distribution. In establishing this feedback, the accelerated population pushes on the upstream plasma and decelerates it before the discontinuity is reached, so that an upstream *precursor* forms, in which the flow speed is monotonically decreasing. At the same time, the cosmic rays press on the downstream gas, slowing it down too. The overall effect is one where the total compression ratio r, from far upstream to far downstream of the discontinuity, actually *exceeds that* (i.e. 4) *of the testparticle scenario*, the case where the canonical power-laws used in [27,37,76] are generated. This situation results from the need of the flow to increase r to adjust for energy and momentum escape [29,31]. If the particle diffusive scale (i.e. mean free path λ) is an increasing function of momentum, as is expected to be the case [4] based on inferences of particle diffusion from the Earth's bow shock and also in hybrid plasma shock simulations [38], then higher energy particles will sample a stronger shock, yielding upward curvature in the nonthermal cosmic ray distribution. This curvature is important for gamma-ray emission models, since it introduces enhancements [5] in the TeV range by factors of 2–3 relative to the EGRET range; such increases can be the difference between detection and non-detection by air Čerenkov experiments like Whipple, CAT, CANGAROO and HEGRA. The curvature and the modification of the flow hydrodynamics depend on each other intimately in a highly non-linear fashion. Typical distributions of particles that are accelerated in *cosmic ray modified* shocks are presented in numerous papers [29,31,47,30,6].

The self-consistent Monte Carlo approach to shock acceleration in [6] includes neutral pion decay emission, bremsstrahlung and inverse Compton emission components. The cessation of acceleration above $E_{max} \sim 1$ TeV - 10 TeV range caused by the spatial and temporal limitations of the expanding SNR shell yields gamma-ray spectral cutoffs that are consistent with the

Whipple TeV upper limits [58,15]. The Monte Carlo approach generates particle diffusion scales that are always much less than the remnant's shock radius (as in [27]) so that the effects of shock curvature can be neglected. This may also render the lack of time-dependence in the technique a less relevant limitation. A prominent feature of the model of [6] is the low value of the electron to proton ratio above 1 GeV, due to a sensible description of the injection of thermal electrons into the acceleration process. This description, which models the way particles diffuse in turbulent plasma environments, guarantees [4] that the electron distribution is steep enough at low energies so as to render the e/p ratio much less than unity above 1 GeV. This determination is entirely consistent with the observation that electrons supply around 2% of the cosmic ray population by number [63], and also blends with limits on the local e/p abundance ratio imposed when modelling the galactic gamma-ray background [45]. This contrasts the situation of [76]. Note that while bremsstrahlung is more efficient than pion decay emission for given cosmic ray electron and proton energies, the emergent bremsstrahlung component can be inhibited if the e/p ratio is low. Future measurements of the unidentified sources by more sensitive experiments in the 1–100 MeV range should constrain the e/p ratio.

While the focus here has been on gamma-rays from shell-type remnants, much can be learned from studying other wavebands also. This has been the approach of Mastichiadis and De Jager [59], who have studied the remnant SN1006. For SN1006, which has not been seen in gamma-rays, they used [59] the recent observations [56] of non-thermal X-rays by ASCA to constrain the energy of electrons and the magnetic field, interpreting the X-ray flux as being of synchrotron origin. This contention (see also Reynolds [68]) assumes that the steep X-ray spectrum is part of a rollover in the electron distribution at energies around 100 TeV. Using microwave and infrared backgrounds appropriate to SN1006, [59] predicted the resulting inverse Compton component in γ-rays, and determined that it would always satisfy the EGRET upper bounds. However, they concluded that TeV upper limits from experiments like Whipple could potentially constrain the ratio $\eta = \lambda/r_g$ of the electron mean free path λ to its gyroradius r_g to values signifying departure from Bohm diffusion (i.e. $\eta \gg 1$), otherwise the TeV flux would exceed that of the Crab nebula. Such a conclusion appears to be borne out by the very recent announcement by Tanimori et al. [78,55] of the detection of SN1006 above 3 TeV by the CANGAROO experiment, with the flux at these energies probably being due to an inverse Compton component. Pinning the X-ray synchrotron spectrum determines $E_{\max}^2 B$ and also a combination of B and the electron density, where $E_{\max}$ is the maximum accelerated electron energy. Through $E_{\max}$, η couples to B so that the gamma-ray inverse Compton flux anti-correlates with both B and $\eta = \lambda/r_g$. This interplay between the wavebands (see also [76]) will play an important role in future model developments for shell-type remnants.

CONCLUSION AND A LOOK AHEAD

The Compton Gamma-Ray Observatory has propelled the study of supernova remnants, plerionic and non-plerionic, into the foreground of gamma-ray astrophysics. The various aspects of the plerions, the shell-type remnants, and the W44 composite discussed in this review serve to underline the diversity of the handful of definitive or candidate γ-ray remnants observed by CGRO. Such a diversity is also reflected in their morphological properties, their optical/IR spectra, environmental densities, etc. It follows that no two sources seem the same so that they must be considered on a case-by-case basis. While the plerions can easily derive their luminosity from the parent pulsar, perhaps the proximity of the non-plerionic sources to dense molecular clouds of various sorts provides a strong clue to the reason for their γ-ray emission. It is clear that if some of the EGRET detections turn out to be of gamma-rays generated in the environs of remnant shells, then gamma-ray emitters must be a minority of remnants, perhaps mostly young, given that they cannot produce ions above around a few TeV in profusion. Remnants that provide cosmic rays up to the knee must consequently be a gamma-ray quiet majority. Alternatively, if fluxes of shell origin are well below EGRET's flux sensitivity, then the notion that shell-type remnants are simultaneously gamma-ray bright and prolific producers of cosmic rays becomes tenable. It has therefore become evident that the Whipple upper limits have not destroyed the hypothesis that shocks in shell-type remnants energize the particles responsible for the gamma-ray emission, but rather have provided a powerful tool for constraining our understanding. Much remains to be explored in this field, in particular the relationship between the clouds and the shock parameters, the degree of ionization of the environment, the precise location of the gamma-ray emission, differentiation between plerion-driven and shock-powered gamma-ray sources, and the maximum energies and relative abundances of the produced cosmic rays. The next generation of both space-based and ground-based gamma-ray telescopes, with better angular resolution and cumulatively-broad spectral range will have a significant impact on this field, particularly in coordination with X-ray and radio observations.

Acknowledgments: We thank our collaborators Alice Harding, Apostolis Mastichiadis, Don Ellison, Steve Reynolds and Isabelle Grenier for many informative discussions about supernova remnants and shock acceleration theory. We also thank Joe Esposito for providing the images used in Figure 4.

REFERENCES

1. Allen, G. E. et al. 1997, *Astrophys. J.* in press.
2. Arendt, R. G. 1989, *Astrophys. J. Supp.* **70**, 181.
3. Aschenbach, B., & Brinkmann, W. 1975, *Astron. Astr.* **41**, 147.

4. Baring, M. G. 1997 in *Proc. Les Arcs Moriond Workshop*, in press (2 reviews).
5. Baring, M. G., Ellison, D. C. & Grenier, I. A., in *The Transparent Universe*, eds. Winkler, C. et al. (ESA, SP-382, ESA, Noordwijk) p. 81.
6. Baring, M. G., et al. 1997, in these Proceedings.
7. Bartlett, L. M. 1994, Ph.D. Thesis, University of Maryland, unpublished.
8. Becker, W., Brazier, K. T. S. & Trümper, J. 1995, *Astron. Astr.* **298**, 528.
9. Blandford, R. D. & Cowie, L. L. 1982, *Astrophys. J.* **260**, 625.
10. Blandford, R. D. & Eichler, D. 1987, *Phys. Reports* **154**, 1.
11. Bork, T. 1989, Ph.D. Thesis, Ludwig-Maximilians-Universität, unpublished.
12. Borkowski, K. J. et al. 1996, *Astrophys. J.* **466**, 866.
13. Brazier, K. T. S., et al., 1996, *MNRAS* **281**, 1033.
14. Brazier, K. T. S., et al., 1997, *MNRAS* submitted.
15. Buckley, J. H., et al., 1997, *Astron. Astr.* in press.
16. Carter-Lewis, D. A., et al. 1997, Proc. 25th Int. Cosmic Ray Conf. (Durban), **3**, 161.
17. Chevalier, R. A. 1977, *Astrophys. J.* **213**, 52.
18. de Jager, O. C., & Harding, A. K. 1992, *Astrophys. J.* **396**, 161.
19. de Jager, O. C., et al. 1996a, *Astrophys. J.* **457**, 253.
20. de Jager, O. C., Harding, A. K., Sreekumar, P., & Strickman, M. 1996, *Astron. Astrophys. Suppl. Ser.* **120**, C441.
21. de Jager, O. C., Harding, A. K., & Strickman, M. 1996, *Astrophys. J.* **460**, 729.
22. de Jager, O. C., & Mastichiadis, A. *Astrophys. J.* **482**, 874.
23. Dermer, C. D. 1986, *Astron. Astr.* **157**, 223.
24. Dickel, J. R., Dickel, H. R., & Crutcher, R. M. 1976, *Pub. Astron. Soc. Pacific* **88**, 840.
25. Drury, L. O'C. 1983, *Rep. Prog. Phys.* **46**, 973.
26. Drury, L. O'C., Duffy, P. & Kirk, J. G. 1996, *Astron. Astr.* **309**, 1002.
27. Drury, L. O'C., Aharonian, F. A. & Völk, H. J. 1994, *Astron. Astr.* **287**, 959.
28. Drury, L. O'C., & Völk, H. J. 1981, *Astrophys. J.* **248**, 344.
29. Eichler, D. 1984, *Astrophys. J.* **277**, 429.
30. Ellison, D. C., Baring, M. G., & Jones, F. C. 1996, *Astrophys. J.* **473**, 1029.
31. Ellison, D. C., & Eichler, D. 1984, *Astrophys. J.* **286**, 691.
32. Ellison, D. C., & Reynolds, S. P. 1991, *Astrophys. J.* **382**, 242.
33. Ellison, D. C. et al. 1994, *Pub. Astron. Soc. Pacific* **106**, 780.
34. Esposito, J. A., et al., 1996, *Astrophys. J.* **461**, 820.
35. Frail, D. A., Goss, W. M., & Whiteoak, J. B. Z. 1994, *Astrophys. J.* **437**, 781.
36. Frail, D. A. et al. 1996, *Astrophys. J.* **464**, L165.
37. Gaisser, T. K., Protheroe, R. J. & Stanev, T. 1997, *Astrophys. J.* in press.
38. Giacalone, J., et al., 1993, *Astrophys. J.* **402**, 550.
39. Gould, R. J. 1965, *Phys. Rev. Lett.*, **15**, 577.
40. Green, D., SNR catalogue, http://www.mrao.cam.ac.uk/surveys/snrs/
41. Harding, A.K., de Jager, O.C., & Gotthelf, E. 1997, Proc. 25th Int. Cosmic Ray Conf. (Durban), **3**, 325.
42. Harrus, I. M., Hughes, J. P., Helfand, D. J. 1996, *Astrophys. J.* **464**, L161.

43. Hester J. J., et al. 1995, *Astrophys. J.* **448**, 240.
44. Higdon, J. C. & Lingenfelter, R. E. 1975, *Astrophys. J. (Lett.)* **198**, L17.
45. Hunter, S. D., et al. 1997, *Astrophys. J.* **481**, 205.
46. Hunter, S. D., Kinzer, R. L., & Strong, A. W. 1997, in these Proceedings.
47. Jones, F. C. & Ellison, D. C. 1991, *Space Sci. Rev.* **58**, 259.
48. Jones, L. R., Smith, A., & Angellini, L. 1993, *MNRAS* **265**, 631.
49. Jung, G. V. 1989, *Astrophys. J.* **338**, 972.
50. Kaaret, P. & Cottam, J. 1996, *Astrophys. J. (Lett.)* **462**, L35.
51. Kennel, C. F. & Coroniti, F. V. 1984, *Astrophys. J.* **283**, 694.
52. Keohane, J. W., Rudnick, L. & Anderson, M. A. 1996, *Astrophys. J.* **466**, 309.
53. Keohane, J. W., et al. 1997, *Astrophys. J.* **484**, 350.
54. Kifune, T. et al. 1995, *Astrophys. J.* **438**, L91.
55. Kifune, T. et al. 1997, to appear in *Towards a Major Atmospheric Čerenkov Detector*, ed. O. C. de Jager (Wesprint, Pochefstroom).
56. Koyama, K. et al. 1995, *Nature* **378**, 255.
57. Lagage, P. O. & Cesarsky, C. J. 1983, *Astron. Astr.* **125**, 249.
58. Lessard, R. W., et al. 1995, Proc. 24th Int. Cosmic Ray Conf. (Rome), **2**, 475.
59. Mastichiadis, A. & de Jager, O. C. 1996, *Astron. Astr.* **311**, L5.
60. Mastichiadis, A. 1996, *Astron. Astr.* **305**, L53.
61. Much, R. P., et al. 1995, *Astron. Astr.* **299**, 435.
62. Much, R. P., et al. 1996, *Astron. Astrophys. Suppl. Ser.* **120**, C703.
63. Müller, D., et al. 1995, *Proc. 24th ICRC (Rome)* **III**, 13.
64. Mukherjee, R., Grenier, I. A., & Thompson, D. J. 1997, in these Proceedings.
65. Naito, T. & Takahara, F. 1994 *J. Phys. G: Nucl. Part. Phys.* **20**, 477.
66. Pelling, R. M. et al. 1987, *Astrophys. J.* **319**, 416.
67. Pollock, A. M. T. 1985, *Astron. Astr.* **150**, 339.
68. Reynolds, S. P. 1996, *Astrophys. J. (Lett.)* **459**, L13.
69. Rho, J. et al. 1994, *Astrophys. J.* **430**, 757.
70. Sakurazawa, K. 1997, Proc. 25th Int. Cosmic Ray Conf. (Durban), **3**, 165.
71. Seward, F. D. 1989, *Space Sci. Rev.* **49**, 385.
72. Stecker, F. W. *Cosmic Gamma Rays*, (NASA SP-249, NASA, 1971)
73. Stephens, B. A., & Badhwar, G. D. 1981, *Astrophys. Sp. Sci.* **76**, 213.
74. Strickman, M., de Jager, O. C., & Harding, A. K. 1996, *Astron. Astrophys. Suppl. Ser.* **120**, C449.
75. Sturner, S. J. & Dermer, C. D. 1995, *Astron. Astr.* **293**, L17.
76. Sturner, S. J., et al., 1997, *Astrophys. J.* in press.
77. Tan, L. C. & Ng, L. K. 1983, *J. Phys. G: Nucl. Part. Phys.* **9**, 1289.
78. Tanimori, T., et al. 1997, IAU Circ. 6706.
79. Thompson, D. J. et al. 1994, *Astrophys. J.* **436**, 229.
80. Thompson, D. J. et al. 1996, *Astrophys. J. Supp.* **465**, 385.
81. Vacanti, G., et al. 1991, *Astrophys. J.* **377**, 467.
82. Weekes, T. C. et al. 1989, *Astrophys. J.* **342**, 379.
83. Willmore A. P., et al., 1992, *MNRAS* **254**, 139.
84. Wootten, H. A. 1977, *Astrophys. J.* **216**, 440.
85. Yoshikoshi, T. 1996, Ph.D. Thesis, Univ. of Tokyo, unpublished.

Diffuse Galactic Continuum Radiation

Stanley D. Hunter,[1] Robert L. Kinzer,[2] and Andy W. Strong[3]

[1] NASA/Goddard Space Flight Center, Greenbelt, MD 20771
[2] Naval Research Laboratory, Washington, D.C. 20375
[3] Max-Planck-Institute für Extraterrestriche Physik, Garching, Germany

Abstract. The diffuse gamma-ray continuum emission, which arises from cosmic-ray protons and electrons interacting with the interstellar matter and low energy photons, is the dominant feature of the gamma-ray sky. This emission has been studied from 0.05 MeV to over 50 GeV using the observations from OSSE, COMPTEL, and EGRET on the Compton Gamma-Ray Observatory. The OSSE observations show the diffuse continuum to be a composite of three independent components: e^+e^- anihilation line plus continuum, a soft low-energy component, and a dominant, hard component. The COMPTEL results are the first measurement of the diffuse emission at intermediate energies and the EGRET results have provided a much clearer image of the spatial and spectral distribution of this emission at higher energies. Although significant contributions from unresolved distributions of point sources cannot be excluded, particularly from the COMPTEL and OSSE data, the spectra from the three instruments, from 1 MeV to about 1 GeV, can be explained with the sum of bremsstrahlung, nucleon-nucleon (π^0 decay), and inverse Compton components. However, above about 1 GeV, the EGRET spectrum is harder than expected from the nucleon-nucleon spectrum derived from an $E_p^{-2.7}$ cosmic-ray proton spectrum observed in the Solar neighborhood.

These observations and results are discussed in terms of the correctness of our interpretations of the observations, what does and what does not conform to the standard picture, the significance of these observations for cosmic-ray physics, and how much of the observed flux is due to point sources.

INTRODUCTION

Prior to the launch of the *Compton Gamma-Ray Observatory* (CGRO) on April 5th 1991, a number of non-simultaneous observations of the Galactic diffuse emission were made with a variety of instruments covering various and generally non-overlapping energy ranges. These include observations in the 0.3–8.5 MeV energy range with the Gamma-Ray Spectrometer on the *Solar Maximum Mission* [1], at higher energies (35 MeV – 150 MeV) with SAS 2 [2,3] and with COS B (50 MeV – 6 GeV) [4,5]. There were no observations made in the intermediate (10 – 50 MeV) energy range. A detailed review of the analyses of these observations above about 1 MeV is given by Bloemen [6].

This paper provides an overview of the recent observations made with the OSSE, COMPTEL, and EGRET instruments on CGRO, covering over six decades in gamma-ray energy (50 keV – > 50 GeV). The BATSE instrument is also sensitive to the diffuse gamma radiation over its energy range from 0.02 MeV to 100 MeV. However, the results of a long term project to extract the diffuse emission from the BATSE observations [7] are not yet available.

The OSSE, COMPTEL, and EGRET observations are providing us with a much clearer understanding of the spectrum and spatial distribution of the Galactic diffuse gamma radiation. This review examines the scientific results from these observations and examines our interpretations: how these results conform, or fail to conform, with the standard picture; the significance for cosmic-ray physics; and the significance of the contribution from point sources.

CGRO OBSERVATIONS AND RESULTS

OSSE – Oriented Scintillation Spectrometer Experiment

The OSSE instrument consists of four identical phoswich spectrometers which are sensitive to gamma rays from about 50 keV to 10 MeV. Passive tungsten slat collimators define a $3.8° \times 11.4°$ FWHM aperture for each detector. A detailed description of the OSSE instrument is given by Johnson et al. [8].

Purcell et al. [9] and Kinzer et al. [10] describe OSSE observations of the central region of the Galactic plane ($b = 0°$) for which the large angle dimension of the OSSE collimator was aligned with the plane of the Galaxy. This alignment maximizes the signal from the extended Galactic ridge emisison while minimizing contamination of plane emission in the background measurements, at $|b| = 10°$.

The high statistical quality of the OSSE observations of the Galactic center allows detailed investigations of various spectral components over the 0.050 to 10 MeV energy range. Kinzer et al. [10] fit various photon models to the spectra by folding the model through the instrument response function, on the assumption of a 5° FWHM width for the Galactic plane emission, and comparing the resultant count spectrum to the observed spectra. The model spectra contained at least four components: a hard cosmic-ray induced component, a soft low-energy component which includes a discrete source contribution, a positron annihilation line at 0.511 MeV, and a positronium continuum annihilation component. The 1.809 MeV ^{26}Al line has an intensity below the sensitivity of a single two-week observation and was not included.

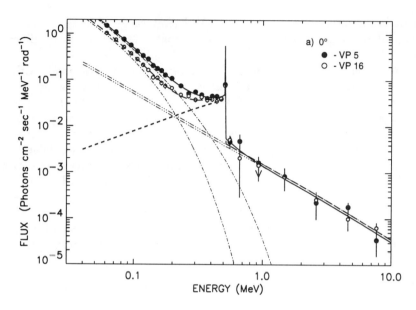

FIGURE 1. OSSE spectra of the diffuse emission from the Galactic center region ($l = 0°$, $b = 0°$, VP 5 and VP 16). The best fit four component model of Kinzer et al. [10] is shown as a solid line. The model components are: exponentially absorbed power-law, dash-doted line; positronium continuum and 2.5 keV wide narrow line at 0.511 MeV, dashed line; and cosmic-ray interaction continuum model of Skibo [11], dash-triple dotted line.

The best fit to the OSSE spectrum from the Galactic center region of the plane (Fig. 1) was obtained by combining the high-energy continuum model of Skibo [11] with an exponentially absorbed power-law model for the low-energy component and models for positron 511 keV annihilation line and the positronium continuum. In the two Galactic center spectra shown in Figure 1, the soft low-energy component fluxes differ by a factor of about 1.7. This strong variation below 0.2 MeV is attributed to one or more bright and highly variable point sources near the Galactic center.

Since the OSSE instrument is non-imaging, it is sensitive to both the diffuse emission and compact sources in the aperture. Purcell et al. [9] used coordinated observations of the Galactic center region made with the imaging SIGMA instrument [12] to separate the discrete source contribution from the diffuse emission observed with OSSE over the energy range 50 keV – 4.5 MeV. Purcell et al. [9] fit a similar model (consisting of a single power law, a photopeak line fixed in energy at 511 keV with a width of 2.5 keV, and a positronium continuum component) to this source subtracted spectrum. The fitted power law spectral index was -2.4 ± 0.1, with a continuum flux of

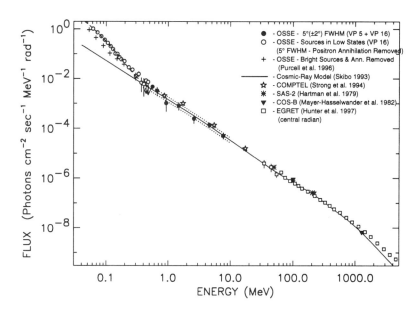

FIGURE 2. Comparison of the OSSE, COMPTEL, and EGRET estimates of the diffuse emission from the Galactic center region without the positron annihilation components and bright source component. The solid line shows the cosmic-ray interaction continuum model of Skibo [11] fit to the OSSE observations for a 5° FWHM Galactic latitude width.

$(1.61 \pm 0.08) \times 10^{-2}$ cm^{-2}s^{-1}MeV^{-1} at 100 keV. This analysis showed that the spectrum below 0.3 MeV towards the Galactic center has roughly the same intensity as the emission at $l = 25°$ and $l = 339°$ after correcting for the discrete sources resolved by SIGMA. This suggests that the residual soft low-energy component of the Galactic diffuse emission is broadly distributed in Galactic longitude and that bright discrete sources contribute significantly to the Galactic center spectrum below 500 keV.

The OSSE spectrum of the plane toward the Galactic center, unfolded on the assumption of a latitude width of 5° FWHM and corrected for the discrete source component, with the positron-annihilation components removed, is shown in Figure 2. Also included are recent results from COMPTEL and EGRET, described below, and earlier findings. The figure shows the best fit model for the soft low-energy component and the cosmic-ray continuum model of Skibo [11]. The Skibo model, as fit to the OSSE data, is shown over the entire energy range. The agrement between leaky-box model of Skibo and the OSSE, COMPTEL, and EGRET data, up to 1 GeV, is fairly good. The discrepancy above 1 GeV and more physical 3-dimensional models are discussed below.

COMPTEL – Imaging Compton Telescope

The COMPTEL instrument is the first medium-energy imaging gamma-ray telescope ever flown on a satellite. It is sensitive to gamma rays in the energy range from 0.75 – 30 MeV, which overlaps the high end of the OSSE energy range and adjoins the low end of the EGRET energy range. Gamma rays in this energy range are produced by cosmic-ray electrons interacting via electron bremsstrahlung and inverse Compton scattering of low-energy photons. Photons in this energy range are difficult to detect; the double Compton scattering detection technique used in the COMPTEL instrument and its large size [13] make it about 20 times more sensitive than prior detectors in this energy range [14].

Derivation of the Galactic diffuse emission spectrum using the COMPTEL data requires fitting a model to the data and is hence sensitive to the handling of the instrumental response and background. The Galactic center spectrum recently derived by Bloemen et al. [15] is compared with earlier results of Strong et al.; in Figure 3a, the early results of Strong et al. [16] are in agreement, but the more recent estimates [17,18] are too high at 1–10 MeV. The most recent values by Strong et al. [19] (Fig. 3b) are lower at 1–10 MeV, but are more reliable than the earlier ones. The higher values in [17,18] were due to inadequate handling of instrumental background.

Interpretation of the diffuse emission spectrum in the medium-energy range has evolved considerably since the launch of CGRO. In 1994 electron bremsstrahlung was assumed to be the dominant production mechanism [20,16]. Later work [18] showed that inverse Compton emission is an important component. The current picture [19] indicates that there may be much

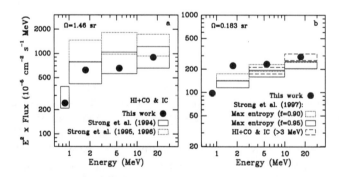

FIGURE 3. (a) COMPTEL broad-band spectra of the inner Galaxy ($300° < l < 60°$, $|b| < 20°$), as recently determined by Bloemen et al. [15] (black dots), compared to previous results of Strong et al. [16–18], fitting the same H I and CO observations and IC intensity model (IC not used in [16]). (b) Comparison with the most recent results of Strong et al. [19] ($330° < l < 30°$, $|b| < 5°$).

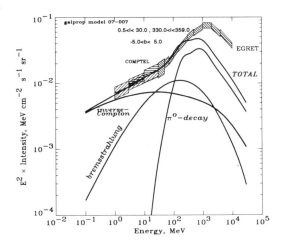

FIGURE 4. The diffuse gamma-ray spectrum for the inner Galaxy, $330° < l < 30°$, $|b| < 5°$, from [19]. ///: from COMPTEL maximum entropy images; ≡: free fit of COMPTEL data to model based on gas + inverse Compton. \\\: EGRET data [34]. Curves show the bremsstrahlung, inverse Compton and π^0-decay components and total for the physical model described in [21]. The injection spectrum used for the electrons is $E^{-2.1}$.

less bremsstrahlung and that inverse Compton emission may be the dominant component of the continuum emission below about 10 MeV, although a contribution from nuclear deexcitation lines [15] can't be ruled out.

The bremsstrahlung component of the medium-energy (1 – 50 MeV) diffuse emission, Figure 4, combined with observations of the diffuse radio synchrotron emission which traces electrons with energies of typically 100 MeV – 10 GeV, can be used to derive the shape of the cosmic-ray electron spectrum below 10 GeV, which is highly uncertain due to Solar modulation. Strong & Moskalenko [21] conclude that bremsstrahlung is less important than earlier estimates and that the electron injection spectrum must be harder, $E^{-2.1}$, than previously thought, $E^{-2.4}$, and that this agrees with recent detailed modelling of the radio synchrotron data.

The latitude distribution of the diffuse emission in the 1 – 3 MeV range has been derived by Strong et al. [17]. The latitude distribution in the 10 – 30 MeV range is shown in Figure 5. The profile is fairly well described by the sum of a narrow (2° FWHM) bremsstrahlung component plus a broader (8° FWHM) inverse Compton component [17]. Within about ±10° of the plane this model agrees fairly well with the observations.

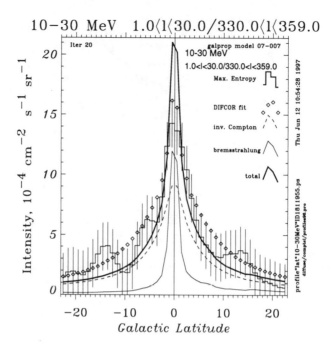

FIGURE 5. COMPTEL latitude profile of the diffuse emission in the Galactic center region compared to a prediction for a cosmic-ray interaction model [19].

EGRET – Energetic Gamma Ray Experiment Telescope

The EGRET instrument is sensitive to gamma rays with energy from about 30 MeV to over 30 GeV, adjoining the COMPTEL energy range. These photons are produced by the interactions of cosmic rays with the interstellar medium via electron bremsstrahlung, nucleon-nucleon processes (π^0 decay), and, to a lesser extent, inverse Compton scattering of low-energy photons.

The greater sensitivity and spatial and energy resolution of the EGRET instrument [22] compared to SAS 2 [23] and COS B [24] has enabled much more detailed spatial and spectral analyses of the diffuse emission to be performed than were possible with these earlier instruments.

The EGRET observations of the diffuse emission from the Galactic plane from the first fifteen months (1991 May 15 – 1992 Nov 17, Phase 1) and the following ten months (1992 Nov 17 – 1993 Sep 14, Phase 2) of EGRET operation have been analyzed by Hunter et al. [25]. These observations provide fairly uniform exposure to the Galactic plane (average exposure = 1×10^9 cm^2 s). They determined the longitude and latitude distributions for 4 broad energy intervals (30–100 MeV, 100–300 MeV, 300–1000 MeV, and 1000–30,000 MeV) averaged, respectively, over $1° \times 4°$ and $20° \times 0.5°$ ($l \times b$) bins (see Figures 2 and 3 of [25]). Spatially resolved, average spectra of the diffuse emission were also determined for 11 energy intervals covering the range from 30 MeV to 30 GeV for $10° \times 4°$ ($l \times b$) bins (see Figure 5 of [25]). The average diffuse gamma-ray spectrum of the inner Galaxy, $300° < l < 60°$, $|b| \leq 10°$, is shown in Figure 6. Spatially resolved spectra have also been determined by Dixon & Samimi [26] using a model independent, wavelet-based analysis technique.

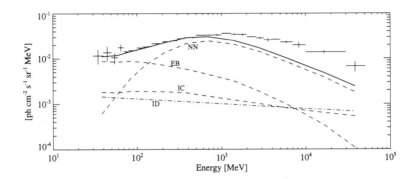

FIGURE 6. The average diffuse gamma-ray spectrum of the inner Galaxy region, $300° < l < 60°$, $|b| \leq 10°$ (0.73 sr) observed with EGRET [25]. The contributions from point sources detected with more than 5σ significance [27,28] have been removed. The data are plotted as crosses where the horizontal line indicates the width of the energy interval and the vertical line the $\pm 1\sigma$ statistical error. The best-fit model calculation (see [25]) plus the isotropic diffuse emission [29] is shown as the solid line. The individual components, nucleon-nucleon (NN), electron-bremsstrahlung (EB), and inverse Compton (IC) of this calculation are shown as dashed lines. The isotropic diffuse emission (ID) is shown as a dash-dot line.

The components of the diffuse emission, electron Bremsstrahlung, nucleon-nucleon, inverse Compton, and the extragalactic (isotropic) diffuse, shown in Figure 6 were derived from a 3-dimensional model of the Galactic interstellar matter distribution on the assumption that the cosmic-rays are coupled to the matter [30]. This model has only two parameters that are considered to be adjustable, the molecular hydrogen mass calibrating ratio (the X-ratio), and the scale of the cosmic-ray coupling to the matter. The spectral components calculated using these parameter values are shown in Figure 6. The best-fit value for the coupling scale was found to be 1.76 ± 0.2 kpc and the value for the X-ratio is given in Table 1 along with other values determined from the EGRET gamma-ray data on several spatial scales, local, individual molecular clouds, Galactic arms, and the Galaxy as a whole. These results, summarized in Table 1, are consistent with a trend for higher values of the X-ratio in the outer Galaxy suggested by non-gamma-ray determinations of the X-ratio, such as Sodrowski et al. [31]. Increasing values of the X-ratio in the outer

Galaxy would be consistant with the metalicity gradient. There is also an indication of a smaller than average value for the X-ratio locally. The energy dependence of the X-ratio over the EGRET energy range has also been investigated [32–35]. An apparent energy dependence would indicate cosmic-ray exclussion of enhancment in molecular clouds. Above 100 MeV there does not appear to be any significant variation with energy. Below 100 MeV the errors are much larger, but are also consistent with no variation.

TABLE 1. The $W_{CO}/N(H_2)$ Calibration Factor (X-ratio)

Region	X-ratio/10^{20} cm^{-2}(K km s^{-1})$^{-1}$	Authors & References
Whole Galaxy	1.56 ± 0.05	Hunter et al. 1997 [25]
Whole Galaxy	1.9 ± 0.2	Strong & Matttox 1996 [34]
Whole Galaxy	1.9 ± 0.3	Strong et al. 1988 [36]
Outer Galaxy	2.7 ± 0.1	Hunter et al. 1997 [25]
Perseus Arm	2.5 ± 0.9	Digel et al. 1996 [35]
Ophiuchus	1.1 ± 0.2	Hunter et al. 1994 [32]
Orion	1.06 ± 0.14	Digel et al. 1995 [33]
Local	0.9 ± 0.14	Digel et al. 1996 [35]

In general, the spatial agreement between the Hunter et al. model and observation is very good, particularly in the directions of the Galactic arm tangents. There are, however, a few areas where the fit is not as good which provide interesting insights into the diffuse emission. At medium latitudes ($3° < |b| < 10°$) toward the Galactic center (Fig. 3 of [25]) the model underpredicts the observed emission. This deficit may be due to underestimation of the inverse Compton intensity in the Galactic center, perhaps owing to undetestimating the density of cosmic-ray electrons or low-energy photons. The longitude profile (Fig. 2 of [25]) shows a small, but, general overprediction in the plane of the outer Galaxy and in isolated longitude regions near the Galactic center. This may indicate that the assumption of dynamic balance between the cosmic-rays and the intersteller matter density in the outer Galaxy must be modified.

The diffuse emission excess above 1 GeV relative to the expected π^0-decay emission (Figs. 2, 4, and 6) [25,34,37] can not be explained by a spectral variation in the X-ratio. The model of Strong & Mattox [34] allowed for a spectral variation of the X-ratio, but found no significant variation (Fig. 7). Comparison of the Hunter et al. model [25], based on the assumption of an energy independent value for the X-ratio, with the EGRET observed flux above 1 GeV found that the discrepancy was proportional to the total predicted intensity at all longitudes rather than peaked towards the Galactic center. This indicates that any explanation of the excess emission must apply equally to both the atomic and molecular hydrogenin the Galaxy.

Hunter et al. suggest four possible explanations for the excess above 1 GeV:

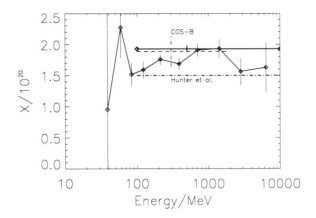

FIGURE 7. Energy dependence of the $N(H_2)/W_{CO}$ calibration factor (X-ratio) in units of $mol\,cm^{-2}\,(K\,km\,s^{-1})^{-1}$ derived by Strong & Mattox [34] for the whole Galaxy in ten energy ranges and for 100–10000 MeV (solid line). The Galactic average values from EGRET [37] (dash-dot line), and COS B [36] (dashed line) are shown for comparison.

1) there may be a possible error in the EGRET calibration, 2) the theory of $p + p \to \pi^0 + X \to \gamma + \gamma$ production may not be adequately understood, 3) there may be a distribution of unresolved point sources, and 4) the shape of the cosmic-ray spectrum that is measured near the Sun may not be representative of the Galactic average spectral shape. Of these, the first is considered unlikely for various reasons and the second, although also unlikely, is being re-examined by Baring & Stecker (private communication).

The third possibility, a distribution of unresolved point sources, has been examined by several authors. These studies include: gamma rays and cosmic rays from supernova explosions and young pulsars [38] and gamma-rays from pulsars [39]. Pulsars, however, are not likely to be the explanation of the high energy excess, which also appears in high latitude spectra.

The fourth possibility, if correct, would require significant alteration to our understanding of the Galactic cosmic-ray distribution. The consequences of an average Galactic cosmic-ray spectrum different from the locally observed spectrum have been investigated by several authors. Mori [40] determined that the $E_\gamma > 100$ GeV excess can be fairly well explained if the average cosmic-ray proton spectrum is $E_p^{-2.45}$, which is very much flatter than the locally observed $E_p^{-2.75}$ spectrum. Mori's "best-fit" spectrum shown in Figure 8 agrees fairly well with the EGRET spectrum above about 500 MeV, however, below this energy the model of Hunter et al. [25], shown in Figure 6, agrees better with the data.

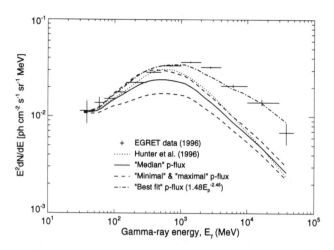

FIGURE 8. Model predictions of the diffuse gamma-ray spectrum based on gamma-ray emissivities from cosmic-ray nuclear interactions based on various assumptions of the proton spectrum [40] compared to the EGRET spectrum [37].

An enhanced inverse Compton component is another possible explanation of the gamma-ray excess above 1 GeV. Chen, Dwyer, & Kaaret [41] found that the high latitude ($29.5° < |b| < 74°$) gamma-ray emission has two components; one is correlated with H I and the other, correlated with the 408 MHz radio emission, was identified as the inverse Compton emission. Chen et al. determined a power-law spectral index of -1.88 ± 0.14 for this emission over the gamma-ray energy range 30–5000 MeV. The spectral index of the inverse Compton emission corresponds to a cosmic-ray electron spectral index of $\alpha_e = 2\alpha_\gamma + 1 = -2.76 \pm 0.28$. This spectrum is harder than expected for the > 10 GeV electrons responsible for inverse Compton in the EGRET range. A harder inverse Compton spectrum was also found by Strong & Mattox [34]. Although this may be due to detection of emission other than pure inverse Compton, it may indicate a large-scale average cosmic-ray spectrum different from that measured locally. This could be possible if the locally measured electron spectrum is dominated by local supernova remnants, and is not typical of the Galactic average spectrum [42].

Other possibilities include: flattening of the extragalactic diffuse spectrum above 1 GeV [43]; hardening of the Galactic diffuse spectrum above 1 GeV by cosmic-ray transport effects in dense molecular clouds [44]; and enhanced cosmic-ray electron bremsstrahlung [42].

Weak evidence of spectral variations of the diffuse emission with Galactic longitude were observed in the EGRET data in the direction of the Galactic anti-center and only at low latitudes ($-2° < b < 2°$) [25]. In this direction the spectrum appears to be steeper above about 4 GeV than at other longitudes or higher latitudes (see Fig. 7 of [25]). This variation of the diffuse gamma-ray emission hints at a possible variation of the cosmic-ray proton spectrum with Galactic radius which might be expected if cosmic-rays are accelerated primarily in the inner Galaxy and diffuse to the outer Galaxy or if the high-energy cosmic rays are confined less well in the outer Galaxy.

The spectral shape of the diffuse emission between about 30 MeV and a few hundred MeV can be used to study the cosmic-ray electron-to-proton (e/p) ratio. In this energy range the dominant production mechanism of the diffuse emission shifts at about 100 MeV from electron (electron bremsstrahlung and inverse Compton) to proton (nucleon-nucleon, π^0-decay) processes. Zhang et al. [45] examined the the COS B data for variation in the spectral ratio I(E_γ > 300MeV)/I(70MeV < E_γ < 300MeV) as a function of Galactic longitude. They suggest that variations in this ratio, typically less than 1 sigma, are most likely due to variations in the e/p ratio. The Spatially resolved spectra ($10° \times 4°$, $l \times b$) derived from the EGRET data [25] do not show any evidence for significant variation in the shape of the diffuse emission spectrum with Galactic longitude below about 4 GeV.

The spectra and distribution of cosmic-ray electrons and protons in the Galaxy have been investigated by several authors and the results compared with the OSSE, COMPTEL, and EGRET data. The electron spectrum derived from a propagation model by Skibo [11] was used by Kinzer et al. [10] and Hunter et al. [25] to derive models of the diffuse emission for analysis of the OSSE and EGRET data, Figures 1, 4, and 6, respectively. The diffuse emission model of Skibo is shown in Figure 2. Strong & Moskalenko [21] have developed a 3-dimensional model, which includes diffusion/convection and transport effects, of the cosmic rays in the Galaxy. This program offers simultaneous predictions of many observable parameters of the cosmic-ray distribution and places significant constraints on their values. A main result of this study is an improved cosmic-ray electron spectrum.

The distribution of cosmic rays in the Galaxy has likewise been examined by several authors using different techniques. Strong & Mattox [34] determined the cosmic-ray emissivity using a radially symmetric model of the Galaxy (Fig. 9). The emissivity in the outer Galaxy has been determined by Digel et al. [35] from a study of the diffuse emission in Cepheus ($100° < l < 130°, -5° < b < +32°$) including the emission from the prominent local molecular clouds in the Cepheus and Polaris flares and the cloud complex associated with Cas A and NGC 7538 in the Perseus arm. Hunter et al. [25] determined the 2-dimensional Galactic cosmic-ray surface density on the assumption that the cosmic-rays are coupled to the interstellar matter. The radial cosmic-ray distribution, derived from this study, for each Galactocentric quadrant is compared in Figure 10 with the radial gradient derived by Strong et al. [36] from the COS B data. The Strong & Mattox emissivity gradient (Fig. 9) shows an indication of a lower cosmic-ray density at the Galactic center and a general decrease in the outer Galaxy. The Hunter et al.

distribution (Fig. 10) also shows this general decrease. In addition, there is a slight density increase in quadrants II and III in the Perseus arm at 13 – 16 kpc from the Galactic center.

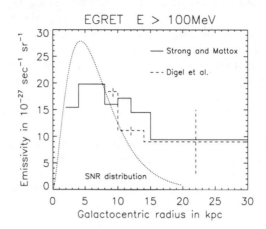

FIGURE 9. Emissivity gradient determined from EGRET data by Strong & Mattox [34], solid line, for the whole galaxy, and by Digel et al. [35] for the outer Galaxy in the direction of Cepheus ($100° < l < 130°, -5° < b < +32°$). The supernova remnant distribution of Case & Bhattacharya [46], dotted line, is more centrally peaked and falls off faster than the cosmic-ray gradient.

CONCLUSIONS

Studies of the Galactic diffuse gamma-ray continuum emission using the instruments on the Compton Gamma Ray Observatory have addressed several important questions about the structure of our Galaxy and the distribution of cosmic rays. Earlier basic interpretations seem now to be confirmed and we have a fairly solid picture of the interstellar matter distribution and the Galactic cosmic-ray density [9,10,16,17,19,25]. The main points of this interpretation include:

- The diffuse Galactic gamma-ray emission has its origin in cosmic-ray electron and proton interactions with the interstellar medium (high energy electron bremsstrahlung and nucleon-nucleon interations leading to π^0 decay) and with low-energy photons (inverse Compton). The Galactic diffuse emission at these energies has a narrow latitude extent along the Galactic plane, with enhancements towards the Galactic center and at the tangent points of the spiral arms.

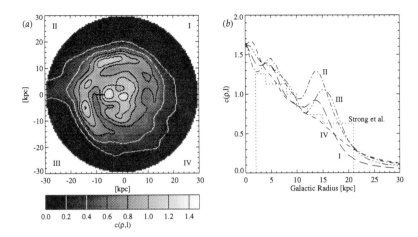

FIGURE 10. (a) The cosmic-ray density relative to the local density, derived on the assumption that the cosmic rays are coupled to the matter [25]. (b) The azimuthal average of the enhancement factor for each Galactocentric quadrant shown in (a). The Galactic average cosmic-ray gradient determined by Strong et al. [36] using the COS-B data is shown as the dotted line.

- The molecular component of the interstellar medium has been clearly shown and the mass calibrating factor (X-ratio) has been fairly well established (Table 1). It appears that the value of the X-ratio is independent of energy (EGRET range), but may vary across the Galaxy.

- The cosmic-ray electron spectrum below about 10 GeV, which is strongly affected by Solar modulation, is tightly constrained by the gamma-ray observations between 10 and 100 MeV where bremsstrahlung is the dominant process. Below 10 MeV, the contribution from inverse Compton emission relative to bremsstrahlung, although somewhat uncertain, offers further constraints on the cosmic-ray electron spectrum. Observations from OSSE and COMPTEL provide evidence that the Galactic diffuse continuum extends down to energies below 100 keV. The spectrum steepens below a few hundred keV (see Fig. 2) but this is likely a point source contribution rather than an indication that the source electron spectrum turns up below a few MeV.

- The density of cosmic-ray electrons and protons in the Galaxy has a general Galactic gradient decreasing towards the outer Galaxy and perhaps peaking at the Galactic center.

- The shape of the diffuse spectrum in the EGRET energy range does not appear to vary with Galactic longitude and/or latitude except near the plane in the Galactic anticenter, where there is weak evidence for steepening.

The instruments on CGRO have, however, also revealed several surprising new aspects of the Galactic emission which show that we don't understand the diffuse emission as well as we might. We may have to rethink some of the basics.

- The high-energy excess above about 1 GeV may indicate that the cosmic-ray proton and/or electron spectrum observed in the local neighborhood is not representative of the Galactic average. A very flat proton spectrum, $E^{-2.45}$, is required to explain the excess if it is due to π^0 decay. A higher density and/or scale height of low-energy photons could explain the excess in terms of increased inverse Compton emission. A unresolved distribution of point sources is another, although unlikely, possibility.

- The cosmic-ray electron spectrum below about 6 GeV can, in principle, be derived from the spectrum of the diffuse emission. This determination is, however, complicated by the uncertainty in the relative contribution of bremstrahlung and inverse Compton emission below about 100 MeV.

- Supernova remnants have been proposed as the source of the cosmic rays, although this has not been confirmed. The cosmic-ray distribution is much flatter than the supernova distribution although there are large uncertainties in the supernova distribution. The flatter cosmic-ray distribution may, however, simply reflect the effects of cosmic-ray propagation.

- The contribution from bright, resolved point sources to the diffuse emission has been accounted for in the EGRET analyses described above. However, the contribution from a distribution of weaker, and hence unresolved, point sources to the diffuse emission observed with all three instruments is not as readily determined. In fact, the distinction between an unresolved distribution of point sources and the diffuse emission is unclear.

These issues present challenges for future work.

REFERENCES

1. Harris, M.J., Share, G.H., Leising, M.D., Kinzer, R.L., & Messina, D.C. 1990, ApJ, 362, 135
2. Fichtel, C.E., et al. 1975, ApJ, 198, 163
3. Hartman, R.C., et al. 1979, ApJ, 230, 597
4. Mayer-Hasselwander, H., et al. 1980, in Ann. NY Acad. Sci. 336, Ninth Texas Symp. on Relativistic Astrophysics, 211
5. Mayer-Hasselwander, H., et al. 1982, A&A, 105, 164

6. Bloemen, Hans (J.B.G.M.), 1989, Annu. Rev. Astron. Astrophys. 27, 469
7. Zhiang, S.N. 1997, private communication
8. Johnson, W.N., et al. ApJS, 86, 693
9. Purcell et al. 1996, A&AS, 120(4), 389
10. Kinzer R.L., et al. 1997, in preparation
11. Skibo, J.G., 1993, Ph.D. Dissertation, University of Maryland
12. Claret, A., et al. 1995, Adv. Sp. Res. 15, 557
13. Schönfelder, V., et al. 1993, ApJS, 86, 657
14. Diehl, R., et al. 1992, in "Data Analysis in Astronomy IV", V. Di Gesù et al. eds., Plenum Press, 201
15. Bloemen, H., et al. 1997, these proceedings
16. Strong, A.W., et al. 1994, A&A, 292, 82
17. Strong, A.W., et al. 1995, Proc. 24th ICRC, Rome, 2, 234
18. Strong, A.W., et al. 1996, A&AS, 120(4), 381
19. Strong, A.W., et al. 1997, these proceedings
20. Bloemen, H. 1993, in "Back to the Galaxy", eds. S.S. Holt, F. Verter (New York) AIP Conf. Proc. 278, 409
21. Strong, A.W., & Moskalenko, I.V. 1997, these proceedings
    College Park, AIP Conf. Proc. 304, 463
22. Thompson, D.J., et al. 1993, ApJ, 86, 629
23. Derdeyn, S.M., et al. 1972, Nucl. Instum. Methods Phys. Res., 98, 557
24. Bignami, G.F., et al. 1975, Space Sci. Instum., 1, 245
25. Hunter, S.D., et al. 1997, ApJ, 481, 205
26. Dixon, D.D., & Samimi, J. 1997, these proceedings
27. Thompson, D.J., et al. 1995, ApJS, 101, 259
28. Thompson, D.J., et al. 1996, ApJS, 107, 227
29. Sreekumar, P., et al. 1997, ApJ, submitted
30. Bertsch, D.L., et al. 1993, ApJ, 416, 587
31. Sodrowski, T.J., et al. 1995, ApJ, 452, 262
32. Hunter, S.D., et al. 1994, ApJ, 436, 216
33. Digel, S.W., et al. 1995, ApJ, 441, 270
34. Strong, A.W., & Mattox, J.R. 1996, A&A Lett. 308, 21
35. Digel, S.W., et al. 1996, ApJ, 463, 608
36. Strong, A.W., et al. 1988, A&A, 207,1
37. Hunter, S.D., et al. 1995, Proc. 24th ICRC, Rome, 2, 182
38. Dorman, L. 1997, these proceedings
39. Pohl, M., Kanbach, G., Hunter, S.D., & Jones, B.B. 1997, these proceedings
40. Mori, M. 1997, ApJ, 478, 225
41. Chen, A., Dwyer, J., & Kaaret, P. 1996, ApJ, 463, 169
42. Pohl, M., & Schlickeiser, R. 1991, A&A, 252, 565
43. Stecker, F.W, & Solomon, M. 1996, 464, 600
44. Skibo, J.G. 1997, these proceedings
45. Zhang, L., Chi, X., Wolfendale, A.W., & Issa, M.R. 1994, in "Second Compton Symposium," College Park, AIP Conf. Proc. 304, 494
46. Case, G., & Bhattacharya, D. 1997, A&A, 120(4), 437

Galactic $e^+ - e^-$ Annihilation Line Radiation

D. M. Smith[1], W. R. Purcell[2], and M. Leventhal[3]

[1] *Space Sciences Laboratory, University of California, Berkeley, CA 94720*
[2] *Dept. of Physics and Astronomy, Northwestern University*
[3] *Department of Astronomy, University of Maryland, College Park*

Abstract. The longest-known and brightest Galactic gamma-ray line is the line at 511 keV from positron annihilation. Despite over two decades of observations, however, the distribution and sources of this line are not well understood. It is only with a steadily increasing database, primarily from OSSE but including other instruments, combined with increasingly sophisticated algorithms to deconvolve image information, that we are now beginning to grasp the shape of the distribution.

I INTRODUCTION - BEFORE COMPTON

A gamma-ray line with energy slightly below 0.5 MeV was first detected from the direction of the Galactic Center (GC) in 1970 [16] with a scintillator experiment having limited energy resolution. In 1977, an observation with a high-resolution germanium spectrometer showed that the line was narrow and at precisely 511 keV, establishing it as positron annihilation radiation [20]. The apparent shift in the lower-resolution scintillator was due to the line being convolved with the three-photon continuum from the decay of orthopositronium. This continuum rises almost linearly with energy, peaks at 511 keV, and drops immediately to zero. It had already been suggested as the cause of the shift [19].

The annihilation line is the longest known and most studied extra-solar gamma-ray line. The history of our understanding of this emission before *Compton* is discussed in depth in the review paper by Tueller [54] in the Proceedings of the first Compton Symposium, and in an earlier paper by Lingenfelter and Ramaty [27]. Recent results from balloon instruments are summarized by Teegarden [51]. Positron annihilation processes and cross sections are most recently discussed in Guessoum, Ramaty and Lingenfelter [11] and Guessoum, Skibo and Ramaty [12].

In brief, after the narrow nature of the line and the presence of the orthopositronium continuum were established, there were two important conclusions drawn before the era of *Compton*. The first was that its flux increases with the aperture of the instrument viewing it, implying that the source is extended on a scale of tens of degrees [1,8,51]. Figure 1 shows the flux as a function of aperture for a large number of missions.

The second and more controversial conclusion was that the source is variable. The principal evidence for this was a series of balloon flights from 1977 through 1986 (see [23] and references therein), which indicated a sharp drop in flux around 1980. Figure 2 shows these data as part of a time history of 511 keV flux measurements, with no correction for field of view. This drop was originally reported by HEAO-3 [42], but a later analysis of those data indicated that the change was not statistically significant [30]. An extended source and rapid change are not compatible for two reasons: the usual light-travel-time argument, and also the time required for positrons to slow down, possibly form positronium, and annihilate in the interstellar medium, which is on the order of 10^5 years [11]. Therefore a two-component model, with an extended source of uncertain shape and a variable point source close to the GC, was the most sophisticated understanding we had of the distribution

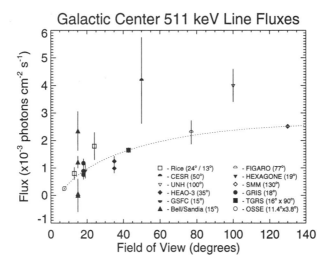

FIGURE 1. GC 511 keV line fluxes vs. detector field of view. Data points taken by balloon instruments are labeled Rice [15,16,14], CESR [1], UNH [9] GSFC [33], Bell/Sandia [19-22], FIGARO [32], HEXAGONE [45] and GRIS [23,10,24]; data points taken by spacecraft are labeled HEAO-3 [29], SMM [42,43,13], TGRS [51], and OSSE [38]. See Smith et al. [48] and Teegarden et al. [52] in these proceedings for new numbers from BATSE (130°) and TGRS.

before the launch of *Compton* [27].

The main candidate for the variable point source was 1E 1740.7-2942, a bright and extremely hard x-ray source about 1° from the GC, which is considered a black-hole candidate because its spectral shape and intrinsic luminosity are identical to those of Cyg X-1 in its hard (low) state. From 1990 to 1992, three discoveries were made which brought attention to this object: a 1-day flare of a bright, broad emission line near 400 keV seen with the SIGMA imager on *GRANAT* [50,3]; a dense molecular cloud in the line of sight of, and possibly surrounding, the object [2,31]; and its radio emission, in the form of a variable core and two well-formed lobes [32]. Ramaty et al. [40] concluded that a large number of electron/positron pairs are created in brief flaring events, and that while some annihilate immediately in the accretion disk, giving the prompt, broad line, others escape via the jets into the molecular cloud, giving variations in the narrow line over a few years.

A number of other transient emission features have been reported which may be associated with positron annihilation, but *Compton* has contributed a number of upper limits and negative results on such features. The review article by M. Harris in these proceedings discusses the controversy in detail.

The extended source was (and still is) interpreted as due in a large part

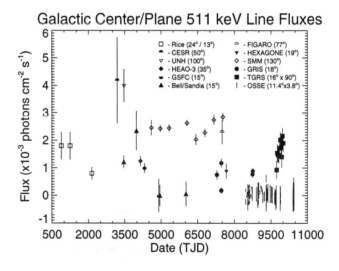

FIGURE 2. 511 keV line fluxes vs. time. Data points taken by balloon instruments are labeled Rice [15,16,14], CESR [1], UNH [9] GSFC [33], Bell/Sandia [19-22], FIGARO [32], HEXAGONE [45] and GRIS [23,10,24]; data points taken by spacecraft are labeled HEAO-3 [29], SMM [42,43,13], TGRS [51], and OSSE [38]. The lowest of the GRIS points was taken in the Galactic plane away from the center. See Smith et al. [48] and Teegarden et al. [52] in these proceedings for new numbers from BATSE (130°) and TGRS.

to positrons from the decay of freshly synthesized radioisotopes. The decays of ^{26}Al and ^{44}Ti, in addition to producing characteristic gamma-rays (see the review by Diehl & Timmes in these proceedings), also produce positrons, as does the decay of ^{56}Co, which is produced in supernovae much more copiously than the others, but which decays quickly enough that only a small and uncertain fraction of the positrons escape the ejecta to annihilate in the ISM. Therefore supernovae, novae, and Wolf-Rayet stars, all of which eject at least some of these isotopes into the ISM, are positron sources. If black holes besides 1E 1740.7-2942 produce positrons, they would not annihilate within a few years (since there would be no coincident molecular cloud), but could create hotspots in the extended emission.

II COMPTON RESULTS BEFORE THIS SYMPOSIUM

By the time of the first Compton Symposium in 1992, Purcell et al. [35] had already compiled and analyzed a number of OSSE pointings around the GC from Phase I, and had already produced the best spatial information available about the annihilation line. Fitting various models, they found that distributions dominated by the disk (CO, visual luminosity, etc.) were clearly rejected, and that the Galactic bulge dominated the emission. Other results presented in that paper included: a value for the positronium fraction (0.96 ± 0.04) which is virtually the same as today's best OSSE value; the first joint modeling of OSSE results with those of another instrument (the *Solar Maximum Mission Gamma-Ray Spectrometer - SMM/GRS*); and the first search for variability by plotting the residuals of each pointing after the model is subtracted (no variability was seen). The SMM/GRS data are particularly complementary to OSSE, since they integrate over a very large field of view (see Figure 1 and Figure 2). These results (with less emphasis on fitting the distribution) were also presented in The Astrophysical Journal Letters [36].

An independent analysis using the Phase I OSSE data together with balloon observations from 1977-1989 was published by Skibo et al. [45]. They concluded that a variable point source was necessary (based mostly on the balloon data), and demonstrated that several physically plausible distributions (COSB gamma-rays, CO, hot plasma, warm gas, pulsars, and novae) failed to fit the combined data well, even when a variable point source was allowed; of the choices, the best fit was from the nova distribution.

In the next report from the OSSE team, at the second Compton Symposium [37], the OSSE 511 keV database contained more pointings; more sophisticated models were fit to the data, although the principal conclusion was still that the bulge dominates the disk; and results from another instrument (the GRIS balloon) were included in the joint fits. In the conclusion it was suggested that the disk component is due to decay of ^{26}Al, while the bulge component is due to decay of ^{56}Co and ^{44}Sc from Type Ia supernovae. Ramaty et al. [41], continuing the work presented in Skibo et al. [45], found their results changed by the inclusion of the additional OSSE data: a point source or extended spheroid added to one of the disk-based distributions was now necessary, a conclusion which agreed with that of Purcell et al. [37].

At the San Antonio meeting of the American Astronomical Society in January 1996, W. Purcell presented model fits to combined data from OSSE, SMM/GRS and other instruments. The model included a Galactic bulge component, a very broad circular halo which carried most of the high flux seen by SMM/GRS, and a narrow disk which was very faint, including just enough flux to be consistent with the 1809 keV radiation from ^{26}Al.

At the third Compton Symposium, the OSSE team presented results from a subset of the data with the long axis of the collimator parallel to the Galactic Plane [18]. It was demonstrated that the positronium continuum and annihilation line had similar profiles in Galactic longitude; in other words, the high positronium fraction (0.97 ± 0.03) remained constant. Tueller et al. [55], in the same proceedings, looked only at those OSSE data with the collimator perpendicular to the Galactic plane and near the GC, to find the center of the bulge component in Galactic longitude. They pointed out that their best fit, $b = (-1.36 \pm 0.33)°$, was more consistent with 1E 1740.7-2942 than with the GC itself. Also at the third Symposium, Smith et al. [47], in a paper discussing BATSE upper limits on annihilation transients, reported that the GC line could be seen with BATSE.

Recently, the analysis of the annihilation distribution with OSSE has taken the biggest step since the publication of the Phase 1 results, with the adoption of model-independent mapping techniques. Purcell et al. [38] presented maps made by two techniques: Basis Pursuit Inversion (BPI) [4] and the Maximum Entropy Method (MEM) [13]. Both maps used OSSE data through Cycle 5 as well as results from HEAO-3, SMM/GRS, the Transient Gamma-Ray Spectrometer (TGRS) on *Wind*, and balloon instruments; only the OSSE data contributed significantly to the detailed structure derived, however. The data from the other instruments, which all have larger fields of view, set the amount of broader, low-surface-brightness emission, which can be invisible to OSSE if it appears in both its source and background pointings. Both maps showed a bright Galactic bulge and faint Galactic disk, consistent with all the previous model-fitting work with the OSSE data, but also showed a feature which none of the modelers could have predicted, and never looked for: a bright spot at roughly $l = -4°, b = +7°$, with about half the flux of the bulge component.

Cheng et al. [5] pursued the MEM procedure in greater detail, introducing a positivity constraint on the map and doing a bootstrap analysis to estimate the significance of the new feature, which was 3.5σ.

III NEW RESULTS

At this Symposium, W. Purcell presented the latest model-independent mapping results, which are being submitted to The Astrophysical Journal [39]. The most important addition is the Cycle 6 data, which consisted of rectangular grids of OSSE pointings at two different collimator angles, to get a nearly uniform exposure over an approximately 20° × 20° field around the GC. This improved the data set considerably, which had been an ensemble of unrelated pointing strategies giving an uneven exposure. The high-latitude feature also appears when the Cycle 6 data are used alone, which is a very important confirmation. A Singular Value Decomposition (SVD) algorithm replaced BPI as the alternative to MEM, and of the non-OSSE data, only SMM/GRS and TGRS were retained. The location of background pointings was not available for the various balloon observations, and may be of importance in their interpretation; furthermore their statistical significance was lower. Finally, new model-fitting was done, with the inclusion of the high-latitude feature, as a check on the other algorithms. Figure 3 shows the SVD map including all data through Cycle 6.

In Purcell et al. [39], the high-latitude feature is found to be extended, and in fact larger than the Galactic bulge feature: ~ 10° vs. ~ 4° FWHM. The exact values and errors vary slightly with which algorithm is used and whether the non-OSSE data are included. The flux of the high-latitude feature is now understood to be comparable to the Galactic bulge flux ($\sim 4 \times 10^{-4}$ ph cm^{-2} s^{-1}), which only looks brighter because the Galactic plane component

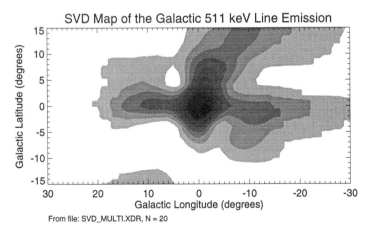

FIGURE 3. Singular Value Decomposition map including OSSE data through Cycle 6. The GC, Galactic plane and high-latitude extension are visible.

is (presumably) also present "underneath" it and because it is more compact. The best fit centroid for the high-latitude feature is approximately $l = -2°$, $b = 9°$, but with a couple of degrees uncertainty due to its extent.

In contrast to the fits presented in San Antonio, the disk component now carries most of the flux seen by the large field-of-view instruments, and a circular halo is not needed in the fit. The disk is no longer narrow in latitude, but broad ($\sim 12°$ FWHM). Two changes made this the better fitting model: the introduction of new OSSE data, and an improvement in the way the SMM/GRS angular response is treated. It cannot be ruled out that there may be both broad and narrow disk emission, in fact it is likely; but it is not required to fit the current data set.

A special evening session was held at this Symposium to discuss the high-latitude feature and related issues. C. Dermer presented a model [7] which invokes a modest starburst of Type II supernovae in the GC region. In this model, hot gas from the ensemble of supernovae breaks out of the galactic disk and erupts into a fountain, carrying to high galactic latitudes both the positrons and the medium required for their annihilation. Indeed, any model which places the emission as far as the GC must also allow for a significant amount of ordinary matter to be transported to high latitudes. Other sources of positrons and energy to drive the "fountain" were suggested by Purcell and others: a pair jet from the black hole at Sgr A*, a pair jet from a solar-mass black hole, or a single gamma-ray-burst-like event within the past 10^6 years.

It is worth stressing that the 511 keV observations do nothing to discourage the interpretation of the high-latitude feature as being in the foreground, and therefore possibly in the Galactic plane [39]. The images are not fine enough to determine if there is a bridge of emission between the bulge and high-latitude sources. At least one candidate foreground source was mentioned briefly by I. F. Mirabel, who noted that Nova Ophiuchus 1977 (V2107 Oph), a black-hole x-ray nova, is consistent with the position of the new feature.

Finally, Purcell et al. [39] also address the possibility of emission from 1E 1740.7-2942. They find that the longitude offset of the bulge is small ($\sim 0.4°$), and not statistically significant. Furthermore, a fit which includes a point source at 1E 1740.7-2942 as well as the bulge, disk, and high-latitude features gives a null result: $(0.5 \pm 5.3) \times 10^{-5}$ ph cm^{-2} s^{-1}, the most sensitive limit yet reported. It is quite possible that the longitude offset in the 511 keV line measured by Tueller et al. [55], and consistent with the position of 1E 1740.7-2942, was in fact due to a contribution from the high-latitude feature, which was unknown at the time and which has a longitude offset in the right direction. Since the Tueller et al. [55] data were taken with long axis of the OSSE collimator perpendicular to the plane, there was a significant response at the latitude of the new feature.

Other results at this meeting included two from GRIS, a balloon-borne germanium spectrometer. Cheng et al. [6], using data at different atmospheric depths and a blocking crystal which covered and uncovered the aperture, find excess 511 keV flux above the extrapolation of the atmospheric emission to zero atmospheric depth; this may be due to an isotropic, or near-isotropic, 511 keV flux from the sky. L. Cheng also mentioned that during this same flight, in which the instrument pointed at the zenith, there was an indication of excess emission as the galactic plane (near $l = 240°$) passed overhead, in addition to the expected excess from the GC. Taken together, these results suggest foreground emission over much of the sky; the Local Bubble and the Gould belt of nearby star-forming regions are possible candidate sources. If the new high-latitude feature from OSSE is local, such hotspots could be fairly numerous; the OSSE feature could be the first to be found simply because it is in the region which has been intensely examined.

An OSSE upper limit to 511 keV radiation from the extraordinarily active black hole candidate GRS 1915+105 is reported in these proceedings [48], and interpreted as a joint limit on the past activity of this object and the positron fraction in its radio jets. Also in these proceedings, Smith et al. [49] give a total annihilation flux measured with BATSE of $1.49 \pm 0.02 \times 10^{-3}$ ph cm^{-2} s^{-1} and discuss its consistency with other wide-field-of-view instruments [44,52], and Teegarden et al. [53] present the latest flux results from TGRS. Ling et al. [26] have also seen the 511 keV line with BATSE using a completely different algorithm and data format (see also Wallyn et al. [56]). It remains to be seen if Earth occultation will provide useful mapping information from BATSE data at 511 keV.

In summary, the current results seem to be drawing our gaze in two directions simultaneously: inward to smaller scales around the GC, in the hope of using structure on the 1° scale to distinguish among models for the Galactic bulge and high-latitude emission, and outwards beyond the GC and the narrow plane, searching for smooth or patchy emission from any and all parts of the sky to reconcile all the wide-field-of-view observations.

REFERENCES

1. Albernhe, F. et al. 1981, A&A, 94, 214
2. Bally, J. & Leventhal, M. 1991, Nature 353, 234
3. Bouchet, L. et al. 1991, ApJ, 275, L45
4. Chen, S., & Donoho, D., 1994, *Proceedings of the 28th Asilomer Conference on Signals, Systems and Computers,* ed. A. Singh, 1, 41
5. Cheng, L. X. et al. 1997a, ApJ, 481, 43
6. Cheng, L. X. et al. 1997b, these proceedings
7. Dermer, C., & Skibo, J., ApJ, submitted
8. Dunphy, P. P., Chupp, E. L., & Forrest, D. J. 1983, in *Positron-Electron Pairs in Astrophysics,* ed. M. L. Burns, A. K. Harding, and R. Ramaty (New York, AIP), p. 237
9. Gardner, B. M. et al. 1982, in *The Galactic Center,* ed. G. R. Riegler &I R. D. Blandford (New York, AIP), p. 144

10. Geherls, N. et al. 1991, ApJ, 375, L13
11. Guessoum, N. et al. 1991, ApJ, 378, 170
12. Guessoum, N. et al. 1997, in *Proceedings of the 2nd Integral Workshop*, ed. C. Winkler, T. J.-L. Courvoisier, & P. Durouchoux (ESA SP-382), p. 113
13. Gull, S. F. & Daniell, G. J. 1978, Nature, 272, 686
14. Harris, M. J. 1990, ApJ 362, L35
15. Haymes, R. C. et al. 1975, ApJ, 201, 593
16. Johnson, W. N. et al. 1972, ApJ, 172, L1
17. Johnson, W. N., & Haymes, R. C. 1973, ApJ, 184, 103
18. Kinzer, R. L. et al. 1996, A&AS, 120, 317
19. Leventhal, M. 1973, ApJ, 183, L147
20. Leventhal, M. et al. 1978, ApJ, 225, L11
21. Leventhal, M. et al. 1980, ApJ, 240, 338
22. Leventhal, M. et al. 1982, ApJ, 260, L1
23. Leventhal, M. et al. 1986, ApJ, 302, 459
24. Leventhal, M. et al. 1989, Nature, 339, 36
25. Leventhal, M. et al. 1993, ApJ, 405, L25
26. Ling, J. et al. 1997, ApJS, submitted
27. Lingenfelter, R. E. & Ramaty, R. 1989, ApJ, 383, 686
28. Lingenfelter, R. E., & Hua, X. 1991, ApJ, 381, 426
29. Mahoney, W. A. et al. 1993, A&AS 97, 159
30. Mahoney, W. A. et al. 1994, ApJS 92, 387
31. Mirabel, I. F. et al. 1991, A&A 251, L43
32. Mirabel, I. F. et al. 1992, Nature, 358, 215
33. Neil, M. et al. 1990, ApJL, 356, L21
34. Paciesas, W. S. et al. 1982, ApJL, 260, L7
35. Purcell, W. R. et al. 1993a, in *Compton Gamma-Ray Observatory*, ed. M. Friedlander, N. Gehrels, and D. J. Macomb (New York, AIP), p. 107
36. Purcell, W. R. et al. 1993b, ApJ, 413, L85
37. Purcell, W. R. et al. 1994, in *The Second Compton Symposium*, ed. C. E. Fichtel, N. Gehrels, & J. P. Norris, (New York, AIP), p. 403
38. Purcell, W. R. et al. 1997a, in *Proceedings of the 2nd Integral Workshop*, ed. C. Winkler, T. J.-L. Courvoisier, & P. Durouchoux (ESA SP-382), p. 67
39. Purcell, W. R. et al. 1997b, ApJ, submitted
40. Ramaty, R. et al. 1992, ApJ, 392, L63
41. Ramaty, R. et al. 1994, ApJS, 92, 393
42. Riegler, G. R. et al. 1981, ApJ, 248, L13
43. Share, G. H. et al. 1988, ApJ, 326, 717
44. Share, G. H. et al. 1990, ApJ, 358, L45
45. Skibo, J. G. et al. 1992, ApJ, 397, 135
46. Smith, D. M. et al. 1993, ApJ, 424, 165
47. Smith, D. M. et al. 1996, A&AS, 120, 361
48. Smith, D. M. et al. 1997a, these proceedings
49. Smith, D. M. et al. 1997b, these proceedings
50. Sunyaev, R. et al. 1991, ApJ, 383, L49

51. Teegarden, B. J. 1994, ApJS, 92, 363
52. Teegarden, B. J. et al. 1996, ApJ, 463, L75
53. Teegarden, B. J. et al. 1997, these proceedings
54. Tueller, J. 1993, in *Compton Gamma-Ray Observatory*, ed. M. Friedlander, N. Gehrels, and D. J. Macomb (New York, AIP), p. 97
55. Tueller, J. et al. 1996, A&AS, 120, 369
56. Wallyn, P. et al. 1997, in *Proceedings of the 2nd Integral Workshop*, ed. C. Winkler, T. J.-L. Courvoisier, & P. Durouchoux (ESA SP-382), p. 109

Galactic Gamma-Ray Line Emission From Radioactive Isotopes

Roland Diehl

Max-Planck Institut für extraterrestrische Physik
Giessenbachstr. 1, D-85740 Garching, Germany

Frank X. Timmes

Astronomy & Astrophysics
University of California, Santa Cruz CA 95064, USA

Abstract. Measurements of the Galactic γ-ray sky in lines attributed to the radioactive decay of ^{7}Be, ^{22}Na, ^{26}Al, ^{44}Ti, ^{56}Ni, and ^{60}Fe are collected, organized, and collated with models for the production sites/mechanisms, spatial distributions, and the chemical evolution of these isotopes.

INTRODUCTION

Radioactivity was discovered a little more than a century ago when Henri Becquerel included potassium and uranium sulfates as part of a photographic emulsion mixture (Becquerel 1896). He soon found that all uranium compounds and the metal itself were "light sources", the intensity was proportional to the amount of uranium present, with chemical combination having no effect. Two years later, Pierre and Marie Curie coined the term "radioactive" for those elements that emitted such "Becquerel-rays". A year later, Ernest Rutherford demonstrated that at least three different kinds of radiation are emitted in the decay of radioactive substances. He called these "alpha", "beta", and "gamma" rays in an increasing order of their ability to penetrate matter (Rutherford 1899, Feather 1973). It took a few more years for Rutherford and others to conclusively show the alpha rays were identical with the nuclei of helium atoms (Rutherford 1905; Rona 1978), and the beta rays were electrons (Becquerel 1900; Badash 1979). By 1912 it was shown that the gamma rays had all the properties of very energetic electromagnetic radiation (e.g., Allen 1911).

We now understand radioactive decay as transitions between different states of atomic nuclei, transitions that are ultimately attributed to electroweak interactions. Measurement of the decay products has grown into an important tool of experimental physics: On Earth, it forms the basis of radioactive dating through high precision isotopic analysis, in tree rings, terrestrial rocks, and meteoritic samples, to name just a few (Rolfs & Rodney 1989). Distant radioactive material throughout the universe may be studied in detail by measuring the material's gamma-ray line spectrum. These gamma-ray lines identify a specific isotope, and the abundance of the distant material directly relates to the measured gamma-ray line intensity (Clayton 1982). These γ-rays are also unaffected by the intervening matter once the radioactive nucleus has left its dense production site and gone into the interstellar medium. The characteristic half-life constitutes an exposure time scale of the sky in a specific γ-ray line. Given reasonable event frequencies and stellar nucleosynthesis yields, the cumulative radioactivity from many events (^{26}Al, ^{60}Fe, and to a lesser extent ^{22}Na and ^{44}Ti), and individual events (^{44}Ti, ^{56}Ni, and perhaps ^{7}Be, ^{22}Na) can be examined.

By the late 1970s various international collaborations had launched experiments on various stratospheric balloons and space satellites and to explore cosmic rays, X-ray, and γ-ray sources (Murthy & Wolfendale 1993). The first gamma-ray line discovery had been reported (Haymes et al. 1975), several diagnostic lines were reported in the years after. Following these pioneering missions the Gamma Ray Observatory was set out to explore the gamma-ray sky for the first time over a wide range of gamma-ray energies, including the regime of nuclear lines from radioactivity. A delightful summary of the history and relationship to the previous missions by Kniffen, Gehrels, & Fishman (1997) complements the description of the Observatory goals, the instrumentation and first achievements by Gehrels et al. (1995). Today collaborative international agencies have numerous ongoing and planned projects to deepen specific studies in the area of gamma-ray astronomy (ref. recent workshop proceedings edited by Winkler et al. (1997) and Kurfess et al. (1997)).

In the following review, attention is chiefly focused on the Milky Way's gamma-ray line emission that originates from the radioactive isotopes ^{7}Be, ^{22}Na, ^{26}Al, ^{44}Ti, ^{56}Ni, and ^{60}Fe.

PRODUCTION OF RADIOACTIVE ISOTOPES

The nucleosynthetic processes in stars responsible for the production of ^{26}Al have been summarized by Clayton & Leising (1987), Prantzos & Diehl (1996), and MacPherson, Davis & Zinner (1995). In stars, ^{26}Al is produced by proton capture reactions, mainly on ^{25}Mg, and is destroyed by e^+ decay, (n,p), (n,α), and (p,γ) reactions. The final yield from any source is temperature sensitive. ^{26}Al can be produced during: (1) hydrostatic hydrogen burning in the core of massive (≥ 11 $M_\odot$) stars; (2) hydrostatic hydrogen burning in the hydrogen shell of low- and intermediate mass (≤ 9 $M_\odot$) stars while on the asymptotic giant branch; (3) explosive hydrogen burning (temperatures $> 2\times10^8$ K) in massive stars and on the surfaces of white dwarfs (i.e., novae); and (4) hydrostatic and explosive carbon and neon burning in massive stars. Any ^{26}Al produced by stars is ejected in part by strong winds (Wolf-Rayet, asymptotic giant branch stars), and in full amount by explosions (supernova and nova). Besides being formed in stars, ^{26}Al can also in principle be produced by spallation reactions of high-energy cosmic rays on a range of nuclei (mainly silicon, aluminum and magnesium), although at substantially lower efficiency.

Most of the different ways to synthesize ^{26}Al, Type II supernovae, Wolf-Rayet stars, AGB stars, and classical novae, can all, at least according to their respective protagonists, produce sufficient quantities Galaxy-wide. So they cannot all make ~ 2 $M_\odot$, or else there would be too much ^{26}Al in the Galaxy. Each prospective source has it's advantages and difficulties. We discuss their nature briefly, addressing a few of the physical aspects involved.

The treatment of convection, and its implications, constitutes one problem area common to several source types. The convective coupling between mass zones in both the oxygen-neon and hydrogen shell burning regions of massive stars can simultaneously bring light reactants and seed nuclei into the hot zone to aid in the synthesis, and through the same process remove the fragile product from the high temperature region where it might otherwise be destroyed. Convective burning in the oxygen-neon shell of a 20 $M_\odot$ star has been modeled in two-dimensions by Bazan & Arnett (1994). They find that large-scale plume structures dominate the velocity field, physically caused by the inertia contained in moving mass cells. As a result, significant mixing beyond the boundaries defined conventionally by mixing-length theory (convective overshoot) brings fresh fuel into the convective region causing local hot spots of nuclear burning. This is very different from the situation encountered in spherically symmetric computations. Additionally, chemical inhomogeneities create gradients of chemical composition which result in additional mixing (semi-convection). Large scale mass motions and local burning conditions are likely to change the quantitative yields of many isotopes from any single massive star. However, any non-monotonic and/or stochastic nature of the nucleosynthetic yields as a function of stellar mass tend to be smoothed out by integration over an initial mass function of the stellar population contributing to the observable gamma-ray emission. Thus, the general features of the integrated yields, as determined from spherically symmetric models, may remain intact.

Rotation may have also have large systematic effects. In massive stars, the amount of mixing in the hydrogen envelope is increased, thus processing more ^{25}Mg into ^{26}Al. The helium core mass may be larger as a result, thus leading to more ^{26}Al in the larger neon-oxygen layers (Meynet & Maeder 1997; Langer et al. 1997). Rotating or not, complete loss of the hydrogen and helium layers, either through mass loss as a single star or through the effects of close binary evolution (Braun & Langer 1995), could affect the quantitative yields of most of the presupernova abundances. Recent improvement of the physical ingredients in Wolf-Rayet ^{26}Al production models provide some consolidation of hydrostatic nucleosynthesis in massive stars (Arnould et al. 1997; Meynet et al. 1997). These studies point out that combinations of metallicity and mass loss can increase the mass of ^{26}Al ejected by ~ 2. The ^{26}Al abundance present in the hydrogen shell is ejected unmodified in the explosion or, in more massive stars during their Wolf-Rayet phase, by a stellar wind.

The ^{26}Al synthesized in the oxygen-neon shell may be enhanced by $\sim 30\%$ due to operation of the neutrino-process (Woosley et al. 1990). Protons liberated by ν-spallation of ^{20}Ne capture on ^{25}Mg to produce ^{26}Al. When the oxygen-neon shells are located closer to the collapsing core, a larger higher ν-flux is encountered. This may enhance the ^{26}Al yield by $\sim 50\%$. The relative importance of the ν-process contribution depends on the ν energy spectrum: the peak μ and τ neutrino temperature is somewhat uncertain (Myra & Burrows 1990), in spite of the SN 1987A measurements (Arnett et al. 1989). Few transport calculations have been carried long enough (at least 3 s) with sufficient energy resolution to see the hardening of the neutrino spectrum that occurs as the proto-neutron star cools.

The genesis of the central compact remnant in core-collapse supernovae bears on several interesting physical problems, some of which may be constrained by radioactivity observations. Below main sequence masses of 19 $M_\odot$, formation of a neutron star appears likely. Modern nuclear equations of state suggest that above 19 $M_\odot$, the neutron star becomes a black hole. The energy of the explosion, placement of the mass cut, and how much mass falls back onto the remnant can all modify this neutron star - black hole bifurcation point. Each of these processes also dramatically affects the mass of the nucleosynthesis products ejected from the inner regions of the supernova. For example, producing more than 10^{-4} $M_\odot$ of radioactive ^{44}Ti is difficult in models, but making much less is easy. Ejection of any ^{44}Ti is sensitive to how much mass falls back onto the remnant. To achieve a nearly constant kinetic energy of the ejecta, the explosion energy in $M \geq 25$ $M_\odot$ presupernova models must be steadily increased in order to overcome the increased binding energy of the mantle. However, even in the 35 and 40 $M_\odot$ stars the density profile may still cause nearly all the ^{44}Ti produced to fall back onto the compact remnant. Unless the explosion mechanism, for unknown reasons, provides a much larger characteristic energy in more massive stars, it appears likely that stars larger than about 30 $M_\odot$ will have dramatically reduced ^{44}Ti yields and leave massive remnants ($M \geq 10$ $M_\odot$) which become black holes. If, however, the explosion is energetic enough (perhaps overly energetic) more ^{44}Ti is ejected.

The genesis and evolution of an accreting white dwarf which eventually becomes a classical novae is essentially unknown, yet has important consequences on the Galactic γ-ray signal (Kolb & Politano 1997). For example, uncertainty in the binary mass distribution function for nova progenitors corresponds to an uncertainty in the fraction of white dwarfs that originate from ≥ 8 $M_\odot$ main-sequence stars. Detection of strong neon lines in nova envelopes has suggested the category "neon-nova" (Starrfield et al. 1996). Derived from about a dozen well-measured nova events, the fraction of metal-enriched neon-novae is typically quoted as 25-30%. The material composition of the nuclear burning region is usually assumed to be a 50-50 mixture of accreted material (solar composition) and dredged up white dwarf material. Typically the white dwarf material is assumed to oxygen-neon-magnesium in mass proportions of 0.3:0.5:0.2 (e.g. Politano et al. 1995). Evolution of a 10 $M_\odot$ star, however, suggests mass ratios 0.5:0.3:0.01 might be more appropriate (Ritossa et al. 1996). If confirmed, this eliminates most of the necessary seed material from which ^{26}Al is synthesized, since the nova yields of ^{26}Al and ^{22}Na are sensitive to the initial ^{25}Mg and ^{20}Ne abundances. Inconsistencies between the mass ejected by neon-novae models and observed total ejected masses (models are too low by an order of magnitude, e.g. Hernanz et al. 1996) show that our understanding of the nova process is incomplete, in spite of the first-order success of the thermonuclear runaway model. If the white dwarf has $M \geq 1.1$ $M_\odot$, then less accreted matter is needed to trigger a classical nova outburst. Correspondingly, the amount of matter is expected to be less. If the inconsistency between model and observed ejected masses is resolved by use of a $M < 1.1$ $M_\odot$ white dwarf, then the ^{26}Al yields are expected to increase due to the lower burning temperatures. If resolution depends upon larger masses being ejected through increased mixing of core material, then higher burning temperatures is expected to decrease of mass of ^{26}Al ejected, as compared to present baseline models in the 1.15-1.35 $M_\odot$ range.

Gamma-ray astronomy seeks to constrain the astrophysical origin site(s) of radioactive isotopes in the Galaxy, and on smaller spatial scales corresponding to regions of coherent star formation. The differences between the derived spatial distributions of the competing sources are often not large enough to provide a crisp discriminant, especially since young population stars relate to several source types (Prantzos & Diehl 1996). Nevertheless, spatial resolution of γ-ray line measurements in sub-degree domains should provide such a test (e.g. AGB stars resulting in a 1.809 MeV glow trailing spiral arms). Are there other measurable quantities that might help distinguish between the competing sources? Yes, one may look for the associated production of other radioactive isotopes. For example, γ-rays from radioactive ^{60}Fe may be a very good discriminant for the contested origin site of Galactic ^{26}Al for two main reasons: (1) Type II supernovae produce comparable amounts of ^{26}Al and ^{60}Fe, while the other potential candidates produce significantly smaller amounts of ^{60}Fe. (2) Even more importantly, the "^{26}Al follows ^{60}Fe" model has consequences that can be measured with present generation γ-ray spectrometers. For individual events, on the other hand, γ-ray measurements can provide specific constraints to detailed models, complementing other physical parameters which are often observed for these in great detail. We separately discuss these two aspects of γ-ray line measurement impacts in the following.

INTEGRATED NUCLEOSYNTHESIS

The HEAO-C discovery of the 1.809 MeV gamma-ray line from radioactive ^{26}Al in the Galaxy (Mahoney et al. 1982), which had been anticipated from theoretical considerations (Clayton 1971, 1982; Ramaty & Lingenfelter 1977), opened new doors in gamma-ray astronomy. Measurements of various radioactive decays in the Milky Way offers a global proof of the idea that heavy element formation occurs continuously and mainly inside stars (Burbidge et al., 1957, Cameron 1957). In particular, radioactive decay times of $\sim 10^6$ y (as for ^{26}Al and ^{60}Fe) probe a time scale which is short compared to galactic evolution, thus testifying to the occurrence of relatively recent nucleosynthesis events throughout the Galaxy. The ^{26}Al and ^{60}Fe isotopes provide this unique observational window, as massive-star nucleosynthesis is expected to produce these trace elements in quantities sufficient to yield γ-ray line fluxes above instrumental threshold sensitivities. For typical yields of $\sim 10^{-4}$ $M_\odot$ per event, approximately 10000 individual events will contribute to the line emission from the Galaxy, resulting in a typical flux of $\sim 10^{-5}$ ph cm^{-2} s^{-1}.

Aluminum 26 in the Galaxy

Eight experiments over the 15 years since the pioneering detection have reported detecting γ-ray emission from radioactive ^{26}Al (see review by Prantzos & Diehl 1996). Instrumental capabilities differ substantially, but the integrated flux measured from the general direction of the inner Galaxy, integrated over latitude and the inner radian in longitude, has been used to roughly compare results. We summarize the flux values in Fig. 1, including the results reviewed by Prantzos & Diehl (1996) plus the recent GRIS measurement (Naya et al. 1996). All measurements are consistent with values $\sim 4 \times 10^{-4}$ ph cm^{-2} s^{-1}/rad, as indicated by the horizontal line. Note that the determination method varies between the instruments, and in particular the flux values for the non-imaging instruments depend on the assumed spatial distribution: All non-imaging instruments essentially assume the same (or equivalent) smooth spatial distribution derived from COS-B measurements of Galactic gamma-rays in the 100 MeV regime. The distribution of 1.809 MeV emission seems significantly different, however. Maximum entropy deconvolution images derived from the COMPTEL Telescope measurements show spatial structure in the emission (Fig. 2). The ridge of the Galactic plane dominates, but there is asymmetry in the emission profile along the disk, and there are several prominent regions of emission such as Vela and Cygnus. All estimates of the absolute ^{26}Al mass in the Galaxy rest on assumptions about the spatial

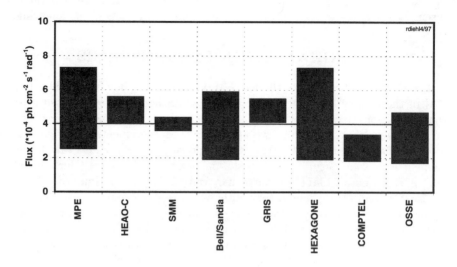

FIGURE 1. Reported 1.809 MeV flux values for the inner Galaxy, from 8 experiments. The length of each vertical stripe is set by the reported uncertainties. A value of $\sim 4 \times 10^{-4}$ ph cm^{-2}s^{-1}rad^{-1} appears plausible given all experiments, although some discrepancies will have to be investigated.

distribution of the sources, as the 1.809 MeV measurements themselves do not carry distance information. The COMPTEL team fitted several models of candidate source distributions to the COMPTEL data, such as CO survey data (Dame et al. 1987), analytical models based on exponential disks, and H II region data evaluated in the context of spiral-arm structure (Taylor & Cordes 1993). When localized regions of emission beyond the inner Galaxy are excluded, then all axisymmetric model fits yield a Galactic mass of 2-3 M$_\odot$ (Diehl et al. 1995, Knödlseder et al. 1996a, Diehl et al. 1997). The extent of spiral-arm emission can be estimated if a composite model of disk emission plus emission along spiral-arms is adopted and compared to the disk-only model. Spiral structure appears significant, including between 1.1 M$_\odot$ and all of the ^{26}Al (Diehl et al. 1997).

Yet, imaging of MeV gamma-rays is far from straightforward, both due to the high instrumental backgrounds, and the complex gamma-ray detection methods. Consistency checks between different techniques have shown that some of the spikyness of the apparent emission in the COMPTEL result can be instrumental (e.g., Knödlseder et al. 1996a). Nevertheless, significant emission from the Cygnus, Carina, and Vela regions appears consolidated. In the Vela region, there was some hope to detect ^{26}Al from one single source, the nearby Vela supernova remnant (Oberlack et al. 1994; Diehl et al. 1995).

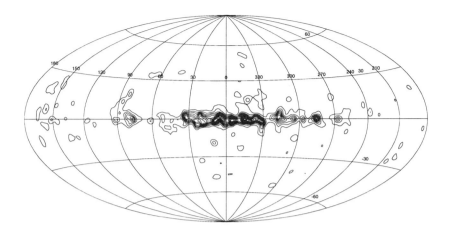

FIGURE 2. COMPTEL 1.8 MeV all-sky image, derived from 5 years of observation through Maximum Entropy deconvolution (Oberlack et al., 1997)

Recent results show the main 1.809 MeV feature is significantly offset from the Vela supernova remnant. A direct calibration of core-collapse supernova nucleosynthesis, which was a tantalizing prospect two years ago (Diehl et al. 1995), does not appear feasible as superimposed emission from other sources at larger distances must be considered significant (Oberlack et al. 1997). Another candidate source in the Vela region is the Wolf Rayet binary system γ^2 Velorum, holding the closest known Wolf Rayet star WR11 (van der Hucht 1988). Recent Hipparcos parallax measurements suggest this binary system is at a distance of 250-310 pc, which is even closer than previous estimates of 300-450 pc (van der Hucht et al. 1997; Schaerer, Schmutz & Grenon 1997). At this close of a distance the absence of a detection from γ^2 Velorum in the COMPTEL data is unexpected, particularly since recent models suggest larger yields for this object (Meynet et al. 1997).

About 80% of the prominent 1809 keV emission associated with the Cygnus region can be understood in terms of the expected ^{26}Al signal from known sources (del Rio et al. 1996). One may be worried about this high percentage, since ^{26}Al decays on a time scale longer than the emission features of supernova remnants and Wolf-Rayet winds used to make this inference. The ^{26}Al attributed to 'seen' sources should be multiplied by a factor ≥ 1 to account for 'unseen' sources. The latest COMPTEL images show structures which appear consistent with the Cygnus superbubble and Cyg OB1, but further analysis is required on this data to investigate the significance of emission from Wolf-Rayet stars in this region.

The Carina region shows the densest concentration of young open clusters along the plane of the Galaxy. Knödlseder et al. (1996b) discuss how this concentration may relate to the observed 1.809 MeV feature, which is spatially more confined than originally estimated from analysis of the Milky Way's spiral structure (Prantzos 1993ab). It is also interesting that some of the 1809 keV image structures which fail to align with spiral arms, do coincide with directions towards nearby associations of massive stars (Diehl 1995; Knödlseder et al. 1997). Patchiness in such a nucleosynthetic snapshot might be expected from the clustering of formation environments of massive stars (Elmegreen & Efremov 1996). If viewed from the outside, the Milky Way might also display the signs of massive star populations in the form of HII regions arranged like beads on a string along spiral arms, such as observed in M31 from H_α emission analysis (Williams et al. 1995), or in M51 from heated dust seen in infrared continuum at 15 μm (Kessler et al. 1996). Interstellar absorption and source confusion prevents such mapping within the Milky Way, unfortunately. Therefore, detailed investigations of the COMPTEL image systematic uncertainty, that is, a quantitative limit to artificial bumpiness of the imaging algorithm, will be important for such interpretation of 1.809 MeV emission. Such concerns also hold for different instruments and future measurements of $\geq$ 20°segments of the Galactic plane with sub-degree resolutions.

In a recent balloon flight, the Gamma Ray Imaging Spectrometer (GRIS) drift scanned the Galactic center region with its $\sim$100° field of view, and detected the 1.809 MeV line at 6.8σ with a flux of 4.8$\pm$0.7 $\times$ 10^{-4} ph cm^{-2} s^{-1}rad^{-1} (Naya et al. 1996). The main surprise of this measurement is the width of the astrophysical line profile, which was significantly broader than the instrumental resolution of the Ge detector, and reported as ΔE = 5.4$\pm$1.4 keV. This line width is much larger than expected from Galactic rotation ($\leq$ 1 keV; Gehrels & Chen 1996), which dominate above the broadening from random motions in the interstellar medium (Ramaty & Lingenfelter 1977). It is presently difficult to understand how such motion could be maintained over the million year time scale of ^{26}Al decay. Thermal broadening by a very hot phase of the interstellar medium ($\sim$ 10^8 K) with long cooling times ($\sim$ 10^5 years), or a kinetic broadening at high average velocities ($\sim$ 500 km/s) seems required. Either case requires extremely low density phases of the interstellar medium on large spatial scales (Chen et al. 1997). Alternatively one can hypothesize massive, high speed, dust grains rich in ^{26}Al to explain the measurement (Chen et al. 1997). Further observations of the line shape details are required to examine any spatial variations in the line broadening. Spectral resolution of $\sim$ 2 keV is required for such a study. Although INTEGRAL may still have insufficient energy resolution to make complete velocity maps of the 1.809 MeV emission along the plane of the Galaxy, the brightest features can probably have their velocity centroids determined well enough to place them on the Galactic rotation curve and thus derive a distance to the features. Surveys which combine velocity information and Galactic latitude extent could then examine the existence and nature of any Galactic "fountains" and "chim-

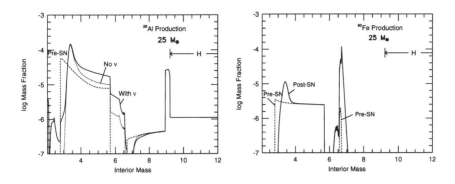

FIGURE 3. Mass profiles of ^{26}Al (left) and ^{60}Fe (right) for a 25 M$_\odot$ model.

neys" from possible "venting" of ^{26}Al into the Galactic halo. In any case, the large line width measured by GRIS, although inconsistent with the HEAO-C line width limit of ≤ 3 keV, needs confirmation, since it could have profound implications to our understanding of the interstellar medium in the Galaxy.

Iron 60 in the Galaxy

Physically, ^{60}Fe should be a good discriminant of different source types generating ^{26}Al, because massive stars produce ^{26}Al and ^{60}Fe in the same regions and in roughly comparable amounts (Fig. 3). Shown are abundances as a function of mass inside a 25 M$_\odot$, solar metallicity, Type II supernova model (Woosley & Weaver 1995) at the end of the presupernova evolution (dashed) and the final, post explosion abundances (solid curves; Timmes et al. 1995). While the ^{26}Al production occurs in the hydrogen shell and the oxygen-neon shell, ^{60}Fe is produced in He shell burning and at the base of the oxygen-neon shell. Most important is that the majority of both ^{26}Al and ^{60}Fe are produced, mainly during the presupernova evolution, between 3-6 M$_\odot$. These two isotopes should have similar spatial distributions after the explosion of the star.

Certainly ^{26}Al and ^{60}Fe are produced in different ratios in stars of different mass (Fig. 4). Massive stars heavier than ~ 25 M$_\odot$ tend to synthesize more ^{26}Al than ^{60}Fe in the Woosley & Weaver models, while the two isotopes are produced in roughly equal amounts below 25 M$_\odot$. An estimate for the injection rate into the Milky Way is the steady-state event rate times the average mass ejected per event. Taking $M(^{26}\text{Al}) \sim 10^{-4}$ M$_\odot$, $M(^{60}\text{Fe}) \sim 4 \times 10^{-5}$ M$_\odot$ and ~ 2 core-collapse supernovae per century, one has $\dot{M}(^{26}\text{Al}) \sim 2.0$ M$_\odot$/Myr and $\dot{M}(^{60}\text{Fe}) \sim 0.8$ M$_\odot$/Myr. More refined chemical evolution calculations

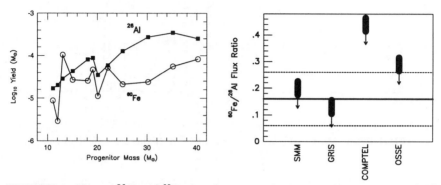

FIGURE 4. Mass of ^{26}Al and ^{60}Fe ejected versus main–sequence progenitor mass (left), assuming no mass loss. ^{26}Al to ^{60}Fe ratio: range predicted from supernova models (line) and measurements.

suggest that Type II supernovae are responsible for 2.2 ± 1.1 M$_\odot$ of ^{26}Al and 1.7 ± 0.9 M$_\odot$ of ^{60}Fe in the Galaxy. These values, when coupled with the two isotopes being synthesized in the same physical region, gives rise to several predictions. When ^{60}Fe is dominated from these sources, and can be unambiguously detected with gamma-ray telescopes, the ^{60}Fe flux map will peak towards the inner Galaxy, the ^{60}Fe emission will follow the ^{26}Al distribution (adding weight to assertions that Type II events make most of the Galactic ^{26}Al), the ^{60}Fe and ^{26}Al hot spots will overlap, and the ^{60}Fe/^{26}Al line flux ratio will be $16 \pm 10\%$: the ^{26}Al image (see Fig. 2) will be the ^{60}Fe image; the ^{60}Fe skymap will have the same morphology, but will be dimmer.

A few recent ^{60}Fe measurements and flux ratios with ^{26}Al are shown in Fig. 4. SMM reported an upper limit of 8.1×10^{-5} ph cm^{-2} s^{-1} for the 1.173 MeV ^{60}Co line over the central radian of Galactic longitude, giving an upper limit of 1.7 M$_\odot$ of ^{60}Fe (Leising & Share 1994). This ^{60}Fe mass is consistent with the expectations given above, and is $\sim 20\%$ of the ^{26}Al flux. Recently, the GRIS (Naya et al. 1997), OSSE (Harris 1997), and COMPTEL (Diehl et al. 1997) teams have reported upper limit ^{60}Fe/^{26}Al ratios. If the stringent GRIS measurements are confirmed, then the initial model estimates for the total Galactic flux ratio might be too large. The qualitative picture of ^{26}Al and ^{60}Fe tracing each other to a good degree of precision still presents an important observational challenge, even more so with a smaller flux ratio.

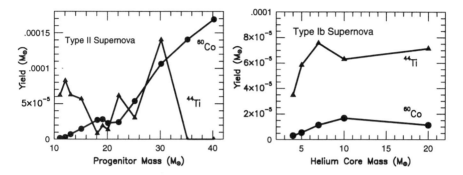

FIGURE 5. ^{44}Ti yields from Type II and Ib supernovae.

SHORTER LIVED RADIOACTIVITIES

Titanium 44

In massive stars, stable ^{44}Ca is produced chiefly, though not exclusively, as radioactive ^{44}Ti in the "α-rich freeze-out". This process occurs when material initially in nuclear statistical equilibrium, and a relatively low density, is cooled rapidly enough that the free α-particles do not have time to merge back into the iron group by the relatively inefficient triple-α reaction. Thus, the distribution of nuclei cools down in the presence of an anomalously large concentration of α-particles (Hix & Thielemann 1996). Representative yields of ^{44}Ti from solar metallicity Type II and Type Ib supernova models are shown in Fig. 5 (Woosley & Weaver 1995; Woosley, Langer & Weaver 1995). Production in Type Ib supernovae is more uniform because the models converge to a common presupernova mass in the narrow range 2.3–3.6 $M_\odot$. All of the models, whose ejecta all have $\sim 10^{51}$ erg of kinetic energy at infinity, predict ^{44}Ti yields between 1-15 $\times 10^{-5}$ $M_\odot$. Typical values are $\sim 3 \times 10^{-5}$ $M_\odot$ for the Type II models, or twice that value for the Type Ib models.

Thielemann, Nomoto & Hashimoto (1996) have also examined the detailed nucleosynthesis of core collapse supernovae, and they find, in general, that larger amounts of ^{44}Ti are ejected. The differences may be due to the nuclear reaction rates employed, or how the explosion is simulated, or the progenitor structure. Thielemann et al. explode their stellar models by depositing thermal energy deep in the neutronized core (which later becomes the neutron star). This gives a larger entropy to the innermost zones than what a momentum (piston) driven explosion would impart and, in principle, eject more material. A larger entropy also ensures a more vigorous alpha–rich freeze out, and thus a larger ^{44}Ti production. Thielemann et al. sum the ejecta from the outside of the star inwards and place the mass cut (which is artificial in all models) at the position where sufficient ^{56}Ni is produced to explain the observations. Woosley & Weaver place the piston at a suitable first approximation to mass cut and then follow an explosion trajectory. It is the difference between injecting momentum or energy in modeling the explosion that leads to the two groups following different adiabatic paths (Aufderheide, Baron & Thielemann 1991). What self-consistent explosion models do, exploding via neutrino heating and multidimensional convection, might be reflective of one dimensional momentum driven explosions, one dimensional energy driven explosions, or some intermediate case. Hence, the differences between the two groups in the amount of ^{44}Ti ejected is an example of the spread one obtains due to uncertainties in modeling the Type II explosion mechanism.

There are 3 γ-ray lines one can use to examine or detect the decay of ^{44}Ti; the 67.9 and 78.4 keV lines from the ^{44}Sc de-excitation cascade, and the 1.157 MeV line as ^{44}Ca decays to it's stable ground state. These transitions are shown in Fig. 6 with their respective spins, parities, and energies. The half-life of ^{44}Ti, used to translate observed γ-ray fluxes into a supernova mass of ^{44}Ti, is surprisingly uncertain (Fig. 6). Values range from 39 to 66 years, with a trend towards around 44 years for methods determining the activity, and scattering around 58 years for methods following the decay curve. Why is the half-life so difficult to measure? The basic reasons are that the number of ^{44}Ti nuclei one can obtain is small, and the half-life is large.

The Cas A supernova remnant is relatively close (2.8–3.7 kpc), young (explosion in 1668–1680) and wide (physical diameter $\sim$ 4 pc), making it one of the premier sites for studying the composition and early behavior of a supernova remnant. It may only be equaled as SN 1987A unfolds. The discovery of 1157 keV gamma-rays from the $\sim$ 300 year-old Cas A supernova remnant (Iyudin et al. 1994; Fig. 7) was a scientific surprise, because supernovae models had indicated $\sim 3 \times 10^{-5}$ M$_\odot$ of ^{44}Ti would be ejected (see Fig. 5), which translates into a γ-ray intensity generally below instrument flux sensitivities. Recent analysis of COMPTEL data supports a 5σ detection of Cas A at $4.8 \pm 0.9 \times 10^{-5}$ ph cm^{-2} s^{-1} in the 1.157 MeV line, implying 1.0-4.0$\times 10^{-4}$ M$_\odot$ of ^{44}Ti (Iyudin et al. 1997). Conversion of the measured flux into the mass limits takes into account the uncertainties in the ^{44}Ti half-life, distance to the event, and the precise time of the explosion.

Data taken with the OSSE instrument, when best-fit to all three ^{44}Ti gamma-ray lines, gives an (insignificant) flux value of $1.7 \pm 1.4 \times 10^{-5}$ ph cm^{-2} s^{-1} for Cas A (The et al. 1996). The uncertainty of systematic errors in the COMPTEL flux determination, coupled with low OSSE upper limits, confused the community for a while, but the two instruments now appear to yield values that are consistent within their respective uncertainties (Iyudin et

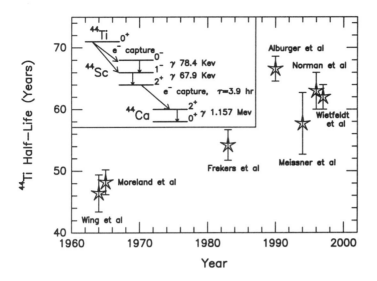

FIGURE 6. ^{44}Ti decay scheme, and measurements of the half life.

al. 1994, 1997; Schönfelder et al. 1996; The et al. 1995, 1996). Measurements of the 68 and 78 keV photons by the Rossi X-Ray Timing Explorer (RXTE) result in a measured flux of $2.87 \pm 1.95 \times 10^{-5}$ ph cm^{-2} s^{-1} (Rothschild et al. 1997). Due to activation of radioactive daughters by cosmic rays, however, the detection of ^{44}Ti lines has been more difficult than expected also for this instrument. Additional measurements are necessary.

From the γ-ray observations one may constrain the ^{44}Ti half-life by assuming the supernova model yields, a distance to Cas A of 3.4 kpc, and the year of the explosion as ~ 1680. This procedure tends to favor the larger ^{44}Ti half-lives.

The abundance of ^{44}Ti and ^{56}Ni as a function of mass inside a 25 M$_\odot$ star is shown in Fig. 8. The mass cut is shown as the solid vertical line. Everything interior to the mass cut becomes part of the neutron star, everything exterior may be ejected, depending on how much mass falls back onto the neutron star during the explosion. Regardless, if ^{44}Ti is ejected, so is ^{56}Ni (Timmes et al. 1996). The isotope ^{56}Ni is made with ^{44}Ti, in the same mass zone regime, and is 3 orders of magnitude more abundant than ^{44}Ti. A large quantity of ^{56}Ni ejected means a bright supernova. How bright? An abundance of ^{44}Ti as large as reported ($\sim 10^{-4}$ M$_\odot$) would probably mean the ejection of at least 0.05 M$_\odot$ of ^{56}Ni. With or without a hydrogen envelope. the supernova would then have a peak luminosity brighter than the 10^{42} erg/s observed for SN 1987A. If no interstellar absorption would attenuate the optical light curve, the Cas A supernova should have had a peak apparent magnitude of

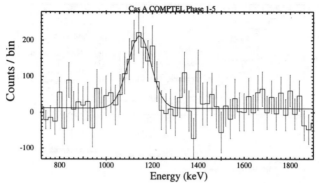

FIGURE 7. ^{44}Ti measurement from Cas A (Iyudin et al. 1997)

-4, easily recognizable on the sky. Cas A was not widely reported as such; some 10 magnitudes of visual extinction is required to make the γ-ray ^{44}Ti measurements consistent with these historical records.

There may indeed have been such a large visual extinction to Cas A at the time of the explosion. At a distance of $\sim$ 3 kpc the optical extinction in the plane of the Galaxy has long been estimated to be 4-6 magnitudes. If Cas A was embedded in a dusty region or experienced significant mass loss which condensed into dust grains before the explosion, the extinction could have been even larger. Measurements of the X-ray scattering halo around Cas A from the ROSAT and ASCA satellites offer some compelling evidence for a

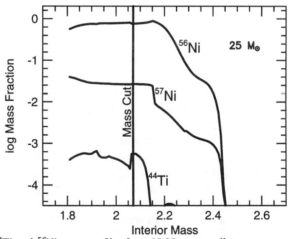

FIGURE 8. ^{44}Ti and ^{56}Ni mass profiles for a 25 M$_\odot$ core collapse supernova model.

larger reddening correction ($N_H=1.8\times10^{22}$ cm^{-2} attributed to an additional material component that must be close to Cas A, and relatively dust-free (Predehl & Schmitt 1995). This corresponds to an additional optical extinction of $A_V=5$ at the time of the supernova, just about what is required. While hypotheses such as the explosion took place in a dense molecular cloud, or a optically thick cloud occulted Cas A at the time of the explosion cannot be ruled out, secondary evidence suggest these scenarios are unlikely. A perhaps more natural hypothesis for this local material is the dusty shell of material ejected prior to the explosion as a Type Ib supernova. The supernova shock wave could have destroyed the dust as it propagated through the debris and the material surrounding the Cas A supernova. This may have resulted in a dimmer supernova, explaining the lack of optical detection, and it would naturally explain the excess neutral hydrogen column density (Hartmann et al. 1997).

The present nature of the compact remnant in Cas A is ambiguous, with the actual mass at the onset of carbon burning being critical. Deep X-ray imaging around Cas A does not reveal the synchrotron nebula one might expect around a neutron star (Ellison et al. 1994). Deep infrared images taken under excellent seeing conditions have set strict magnitude limits on the presence of a stellar remnant (van den Bergh & Pritchet 1986; Fesen & Becker 1991). The region around the center of Cas A simply appears void of any detectable stars. So observationally, a black hole may have formed. This would make the ejection of a lot of ^{44}Ti more difficult, but not an impossibly rare event. The black hole mass would need to be not too far above the critical gravitational mass, probably less than 2 $M_\odot$.

Beryllium 7 and Sodium 22

Classical novae are the most plausible sources for detection of the 478 keV gamma-ray line from the decay of radioactive ^{7}Be, and the 1.275 MeV gamma-ray line from the decay of ^{22}Na. Other sources such as red giant stars and supernovae also synthesize ^{7}Be, but the isotope cannot be transported with sufficient rapidity into regimes that are transparent to γ-rays. The explosion of a massive star also produces ^{22}Na in potentially detectable amounts, but classical novae remain a favored candidate due to their relative proximity (known Galactic events) and their relative frequency of occurrence ($\sim$ 30/year in the Galaxy). Each of these two isotopes probe different phases of the thermonuclear runaway model. Both have also been a target of γ-ray astronomy for quite some time (Hoyle & Clayton 1974).

Nuclei of mass A=7 are formed predominantly through the ^{3}He$(\alpha,\gamma)^7$Be reaction (Fig. 9), whether in classical novae explosions, hydrostatic red giant envelopes, or primordial nucleosynthesis scenarios. In novae, the ^{3}He seed material originates mainly from the accreted material and to a much lesser

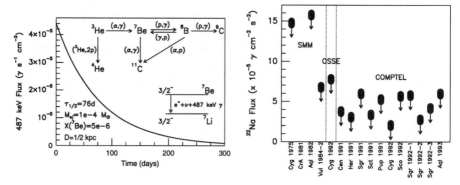

FIGURE 9. ^{7}Be decay scheme and production processes (left), and ^{22}Na measurements from individual nova (right).

extent from incomplete hydrogen burning (first two steps of the pp chain). Competing with this reaction in the consumption of ^{3}He is ^{3}He(^{3}He,2p)^{4}He reaction. This third step of the pp chain is always faster, except at very low ^{3}He abundances. As such, the final ^{7}Be yields from novae display a logarithmic, not a linear, dependence on the initial ^{3}He content (Boffin et al. 1993). A solar ^{3}He mass fraction in the accreted envelope cannot generate a ^{7}Li (^{7}Be) abundance that greatly exceeds the solar mass fraction.

After ^{7}Be is produced, it is destroyed mainly by proton captures to form ^{8}B. This destruction is quite efficient below temperatures of $\sim 10^8$ K and densities above ~ 100 g/cm^3. At hotter temperatures the photodisintegration of ^{8}B reduces the overall destruction somewhat. Including the leakage of ^{8}B to ^{9}C is important, since it reduces the ^{7}Be abundances by allowing a flow out of any ^{7}Be-^{8}B equilibrium (Boffin et al. 1993). Once the nuclear reactions cease any remaining ^{7}Be will beta decay to ^{7}Li with a half-life of 76 days, and produce the 487 keV photon that is a target of γ-ray astronomy. Assuming a nova ejects a total mass of 10^{-4} M$_\odot$, of which the ^{7}Be mass fraction is 5×10^{-6}, the 487 keV flux is shown as a function of time in Fig. 9 for a distance to the novae of 0.5 kpc. These values may not be typical and preferences for other values are easily accommodated; the flux scales linearly with the total ejected mass and ^{7}Be mass fraction, while scaling with the inverse square of the distance. For the conditions shown, a space-borne spectrometer with a sensitivity of 10^{-5} ph cm^{-2} s^{-1} will be able to study such novae for about 3 months.

Convective processing and mixing timescales in nova envelopes are critical issues. Significant γ-ray signals from ^{7}Li can only occur when convection can transport freshly synthesized ^{7}Be into a cooler and less dense regions. Spherically symmetric models of classical novae which incorporate time-dependent convection and the accretion phase tend to give ^{7}Be yields between 10^{-12}-10^{-10} M$_\odot$ depending on the mass and composition of the white dwarf and accreted material (Politano et al. 1995). Two-dimensional hydrodynamic models that examine the evolution of a fully convective hydrogen-rich envelope during the earliest stages of the thermonuclear runaway are beginning to be calculated, and should help clarify some of these uncertainties in the future (Shankar & Arnett 1994; Glasner, Livne, & Truran 1997).

There have been no reported detections of 478 keV line emission from either the cumulative effects of many novae (∼100) near the Galactic center region, or individual novae. However, SMM reported a 3σ upper limit flux of 1.7×10^{-4} ph cm^{-2} s^{-1}, assuming a point source of constant intensity at the Galactic Center (Harris, Leising & Share 1988). They also reported 3σ upper limit fluxes of 2.0×10^{-3} ph cm^{-2} s^{-1} from Nova Aql 1982, 8.1×10^{-4} ph cm^{-2} s^{-1} from Nova Vul 1984, and 1.1×10^{-3} ph cm^{-2} s^{-1} from Nova Cen 1986. These results imply upper limit ^{7}Be abundances that are about an order of magnitude above even the most optimistic theoretical expectations.

The radioactive isotope ^{22}Na is synthesized in classical novae at the interface between the white dwarf and the accreted envelope. How much carbon, oxygen, magnesium and neon is dredged up from the white dwarf, or accreted from the binary companion is a critical issue for determining the strength of any ^{22}Na γ-ray signal. The chief nuclear flows that create and destroy ^{22}Na are discussed in detail by Higdon & Fowler 1987 and Coc et al. 1995, along with the uncertainties in several key reaction rates. As the ejecta cools, expands, and nuclear burning ceases (∼ 7 days), ^{22}Na decays with a half-life of 2.60 years from the $J^\pi=3^+$ state to a short-lived excited state of ^{22}Ne, emitting a 1.275 MeV gamma-ray line in the process.

Gamma-ray line emission at 1.275 MeV from classical novae in the Milky Way remains to be positively detected. SMM, OSSE, and COMPTEL have, however, reported upper limits for several individual nova (Leising 1993; Iyudin et al. 1995; Fig 9). The 2σ upper limit fluxes reported by COMPTEL puts tight constraints on thermonuclear runaway models of classical novae. For neon-novae, which are expected to eject the largest ^{22}Na masses, the average COMPTEL upper limit flux of 3×10^{-5} ph cm^{-2} s^{-1} translates into an upper limit on the ejected ^{22}Na mass of 3.7×10^{-8} M$_\odot$ (Iyudin et al. 1995).

Spherically symmetric models of classical novae which incorporate a more self-consistent treatment of the hydrodynamic evolution and refined initial compositions tend to give ^{22}Na yields of ∼10^{-8} M$_\odot$ for the most favorable neon-novae cases (Starrfield et al. 1996). In general, these recent improvements to the modeling have reduced the ^{22}Na yield by about an order of magnitude compared to the older models (Starrfield et al. 1992) that lacked these refinements and were at distinct odds with COMPTEL measurements.

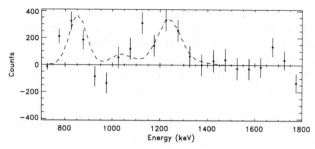

FIGURE 10. SN1991T spectrum as measured by COMPTEL, after subtraction of a background model (Morris et al., 1995, and these proceedings).

It may still be that present nova models are biased from the observations of a few exceptional events, and that the typical neon-novae ejects much less mass into the interstellar medium than the exceptional events.

Nickel 56

A goal of observational γ-ray astronomy, since at least Clayton, Colgate & Fishman in 1969, has been detection of line radiation from the decay of radioactive ^{56}Ni and ^{56}Co produced in Type Ia supernovae. Type Ia events are favored over the other supernova classes because they produce a lot of ^{56}Ni (~ 0.6 M$_\odot$; Thielemann et al. 1986), and they expand rapidly enough to allow the γ-rays to escape before all the fresh radioactivity has decayed. Type Ia supernovae are favored over the other classes because they produce a lot of ^{56}Ni. Even so, detection of these events has been difficult to achieve. Type Ib supernovae synthesize 5 to 10 times less ^{56}Ni than a typical Type Ia, but they also expand rapidly. The best studied supernova of all in radioactive line emission was the Type II supernova SN 1987A (e.g., Arnett et al. 1989), but such studies were possible only because of its proximity. The γ-ray line detection of any Type II supernova outside the Local Group is very improbable.

Only one Type Ia supernova has been seen in γ-rays, SN 1991T in NGC 4527 (Morris et al. 1995). This galaxy is ~ 17 Mpc distant and in the direction of the Virgo cluster. The supernova was unusually bright at maximum (0.7 M$_\odot$ of ^{56}Ni; Höflich et al. 1996) and the light curve evolved unusually slow. Detection of high velocity ($\sim$13000 km/s) iron and nickel in the outer layers of SN 1991T favors models where the subsonic flame front propagates larger distances from the white dwarf core before making the transition to a detonation. These types of delayed-detonation models are also consistent with the velocity profile of most of the other ejecta (silicon, calcium) seen in SN 1991T. Shigeyama et al. (1993) suggest that detection in the early light curve of the 812 keV γ-ray line from the decay of ^{56}Ni would be direct evidence for delayed-detonation models, as the line cannot be seen when the ^{56}Ni is embedded deeper in other categories of Type Ia models.

The tentative COMPTEL detection of the ^{56}Co decay γ-rays (Morris et al. 1995, Fig. 10) indicates this isotope was present in the outer envelope, and thus supports extensive mixing scenarios. At a distance of 13 Mpc, however, the COMPTEL measurement converts into a surprisingly large ^{56}Ni mass of 1.3 M$_\odot$. This requires that almost all of the Chandrasekhar mass white dwarf must be turned into radioactive ^{56}Ni. The OSSE upper limits (Leising et al. 1995) may indicate that the ^{56}Co line flux derived by Morris et al. (1995) is optimistic, although the detection itself seems significant (Morris et al., these proceedings). More detections of Type Ia supernovae in ^{56}Ni are required to clarify how typical SN 1991T was. COMPTEL and OSSE will hopefully remain ready to observe any nearby events (for estimates see below).

ASSOCIATED ASTROPHYSICAL CHALLENGES

Meteoritic Grains

Although the nucleosynthetic processes occurring in different stars generally result in a wide range of isotopic compositions, by far most of the material from the many stellar sources that contributed to the proto-solar cloud was thoroughly processed and mixed, which resulted in the essentially isotopically homogeneous solar system we know today. However, a small fraction of the original material, in the form of presolar dust grains, survived solar system formation and was trapped in primitive meteorites. From their highly unusual isotopic compositions, compared to that of the solar system as a whole, these presolar grains are inferred to have formed in circumstellar atmospheres or, in some cases, in nova or supernova explosions. Because their compositions reflect the isotopic and chemical signatures of their sources, presolar grains provide information about stellar evolution and nucleosynthesis, mixing processes in stars, the physical and chemical conditions of stellar atmospheres, and the chemical evolution of the Galaxy (Anders & Zinner 1993; Ott 1993).

There are several interesting parallels between ^{26}Al, ^{60}Fe and ^{44}Ti in γ-ray astronomy and the laboratory study of presolar meteoritic grains. Enhanced ^{26}Mg/^{24}Mg ratios in the Ca-Al rich inclusions of the Allende meteorite were first evidence for live ^{26}Al in the early solar system (Lee, Papanastassiou, & Wasserburg 1977). Present compilations (MacPherson et al. 1995), with over 1500 data points derived from various types of meteoritic samples, confirm that the enhancements in ^{26}Mg are observed in these old and stable inclusions of solar system material, and homogeneously show an isotopic ratio ^{26}Al/^{27}Al of $\sim 5\ 10^{-5}$. This is interpreted as *in situ* decay of ^{26}Al in those early droplets (MacPherson et al. 1995), mainly from the strong correlation of ^{26}Mg excess with aluminum abundances of the samples. This is direct proof of an injection

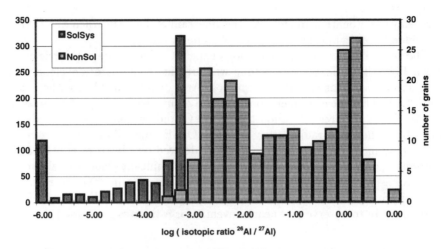

FIGURE 11. Histogram of inferred ^{26}Al/^{27}Al for interstellar grains of various types (wide hatched columns) and solar system objects (narrow filled), after MacPherson et al. (1995).

of (at least) ^{26}Al into the solar nebula shortly before solar system formation. The inferred ^{26}Al/^{27}Al ratio of $\sim 5\times10^{-5}$ is substantially larger than the ratio of 2-3$\times 10^{-6}$ estimated from γ-ray measurements (Diehl et al. 1995).

Figure 11 compares the inferred initial ^{26}Al/^{27}Al ratio distributions of presolar grains (corundum, graphite and silicon carbide) with the total population of data for aluminum-rich material (mostly Ca-Al rich inclusions) that represents solar system aluminum. There is very little overlap between the two populations. This reassures that these types of meteoritic samples constitute different observational windows which may not be expected to correlate, but rather carry imprints of the specific grain formation event, or early solar system processes, respectively (see MacPherson et al. 1995).

An indication that live ^{60}Fe also existed when the meteorites solidified is found from isotopic analysis of the Chernvony Kut meteorite (Shukolyukov & Lugmair 1993). The measured excess of ^{60}Ni, after alternative modes of production such as spallation and (n,γ) reactions on ^{59}Co could be eliminated, leads to a ^{60}Fe/^{56}Fe ratio at the time of iron-nickel fractionation of $\sim 7.5\times10^{-9}$. This is consistent with the inferred 10 million year hiatus before the formation of the Ca-Al rich inclusions in the Allende meteorite, which has a much larger ^{60}Fe/^{56}Fe ratio at the time of fractionation of $\sim 1.6\times10^{-6}$ (Birck & Lugmair 1988). However, the possibility that some of the ^{60}Ni in the Ca-Al rich inclusions is fossil rather than from the in situ decay of ^{60}Fe cannot be excluded. Large excesses in ^{44}Ca have been observed in four low-density graphite grains and five silicon carbide grains of type X extracted from the Murchison meteorite; these are shown to be related to radioactive decay of ^{44}Ti (see Fig.

6) (Nittler et al. 1996; Hoppe et al. 1996). Because ^{44}Ti is only produced in supernovae, these grains must have a supernova origin. Moreover, the silicon, carbon, nitrogen, aluminum, oxygen, and titanium isotopic compositions of these large grains (>1 micron) require a Type II supernova source. This is strong evidence that these grains are supernova condensates and provides evidence for deep and heterogeneous mixing of different supernova regions, including the nickel core.

Extensive high-precision measurements of other meteorites discovered many other anomalous ratios attributed isotopes that were still radioactive during the decoupling of the solar nebula from the interstellar medium 4.6 Gy ago (e.g., Harper 1996). Combined arguments lead to hypotheses of the solar system formation being related to a nearby event such as a supernova or an AGB star (Cameron 1993; Wasserburg et al. 1995). The solar nebula experienced enrichment above interstellar abundances in relatively shortlived radioactive species, either from the triggering events (if they occurred), or from some previously unconsidered process during formation of the sun.

Shu et al. (1997) discuss the idea that the chondrules and Ca-Al rich inclusions found in chondritic meteorites might have been formed as solids entrained and melted in the bipolar wind that results from the interaction of the accreting proto-solar nebula and the magnetosphere of the young protosun. Aerodynamic sorting and a mechanical selection for molten droplets that rain back onto the disk at planetary distances explain the size distributions and patterns of element segregation that we observe in carbonaceous and ordinary chondrites. Cosmic ray ions generated in the flares that accompany the general magnetic activity of the inner region may irradiate the precursor rocks before they are launched in the bipolar wind. Under certain scaling assumptions for the efficiency of the process in protostars, Shu et al. find that cosmic-ray bombardment can generate the short-lived radionuclides ^{26}Al, ^{41}Ca, and ^{53}Mn at their inferred meteoritic levels. Simply put, this mechanism captures material trying to get on the sun, heats it up, irradiates it with cosmic rays, and dumps it back onto the disk farther out (Glanz 1997). It is far from clear, however, that such low-energy cosmic rays can explain most of the complex isotopic patterns found in chondrites. Additional measurements are necessary to see if the morphological and isotopic anomalies are best explained by such a single mechanism.

Positrons from Radioactivities

Radioactive decays can generate positrons whenever the energy level of the daughter atom is more than the 1.022 MeV pair production threshold below the energy level of the parent. Some interesting amounts are positrons and produced from the β^+ decay of ^{56}Co, ^{57}Co, ^{26}Al, ^{44}Sc, and the distinct nova products ^{13}N and ^{18}F. Annihilation produces two 511 keV photons if the positron-electron spins point in opposite directions, or 3 photons in a continuous energy spectrum if the spins are parallel. This 3 photon annihilation process usually involves formation of positronium, which then decays before being collisionally destroyed (more likely in the low densities of interstellar space). The fraction of positronium carries information about the annihilation environment. A cold, neutral environment results in positronium fractions of 0.9-0.945 (Bussard, Ramaty & Drachman 1979; Brown, Leventhal & Mills 1986), while larger positronium fractions tend to indicate annihilation in warm (5×10^3 K or hotter) environments. Unbound positrons may have kinetic energies up to ~MeV, and thus may be relativistic when they are produced. Deceleration and thermalization are more likely than annihilation-in-flight, so that the positron lifetime in interstellar space before annihilation can be $\sim 10^5$ y. The thermalization process yield intrinsically narrow 511 keV linewidths that are related to the annihilation environment rather than the positron production environment.

In addition to the annihilation γ-ray signals from radioactive decays, there are also such signals from pair plasma disks around accreting compact remnants, plasma jets from dynamo action of accreting compact sources, and γ-γ-reactions in strong magnetic fields. How are these signals differentiated? Positrons from radioactive decays usually occur in in intrinsically diffuse environments with a long thermalization lifetime in the interstellar medium. Positrons ejected from processes surrounding compact objects, however, are usually annihilated nearby resulting in a recognizable more localized (and time variable) source of annihilation photons. These regions also have a smaller positronium fraction than regions where radioactive decays usually occur.

Almost all of the Galactic-disk annihilation luminosity may be explained by radioactive sources (Ramaty, Skibo & Lingenfelter 1994). OSSE measurements of the Galactic bulge and disk annihilation components (Kinzer et al. 1996; Purcell et al. 1997) can be analyzed in terms of plausible spatial distribution models that aim to separate the disk radioactivity component from the Galactic bulge component. The annihilation productions are 10^{43} e$^+$s^{-1} for the disk and 2.6×10^{43} e$^+$s^{-1} for the bulge. About 20% of the disk component can be assigned to ^{26}Al, based on the 1.809 MeV skymaps (Ramaty et al. 1994), and suggests a large contribution from sources related to an old stellar population, such as novae, Type Ia supernovae, or compact X-ray sources. Kinzer et al. (1996) determine a positronium fraction between 0.94 and 1.0 in the inner Galaxy, suggesting that the contribution from compact sources should be small. However, this constraint strongly depends on the environment of the compact sources; Ramaty et al. (1994) point out that the entire bulge component luminosity could be explained from 1E1740.7-2942 positron production alone, if positrons are not rapidly annihilated in a target close to that compact source. Hernanz et al. (1996) estimate from their simulations of classical novae that the peak 511 keV emission reaches $\sim 10^{-2}$ (D/1kpc)^2 ph cm^{-2} s^{-1} for a short period about 7 hours after the outburst. This would make detections out to distances of the Galactic Center feasible, if timing of the observations would be fortunate. Overall, however, the nova contribution to the diffuse 511 keV glow of the Galaxy is expected to be low (Lingenfelter & Ramaty 1989).

Galactic Nucleosynthesis and Supernova Rates

Direct measurement of the Milky Way's supernova rate is notoriously difficult. Various methods have been attempted, such as O-B2 star counts within 1 kpc of the Sun, radio supernova remnant counts, γ-ray fluxes from the decay of ^{44}Ti, neutrino burst detections, and pulsar birth rate estimates. These indirect methods only yield upper limits with large error bars. Galactic supernova rates have therefore been based on extragalactic supernova searches. These estimates depend upon morphological type, Hubble constant, completeness of the surveys, determination of reliable control times, and the uncertain luminosity of the Galaxy. The total blue luminosity, Hα and far-infrared fluxes of the Milky Way remain uncertain, in contrast to most Local Group galaxies. Nevertheless, using the extragalactic estimates with a total Galactic blue luminosity of 2.3×10^{10} L$_\odot$, a Hubble constant of 75 km/s/Mpc, and a Sbc Galactic morphology, the Galactic core-collapse supernova rate is estimated to be 2.4-2.7 per century while the Type Ia rate is estimated to be 0.3-0.6 per century (van den Bergh & McClure 1994; Tammann, Löffler, & Schröder 1994). Thus, the total rate is about 3/century and the ratio of core collapse to thermonuclear events is about 6.

There are 6 known local supernovae: Lupus (SN 1006) is considered to have been a Type Ia event, Cas A (SN 1680) and Tycho (SN 1572) which were most likely Type Ib supernovae, and the Crab (SN 1054), SN 1181, and Kepler (SN 1604) were probably Type II events. This listing may be complete to ~ 4 kpc. For an exponential disk with a 4 kpc scale length extending between 3-15 kpc, about 9% of all Galactic OB stars will be located within 4 kpc of the Sun. At a radial distance of 8.5 kpc, the number of supernova with massive progenitors within 4 kpc of the Sun may be estimated as $\sim 0.09\times0.85\times3 \sim 2.3$ per millennium (van den Bergh & McClure 1994). This is close to the observed number that are known to have occurred within 4 kpc of the Sun during the last 2000 years. However, the statistical uncertainty in the frequency with which supernova of different types occur in galaxies of different Hubble class is large, and this agreement may be fortuitous.

The total ^{44}Ti line flux originating from the central regions of the Milky Way was sought by the large field of view spectrometers aboard the HEAO3 and SMM satellites. Analysis of the data taken with HEAO3 gave an upper limit of 2×10^{-4} ph cm^{-2} s^{-1} on the Galactic 67.9 & 78.4 keV emission (Mahoney et al. 1992). Searching through nearly 10 years of data taken by SMM gave an upper limit of 8×10^{-5} ph cm^{-2} s^{-1} for the inner 150° of the Milky Way in the

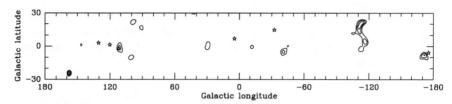

FIGURE 12. A first COMPTEL map of the Galactic plane in the 1.157 MeV line from ^{44}Ti (Dupraz et al. 1997), using the first 3 years of data. Star symbols mark locations of historical supernovae 1181, Tycho, Cas A, Kepler, Lupus, and Crab (from left to right). The Cas A signal in these data corresponds to 3.5σ significance, likelihood contour levels are in 1σ steps for known sources, $\sim 0.75\sigma$ for search of new sources.

1.157 MeV line (Leising & Share 1994), the imaging COMPTEL instrument sets upper limits below 2×10^{-5} ph cm^{-2} s^{-1} for the known historical events (Dupraz et al. 1997).

A first COMPTEL 1.07-1.25 MeV background subtracted skymap is shown in Fig. 12 (Dupraz et al. 1997). Even though only part of the data have been used in this analysis, and the potentially more sensitive background modelling from adjacent energies has not been employed, first conclusions can be drawn: Cas A is the only historic event detected (at $\sim 5\sigma$ now, see above). Upper limits to the ^{44}Ti line flux from Tycho and Kepler are about 1/2 of the Cas A signal. Even with the longest ^{44}Ti half-lives measured, Lupus, Crab, and SN 1181 are far too old to have significant amounts of ^{44}Ti remaining to be be detected. No previously unknown event is found, even though COMPTEL's sensitivity should provide complete sampling of the last century beyond the distance of the Galactic center.

Core collapse supernovae and Chandrasekhar mass thermonuclear supernovae tend to account for $\sim$1/3 of the solar ^{44}Ca abundance, with most of the production attributable to Type II events. The mass of ^{44}Ca produced as itself is comparable to the ^{44}Ca synthesized from the radioactive decay of ^{44}Ti. Even at this $\sim$1/3 of solar level, the sky should contain several mean Type II supernovae remnants bright enough to be seen in their ^{44}Ti afterglow. However, no search (HEAO 3, SMM, COMPTEL, OSSE) has produced a detection of ^{44}Ti line emission from any young, previously unknown, and visually obscured remnants. We simply do not see the expected number of supernovae remnants emitting gamma-rays from the decay of radioactive ^{44}Ti. After eliminating alternative modes of increasing the ^{44}Ca production (e.g., larger yields), the conclusion becomes almost inescapable. The solar abundance of ^{44}Ca is apparently due to rare events with exceptionally high ^{44}Ti yields. This conclusion was already derived from the SMM upper limits to ^{44}Ti gamma-rays (Leising & Share 1994). Rare events could mean Type II events where the explosion energy is large enough to minimize the mass of ^{44}Ti that falls back onto the compact remnant, or Type Ia events that manage to explode from a star with high central density. Mapping the Galaxy in ^{44}Ti should help to resolve the unknown rate of these rare events; they would still mark bright spots in the 1157 keV sky. The COMPTEL survey from the first three mission years (Dupraz et al. 1997) still appears inconclusive in this respect.

Integrated nucleosynthesis measurements, such as the 1.809 MeV ^{26}Al observations, can be a useful measure of the Galactic supernova rates. If a dominating origin of ^{26}Al from massive stars is adopted (Prantzos & Diehl 1996), then the γ-ray data and nucleosynthesis yields provide independent measures of the massive star formation rate in the Galaxy. The core-collapse supernova rate determined in this way is consistent with various classical rate determination methods, such as on H_α measurements and supernova records in "similar" distant galaxies (Timmes et al. 1997). Crucial to such analyses is the integrated radioactive mass inferred from the γ-ray measurements and the spatial distribution of the emission in the Galaxy. Given the wide range of 0.4-3.5 $M_\odot$ from present data, γ-ray measurements cannot help resolve the various systematic uncertainties present in the classical methods (e.g. galaxy type, Hubble constant, sample completeness). Future measurements of the position and γ-ray line shape may provide the basis for a 3-dimensional deconvolution of the apparent emission (Gehrels & Chen 1996). Note that measurement of a 6 keV wide 1.809 MeV line by Naya et al. (1996) suggests that any 0.1-0.9 keV line shifts expected from Galactic rotation may not allow a direct mapping of the emission.

Other Galaxies

In undertaking searches for γ-ray line signals from supernovae, it is useful to have some guidelines as to what might be expected. With recent catalogs of supernova events, and distances based upon some assumptions regarding the peak absolute magnitude of supernovae, estimates of the detectable event rate as a function of γ-ray telescope sensitivity can be made. Vigorous ground based programs that search, discover, classify, and catalog supernovae are essential prerequisites in studying the emission in detail, and hence, studying the explosion in detail.

Data on more than 1000 extragalactic supernovae are given in the Sternberg Astronomical Institute Supernova Catalogue (Tsvetkov, Pavlyuk, & Bartunov 1997), and the Asiago Supernova Catalogue (Cappellaro Barbon, & Turatto 1997). About 50% of the supernova in either catalog lack, or have an uncertain, supernova type identification; but certainly the brightest and more recent events do have supernova types assigned. All Type Ia events are assumed to have a peak absolute bolometric magnitude of M=-19.0. This standard candle assumption may be challenged on both observational (e.g., Saha et al. 1996) and theoretical (e.g., Höflich et al. 1996) grounds, but it remains a useful first

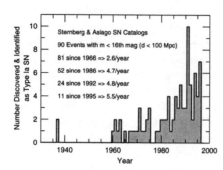
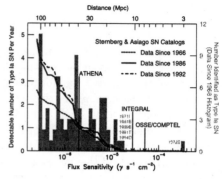

FIGURE 13. Number of events discovered and identified as Type Ia supernovae vs the year of the explosion (left). Flux sensitivity histogram of 1966–1996 Type Ia events (right).

approximation. Distances to individual supernovae are then calculated from the catalogued magnitudes by the standard formula $\log_{10} D = (m - M + 5)/5$, with no reddening corrections applied.

All the various Type Ia models tend to produce peak 847 keV line fluxes in the $2\text{-}10 \times 10^{-5}$ ph cm^{-2} s^{-1} range at $D_p = 10$ Mpc, reaching this maximum $\sim$ 75 days after the explosion (Chan & Lingenfelter 1991; Khokhlov, Müller, & Höflich 1993). The peak 1.238 MeV line flux slightly is smaller at the same distance (smaller branching ratio), and tends to lie in the $1.5\text{-}6 \times 10^{-5}$ ph cm^{-2} s^{-1} range. Here we simply assume all Type Ia supernovae have a peak gamma-ray flux of $F_p = 3 \times 10^{-5}$ ph cm^{-2} s^{-1} at $D_p = 10$ Mpc. Peak gamma-ray fluxes are then calculated by the flux ratio $F = F_p (D_p/D)^2$. Catalog completeness, volumes sampled, and limitations of the assumptions are important concerns. These are discussed in detail by Timmes & Woosley (1997), who also report a similar analysis for core-collapse supernovae.

A total of 90 Type Ia supernovae brighter than 16th apparent magnitude are listed in these catalogs (Fig. 13). As long-term search, discovery and classification programs continue to mature, the significance of an apparent average detection and classification rate of roughly 5/year may be evaluated. These implied rates are lower bounds for several reasons; large visual extinctions by dust from the parent galaxy may hide events, search programs are becoming more efficient, and unclassified events may include some Type Ia supernovae.

The fluxes for 81 events discovered and identified as Type Ia supernova since 1966 are histogrammed in Fig. 13 (individually listed are the 6 brightest Type Ia supernovae). The apparent magnitudes may be converted into distances and peak γ-ray flux values, making the above standard candle assumptions (see top and bottom ordinates). Integrating the histogram and normalizing to the time frames' average Type Ia supernova rate gives the γ-ray event rate versus flux curve. This can be compared with instrumental sensitivities. An instrument with a flux sensitivity of $\sim 2\times10^{-6}$ ph cm^{-2} s^{-1} to 5000 km/s broadened lines should easily observe Type Ia events in the Virgo (18.2 Mpc), Fornax (18.4 Mpc), and possibly the Hydra (41 Mpc) galaxy clusters at a rate of 1/year (roughly 10-30% of all Type Ia events within 100 Mpc). INTEGRAL's spectrometer sensitivity of $\sim 6\times10^{-6}$ ph cm^{-2} s^{-1} (Winkler 1995), for a 10 day observation, could detect all Type Ia supernovae out to ~ 20 Mpc and allow a rate near 1 Type Ia event/year. Present instrument broad line flux sensitivities larger than 1.5×10^{-5} ph cm^{-2} s^{-1} are probably limited to detecting Type Ia supernova within $\sim$10 Mpc. These rate estimates should be regarded as rough guidelines, to aid, for example, in the design of future instruments.

CONCLUDING REMARKS

The search for an overarching paradigm which encompasses measurements of γ-ray emission from radioactive isotopes, the solar abundances, presolar meteoritic grains, stellar nucleosynthesis calculations, and Galactic chemical evolution has seen major advances during the time of the Compton Observatory mission. Yet it remains unfulfilled. Our goal is to understand the origin of every isotope in every location (and time) in the universe. Next steps towards this goal will be to determine the Galactic ^{26}Al and ^{60}Fe amounts, to put the connection of these isotopes with massive stars on solid grounds, and to find and understand exeptional events which eject ^{44}Ti, ^{60}Fe, ^{22}Na, helping to enlighten the variety of nova and supernova event types. We shall use this understanding, as it develops, as a diagnostic for the physical processes and evolution of the universe and its contents. New instrumentation and analysis methods, and theory advances, and their often conflicting constraints will yield exciting adventures in those future studies.

Acknowledgements Useful discussions with many colleagues, in particular Dieter Hartmann, Mark Leising, and Stan Woosley, are reflected in this article. This work has been supported by the Max Planck Gesellschaft (RD) and a Compton Gamma Ray Observatory Postdoctoral Fellowship (FXT).

REFERENCES

1. Allen, S. J., 1911, Phys. Rev., 34, 296
2. Anders E. & Zinner E., 1993, Meteoritics, 28, 490
3. Arnett, W. D., et al., 1989, ARAA, 27, 629
4. Aufderheide, M., Baron, E., & Thielemann, F. -K. 1991, ApJ, 370, 630

5. Badash, L. B., 1979, Radioactivity in America
6. Bazan, G., & Arnett, D., 1994, ApJ, 433, L41
7. Becquerel, H., 1896, Compt. Rend., 122, 420
8. Becquerel, H., 1900, Compt. Rend., 130, 809
9. Birck, J. L., & Lugmair, G. W., 1988 Earth Plan. Sci. Lett., 90, 131
10. Boffin, H. M. J., et al., 1993, A&A, 279, 173
11. Brown, B. L., Leventhal, M., & Mills, A. P., Jr., 1986, Phys. Rev. A, 33, 2281
12. Burbidge, E. M., et al., 1957 Rev. Mod. Phys., 29, 547
13. Bussard, R. W., Ramaty, R., & Drachman, R. J., 1979 ApJ, 228, 928
14. Cameron, A. G. W., 1957, Chalk River Report, CRL-41
15. Cameron, A. G. W., 1993, in Protostars and Planets III, 47
16. Cappellaro, E., Barbon, R., & Turatto, H., 1997, www.pd.astro.it/supern
17. Chan, K., & Lingenfelter, R., 1991, ApJ, 368, 515
18. Chen, W., et al., 1997, ESA-SP 382, 105
19. Clayton, D. D., 1971, Nature, 234, 291
20. Clayton, D. D., 1982, in: Essays in Nuclear Astrophysics, 401
21. Clayton, D. D., Colgate, S. A., & Fishman, G. 1969, ApJ, 220, 353
22. Clayton, D. D. & Leising, M. D. 1987, Phys. Rep., 144, 1
23. Coe, A., et al., 1995, A&A, 299, 479
24. Dame, T. M., et al., 1987, ApJ, 322, 706
25. Del Rio, E., et al., 1996, A&A, 315 237
26. Diehl, R., et al., 1995, A&A, 298, 445
27. Diehl, R., et al., 1997, these proceedings
28. Dupraz, C., et al., 1997, A&A, in press
29. Ellison, D. C., et al., 1994, PASP, 106, 780
30. Elmegreen, B., G., & Efremov, Y. N., 1996, ApJ, 466, 802
31. Feather, N., 1973, Lord Rutherford, London
32. Fesen, R. A., & Becker, R. H., 1991, ApJ, 371, 621
33. Gehrels, N., & Chen, W., 1996, A&A Supp., 120, 331
34. Gehrels, N., et al., 1993, Sci. Amer., 269, 68
35. Glasner, S. A., Livne, E., & Truran, J. W., 1997, 475, 754
36. Glanz, J., 1997, Science, 276, 1789
37. Harper, C. E., 1996, ApJ, 466, 1026
38. Harris, M. J., 1997, these proceedings
39. Harris, M. J., Leising, M. D., & Share, G. H. 1991, ApJ, 375, 216
40. Hartmann, D. H., et al., 1997 in Nucl. Phys. A, in press
41. Haymes, R. C., et al., 1975, ApJ, 201, 593
42. Hernanz, M., et al., 1996, ApJ, 465, L27
43. Hix, W. R., & Thielemann, F. K., 1996, ApJ, 460, 869
44. Höflich, P., et al., 1996, ApJ, 472, 81
45. Hoppe, P, et al., 1996, Science, 272, 1314
46. Hoyle, F., & Clayton, D. D., 1974, ApJ, 191, 705
47. Iyudin, A. F., et al., 1994, A&A, 284, L1
48. Iyudin, A. F., et al., 1995, A&A, 300, 422
49. Iyudin, A. F., et al., 1997, ESA SP-382, 37

50. Kessler, M. F., et al., 1996, A&A, 315, L27
51. Khokhlov, A., Müller, E., & Höflich, P., 1993, A&A, 270, 223
52. Kinzer, R. L., et al. 1996 A&A Supp., 120, 317
53. Kniffen, D. A., Gehrels, N., & Fishman, G., 1997, these proceedings
54. Knödlseder, J., et al. 1996a, A&A Supp., 120, 327
55. Knödlseder, J., et al. 1996b, A&A Supp., 120, 335
56. Knödlseder, J., et al. 1997, A&A, in press
57. Kolb, U., & Politano, M., 1997, A&A, 319, 909
58. Langer, N., et al., Nuc. Phys. A, in press
59. Lee, T., Papanastassiou, D. A., & Wasserburg, G. J. 1977, ApJ, 211, L107
60. Leising, M. D., 1993, A&A Supp., 97, 299
61. Leising, M. D. & Clayton, D. D., 1987, ApJ, 323, 159
62. Leising, M. D., et al., 1995, ApJ, 450, 805
63. Leising, M. D., & Share, G. H., 1994, ApJ, 424, 200
64. Lichti, G. G., et al., 1994, A&A, 292, 569
65. Lingenfelter, R. E., & Ramaty, R. 1989, ApJ, 343, 686
66. MacPherson, G. J., Davis, A. M., & Zinner, E. K. 1995, Meteoritics, 30, 365
67. Mahoney, W. A., et al., 1982, ApJ, 262, 742
68. Mahoney, W. A., et al., 1992, ApJ, 387, 314
69. Meynet, G., et al., 1997, A&A, 320, 460
70. Morris, D. J., et al., 1995, N.Y.Acad.Sci., 759, 397
71. Murthy, P. V. R., & Wolfendale, A. W., 1993, "Gamma-ray astronomy"
72. Myra, E. S. & Burrows, A., 1990, ApJ, 364, 222
73. Naya, J. E., et al., 1996, Nature, 384, 44
74. Naya, J., et al., 1997, these proceedings
75. Nittler, L. R. et al., 1996, ApJ, 462, L31
76. Ott, U., 1993, Nature, 364, 25
77. Oberlack, U., et al, 1997, A&A, in preparation
78. Politano, M., et al., 1995, ApJ, 448, 807
79. Prantzos, N., 1993a, ApJ, 405, L55
80. Prantzos, N., 1993b, A&A, 97, 119
81. Prantzos, N. & Diehl, R., 1996, Phys. Rep., 267, 1
82. Predehl, P., & Schmitt, J. H. M. M., 1995, A&A, 293, 889
83. Purcell, W. R., et al., 1997, these proceedings
84. Ramaty, R. & Lingenfelter, R. E., 1977, ApJ, 213, L5
85. Ramaty, R., Skibo, J. G., & Lingenfelter, R. E. 1994, ApJS, 92, 393
86. Ritossa, C., Garcia-Berro, E., & Iben, I., Jr., 1996, ApJ, 460, 489
87. Rolfs, C. E., & Rodney, W. S., 1988, Cauldrons in the Cosmos, U. Chicago
88. Rona, E., 1978 How it came about: radioactivity, atomic energy
89. Rothschild, R. E., et al., 1997, these proceedings
90. Rutherford, E., 1899, Phil. Mag., 47, 109
91. Rutherford, E., 1905, Radio-Activity, Cambridge U. Press
92. Rutherford, E., 1919, Phil. Mag., 37, 581
93. Saha, A., et al., 1996, ApJS, 107, 693
94. Schaerer, D., Schmutz, W., & Grenon, M., 1997, ApJ, 484, L153

95. Shankar, A., & Arnett, D., 1994, ApJ, 433, 216
96. Schönfelder, V., et al., 1996, A&A Supp., 120, 13
97. Shigeyama, T., et al., 1993, A&A Supp, 97, 223
98. Shu, F. H., et al., 1997, BAAS, 190, 4904
99. Shukolyukov, A., & Lugmair, G. W., 1993 Science, 259, 1138
100. Starrfield, S., 1996, PASP 99, 242
101. Starrfield, S., et al., 1992, ApJ, 391, 71
102. Tammann, G. A., Löffler, & Schröder, 1994 ApJS, 92, 487
103. Taylor, J. H., & Cordes, J. M., 1993, ApJ, 411, 674
104. The, L. S., et al., 1995, ApJ, 444, 244
105. The, L. S., et al., 1996, A&A Supp., 120, 357
106. Thielemann, F. -K., Nomoto, K., & Yokoi, Y. 1986, A&A, 158, 17
107. Thielemann, F. -K., Nomoto, K., & Hashimoto, M. A. 1996, ApJ, 460, 408
108. Timmes, F. X., et al., 1995, ApJ, 449, 204
109. Timmes, F. X., et al., 1996, ApJ, 464, 332
110. Timmes, F. X., Diehl R., & Hartmann D.H., 1997, ApJ, 479, 760
111. Timmes, F. X., & Woosley, S. E., 1997, ApJ, in press
112. Tsvetkov, D., Pavlyuk, N., & Bartunov, O., 1997, www.sai.msu.su/groups/sn
113. van den Bergh, S., & McClure, R. D., 1994, ApJ, 425, 205
114. van den Bergh, S., & Pritchet, C. J., 1986, ApJ, 307, 723
115. van der Hucht, K. A., et al., 1988, A&A, 199, 217
116. van der Hucht, K. A., et al., 1997, New Astron., 2, 245
117. Wasserburg, G. J., et al., 1995, ApJ, 440, L101
118. Williams, B. F., Schmitt, M. D., & Winkler, P. F., 1995, BAAS, 186, 4911
119. Winkler, C., 1995, Exp. Astron., 6, 71
120. Woosley, S. E., et al., 1990, ApJ, 356, 272
121. Woosley, S. E., Langer, N., & Weaver, T. A., 1995, ApJ, 448, 315
122. Woosley, S. E., & Weaver, T. A., 1995, ApJS 101, 181

Galactic Nuclear Deexcitation Gamma-Ray Lines

H. Bloemen[*][†] & A.M. Bykov[‡][*]

[*]*SRON-Utrecht, Utrecht, The Netherlands*
[†]*Leiden Observatory, Leiden, The Netherlands*
[‡]*A.F. Ioffe Institute, St. Petersburg, Russia*

Abstract. COMPTEL is finding evidence for nuclear-interaction γ-ray lines in the galactic MeV radiation, although conclusive line identifications are still lacking. In addition to Orion, first indications are now seen in emission from the inner Galaxy as well. The data analysis is complicated by the extended nature of the observed emission and by the fact that the lines can be expected to be broad and structured and possibly even gravitationally redshifted. The emission from Orion has stimulated studies of γ-ray line production near OB associations that are enriching and energizing their environment. The inner-Galaxy emission, apparently extending well out of the galactic plane, suggests that other sources may have to be considered as well, such as possibly compact objects. We review the current status.

1. INTRODUCTION

In addition to the intense $e^{\pm}$-annihilation 511 keV emission from the inner Galaxy (see review [102] in this volume), all astrophysical (non-solar) nuclear γ-ray lines that are firmly detected so far originate from the decay of long lived radionuclei produced in nucleosynthesis processes. This includes the lines at 0.847 and 1.238 MeV from ^{56}Co, at 1.157 MeV from ^{44}Ti, and at 1.809 MeV from ^{26}Al (see review [33] in this volume). Another category of γ-ray lines results from energetic ($\sim$ 2–100 MeV/nucleon) particle interactions. These are well known from solar flares (e.g. [26,69] and review [96] in this volume), but no compelling evidence was seen from astrophysical sources prior to the Compton Observatory (CGRO), although possible detections have been reported [46,53,58]. Such (astrophysical) nuclear-interaction lines form the subject of this review — COMPTEL is providing a new trigger to the field.

An extensive evaluation of candidate γ-ray lines from nuclear interactions was presented by Ramaty, Kozlovsky & Lingenfelter (1979) [89]. The

γ-ray line emission originates in a variety of ways, such as through direct excitation of nuclear levels, through the production of excited secondary nuclei, and through the production of radioactive species which decay into excited levels of other nuclei. A wealth of spectral structure can be produced, depending on the compositions of the energetic particles and the ambient medium and on the energy spectrum of the particles. The main candidate lines (in case of roughly solar composition) are from ^{12}C (4.44 MeV) and ^{16}O (6.13, 6.92, and 7.12 MeV) in the 3–7 MeV regime and from heavier nuclei such as ^{20}Ne (1.63 MeV), ^{22}Ne (1.28 MeV), ^{24}Mg (1.37 MeV), and ^{28}Si (1.78 MeV) at 1–2 MeV. Lines from energetic nuclei are distinguishable from those of ambient nuclei (excited by protons and α-particles) because they are Doppler broadened and can contain line splitting features due to the fact that the emission is not isotropic in the rest frame of the nuclei [23,93,56].

Gamma-ray spectroscopy of nuclear deexcitation lines thus provides a potentially important tool to study violent phenomena in active galactic environments, ranging from the interstellar medium (e.g. near associations of massive stars) to the direct vicinity of compact objects. It was certainly not obvious and not predicted, however, that CGRO should be able to detect such lines. Early modelling of γ-ray line production in the interstellar medium (ISM) [47,65,89], necessarily based on indirect and highly uncertain estimates of low-energy (<100 MeV/nucl) cosmic rays, was not promising for CGRO observations. But more localized diffuse source regions involving massive stars and associated stellar winds and supernova explosions, which are in retrospect good candidates (§3), were not considered. A very different potential production environment considered prior to CGRO is the accretion flow onto a neutron star, proposed more than 25 years ago [100] and further explored in a number of subsequent studies [88,62,13,1,11,4]. But it seemed unlikely that line emission from individual objects would be detectable by CGRO. Again in retrospect, however, observing the collective signal of a large ensemble may be possible, although the expectations are strongly model dependent (§3).

Needless to say, COMPTEL's first observations of Orion [8], showing enhanced emission at 3–7 MeV (and thus possibly indicating the presence of ^{12}C and ^{16}O lines), came as a pleasant suprise, but also as a worry for the observers because the measurements are difficult and the predictions were simply not promising. But adding more and more observations the evidence is growing [9] and plausible theoretical scenarios have been proposed in the meantime (§3). In addition, evidence is now seen from the inner Galaxy as well [10].

2. OBSERVATIONS

Before the CGRO mission, observations of the galactic γ-ray emission in the 1–10 MeV regime were focussed on the inner radian, but even here any steady emission from the Milky Way was marginally detectable if seen at all (apart from the ^{26}Al 1.8 MeV line — see review [33]) and any imaging information was basically lacking [41,63,74,59,98,82,111]. An exception was formed by SMM (although without imaging information), which has provided a detailed spectrum of the integrated inner-Galaxy emission. Nevertheless, only strong (broad) line features with fluxes of typically a few times 10^{-4} γ cm^{-2} s^{-1} rad^{-1} would have shown up if present, as can be deduced from the SMM studies presented in e.g. [61,42,60].

Possibilities for γ-ray spectroscopy with CGRO are provided by the COMPTEL [99] and OSSE [54] instruments. The COMPTEL Compton telescope enables wide-field ($\sim 30°$ radius) imaging with an angular resolution of typically 1°– 3° in a spectral band that ranges from about 0.75 MeV to 30 MeV. In combination with an energy resolution ranging from about 10% FWHM at the lowest energies to about 5% FWHM around 10 MeV, the instrument offers the best possibilities so far for spatially resolved γ-ray spectroscopy. Good examples are the ^{26}Al (e.g [32,73]) and ^{44}Ti [51,52,36] studies. OSSE covers the energy range 50 keV – 10 MeV. Its collimators provide a field of view of 3.8° × 11.4° FWHM. Background measurements have to be obtained by periodic off-set pointings, which limits OSSE's capability to observe extended emission. This is unfortunate because the γ-ray line emission that may be seen by COMPTEL, as described below, is of extended nature.

2.1 The Orion/Monoceros region

The Orion Complex, which is the nearest region of giant molecular clouds and active massive star formation at a distance of $\sim$ 450 pc (e.g. [38,14]), is a well-known source of high-energy γ-rays ($\gtrsim$ 100 MeV), observed by COS-B [7] as well as by EGRET [34]. The γ-ray emission approximately follows the distribution of the molecular gas as traced by CO observations (including the adjacent so called Mon R2 cloud at a distance of $\sim$800 pc) and can be attributed to cosmic-ray (CR) interactions with the gas nuclei through $\pi°$ decay (dominant) and bremsstrahlung, like most of the high-energy γ radiation from the Milky Way up to $\sim$ 1 GeV (e.g. [6,49]). The inferred flux of CR protons (with energies between several hundreds of MeV and a few tens of GeV) is very similar to that directly measured near Earth, which holds for the Galaxy at large as well (see e.g. review [6]).

The Orion/Monoceros region was in the field-of-view of COMPTEL during several observation periods of the CGRO Phase-I sky survey in 1991–1992. Given the high-energy γ-ray observations and the measured MeV flux of the Milky Way towards the inner Galaxy, it was clear that any diffuse bremsstrahlung emission from the Orion clouds could not be expected to be seen by COMPTEL. However, intense emission was detected (at least an order of magnitude higher than expected), but only in the energy range 3–7 MeV [8], with a flux of $\sim 10^{-4}$ γ cm^{-2} s^{-1}. The emission was seen to peak between

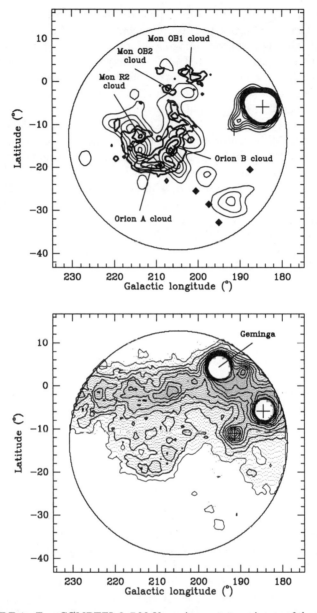

FIGURE 1. *Top:* COMPTEL 3–7 MeV maximum-entropy image of the Orion/Monoceros region [9] (thin contours). CO observations of Orion and Monoceros [31] are superimposed (thick contours; not convolved with the COMPTEL PSF). Adjacent to the strong Crab source, the AGN PKS 0528+134 is seen (both indicated by + signs). EGRET has detected 4 other AGN source candidates in this field (diamonds), which might be responsible for the emission at $\sim (195°, -25°)$. *Bottom:* EGRET map (>100 MeV) of the same region (Phase 1–3 data).

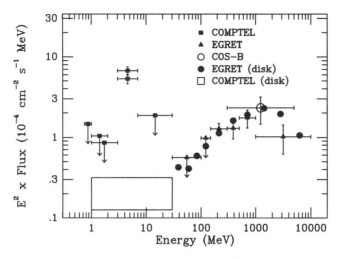

FIGURE 2. Broad-band flux spectrum of the Orion/Mon R2 complex ($\times\ E^2$), determined by CO model fits [9], illustrating the excessive 3–7 MeV emission compared to COMPTEL observations at neighbouring energies, to observations by COS-B [7] and EGRET [34] above 30 MeV, and to expected flux values based on typical emissivity values of the galactic disk, as derived from observations by COMPTEL [10] (approximate 1–30 MeV average — see §2.2) and EGRET [103]. All upper limits are at the 2σ level. For the 3–7 MeV range, the slightly higher 'model-independent' flux [9] is shown as well.

the two main molecular cloud components (Orion A & B), roughly coinciding with the OB1-bcd associations, and extended in the direction of Mon R2.

It seemed most natural to attribute the emission to the 4.44 and 6.13 MeV lines of ^{12}C and ^{16}O, for which some evidence was indeed seen in raw count spectra. No stringent constraints on the line widths could be obtained, although a dominantly narrow-line origin seemed unlikely. If due to a low-energy CR component (with standard CR composition), pervading the clouds, it could readily be inferred from [89] that the energy density of this CR component had to be two orders of magnitude higher than that of the canonical ($\gtrsim 100$ MeV/nucleon) cosmic rays (~ 1 eV cm^{-3}). This would violate the CR ionization rate of the clouds by a similar magnitude (and make CR heating a major contributor to the infrared luminosity of Orion [30]). In addition, this seemed difficult to reconcile with the fact that the particle density above a few 100 MeV/nucleon should not be enhanced above the average Galactic value because no evidence for any significant enhancement of high-energy γ-ray emission is seen, as noted above. It seemed therefore most likely that enhanced abundances of fast C and O nuclei are responsible (producing a strong broad-line component), which would alleviate these problems. But much more detailed evaluations of these aspects are available now (§3.1). One way or another, the basic conclusion that could be drawn from this initial detection was

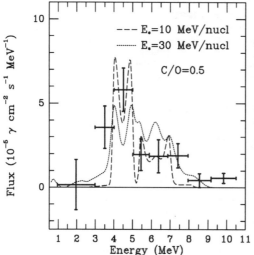

FIGURE 3. COMPTEL spectrum of the Orion/Mon R2 complex [9]. Statistical $\pm 1\sigma$ error bars are shown. Model spectra from the deexcitation of energetic C and O nuclei (pure broad-line scenario) are superimposed, including the line-splitting effect [23] (arbitrarily scaled). The energy resolution of COMPTEL is not accounted for. The particle spectra are assumed to have breaks at $E_\star = 10$ and 30 MeV/nucleon.

that a large flux of low-energy accelerated particles appears to be present in Orion, if the observed γ-ray emission is indeed of diffuse line origin.

After this first report, it took almost two years until the Ori/Mon area was well within the COMPTEL field-of-view again. A series of new observations has became available in the meantime. All data obtained until April 1996 have now been combined [9], which has increased the effective exposure by a factor of ~ 4. The 3–7 MeV radiation is observed at the 9σ significance level with a flux of $\sim 1.3 \times 10^{-4}$ γ cm^{-2} s^{-1}. For a distance of 450 pc, this corresponds to a luminosity of $\sim 3 \times 10^{39}$ photon s^{-1}, or $\sim 3 \times 10^{34}$ erg s^{-1}. A 2σ flux upper limit of 6×10^{-5} γ cm^{-2} s^{-1} for the 1–3 MeV range was derived. In addition, it can be firmly concluded now that one single source cannot explain the measured intensity distribution: the emission extends over the complex (the Orion A & B and Mon R2 clouds), but it may actually be due to a few localized source regions (Fig. 1). An EGRET map of the high-energy (>100 MeV) γ-ray emission is shown for comparison, emphasizing the pronounced appearance of the Ori/Mon complex in the COMPTEL observations. Some evidence (4.4σ) for emission from the Mon OB1/OB2 area is appearing as well. Although the global correlation with the Ori/Mon clouds is remarkable, the emission may actually be anti-correlated with the molecular gas on smaller scales because of the probably implied high CR ionization rates, if of diffuse origin. Such an anti-correlation is suggested by Fig. 1, but no firm conclusion can be drawn yet.

The excessive 3–7 MeV flux is further illustrated in Fig. 2, which shows a broad-band spectrum including the high-energy γ-ray data. Fig. 3 shows that the emission is apparently spread over the 3–7 MeV range, suggesting a dominantly broad-line origin from energetic C and O nuclei if indeed due to nuclear lines. A preliminary more detailed spectrum was derived as well, which indicates that the emission is not due to narrow lines at e.g. 4.44 and 6.13 MeV. A fully deconvolved narrow-bin spectrum could not be obtained yet, as described in [9]. Also Fig. 3 was produced by treating all energy bins separately. The 3–4 MeV part of the spectrum can be expected to be influenced, because the measured photon energy is below the nominal line energy (e.g. 4.44 MeV) for a significant fraction of the events. At this stage, however, the impact is not obvious: the 3–4 MeV flux may actually be underestimated rather than overestimated [9], which would clearly have an important impact on the interpretation (§3). This uncertainty orginates from the background modelling, particularly in case of extended emission.

OSSE has so far not detected Orion [70,45], but the findings are consistent with the COMPTEL results . The most recent OSSE flux estimate for broad-line emission is $(1.5 \pm 1.0) \times 10^{-4}$ γ cm^{-2} s^{-1} [45]. In addition, the OSSE observations confirm that a narrow-line origin is unlikely and that the emission must be extended rather than point-like.

No evidence for any significant ^{26}Al 1.8 MeV emission from Orion OB1 is seen so far [72], as can be expected (see e.g. [84]) in view of the fact that at least the OB1bcd subgroups are still very young ($\lesssim 5 \times 10^6$ yr; e.g. [14] and references therein). The older OB1a subgroup, with an estimated age of $\sim 12 \times 10^6$ yr (and located about 10° away from the Orion A&B clouds) would be the prime candidate.

2.2 The inner Galaxy

Prior to CGRO, as noted above, even for the inner Galaxy only very little spectral information on the 1–10 MeV emission was available and any imaging information was basically lacking. COMPTEL has provided first maps [106], but until recently spectral analyses were limited to 3 or 4 broad energy bands and focussed on the diffuse continuum emission [104,105,107]. It has become clear from these studies that an important diffuse inverse-Compton (IC) component may be present (with a wide latitude distribution), but the relative contributions of diffuse bremsstrahlung and IC radiation are not well established yet and a significant contribution from unresolved sources and γ-ray lines cannot be excluded.

Results are now available from first attempts to apply a much finer spectral binning to the integrated emission from the inner Galaxy, using all observations that were obtained during the first 5 years of the mission [10]. Apart from the fact that this finer binning will set more stringent constraints on the origin of the continuum emission in future work, it has revealed evidence for deviations from a smooth continuum spectrum. This is illustrated in Figs. 4–6. Spectral excesses seem to be present between ~ 1.3 MeV and the ^{26}Al 1.8 MeV line and at 3–7 MeV. As for Orion, the details are preliminary because

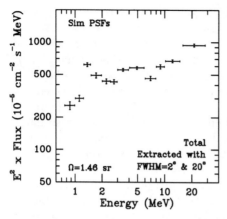

FIGURE 4. COMPTEL spectrum of the inner Galaxy [10] ($|\ell| < 60°$; $|b| < 20°$), obtained from fitting simultaneously models with Gaussian latitude distributions of FWHM = 2° and 20° (constant in intensity over the longitude range $\ell = -60°$ to 60° and zero outside). All events between 1.7 and 1.9 MeV are excluded from the 1.5–2 MeV bin and the response tail of the 1.8 MeV line is modelled out. So the 1.5–2 MeV bin actually covers the energy range 1.5–1.7 + 1.9–2 MeV. Statistical $\pm 1\sigma$ error bars are shown.

the spectral deconvolution is incomplete (although the response tail of the 1.8 MeV line towards lower energies is corrected for). The excess emission appears to have a rather wide latitude distrution, as illustrated in Fig. 5, but the precise extent needs further study. The flux in each excess ($\sim$ 1.3–1.7 and 3–7 MeV) is typically 2×10^{-4} γ cm^{-2} s^{-1} for the inner *two* radians, although not well determined because of the uncertain continuum level underneath. Fig. 6 presents a preliminary more detailed spectrum and places the COMPTEL spectrum in a broader perspective.

The spectral features were not seen so far by e.g. SMM and OSSE, but this may not be surprising because the excesses are broad and relatively weak. A

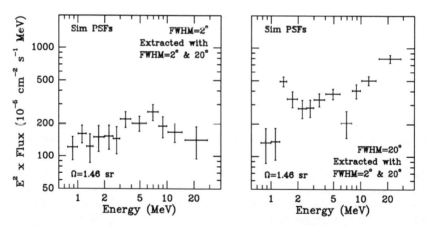

FIGURE 5. COMPTEL inner-Galaxy spectra of the narrow- and wide-latitude components separately [10] ($|\ell| < 60°$; $|b| < 20°$). This separation may of course not properly represent the true sky, but it cleary indicates that a significant fraction of the excesses emission seen in Fig. 4 must have a rather wide latitude extent. All 1.7–1.9 MeV events are excluded and the 1.8 MeV tail is modelled out. Statistical $\pm 1\sigma$ error bars are shown.

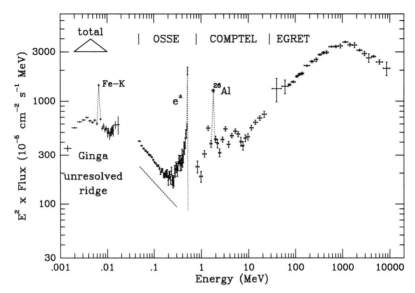

FIGURE 6. The preliminary COMPTEL narrow-bin spectrum of the inner Galaxy [10], together with observations by *Ginga* [117], OSSE [85,86], and EGRET [49] (statistical error bars are shown). This figure is mainly meant to put the COMPTEL result in a broader perspective — the spectral details, such as e.g. the structure *within* the 3–7 MeV range, should be considered preliminary [10]. The (ℓ, b) integration for the COMPTEL and EGRET data points ($-60° < \ell < 60°$; $|b| < 10°$) differs from that of the *Ginga* and OSSE points (inner radian), but this does not disturb the picture significantly because all emission is seen to be concentrated within $\pm 40°$ from the galactic centre. The dotted line near 100 keV represents the best-fit spectrum to two OSSE measurements after subtraction of point sources seen in simultaneous SIGMA observations [86].

study of SMM data [43], assuming the presence of two broad lines at 4.44 and 6.13 MeV (of width 0.8 and 1 MeV), has provided a combined upper limit of 3.0×10^{-4} γ cm^{-2} s^{-1} for the inner radian. The intensity distribution was assumed to follow the CO distribution. Using OSSE data, an upper limit of 2.2×10^{-4} γ cm^{-2} s^{-1} was found for the inner radian in the same scenario [44]. Both values should probably be multiplied by a factor of 1.5 or so to compare with the COMPTEL flux estimate for the inner two radians. Although the upper limits are above this flux estimate, a confirmation by OSSE seems possible. On the other hand, COMPTEL observes a broader spectral excess, extending from about 3 to 7 MeV, and finds that an important fraction of the enhanced 3–7 MeV emission (and the 1.3–1.7 MeV excess) seems to have a rather wide latitude extent, which would be difficult to observe with OSSE. The broad feature at 1.3–1.7 MeV cannot be directly compared with published upper limits from SMM and OSSE data, but typical upper limits of $(1-2) \times 10^{-4}$ γ cm^{-2} s^{-1} are derived for *narrow*-line fluxes from the inner radian at approximately these energies [61,42,60,44].

3. IMPLICATIONS AND MODELS

The spectral features seen by COMPTEL clearly suggest the presence of nuclear deexcitation lines as a possible explanation, but the observational constraints are severe. We review here different model scenarios for γ-ray line production. A diffuse origin related to starforming regions is discussed in §3.1. The possible role of compact objects is addressed in §3.2.

Two other interpretations have been suggested for the 3–7 MeV emission from Orion. Pohl [83] proposed the presence of a γ-ray emitting AGN (blueshifted $e^{\pm}$ annihilation line). This would not explain the extended nature of the emission that is clearly seen now, but an AGN contribution cannot be excluded. Dogiel et al. [35] proposed bremsstrahlung from an enhanced low-energy CR electron population ($E_e < 10$ MeV), but emission below 3 MeV seems hard to avoid if the entire 3–7 MeV flux would be due to bremsstrahlung. Also spectral structure in the 3–7 MeV regime would of course not be expected.

3.1 Starforming regions

Orion versus inner Galaxy

If all giant molecular clouds (GMCs) in the Galaxy would be 3–7 MeV γ-ray sources like Orion, assuming fluxes proportional to the gas mass (most of the galactic molecular gas is in GMCs with masses $\gtrsim 10^5$ M$_\odot$, like Orion — see e.g. [115]), then the expected 3–7 MeV flux from the inner two radians of the Galaxy would be roughly $(2 - 3) \times 10^{-3}$ γ cm^{-2} s^{-1}. This is at least 5 times higher than the *total* 3–7 MeV flux that is actually observed from the galactic narrow ridge in the inner Galaxy, namely $\sim 4 \times 10^{-4}$ γ cm^{-2} s^{-1} (Fig. 5 — left) [10]. In comparison with the *excess* 3–7 MeV flux that may be present in the narrow-ridge emission, which is $\lesssim 10^{-4}$ γ cm^{-2} s^{-1} (Fig. 5 — left), the expected flux is higher by a factor $\gtrsim 20$. So the Orion γ-rays are not produced by an intense (and enriched) low-energy CR component with a homogeneous distribution throughout the Galaxy. In fact, the expected flux would be higher by an additional factor of ~ 2 because emission from the atomic gas would have to be included as well of course.

Comparing the Orion 3–7 MeV γ-ray luminosity of $\sim 3 \times 10^{39}$ photon s^{-1} with the above value of the excess 3–7 MeV flux that may be present in the narrow-ridge emission, we find that $\lesssim 200$ Orion-like γ-ray sources can be present in the inner Galaxy. Williams & McKee [115] estimate that only about 1/8 of the clouds with masses like Orion contains an OB association of luminosity comparable to that in Orion and derive indeed a number of ~ 200 Orion-like complexes in the Galaxy. About 50 more luminous complexes are expected like the Rosette/Mon OB2 complex (at a distance of ~ 1500 pc), from which 3–7 MeV γ-rays may have been seen [9] (see §2.1).

However, in a scenario that associates the γ-rays with starforming activity, one may argue that on average only a fraction $\lesssim 20\%$ of the candidate complexes is actually ON, i.e. really active. This follows from the required energetics, as discussed below, which suggests that the γ-ray production is of intermittent nature with $\tau_{\rm ON} \lesssim 10^5$ yr. As it seems most natural to associate the 'flaring' with supernova (SN) explosions of massive stars [and the preceding Wolf-Rayet (WR) phase], which can be expected to occur on average once every $\sim 5 \times 10^5$ yr ($= \tau_{\rm OFF}$) in an association like Orion OB1 [14], one obtains $\tau_{\rm ON}/\tau_{\rm OFF} \lesssim 0.2$. Based on the result of Williams & McKee [115], one would therefore expect not more than ~ 50 Orion(+Rosette)-like γ-ray sources in the Galaxy, with a total flux from the inner Galaxy of $\lesssim 3 \times 10^{-5}\,\gamma\,{\rm cm}^{-2}\,{\rm s}^{-1}$. Similar numbers are obtained when the implications of this scenario for the production of light elements (Li, Be, B) are considered [25,87], as discussed at the end of this section.

We will see below that the energy output of each individual SN is most likely not the main driving force for the flaring. It is more likely that the flaring is largely regulated by the replenishment of fresh nuclei by the SN and by the preceding WR wind, for which the accelerator machine was already installed at an earlier stage (newly powered by the stellar wind and SN).

The above illustrates that the detection of Orion and an interpretation in terms of diffuse emission do certainly not imply that the inner Galaxy has to be a bright source of 3–7 MeV γ-rays as well. On the contrary, the expected flux seems only marginally detectable by COMPTEL and is approximately an order of magnitude lower than the upper limits set by SMM [43] and OSSE [44]. As stronger emission appears to be seen (including a component that is rather wide in latitude, which we have not addressed above), it seems necessary to consider other types of sources as well.

Modelling of diffuse γ-ray lines

The basic scenario that links the Orion γ-rays to starforming activity was discussed by Bykov & Bloemen [18]. They suggested that the observations may be explained by broad lines originating from energetic nuclei that are accelerated from the enriched hot media in regions of massive star formation, interacting with ambient dense gas in clouds and shells. The particle acceleration results from a powerful energy release in the form of shocks and violent plasma motions which occur in a bubble created by stellar winds (during the OB phase and WR phase) and SNe [22,20,17]. The elemental abundances of the hot rarified plasma inside such a bubble can be very different from the standard cosmic abundances due to the injection of matter enriched with heavy elements from stellar winds and SNe of massive stars. The stochastic MHD fluctuations and multiple shocks will create a nonthermal particle population with a very hard spectrum of low-energy ions.

The distribution function of the nonthermal ion energies in this model is approximately $\propto E^{-1}$ up to a break energy E_*. This break energy achieves its stationary value $E_* \sim 0.5 m_i u^2 \eta^{-1}$ on acceleration time scales τ_a of about a few times 10^4 yr for the model described in [18]. Here m_i is the ion mass and u is the amplitude of the large-scale bulk velocity of the plasma. The efficiency of ion injection η is $\sim 10^{-3}$, if the injection is due to shocks (e.g. [40]). This implies that a break energy E_* of about 10–30 MeV/nucleon is expected. Modelling that accounts for nonlinear effects (i.e. the back reaction of the nonthermal particles on the plasma motions), which becomes necessary after a few times τ_a, has shown that the hard low-energy branch keeps this spectral shape at least until $\tau_e \sim (R/l)^2 \tau_a$, where R is the bubble scale and l is the correlation length of the MHD large-scale plasma motions [21]. The time scale τ_e is about a few million years for the model in [18]. The spectral shape above the break energy is typically very steep (e.g. [17]), which explains the lack of excessive π^0-decay emission from Orion discussed in §2 (indeed requiring a break energy below $\sim$100 MeV/nucleon [109]).

The nonthermal ions accelerated in the bubble then penetrate into dense matter of the surrounding shell and the neighbouring clouds. They encounter here Coulomb and nuclear interactions, which leads to γ-ray production and nucleosynthesis. Due to scattering on small-scale magnetic fluctuations, the nonthermal ions propagate diffusively into the dense medium, with mean free path $\lambda = 10^{16} \lambda_{16}$ cm (at energies of some 10 MeV/nucleon). The penetration depth of the nonthermal ions irradiating the cloud can be estimated as $\Delta \sim \lambda_{16}^{0.5} n_{100}^{-0.5}$ pc, where the particle number density inside the cloud n_{100} is in units of 100 cm^{-3} ($E_* = 30$ MeV/nucleon is assumed here; $\Delta \propto E_*^a$, where typically $1 \lesssim a \lesssim 1.5$). This implies that approximately a few percent of the Orion cloud mass, located near the hot bubble boundary, is actually involved.[1] The modelling of cloud ionisation in such a scenario is discussed in [12].

If complexes like Orion are indeed producing nuclear γ-rays from energetic ions accelerated in the vicinity of OB associations, then the γ-ray line spectra provide a potentially very interesting tool to study the isotopic composition of the ejecta from massive stars, which was immediately realized [18,25,90]. Extensive massive star evolutionary models that are suitable for detailed comparisons are available now (e.g. [116,66,110]). Ramaty et al. [90,92,87] have generated spectra for a variety of energetic particle compositions in order to investigate whether or not the absence of emission from Orion below 3 MeV sets constraints on the origin of the 3–7 MeV emission. They find that at least a solar system composition [2] and a CR source composition [67] appear to be in disagreement. The composition has to be more strongly enriched in

---

[1] It is important to account for this structure of the nonthermal ion and the cloud ionisation rate distributions when comparing the X-ray emission due to bremsstrahlung of knock-on electrons [35] and the K_α line emission [94] with ROSAT observations (discussed below). This might substantially alleviate the K_α inconsistency pointed out in [94].

C–O relative to Ne–Fe.[2] The composition of the extreme late phase wind of a carbon-rich WR star (type WC) was found to be in good agreement. A composition determined by pick-up ions from the breakup of dust would be consistent as well [92], which was also proposed by Ip [50] in connection to the possible acceleration by T-Tauri protostellar winds. Parizot et al. [78] have recently studied the γ-ray line production for wind compositions of massive stars ($\gtrsim 40$ M$_\odot$) when integrated over the entire WR phase (typically $\sim 4 \times 10^5$ yr, with a sharp concentration in $\sim 10^5$ yr [97]), using massive star evolutionary models of [97,66]. This integration is important because the WC mass ejected in the final episode is negligible in comparison to the total mass of the wind. For a massive star, also the mass of the SN ejecta is small compared to the total mass loss during its life. They find that the COMPTEL upper limit on the (1–3 MeV)/(3–7 MeV) flux ratio is consistent with the mean-wind composition. Also the composition of the accumulated ejecta of an entire OB association would be consistent with the flux-ratio upper limit, even after integration over $\sim 10^7$ yr (determined by stellar masses $\gtrsim 20$ M$_\odot$). So COMPTEL is not highly constraining the composition origin yet.

The energy requirements are rather stringent [50,30,90,23,78]. The required power is determined by the break energy E_* (decreasing with E_* for a given γ-ray flux) and by the composition of the particles (the γ-ray emission is energetically most efficient for high abundances of energetic ions). An evaluation of the power deposit for different compositions has been carried out by Ramaty et al. [90,92,87] and more recently by Parizot et al. [78]. The power must be at least a few $\times 10^{38}$ erg s^{-1}, if an optimal composition is assumed (also selective injection into the acceleration process may play a role). Estimates ranging from $\sim 10^{39}$ erg s^{-1} (for $E_* \sim 30$ MeV/nucleon) to $\sim 3 \times 10^{39}$ erg s^{-1} ($E_* \sim 10$ MeV/nucleon) are obtained for the composition of wind material accumulated during the entire WR phase of a massive star ($\gtrsim 40$ M$_\odot$) or for that of an entire OB association [78]. This energy consumption rate adds up to roughly 10^{51} erg or more in 10^5 yr (model dependent — see e.g. examples in [77]), which indicates that a single SN (with a kinetic energy output of $\sim 10^{51}$ erg, of which typically $\sim 10\%$ may be available for particle acceleration) is most likely not responsible for the Orion γ-ray emission.[3] Moreover, a particle spectrum with an approximately single power-law shape would be

---

[2] Evidence for C–O enrichment relative to protons and α-particles follows directly from the fact that no evidence for narrow lines is seen, as discussed in §2, but this important observational constraint is not fully conclusive yet due to the incomplete deconvolution of the COMPTEL data at this stage.

[3] The single SN origin discussed by Cameron et al. [24], Fields et al. [37], Tatischeff et al. [108], and Parizot et al. [81] involves the direct production of γ-ray lines from the fast moving and highly enriched SN ejecta from a massive (WC) star, i.e. without statistical acceleration. Such a γ-ray source would be quite short lived (estimates range from a few tens to a few thousand years), so that very recently a SN should have occurred in Orion. The same holds for the SN scenario including shock acceleration discussed in [79]. The fast moving ejecta may serve as an injection mechanism [17], in addition to shocks.

expected (see e.g. [55]), i.e. without the required break. The same holds for acceleration in a single strong stellar-wind shock, as considered in [71,80]. A single WC wind origin (including time dependence) is discussed in [80] and is anyway found to be not sufficiently efficient to explain γ-ray lines from Orion. So from the point of view of energetics, the collective effect of winds and SNe from an association of stars (as in the scenario [18] discussed above) seems needed.

Although a single WR/SN event appears to be energetically insufficient to produce the observed γ-rays from Orion, the occurence of such an event does probably play a distinct role. It is likely that the γ-ray production is largely regulated by the intermittent replenishment of fresh nuclei by a WR wind (and SN), which are then accelerated by already existing shocks and large-scale plasma motions, as in the scenario discussed above (with new power being provided by the stellar wind and SN). As no WR stars are found in Orion, it seems that a SN must have occurred $\lesssim 10^5$ yr ago, for which evidence is indeed observed [16]. Also indications for SN activity $\sim (2-4) \times 10^6$ yr ago are seen [29,95]. The energy that is required to accelerate the injected ions would be provided by the total energy output of the Orion OB1abc subgroups, which is estimated to be at least $\sim 10^{52}$ erg from simulations of the stellar content of Orion OB1 [14]. If the Orion OB1 association has indeed created the so called Orion-Eridanus bubble, with an estimated dynamical age of $(2-6) \times 10^6$ yr, then $\sim 2 \times 10^{52}$ erg must have been provided [15]. A promising detailed evaluation of this scenario was recently given by Parizot [77].

As illustrated in Fig. 3, the effect of line splitting proposed by Bykov et al. [23] can be very pronounced (see also Ramaty et al. [93] and Kozlovsky et al. [56]), but a fully deconvolved spectrum will be needed to study and utilize this effect. Ramaty et al. [93,56] have argued that the blue wings of lines can be suppressed if the particles interact primarily on the side of a molecular cloud facing us. It seems to us, however, that this anisotropy effect is lost in thick-target cloud interactions. As the average column density of the Orion clouds does not exceed a few times 10^{22} cm^{-2}, multiple fast ion scatterings are required to have enough grammage. If no scattering would occur, then the required power would be very large; in order to produce a significant anisotropy in the scenario proposed by Ramaty et al. [93,56] (with $E_*=100$ MeV/nucleon), a power of $> 10^{40}$ erg s^{-1} seems needed.

X-rays

There is a controversy on the associated X-ray emission that can be expected from Orion. Dogiel et al. [35] calculated the bremsstrahlung from knock-on electrons which are produced by the ions generating the γ-ray lines. They find that this will exceed the upper limit they derive from ROSAT data at 0.5–2 keV ($\sim 6 \times 10^{33}$ versus $\sim 1.3 \times 10^{33}$ erg s^{-1}, excluding point sources).

However, Ramaty et al. [94] find that the contribution of the knock-on process to the total X-ray production by energetic ions is negligible in comparison with inverse bremsstralung and that the X-ray emission from both processes is not inconsistent with the ROSAT data. On the other hand, Ramaty et al. point out that the K_α line emission following electron capture onto fast O nuclei may be in conflict with the ROSAT data, although they note that the problem is at least alleviated if the γ-rays are produced in the partly ionized cloud boundaries. See also footnote 1.

Light elements

The possible presence of large fluxes of low-energy accelarated C and O ions in the Galaxy has stimulated new studies on the origin of light elements (Li, Be, B) [25,17,87,92,112,113] and on the extinct radio-isotopes existing at the time of the formation of the solar system [27,64,24,28,92,3]. This work is beyond the scope of our review, but it provides an interesting link between the γ-rays from Orion and the expected γ-rays from the Galaxy at large. Namely, the number of Orion-like γ-ray sources in the Galaxy that is expected on the basis of the scenario discussed in this section ($\lesssim 50$) is consistent with the number (~ 50) derived from the current rate of boron production in the Galaxy by Cassé et al. [25] (for an energy deposit by the particles in Orion of 10^{39} erg s^{-1}). Ramaty [87] estimated a larger total number of 200–400, which is mainly due to the assumed lower energy deposit by a factor of 2–4. So a nicely consistent picture appears to emerge.

3.2 Compact objects

The presence of spectral excesses in the inner-Galaxy wide-latitude component (§2.2) would be difficult to understand in the scenario discussed above. In addition, significant emission at 3–4 MeV, which may be seen towards the inner Galaxy as well as Orion, cannot be readily explained. If indeed nuclear deexcitation lines are seen in the wide-latitude radiation, then the most plausible alternative explanation seems to be that the collective signal of an ensemble of relativistic objects is being detected, in particular accreting neutron stars (NSs). We summarize here some basic aspects and discuss Orion in this context.

In principle, accretion power can easily provide the energetics. The wide latitude extent would be understood if low-mass X-ray binaries (LMXBs), having a large scale height [76], play an important role. On the other hand, COMPTEL cannot distinguish between a scenario in which a large galactic population of sources produces the emission and a scenario in which only a dozen or so relatively bright ones is responsible (but below the detection limit, i.e. fluxes up to $\sim 2 \times 10^{-5}$ γ cm^{-2} s^{-1}). The line emission can be gravitionally redshifted and thus extend into the 3–4 MeV regime, depending upon the location of the production region. The first pionering idea to study redshifted γ-rays as a probe of the gravitational potential of NSs was put forward already more than 25 years ago [100]. The development of early ideas on γ-ray lines from individual NSs is reviewed in [62] and more recent work is described in [4].

Under the assumption that most of the accretion power is radiated in X-rays, the ratio between the luminosity L_γ of a γ-ray line at energy E_γ (not accounting for Compton attenuation) and the integrated X-ray luminosity L_X can be written as $\zeta \equiv L_\gamma/L_X \approx E_\gamma Q\eta/\phi\mu \approx 10^{-2}Q\eta$, where Q is the average number of line photons emitted per accreting nucleus, η is the abundance of the accretion-flow nuclei emitting the γ-ray line, $\mu \approx 1 + \Sigma A_j\eta_j$ is the mean atomic weight of the accreting matter, and ϕ is the gravitational potential at the stopping surface. It follows from Fig. 6 that the ζ value of the observed line features is of the order 10^{-3} ($L_\gamma \sim 10^{36}$ erg s^{-1} and $L_X \sim 10^{39}$ erg s^{-1} — the total X-ray luminosity of the inner Galaxy). So in case of non-enhanced ion abundances ($\eta \sim 10^{-3}$), a Q value of ~ 100 would be implied.

The efficiency of the γ-ray line production (i.e. the value of Q) depends strongly upon the way accreting matter is slowed down near the NS surface. The two basic possibilities are collisional Coulomb stopping in the NS atmosphere (accompanied by strong nuclear destruction [1,4]) and collisionless stopping due to collective effects and radiation drag forces (see e.g. [39] for a review). A comprehensive study of γ-ray line emission in the Coulomb stopping scenario [4] reached the conclusion that the yields are small ($Q < 0.1$). Moreover, the 2.22 MeV neutron capture line should strongly dominate the nuclear deexcitation lines [5]. Collisionless stopping would therefore clearly be preferred by the observations. Although Q can be much higher in this scenario, a value as large as ~ 100 would set important constraints [19]. Clearly, the efficiency requirement would be less severe if the accreted matter would be enriched with metals (e.g. from a highly stripped dwarf), in which case it may be even possible to detect individual objects (near Eddington, at least in the outburst stage).

Three X-ray binaries (LMXBs) are known in the Orion/Monoceros field shown in Fig. 1 [75], but these are located near the galactic plane: A0620-003 at $(\ell, b) \sim (210.0, -6.5)$, A06556-072 at about $(220.2, -1.7)$, and 4U 0614+09 at about $(200.9, -3.4)$. Only the latter concides with an excess in the map, namely in the Mon OB1/OB2 area. The former is a well known example of the galactic black hole candidates, which might be sources of γ-ray lines due to nonthermal MHD acceleration processes in the accretion disk [68] or due to nuclear excitations during radiative stopping of accreted matter in the outburst phase [19].

Another possibility to be considered is that the correlation of the 3–7 MeV emission with the CO distribution shown in Fig. 1 is mimicked by an ensemble of isolated NSs undergoing Bondi-Hoyle accretion (e.g. [57]) from the molecular clouds. The enhanced γ-ray emission from Orion relative to that from the galactic molecular cloud ensemble (this is the factor $\gtrsim 20$ discussed in §3.1) may then be associated with the fact that the accretion rate depends strongly on the NS velocity ($\propto v^{-3}$) and is thus influenced by e.g. the differing gravitational fields in the inner Galaxy and in the local galactic environment, but this needs careful study. In this scenario, an accretion power in excess of 10^{36} erg s^{-1} is dissipated in the Orion complex. In order to be an efficient γ-ray line source, an isolated NS should have a high column density in the accretion column (at least a few times 10^{24} cm^{-2}) and a high magnetic field ($> 3 \times 10^{11}$ G) [19]. This implies that most of the accretion power, released just above the NS surface, has to be reprocessed to soft X-rays which are efficiently absorbed within the dense Orion clouds (for comparison, the observed infrared luminosity from the Orion complex exceeds 10^{39} erg s^{-1}, e.g. [114], of which a small fraction might be reprocessed accretion power). The anisotropy of the soft emission in this scenario would further reduce the X-ray flux that can be expected to be seen. The 3–7 MeV emission would not be strongly affected by the accreting matter and a low (1–3 MeV)/(3–7 MeV) flux ratio may be basically understood without a strong C–O overabundance due to more efficient Compton attenuation of the γ-ray lines below 2 MeV [19].

No firm conclusions on the contribution from accreting objects can be drawn at this stage, but further study seems definitely warranted. Careful modelling of the expected X-ray emission for candidate γ-ray line emitters among the accreting-NS scenarios seems particularly important.

4. CONCLUSIONS

COMPTEL's finding of spectral structure in the MeV regime suggests the presence of nuclear-interaction γ-ray lines, although conclusive proof is still lacking. First evidence was observed from the Orion complex and indications are now seen in emission from the inner Galaxy as well. The fact that *broad* lines appear to be seen, i.e. involving energetic ions, has far reaching consequences and makes detailed spectroscopy of the lines an extremely powerful technique [23,56]. An origin from energetic ions accelerated in the vicinity of OB associations seems plausible, which would imply that γ-ray line spectra provide a potentially very interesting tool to study the acceleration and the isotopic composition of the wind and SN ejecta from massive stars. But there are indications that the emission extends into the 3–4 MeV regime and that the latitude distribution in the inner Galaxy is rather wide in comparison to what would be expected from sources associated with recent star forming activity. It seems that another category of sources is contributing and that accreting compact objects may play a role here.

Acknowledgements: A.M.B acknowledges support by RBRF grant 95-02-04143a and by the Netherlands Organisation for Scientific Research (NWO).

REFERENCES

1. Aharonian F.A. & Sunyaev R.A., 1984, MNRAS 210, 257
2. Anders E. & Grevesse N., 1989, Geochim. Cosmochim. Acta 53, 197
3. Bateman N.P.T., Parker P.D. & Champagne A.E., 1996, ApJL 472, L119
4. Bildsten L., Salpeter E.E. & Wasserman I., 1992, ApJ 384, 143
5. Bildsten L., Salpeter E.E. & Wasserman I., 1993, ApJ 408, 615
6. Bloemen H., 1989, Ann. Rev. A&A 27, 469
7. Bloemen H., et al., 1984, A&A 139, 37
8. Bloemen H., et al., 1994, A&A 281, L5
9. Bloemen H., et al., 1997, ApJ 475, L25
10. Bloemen H., et al., 1997, these proceedings
11. Böhringer H., Morfill G.E. & Zimmermann H.U., 1987, ApJ 313, 218
12. Bozhokin S. & Bykov A.M., 1994, Astron. Lett. 20, 593
13. Brecher K. & Burrows A., 1980, ApJ 240, 642
14. Brown A.G.A., de Geus E.J. & de Zeeuw P.T., 1994, A&A, 289, 101
15. Brown A.G.A., Hartmann D. & Burton W.B., 1995, A&A, 300, 922
16. Burrows D.N., et al., 1993, ApJ 406, 97
17. Bykov A.M., 1995, Space Sci. Rev. 74 (3/4), 397
18. Bykov A.M. & Bloemen H., 1994, A&A 283, L1
19. Bykov A.M. & Bloemen H., 1997, subm.
20. Bykov A.M. & Fleishman G.D., 1992, MNRAS 255, 269
21. Bykov A.M. & Pushkin N.A., 1997, in prep.
22. Bykov A.M. & Toptygin I.N., 1990, Sov. Phys. JETP 71, 702
23. Bykov A.M., Bozhokin S. & Bloemen H., 1996, A&A 307, L37
24. Cameron A.G.W., et al., 1995, ApJ 447, L53
25. Cassé M., Lehoucq R. & Vangioni-Flam E., 1995, Nature 373, 318
26. Chupp E.L., et al., 1973, Nature 241, 333
27. Clayton D.D., 1994, Nature 368, 222
28. Clayton D.D. & Jin L., 1995, ApJ 451, 681
29. Cowie L.L., Songaila A. & York D.G., 1979, ApJ 230, 469
30. Cowsik R. & Friedlander M.W., 1995, ApJ 444, L29
31. Dame T.M., et al., 1987, ApJ 322, 706
32. Diehl R., et al., 1995, A&A 298, 445
33. Diehl R. & Timmes F.X. 1997, these proceedings
34. Digel S.W., Hunter S.D. & Mukherjee R., 1995, ApJ 441, 270
35. Dogiel V.A., Freyberg M.J., Morfill G.E. & Schönfelder, 1997, Proc. 25th Int. Cosmic Ray Conf., OG 3.4
36. Dupraz C., et al., 1997, A&A 324, 683
37. Fields B.D., et al., 1996, ApJ 462, 276
38. Genzel R. & Stutzki J., 1989, Ann. Rev. A&A 27, 41
39. Ghosh P. & Lamb F.K., 1991, in Neutron Stars: Theory and Observation, eds. J. Ventura & D. Pines, Kluwer, Dordrecht, p. 363
40. Giacalone J., et al., 1993, ApJ 402, 550
41. Gilman D., et al., 1979, ApJ 229, 753

42. Harris M.J., et al., 1990, ApJ 362, 135
43. Harris M.J., Share G.H. & Messina D.C., 1995, Ap 448, 157
44. Harris M.J., et al., 1996, A&AS 120, C343
45. Harris M.J., et al., 1997, A&A, in press
46. Haymes R.C., et al., 1975, ApJ 201, 593
47. Higdon J.C., 1987, Proc. 20th Int. Cosmic Ray Conf., 1, 160
48. Higdon J.C. & Lingenfelter R.E., 1996, A&AS 120, C349
49. Hunter S., et al., 1997, ApJ 481, 205
50. Ip W.-H., 1995, A&A 300, 283
51. Iyudin A., et al., 1994, A&A 284, L1
52. Iyudin A., et al., 1997, in The Transparent Universe, eds. C. Winkler et al., ESA SP-382, p. 37
53. Jacobson A.S., et al., 1978, in Gamma Ray Spectroscopy in Astrophysics, NASA TM-79619, eds. T.L. Cline & R. Ramaty, p. 228
54. Johnson W.N., et al., 1993, ApJS 86, 693
55. Jones F.C. & Ellison D.C., 1991, Space Sci. Rev. 58, 259
56. Kozlovsky B., Ramaty R. & Lingenfelter, 1997, ApJ 484, 286
57. Lamb F.K., 1989, in Timing Neutron Stars, eds. H. Ögelman & E.P.J. van den Heuvel, Kluwer, Dordrecht, p. 649
58. Lamb R.C., et al., 1983, Nature 305, 37
59. Lavigne J.M., et al., 1986, ApJ 308, 370
60. Leising M.D. & Share G.H., 1994, ApJ 424, 200
61. Leising M.D., et al., 1988, ApJ 328, 755
62. Lingenfelter R.E., Higdon J. & Ramaty R., 1978, in Gamma Ray Spectroscopy in Astrophysics, NASA TM-79619, eds. T.L. Cline & R. Ramaty, p. 252
63. Mandrou P., Bui-Van A., Vedrenne G. & Niel M., 1980, ApJ 237, 424
64. Marti K. & Lingenfelter R.E., 1994, AIP Conf. Proc. 327, 549
65. Meneguzzi M. & Reeves H., 1975, A&A 40, 91
66. Meynet G., et al., 1994, A&AS 103, 97
67. Mewaldt R.A., 1983, Rev. Geophys. Space Phys. 21, 295
68. Miller J.A. & Dermer C.D., 1995, A&A 298, L13
69. Murphy R.J., et al., 1991, ApJ, 371, 793
70. Murphy R.J., et al., 1996, ApJ 473, 990
71. Nath B.B. & Biermann P.L., 1994, MNRAS 270, L33
72. Oberlack U., et al., 1995, Proc. 24th Int. Cosmic Ray Conf. 2, 207
73. Oberlack U., et al., 1996, A&AS 120, C311
74. O'Neill T., et al., 1983, Proc. 18th ICRC 9, 45
75. van Paradijs J., 1995, in X-ray Binaries, eds. W.H.G. Lewin, J. van Paradijs & E.P.J. van de Heuvel, Cambridge Univ. Press, p. 536
76. van Paradijs J. & White N.E., 1995, ApJL 447, L33
77. Parizot E.M.G., 1997, A&A, subm.
78. Parizot E.M.G., Cassé M & Vangioni-Flam E., 1997, A&A, subm.
79. Parizot E.M.G., Ellison D.C., Lehoucq R. & Cassé M, 1997, in The Transparent Universe, eds. C. Winkler et al., ESA SP-382, p. 43
80. Parizot E.M.G., Lehoucq R., Cassé M & Ellison D.C., 1997, in The Transpar-

ent Universe, eds. C. Winkler et al., ESA SP-382, p. 93
81. Parizot E.M.G., Lehoucq R., Cassé M & Vangioni-Flam E., 1997, in The Transparent Universe, eds. C. Winkler et al., ESA SP-382, p. 97
82. Peterson L.E., et al., 1990, 21st ICRC 1, 44
83. Pohl M., 1996, A&AS 120, C457
84. Prantzos N. & Diehl R., 1996, Phys. Rep. 267, 1
85. Purcell W.R., et al., 1995, Proc. 24th Int. Cosmic Ray Conf. 2, 211
86. Purcell W.R., et al., 1996, A&AS 120, C389
87. Ramaty R., 1996, A&AS 120, C373
88. Ramaty R., Borner G. & Cohen J.M., 1973, ApJ 181, 891
89. Ramaty R., Kozlovsky B. & Lingenfelter R.E., 1979, ApJS 40, 487
90. Ramaty R., Kozlovsky B. & Lingenfelter R.E., 1995, ApJ 438, L21
91. Ramaty R., Kozlovsky B. & Lingenfelter R.E., 1995, Ann. NY Acad. Sci. (Proc. 17th Texas Symp.) 759, 392
92. Ramaty R., Kozlovsky B. & Lingenfelter R.E., 1996, ApJ 456, 525
93. Ramaty R., Kozlovsky B. & Lingenfelter R.E., 1997, in The Transparent Universe, eds. C. Winkler et al., ESA Publications Division, p. 75
94. Ramaty R., Kozlovsky B. & Tatischeff V., 1997, these proceedings
95. Reynolds R.J. & Ogden P.M., 1979, ApJ 229, 942
96. Ryan J. & Share G.H., 1997, these proceedings
97. Schaller G., Schaerer D., Meynet G. & Maeder A., 1992, A&AS, 96, 269
98. Schönfelder V., von Ballmoos P. & Diehl R., 1988, ApJ 335, 748
99. Schönfelder V. et al., 1993, ApJS, 86, 657
100. Shvartsman V.F., 1970, Astrophysics 6, 56; 1970, Astrofizika, 6, 123
101. Skibo J.G., Ramaty R. & Purcell W.R., 1996, A&A Suppl. Ser. 120, C403
102. Smith D.M., Purcell W.R. & Leventhal M., 1997, this volume
103. Strong A.W. & Mattox J., 1996, A&A 308, L21
104. Strong A.W., et al., 1994, A&A 292, 82
105. Strong A.W., et al., 1996, A&AS 120, C381
106. Strong A.W., et al., 1997, in The Transparent Universe, eds. C. Winkler et al., ESA SP-382, p. 533
107. Strong A.W., et al., 1997, these proceedings
108. Tatischeff V., et al., 1996, ApJ 472, 205
109. Tatischeff V., Ramaty R. & Mandzhavidze N., 1997, these proceedings
110. Thielemann F.K., Nomoto K. & Hashimoto M., 1996, ApJ 460, 408
111. Tueller J., et al., 1992, in The Compton Observatory Science Workshop, NASA Conf. Publ. 3137, p. 438
112. Vangioni-Flam E., Cassé M., Olive K., Fields B., ApJ 468, 199
113. Vangioni-Flam E., Cassé M., Ramaty R., 1997, in The Transparent Universe, eds. C. Winkler et al., ESA SP-382, p. 123
114. Wall W.F., et al., 1996, ApJ 456, 566
115. Williams J.P. & McKee C.F., 1997, ApJ 476, 166
116. Woosley S., Langer N. & Weaver T.A., 1993, ApJ 411, 823
117. Yamasaki N.Y, et al., 1996, A&AS 120, C393

EXTRAGALACTIC GAMMA RAY ASTRONOMY

High Energy Emission from Starburst Galaxies and Clusters

Yoel Rephaeli* and Charles D. Dermer[†]

*Physics Department, Stanford University, Stanford, CA 94305 USA
[†]Naval Research Laboratory, Code 7653, Washington, DC 20375-5352 USA

Abstract. Detection of diffuse high energy X-ray and γ-ray emission from starburst galaxies and clusters of galaxies will significantly advance the study of these systems. We review the results of OSSE observations of the nearby starbursts NGC 253 and M82. Compton scattering of relativistic electrons by the far IR and cosmic microwave background radiation fields can account for the observed emission from NGC 253. The predicted level of this emission in M82 is close to the upper limit set by OSSE. Based on OSSE and recent *ISO* observations of ultraluminous IR galaxies, including Arp 220, it can be inferred that these galaxies are powered largely by starburst activity. Nonthermal emission has not yet been detected from clusters, including the Coma cluster which was the target of a two-week OSSE observation. We review current lower limits on intracluster magnetic fields and upper limits on relativistic electron energy densities in the clusters already observed at high energies. Improved spectra of NGC 253, M82, and the Coma cluster are expected soon from recent observations of these systems by the PCA and HEXTE experiments aboard the *RXTE* satellite.

1. INTRODUCTION

High energy X-ray (HEX) and soft γ-ray emission, at energies higher than typically emitted from diffuse plasmas ($\epsilon > 10$ keV), is of interest as a diagnostic tool in virtually all galactic systems. Generally, HEX emission has not yet been extensively observed from any galactic class; sources detected so far are mostly active galactic nuclei (AGNs). Observations of starburst galaxies (SBGs), IR-luminous galaxies, and clusters of galaxies are perhaps of particular interest because HEX emission quite likely emanates from extended central regions in these systems. The spatial distribution of this emission in SBGs may prove to be the key to determining their nature: an extended emission region would imply that the high energy activity is not powered by a compact nucleus and, therefore, that these galaxies are not scaled-down manifestations of the AGN phenomenon. Rather, it would constitute direct evidence that the IR and HEX outputs of these galaxies are closely related to enhanced star formation activity. This will pave the way to detailed, multispectral investigations of the SB regions.

More generally, interest in X-ray emission from SBGs also stems from the possibility that conditions in their central regions are representative of conditions in young galaxies and the parent galactic population of AGNs and quasars. Related to this is the hypothesis that SBGs – as well as young galaxies – may contribute appreciably to the cosmic X-ray background [43,32]. Moreover, determining the relativistic particle densities and magnetic fields in clusters may shed some light on the values of these quantities in the intergalactic space proper, and possibly also on the role of cosmological seed fields.

Note that a distinction is made here between SBGs and the more general class of IR-luminous galaxies. The latter, loosely defined class is characterized solely by a high far IR (FIR) luminosity that is nominally $\geq 10^{11}$ $L_\odot$. This class consists of various types of galaxies, including SBGs, AGNs, and ultraluminous infrared galaxies (ULIGs; $L_{\rm IR} > 10^{12} L_\odot$). The study of HEX emission from each of these galaxy types is still in its infancy, due mainly to insufficient sensitivity of previous satellite experiments. For the first time, the *Compton Observatory* and the *Rossi X-Ray Timing Explorer* satellites have provided us with basic spectral information on a few nearby non-AGN galaxies. As it happens, these are mostly SBGs; we review these results, and summarize recent observations of other IR-luminous galaxies observed with OSSE.

In clusters with diffuse intracluster (IC) radio emission, the detection of HEX emission may uniquely identify the emission mechanism, yielding new information on the physical properties of the IC gas and thereby providing important insights on its origin and evolution. Specifically, the combination of radio and HEX measurements directly yields the mean strength of magnetic fields and the relativistic electron energy density in the IC space. Adequate spatial resolution of both these emissions may also resolve the issues of origin and propagation mode of the electrons. Extreme possibilities are either that the high-energy electrons originate from a strong radio source at the cluster center or that the electron sources are distributed (or perhaps the electrons are reaccelerated) throughout the cluster.

In this review we briefly discuss the motivation for HEX observations of IR galaxies and clusters, present the results of measurements of a few such systems, and discuss their implications.

2. STARBURST AND IR-LUMINOUS GALAXIES

Infrared emission from galaxies with high IR luminosities is generally thought to be produced either by warm interstellar dust heated by an abundant population of massive young (OB) stars, or by optical/UV and high-energy accretion-disk radiation reprocessed by surrounding gas and dust (for a recent review of IR-luminous galaxies, see [36]). In several nearby spatially-resolved galaxies, such as Arp 220, Mrk 231, and NGC 1068, the IR radiation is evidently emitted from a spatially extended, centrally located region at least a few hundred pc in size. In galaxies whose IR radiation is reprocessed nuclear emission, luminous HEX emission is also expected, and the latter type of galaxy would constitute a sub-class of the AGN category. Because of the obscuration of optical and UV emission and the penetrating power of HEX radiation, observations of IR galaxies with hard X-ray and soft γ-ray telescopes will help to discriminate between these types of galaxies.

We first focus attention on IR luminous galaxies whose activity is powered by young stars, i.e. SBGs. It is thought that the enhanced star formation rate in SBGs is triggered by galactic mergers, implying that these galaxies were more abundant in the earlier universe. Clearly, the class of SBGs is not a distinct one, even though most are either non-Seyfert Markarians [39] or normal spirals [18]. SBGs have been studied predominantly in the radio, infrared, and optical regions of the spectrum. The basic characteristics of SBGs are an enhanced star formation activity in the central ~ 1 kpc region, with copious emission of reprocessed stellar light and IR emission from heated interstellar dust.

The abundance of young stars is accompanied by a high rate of SNe and a large number of SN remnants. Associated with the SN activity are shock heating of the interstellar gas, and acceleration of particles to relativistic energies. Low energy protons ionize and heat the gas, whereas relativistic electrons lose energy mainly by synchrotron and Compton processes, leading mainly to radio and X-ray emission. Compton scattering off the intense FIR radiation field in the central region of the galaxy boosts up photon energies to the X-ray and γ-ray regions [37,31]. Emission at energies of up to a few tens of keV may also be produced in low metallicity X-ray binary systems [13]. In addition, X-rays are emitted from SNe and their remnants, and SN shocks collectively drive a wind which may result in a hot X-ray producing galactic halo [3]. Irrespective of the exact nature of the X-ray emitting mechanism, the IR and X-ray outputs of a SBG are expected to be directly correlated due to being largely triggered by the same parent population of massive stars.

In contrast, luminous HEX emission with a characteristic spectrum and a point-source morphology is expected if the enhanced IR output is reprocessed UV and HEX nuclear radiation. In the 2-10 keV range, Seyfert galaxies display a mean photon spectral index of ~ 1.7, which is explained by an accreting massive black hole which emits HEX radiation with a spectral index ≈ 1.9, upon which is imprinted a Compton reflection component. Observations show that the spectra characteristically continue as power laws to soft γ-ray energies, above which they are cut off at $\epsilon \approx$ several hundred keV (see the review by Johnson et al. [15] in these proceedings). A large compactness of the emission site, deduced from the luminosity and variability time scale of the emission, would provide strong evidence for an accretion-disk origin of the enhanced IR luminosity. This is not expected from an extended SBG region. The temporal signature is therefore key to determining a compact-source origin of the emission.

2a. Observations

Observations of SBGs have been made mostly at low X-ray energies; best studied are the nearby galaxies NGC 253 and M82. These archetypical SBGs were observed with all the major X-ray satellites, including *Einstein* [9], *Ginga* [40,24], *Rosat* [2], and *ASCA* [23]. The low energy X-ray (≤ 10 keV) emission from these galaxies is quite substantial, with X-ray luminosities $\sim 10^{41}$ erg/s. The spectra are best-fit with both thermal and power-law components, and the emission extends beyond the optical sizes of these galaxies.

The most detailed analysis of the low-energy emission from a SBG is that of the *Rosat* and *ASCA* measurements of M82 by Moran & Lehnert [23]. A power-law component with a photon index ~ 1.7 dominates the emission, with two additional thermal components, at 3×10^6 K and 6×10^6 K, required to best-fit the combined spectrum. Photoelectric absorption of the first two components is significant, at a level corresponding to $N_H \sim 10^{22}$ cm^{-2}. The size of the emission region decreases with energy; nonthermal emission is mainly from within the central kpc, while the low temperature component comes from a region extending to at least 5 kpc along the disk minor axis.

Independent evidence for significant thermal and nonthermal emissions comes from the presence of spectral X-ray lines and diffuse radio emission. The lines indicate the presence of hot gas within SN remnants and throughout the disk and halo, whereas radio emission is a direct indication for the presence of relativistic electrons. Irrespective of the exact nature of the thermal emission (for example, whether it is partly produced also in binary systems), it is the nonthermal electrons which give rise to HEX and soft γ-ray emission.

The realization that IR-luminous galaxies are likely also to be luminous HEX sources prompted an analysis [31] of the HEAO-1 A4 survey observations of a sample of 51 bright IRAS galaxies. Combining also *Einstein* and HEAO-1 A2 observations, a marginal detection was deduced which, when fitted with a power-law, led to a mean $0.5 - 160$ keV spectrum with an index of 1.47 ± 0.26. However, this result was heavily weighted by the low energy data, and essentially no evidence was found for emission at > 10 keV. Note also that this sample of 51 bright IRAS galaxies was broadly selected to include all local non-quasars with FIR luminosities $\geq 10^{11}$ $L_\odot$, so it clearly included some other AGNs. (For more on the correct classification of bright IRAS galaxies, see Moran *et al.* [22].)

The two nearby SBGs NGC 253 and M82 were observed with OSSE [1]. NGC 253 was detected at a 4.4 σ significance level in the 50 - 200 keV band,

and an upper limit was set on emission from M82. With very little spectral information and no spatial resolution, the OSSE measurement of NGC 253 allows only general characterization of the emission. OSSE observations of the ULIGs Arp 220, Mrk 273, and Mrk 231 also resulted in upper limits [5,6], although the upper limits on Mrk 231 are not very constraining because of the short live-time on this source. Nevertheless, important implications concerning the nature of the IR luminosity could be drawn from the OSSE observations for Arp 220 because the upper limit to the γ-ray luminosity is far below what might have been expected based on its IR luminosity. No SBGs or IR-luminous galaxies (other than quasars such as 3C 273) have been detected with the COMPTEL and EGRET experiments on *CGRO* [38,20].

2b. Interpretation

Mechanisms which can possibly result in extended emission at energies in the OSSE range are superposed nonthermal emission from SNe and their remnants, emission from low-metallicity massive X-ray binaries, and Compton scattering throughout the disk and halo. Emission from SNe and binary systems is confined to the disk, whereas Compton scattering of electrons by the FIR and CMB fields is expected to extend beyond the disk region. Indeed, radio emission has been observed from the disk and an inner halo region extending a few kpc above the optical disk in both NGC 253 and M82. Bhattacharya *et al.* [1] have estimated that only $\sim$ 20% of the observed flux from NGC 253 can be accounted for by Compton scattering of relativistic electrons by the FIR radiation field in the disk region. In a more quantitative and complete analysis [12] of the observed disk and halo radio emission in this galaxy, it was determined that all the observed flux can in fact be due to Compton scattering off (both) the CMB and FIR fields. While most of the emission in the central disk region is due to scattering off the FIR radiation, the scattering of relativistic electrons in the inner halo by the CMB contributes most of the emission ($\sim$ 70% at 100 keV). Thus, by ignoring the halo emission, Bhattacharya *et al.* underestimated the Compton HEX emission from NGC 253.

In the model adopted by Goldshmidt & Rephaeli [12], electron propagation in the disk and halo is a combination of diffusion and convection; a similar model [19] successfully reproduces the spectral and spatial properties of radio halos of edge-on galaxies. While propagating out of the central disk region (thickness $\sim$ 300 pc) where they are accelerated, electrons lose energy by radiative losses and by ionization and electronic excitations in the neutral and ionized gas, respectively. The electron energy spectrum steepens as they propagate outward. Radiative losses dominate at energies $\gtrsim$ 1 GeV, and most scatterings are with photons of the FIR field, which is $\sim$ 20 more intense than the CMB within the central disk region. Compton scattering off the CMB and

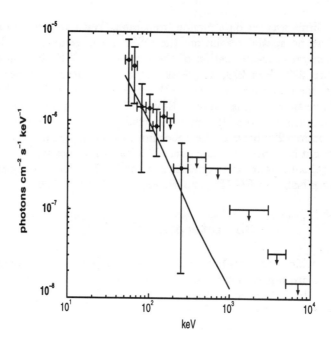

FIGURE 1. OSSE measurements of NGC 253 [1], and model fit [12] (solid line).

synchrotron emission dominates outside this region. The wider spatial distribution of lower energy electrons implies that the CMB contribution increases with distance from the disk. Some 90% of the emission at 50 keV originates from a region $\sim$ 3 kpc thick [12].

Fitting the OSSE data yields a photon index of $\sim$ 1.7 in the halo and $\sim$ 1.5 in the disk [12]. These values are consistent with the observed range of radio spectral indices [4]. The mean magnetic field value deduced from radio observations, together with the assumption of equipartition of energy between the field and relativistic protons – whose energy density is assumed to be a hundred times that of the electrons – is 9 μG. This value is realistic, though on the high side for galactic disks, as is also the implied relativistic electron density. The best-fit spectrum, shown in Figure 1, yields a total luminosity of $\sim 10^{41}$ erg s^{-1} in the 10 - 100 keV band.

Attributing the observed emission to a population of SN Ia at peak brightness would require the very high SN rate of $\sim$ 1 yr^{-1}, at least an order of magnitude higher than the rate deduced observationally [42]. The contribution of these SNe is constrained by the fact that the associated 847 keV line was not observed by OSSE [1]. This consideration makes it also unlikely that the emission is from SN remnants. Even if we assume a mean 10 - 100 keV luminosity of 10^{37} erg s^{-1} per SN remnant, 10^4 remnants would be implied, and with an assumed upper bound of 10^4 yr X-ray active lifetime, a SN rate of $\gtrsim 1$ yr^{-1} would be required. It also does not seem very plausible that the emission is from a population of low-metallicity, massive X-ray binaries, whose emission is thought to extend to ~ 60 keV. One would not expect such systems in the dust (and metal) rich environment of a SBG. Even if it is assumed that their spectra extend to the OSSE band, with a luminosity comparable to those of the few known systems in the LMC (no such systems are known in the Galaxy), then more than a hundred such systems would be required to account for the observed emission. This is probably unrealistic, judging by the rarity of these binaries.

Identification of the emission from NGC 253 as due to Compton scattering would therefore seem very reasonable, given the difficulties with alternative origins. After all, extended radio emission constitutes direct evidence for relativistic electrons over a large region in the disk and halo. The estimate of the FIR radiation field is also quite secure; recall, though, that most of the HEX emission originates in the inner halo, and is due to scattering of the electrons by the CMB. Thus, any uncertainty that might be associated with the FIR energy density has little effect on the estimate of the overall Compton flux. Nonetheless, while the data are suggestive of a Compton origin, this is only a preliminary interpretation. The emission has to be spatially resolved before a diffuse origin in the disk (and also in the inner part of the halo; see [12]) can be determined. On the other hand, appreciable temporal variation will be decisive evidence for an AGN-like source for the high-energy emission.

It is instructive to compare the results of OSSE observations of NGC 253 and M82. While NGC 253 was detected, observations of M82 yielded only upper limits [1]. Yet these two galaxies are quite similar, about equally distant, and were observed for roughly the same time by OSSE. Moreover, diffuse radio emission has been detected also from M82, and – as we have noted – analysis of ASCA observations indicates the presence of a nonthermal component [23] (at energies ≤ 10 keV). In order to understand these results, the level of HEX emission from M82 was estimated in the context of the same model [12] that was applied to NGC 253. Doing so, Goldshmidt [10] determined that the predicted emission from M82 is very close to the OSSE upper limit, which is itself comparable to the flux detected from NGC 253. As expected, therefore, there does not seem to be a qualitative difference between the HEX properties of these two galaxies.

A different approach [5,6] was used to attempt to identify the origin of the far-IR radiation from OSSE observations of the ULIGs Arp 220, Mrk 273, and Mrk 231. According to the standard scenario for quasar evolution [35], the IR radiation from ULIGs is accretion-disk radiation reprocessed by surrounding dust and gas. Gamma-ray observations provide the best method for detecting direct nuclear emission from a dust-enshrouded AGN. Values of the column density to the central nucleus are limited by CO observations and dynamical estimates of gas mass. Assuming that the buried AGN in these sources emit a standard Seyfert spectrum, the observed IR luminosities were used to normalize the luminosities of the X-and-γ component. Upper limits on the emission from Arp 220 obtained with OSSE – shown in Figure 2 – imply that $\lesssim 20\%$ is produced by AGN activity. By inference, the bulk of the IR luminosity therefore originates in non-AGN starburst activity.

This conclusion is in accord with recent ISO observations which suggest that in some ULIGs, including NGC 3256 and NGC 6240 as well as Arp 220, most of the IR emission is powered by massive stars [21]. Further γ-ray observations of these and other IR galaxies are needed to improve limits and ensure that the lack of detection is not due to having observed the galaxies in states of low gamma-ray activity. The preliminary results obtained with OSSE are important because they challenge our present understanding of the origin of the IR luminosity in ULIGs, and the relationship between ULIGs and quasars.

3. CLUSTER HEX EMISSION

The process which produces HEX radiation over kpc scales in SBGs may also give rise to similar emission over Mpc scales in clusters of galaxies. In

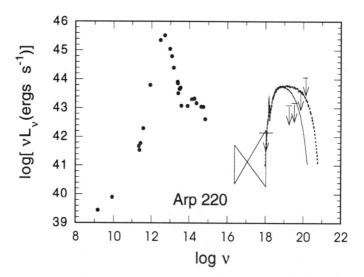

FIGURE 2. Multiwavelength νL_ν spectra of Arp 220 with OSSE upper limits (for references to data, see [6]). The Monte Carlo simulations show the fluxes expected if AGNs with X-and-γ-ray luminosities equal to 10% of the measured 8-1000 μm IR luminosity are buried behind a torus with a neutral hydrogen column density of 10^{24} cm^{-2}. The solid and dotted curves refer to source spectra with 100 and 400 keV cutoffs, respectively.

clusters, relativistic electrons leak out of radio galaxies and scatter photons of the CMB. As mentioned in the Introduction, HEX observations of clusters are of considerable interest because of the additional insight that can be gained on physical conditions in the IC environment.

Diffuse radio emission over regions typically ~ 1 Mpc in size has already been detected in ~ 10 clusters. Spectral (energy) indices are in the range 1.2 - 1.7, and 0.01 - 10 GHz luminosities are 6×10^{40} - 7×10^{41} erg/s (using a value of 50 km s^{-1} Mpc^{-1} for the Hubble constant). There are reasons to believe that diffuse IC radio emission may even be a more common phenomenon, though not clearly apparent in many clusters because its detection requires very careful subtraction of galactic emissions [28]. These measurements imply the presence the IC space of electrons with energies in the interval ~ 1 - 100 GeV, assuming magnetic fields in the range ~ 0.1 - 1 μG. Compton scattering of these electrons by the CMB results in radiation in the ~ 2 keV - 20 MeV band. Detailed calculations of the predicted properties of HEX emission from clusters were made by Rephaeli [26,27]. For a given measured radio flux, the predicted level of HEX emission depends on the value of the *mean, volume-averaged magnetic field B*. If $B > 3$ μG, which is comparable to or higher than a characteristic galactic value, then HEX emission in the 40 - 100 keV range will typically be $\lesssim 10^{-6}$ cm^{-2} s^{-1} keV^{-1} from clusters at distances of a few hundred Mpc. Based on considerations having to do with the likely origin of the fields [28], and estimates from Faraday rotation measurements [17,41,14,11], we expect that $B \leq 1$ μG. Therefore, we can reasonably expect HEX emission from clusters at such distances to be comparable to or higher than the above value.

Only few attempts to detect clusters at energies $\gtrsim 20$ keV have been made. *HEAO-1* A4 survey data from 6 Abell clusters with known radio halos (A401, A1367, Coma, A2255, A2256, A2319) have been analyzed [33,29], and two clusters – Coma and A2319 – were detected at energies of up to ~ 30 keV, with upper limits on the emission at higher energies. The detected emission is most likely from hot gas in these clusters. The bounds on nonthermal emission from all the clusters resulted in lower limits, $B \sim 0.1$ μG, and upper limits on the electron energy densities, generally in the range $\rho \sim$ few$\times 10^{-13}$ erg cm^{-3}. We emphasize that the bounds are on the *mean* values of these quantities. Since no spatial variation was assumed in deriving these results, *the central values of B and ρ may be significantly higher*. For example, in the cores of clusters, values of B as high as a few μG are consistent with these limits.

The A4 cluster observations, which were part of the HEAO-1 all-sky survey, were not sufficiently long for detection of the expected weak level of emission. To measure HEX emission from even a nearby cluster, a dedicated, deep observation is required. The first such observation of the Coma cluster was made by OSSE for a total of two weeks, but with only 4.4 days of on-source time. This observation too did not result in measurement of the sought emission, but an improved lower limit was set on the mean IC magnetic field of $\gtrsim 0.1$

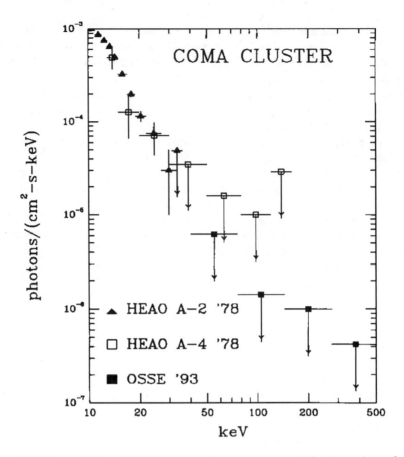

FIGURE 3. OSSE and HEAO-1 A2 & A4 measurements of the Coma cluster [34].

μG [34]. The OSSE data points, along with the results of HEAO-1 A2 and A4 observations, are shown in Figure 3. EGRET upper limits presumably result in a mean magnetic field $\gtrsim 1$ μG [8], but depend on an extrapolation of the electron spectrum to energies where the synchrotron emission is not detected.

The Coma cluster was recently observed by the PCA and HEXTE experiments aboard the *RXTE* satellite. A preliminary analysis of some of the data yields a high quality spectrum at energies up to ~ 25 keV, with emission detected by HEXTE up to at least ~ 40 keV. At this energy the flux is at a level $\sim 10^{-5}$ cm^{-2} s^{-1} keV^{-1}. Analysis [30] of the complete data set is in progress. The main goal, of course, is to determine whether the best-fit model includes a power-law spectral component.

Prior to the launch of *CGRO*, a lower limit of 1.5 μG was placed [7] on the mean magnetic field in the Virgo cluster from high-energy radiation flux upper limits from M87. A long pointing in the Virgo region led to a strong OSSE detection of emission from a direction compatible with both the locations of NGC 4388 and M87 [16]. The Sigma telescope on the *GRANAT* has better imaging capablity than OSSE, and only the Seyfert 2 galaxy NGC 4388 has been reported [25] as a Sigma source, suggesting that most of the HEX emission is associated with NGC 4388. Observations with *RXTE* could determine whether a detection of M87 was obtained with OSSE, and lead to a new value or improved limit for the mean Virgo cluster magnetic field.

Finally, we note that no clusters of galaxies have been detected with the COMPTEL and EGRET experiments on *CGRO* [38,20].

4. SUMMARY

Of the SBGs and ULIGs observed with OSSE on the *Compton Observatory*, only NGC 253 has so far been detected. Compton origin of the emission from NGC 253 seems quite likely, but with little spectral and no spatial or temporal information, this interpretation is not yet secure. Estimates for the HEX emission from M82 show that it is near the sensitivity threshold of the OSSE experiment. The upper limits to the γ-ray emission observed from ULIGs, in particular Arp 220, indicate that the bulk of the large IR luminosities originate from starburst activity, rather than from a buried AGN.

The predicted HEX luminosity of a cluster is as high as $\sim 10^{43}$ erg s^{-1}. The detection of this emission from clusters is, however, more challenging than its detection from the nearby SBGs, because there simply are no nearby clusters with significant level of diffuse radio emission. Whereas the 10 - 100 keV luminosity of NGC 253 is more than ~ 100 times lower than that of Coma, its flux is still ~ 15 times higher. This explains why the emission was detected from NGC 253 but not from the Coma cluster, which was observed with about the same exposure by OSSE [34].

Considerable progress may come soon: NGC 253, M82, and the Coma cluster have all been observed by the PCA and HEXTE experiments aboard *RXTE*, with NGC 253 observed also by SAX. The improved overall HEXTE sensitivity with respect to that attained in the OSSE measurements of these sources will likely lead to definite deductions on the nature of the two nearby SBGs and other IR luminous galaxies and clusters.

REFERENCES

1. Bhattacharya, D. *et al.*, 1994, *ApJ* **437**, 173 (1994).
2. Boller, Th. *et al.*, *A&A* **261**, 57 (1992).
3. Bookbinder, J., Cowie, L.L., Krolik, J.H., Ostriker, J.P., and Rees, M., *ApJ* **237**, 647 (1980).
4. Carilli, C.L., Holdway, M.A., Ho, P.T.P., and De Pree, C.G., *ApJ* **399**, L59 (1992).

5. Dermer, C.D., Shier, L.M., Sturner, S.J., McNaron-Brown, K., and Bland-Hawthorn, J., *Proceedings Second INTEGRAL Workshop: The Transparent Universe*, eds. C. Winkler *et al.* (ESA: Noordwijk), ESA SP-382, p. 447 (1997).
6. Dermer, C.D., Bland-Hawthorn, J., Chiang, J., and McNaron-Brown, K. *ApJ* **484**, L121 (1997).
7. Dermer, C. D., and Rephaeli, Y., *ApJ* **329**, 687 (1988).
8. Enßlin, T. A., and Biermann, P. L., *A&A*, submitted (1997).
9. Fabbiano, G., *ApJ* **330**, 672 (1988).
10. Goldshmidt, O., Ph.D. thesis, Tel Aviv University (1995).
11. Goldshmidt, O., and Rephaeli, Y., *ApJ* **411**, 518 (1993).
12. Goldshmidt, O., and Rephaeli, Y., *ApJ* **444**, 113 (1995).
13. Griffiths, R.E., and Padovani, P., *ApJ* **360**, 483 (1990).
14. Hennessy, G.S., Owen, F.N., and Eilek, J.A., *ApJ* **347**, 144 (1989).
15. Johnson, W.N., *et al.*, these proceedings.
16. Kurfess, J. D., private communication (1997).
17. Lawler, J. M., and Dennison, B., *ApJ* **252**, 81 (1982).
18. Lawrence, A., Walker, D., Rowan-Robinson, M., Leech, K.J., and Penston, M.V., *MNRAS* **219**, 687 (1986).
19. Lerche, I., and Schlickeiser, R., *ApJ* **239**, 1089 (1980).
20. Lin, Y. C. *et al.*, *ApJ* **416**, L53 (1993).
21. Lutz, D. *et al.*, *A&A* **315**, L137 (1996).
22. Moran, E. C., Halpern, J. P., and Helfand, D. J., *ApJS* **106**, 341 (1996).
23. Moran, E.C., and Lehnert, M.D., *ApJ* **478**, 172 (1997).
24. Ohashi, T. *et al.*, *ApJ* **365**, 180 (1990).
25. Paul, J. *The Gamma Ray Sky with Compton GRO and SIGMA*, eds. M. SIgnore, P. Salati, G. Vedrenne (London: Kluwer), p. 15 (1995).
26. Rephaeli, Y., *ApJ* **212**, 608 (1979).
27. Rephaeli, Y., *ApJ* **227**, 364 (1979).
28. Rephaeli, Y., *Comm. Ap.* **12**, 265 (1988).
29. Rephaeli, Y., and Gruber, D.E., *ApJ* **333**, 133 (1988).
30. Rephaeli, Y., and Gruber, D. E., in preparation (1997).
31. Rephaeli, Y., Gruber, D.E., Persic, M., and MacDonald, D., *ApJ* **380**, L59 (1991).
32. Rephaeli, Y., Gruber, D.E., Persic, M., *A&A* **300**, 91 (1995).
33. Rephaeli, Y., and Gruber, D.E., and Rothschild, R.E., *ApJ* **333**, 133 (1988).
34. Rephaeli, Y., Ulmer, M., and Gruber, D.E., *ApJ* **429**, 554 (1994).
35. Sanders, D. B. *et al.*, *ApJ* **347**, 29.
36. Sanders, D.B., and Mirabel, I.F., *Ann. Rev. Astron. Ap.*, **34** 749 (1996).
37. Schaaf, R., *et al.*, *A&A* **336**, 722 (1989).
38. Schönfelder, V. *et al.*, *A&AS* **120**, 13 (1996).
39. Soifer, B.T., Sanders, D.B., Neugebauer, G., Danielson, G.E., Lonsdale, C.J., Madore, B.F., and Peterson, S.E., *ApJ* **303**, L41 (1986).
40. Tsuru, T., *et al.*, *Publ. Astron. Soc. Japan* **42**, L75 (1990).
41. Vallee, J.P., Broten, N.W., and MacLeod, J.M., *ApJ* **25**, 181 (1987).
42. Van Buren, D., and Greenhouse, M.A., *ApJ* **431**, 640 (1994).
43. Weedman, D.W., *Star Formation in Galaxies*, NASA Conf. Publ. 2466, p. 351 (1987).

Seyferts and Radio Galaxies

W. Neil Johnson[1], Andrzej A. Zdziarski[2], Greg M. Madejski[3],
William S. Paciesas[4], Helmut Steinle[5] and Ying-Chi Lin[6]

[1] *Naval Research Laboratory, Code 7651, Washington, DC 20375-5352, USA*
[2] *N. Copernicus Astronomical Center, Bartycka 18, 00-716 Warsaw, Poland*
[3] *NASA/Goddard Space Flight Center, Code 661, Greenbelt, MD 20771, USA*
[4] *Univ. of Alabama, Huntsville, AL 35899, USA*
[5] *Max-Planck-Institut für extraterrestrische Physik, Postfach 1603,
85740 Garching, Germany*
[6] *Stanford Univ., Stanford, CA 94305, USA*

Abstract. Observations with *Compton Gamma Ray Observatory* have provided key insights to the high energy emission mechanisms in Active Galactic Nuclei (AGN) and clearly separate them into a class of Blazars, with apparent non-thermal jet emissions often peaking in the EGRET energy band, and into the Seyfert class with emissions more closely related to the accretion disk presumably energizing the nuclear black hole. OSSE measurements indicate that the high energy emission from Seyfert AGN has a high energy cutoff below $\sim 100-200$ keV and no Seyferts have been detected by COMPTEL or EGRET. Here, we review the X-ray and γ-ray observations of Seyferts and Radio Galaxies and the broad band emission models that describe the observed spectral characteristics. A comparison of Seyfert 1, Seyfert 2, and radio galaxies can be made in the context of the AGN unification models.

1. Introduction

The *Compton Gamma Ray Observatory* (*CGRO*) has contributed substantially to our understanding of Active Galactic Nuclei. Perhaps the most obvious and widely known result is that there are clearly two classes of γ-ray – emitting AGN in the sky: these are the jet-dominated blazars, and the more ordinary, radio-quiet Seyferts and quasars. This is well illustrated on Figure 1, from Dermer & Gehrels [6]. The marked differences between the two classes are seen even in the the radio and optical data: blazars are dominated by compact, milliarcsecond cores, and the radio and optical emission is strongly polarized. The compact radio cores show superluminal expansion. With the rapid and large-amplitude variability observed in the GeV γ-rays, we now think that the entire emission arises in a jet pointing close to the line of sight, where the emission is strongly enhanced due to the relativistic Doppler effect. The radio and optical emissions are most likely due to the synchrotron process, while the hard X-rays and γ-rays are due to Comptonization, either the internal synchrotron radiation, or photons external to the jet; the details are discussed in the paper by Hartman et al. ([12], these proceedings). The blazar luminosity often peaks in the γ-ray band above 1 MeV and all *CGRO* instruments have detected the high energy emissions from several blazars.

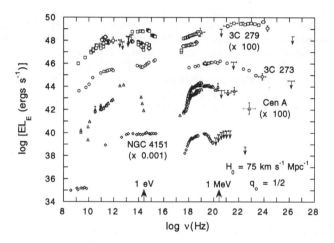

FIGURE 1. Broad band spectra for various AGN as compiled by Dermer & Gehrels [6]. High energy Cen A data is from Steinle et al. [44]

In the case of radio quiet objects, the overall electromagnetic spectra are quite different. Within the current sensitivity, these spectra do not extend beyond a few hundred keV. Instead, they appear to cut off, such that there are no observed photons at or beyond the 511 keV annihilation line; all recent γ-ray detections and measurements of spectra of these objects have been made by OSSE and BATSE. There are essentially no detections of radio-quiet AGN in the COMPTEL or EGRET energy ranges. In many cases, the hard X-rays and soft γ-rays vary on a relatively short time scale, implying a compact emission region (see, e.g., [30]). This implies that they must be produced very closely to the central engine, and are probably the primary form of radiative power in radio-quiet AGN.

In the following sections, we attempt to summarize what we have learned from γ-ray observations of Seyfert Galaxies. We include here radio galaxies with Seyfert nuclei (also called radio-loud Seyferts), which represent a class intermediate between radio-quiet Seyferts and Blazars. In particular, we present an overview of the *CGRO* observations, and put these observations into a multi-wavelength perspective. We discuss briefly the Seyfert Unification schemes, and summarize the current status of the theoretical models that can describe the X-ray – γ-ray emission.

2. CGRO Observations of Seyfert AGN

OSSE Observations. OSSE's small field of view requires the selection of specific sources for study. A list of potential AGN targets was created by selecting bright X-ray objects in the *EXOSAT* catalog [50] and in *HEAO* catalogs [41,36]. In the first 4.5 years of *CGRO* operations (May, 1991 – Dec, 1995), OSSE observed 37 Seyfert class AGN. Of these objects, 24 Seyfert 1 AGN and 13 Seyfert 2s were studied in 115 observations. Tables 1 and 2 summarize these Seyfert observations. The radio galaxy, Cen A, has been included in Table 2 since it has some Sy 2 characteristics. The OSSE column in the tables reports the average flux detected from the sum of all observations of each source in the 50–150 keV enegy band. Upper limits are reported at the 2σ level. OSSE has observed all twelve AGN reported by *HEAO-1* [41] and detects six of them at 4σ or greater. There is no evidence for emission from the *HEAO-1* sources MRK 335 and MCG +5-23-16. A total of 26 Seyferts have been detected by OSSE at a significance of $\sim 3\sigma$ or greater, and 18 Seyferts detected at a significance of 5σ or greater. These 18 sources, in order of decreasing significance, are Cen A, NGC 4151, NGC 4388, IC 4329A, NGC 4945, MCG +8-11-11, NGC 5548, NGC 7172, NGC 4507, 3C 120, NGC 5506, MCG –5-23-16, MCG –6-30-15, 3C 390.3, NGC 3227, NGC 7469, MRK 509, and 3C 111.

BATSE Observations. Because of sensitivity limitations, detailed studies of individual AGNs with BATSE are limited to a few bright sources, among which NGC 4151 is the only Seyfert galaxy. However, the sensitivity can be improved significantly by using long integration times. Preliminary results indicate that, for relatively isolated sources such as most AGN, integration times of order 1000 days are sufficiently free of systematic errors to permit survey studies of classes of sources. Bassani et al. [4] studied a sample of 17 type 2 Seyfert galaxies with BATSE, of which 5 were significantly detected and another 5 marginally detected. The high detection ratio (roughly 50%) supports the inference from the unified model that Seyfert 2s should be strong hard X-ray emitters. Malizia et al. [29] used BATSE to study the AGN in the Piccinotti sample [36], which consists predominantly of Seyferts (30 out of 36 objects). Among the 30 Seyferts, 20 were detected by BATSE at a significance of more than 5σ. Tables 1 and 2 summarize the BATSE Seyfert detections.

COMPTEL Observations. During the first 15 months of the *CGRO* mission, COMPTEL and EGRET conducted an all-sky survey. The COMPTEL data from this period has been examined for emission from 26 X-ray selected Seyfert galaxies by Maisack et al. [25,27]. No significant emission from any of the objects was detected. Tables 1 and 2 include the COMPTEL limits from Maisack et al. [27] in the 1–3 MeV energy band.

EGRET Observations. No Seyfert galaxies have been detected by EGRET. In order to make sure that Seyfert galaxies as a class are indeed not EGRET sources, Lin et al. [21] carried out a systematic search for Seyfert galaxies in the EGRET all-sky survey data. The source candidates were first

TABLE 1. Seyfert 1 AGN Observed by CGRO

Source	Seyfert Class	OSSE Flux[a]	BATSE Detection[b]	COMPTEL Upper Limit[c]	EGRET Upper Limit[d]
3C 111	1	2.7 ± 0.5	–	0.6	0.9
3C 120	1	2.5 ± 0.3	5.3	0.8	0.9
3C 390.3	1	2.5 ± 0.3	–	–	0.7
3C 445	1	–	12.9	1.0	–
ARK 120	1	–	–	1.3	–
ESO 141-G55	1	2.4 ± 0.7	nsf	3.2	1.5
ESO 198-G24	1	–	4.6	–	–
FAIRALL 9	1	< 1.4	4.1	1.2	–
H 0557-385	1	–	nsf	–	–
H 0917-074	1	–	6.0	–	–
H 1846-786	1	–	3.3	–	–
IC 4329A	1	7.5 ± 0.4	21.6	1.2	–
III ZW 2	1	1.5 + 0.7	3.6	1.6	–
MCG +8-11-11	1	3.7 ± 0.3	–	1.8	1.1
MCG -2-58-22	1.5	3.3 ± 0.8	7.1	1.5	0.9
MCG -6-30-15	1	4.1 ± 0.6	3.5	2.1	1.0
MRK 1152	1.5	–	5.6	–	–
MRK 279	1	1.4 ± 0.5	–	–	0.7
MRK 335	1	< 1.5	–	2.5	0.9
MRK 464	1	< 2.0	–	–	–
MRK 509	1	3.6 ± 0.6	13.8	1.1	0.8
MRK 590	1.2	< 2.8	5.2	–	0.7
MRK 841	1	3.1 ± 0.9	–	–	0.9
NGC 3227	1.5	3.2 ± 0.5	17.5	1.9	–
NGC 3783	1	4.0 ± 1.2	16.5	0.7	1.2
NGC 4051	1	–	–	1.0	0.5
NGC 4151	1.5	24.5 ± 0.4	81.6	1.0	0.5
NGC 4593	1	< 2.9	7.2	0.7	1.2
NGC 5506	1.9	5.0 ± 0.6	18.5	1.0	–
NGC 5548	1.5	4.8 ± 0.5	12.8	3.0	1.0
NGC 6814	1	2.8 ± 0.5	–	0.9	–
NGC 7213	1	2.0 ± 0.7	4.7	–	1.0
NGC 7314	1.9	1.6 ± 0.7	nsf	2.3	–
NGC 7469	1	3.2 ± 0.5	5.7	1.4	1.1

[a] Flux in 50 – 150 keV band in units of 10^{-4} photons cm^{-2} s^{-1}
[b] Statistical significance in σ of BATSE detection in 20 – 100 keV band [29]. No significant flux detected for source is indicated by "nsf".
[c] Flux Limit (2σ) in 1 – 3 MeV band in units of 10^{-4} photons cm^{-2} s^{-1} [27]
[d] Flux Limit (2σ) for E > 100 MeV in units of 10^{-7} photons cm^{-2} s^{-1}

TABLE 2. Seyfert 2 AGN Observed by CGRO

Source	Seyfert Class	OSSE Flux[a]	BATSE Detection[a]	COMPTEL Upper Limit[a]	EGRET Upper Limit[a]
Cen A	2 [b]	42.5 ± 0.4	68[c]	1.0 ± 0.3[d]	1.8 ± 0.4[e]
ESO 103-G55	2	–	5.8	–	–
FAIRALL 49	2	–	nsf	–	–
MCG +5-23-16	2	< 1.8	–	1.1	–
MCG -5-23-16	2	3.8 ± 0.5	6.6[f]	–	–
MRK 3	2	1.8 ± 0.4	–	–	–
MRK 348	2	< 2.8	–	–	1.0
MRK 463	2	–	6.7[f]	–	–
MRK 78	2	–	4.7[f]	–	–
NGC 1068	2	< 1.1	–	1.4	0.8
NGC 1275	2	1.8 ± 0.5	–	–	0.9
NGC 2110	2	3.6 ± 0.7	3.1[f]	–	–
NGC 2992	2	< 1.2	7.1	–	–
NGC 4388	2	9.1 ± 0.4	14.2[f]	–	–
NGC 4507	2	5.2 ± 0.6	13.2[f]	–	–
NGC 4945	2	10.9 ± 0.7	4.6[f]	–	–
NGC 526A	2	3.2 ± 0.8	12.9	–	–
NGC 7172	2	6.3 ± 0.6	12.1	–	–
NGC 7582	2	2.9 ± 0.8	9.1	2.0	–

[a] see Table 1 for column descriptions
[b] Radio Galaxy with Sy 2 like nucleus with strong optical extinction
[c] Zhang [60]
[d] Detection, sum of all COMPTEL low state data, Steinle et al. [44]
[e] Detection, Nolan et al. [34]
[f] BATSE data from Bassani et al. [4]

selected according to their X-ray fluxes in two catalogs of X-ray-detected AGN: the *EXOSAT* Catalog [50] in the 2–10 keV energy range and the *Ginga* Catalog [28] in the 2–30 keV energy range. Additional sources were added to the list for their reported historical low-energy γ-ray detections [3] or observations by OSSE. The resulting list of 22 Seyfert galaxies (see Tables 1 and 2) were searched for in the EGRET all-sky survey data. As reported by Lin et al. [21], no significant detection was found for any of them. The EGRET limits presented in Tables 1 and 2 represent an improvement over the original results. More data have been included (May, 1991 – October, 1995) and the EGRET analysis software and the background model [13] have been improved. Tabulated in the table are the 2-σ flux upper limits for $E > 100$ MeV for the same list of 22 candidates as used in Lin et al. Again there is no evidence of flux excesses for any of these 22 Seyfert galaxies in this more sensitive data set.

3. Gamma Ray Spectra of Seyfert and Radio Galaxies

The average OSSE spectrum for each Seyfert AGN was formed by a livetime-weighted sum of all observations. The spectra of 17 of these Seyferts were of sufficient significance to perform spectral characterization. A variety of photon models characterized by either thermal or non-thermal natures were applied to the individual Seyferts and to the average Seyfert class spectra.

With the exception of the brightest sources, Cen A and NGC 4151, all of the spectra of individual sources were reasonably well described by simple power-law models or exponentially truncated power-law models. NGC 4151 and Cen A are discussed separately below. The statistical precision of the measurements of the typical Seyfert did not permit discrimination between the models. Figure 2 shows a plot of the best power-law photon indices for sources with a detection significance of $\geq 5\sigma$ in the $50-150$ keV energy range. Cen A and NGC 4151 have been excluded from the figure since a power-law does not provide a good fit to their data. The average index for this set is 2.4 with a variance of 0.3, which is clearly softer than the typical intrinsic *Ginga* power-law index, $\Gamma \sim 1.9$, in the X-ray band [33].

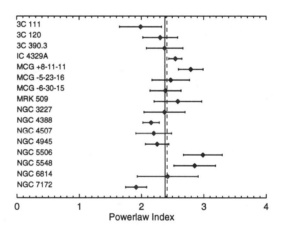

FIGURE 2. Comparison of the best fit power-law photon indices of OSSE Seyfert observations with detection significance $\geq 5\sigma$. The horizontal bars represent the 1σ uncertainties in the best fit index for each source. The solid vertical line is the weighted mean index and the dashed vertical line is the average. Variance around the mean is 0.3.

3.1 NGC 4151

The brightest radio-quiet AGN in the OSSE sky is the Seyfert NGC 4151. Initially, NGC 4151 was believed to be unique, not fitting with the Seyfert

1 / Seyfert 2 Unification paradigm (see below), but the recent results using OSSE measurements along with X-ray data from *Ginga*, *ROSAT* and *ASCA* [57,15] show that NGC 4151 is actually more similar to other radio quiet AGN than we initially thought; some ambiguity had to do with the fact that this source is also heavily absorbed in X-rays. Since it is so bright, the statistical significance of the γ-ray data place significant constraints on the modeling of the source emission mechanisms. The OSSE observations have been reported by Masaick et al. [26] and Johnson et al. [15]. Through October, 1996, OSSE observed NGC 4151 11 times. The long-term average flux from the source in the 50–150 keV band has varied by no more than a factor of 2 (2–4 $\times 10^{-3}$ photons cm^{-2} s^{-1}). On shorter time scales, day-to-day variability was detected by OSSE up to a magnitude of $\pm 25\%$ of the observation average.

The relative constancy of NGC 4151 seen by OSSE is confirmed by long-term BATSE monitoring. Using the Earth occultation technique with BATSE data, Parsons et al. [35] have produced continuous light curves of NGC 4151 in two energy bands (20–70 and 70–200 keV) covering nearly six years with one-week time resolution. On these timescales, the source shows remarkably little variability. They find no obvious long-term trends or evidence for multiple source states, and the scatter of the data about the mean indicates an intrinsic source variability of about 20%.

Simple spectral models have been fit to the 11 individual OSSE observations. Models with softening spectra at higher energies (for example, Comptonization models) provide the best fits. Simple power law models fit to the individual observatons produce power law number indices, $\Gamma \approx 2.5 - 2.9$, but with unacceptable χ^2 for many of the data sets. As shown in Table 3, the best power law fit to the total spectrum produced an unacceptable $\chi^2 = 119$ for 49 degrees of freedom (d.o.f.). A simple exponential model is a fair description of the observed spectra with e-folding energies in the ~ 30–45 keV range. The simple exponential spectrum fit to the total of all observations, however, is found to be unacceptable ($\chi^2 = 106/49$ d.o.f.).

TABLE 3. Spectral fits to average of all OSSE NGC 4151 data in 50–800 keV band.

Model	I^a	Γ	E_c or kT	τ	χ^2/d.o.f.
Power Law (PL)	1.72 ± 0.03	2.70 ± 0.04			119/49
Exponential	1.93 ± 0.03		38 ± 1		106/49
PL-Exp[b]	1.83 ± 0.03	$1.60^{+0.24}_{-0.26}$	96^{+28}_{-19}		34/48
Therm. Comptonization[c]	1.82 ± 0.03		44^{+7}_{-5}	2.1 ± 0.3	39/48

[a] Flux value at 100 keV in units of 10^{-5} photons cm^{-2} s^{-1} keV^{-1}
[b] Model is $\Phi(E) = I(E/100)^{-\Gamma} \exp^{-(E-100)/E_c}$ photons cm^{-2} s^{-1} keV^{-1}
[c] Following Sunyaev & Titarchuk [47]

The simplest model which provides good fits to all data sets is the exponentially cutoff power law model (PL-exp), $\Phi(E) \propto E^{-\Gamma} \exp(-E/E_c)$, which

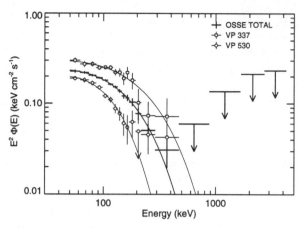

FIGURE 3. The average spectrum of NGC 4151 from all the OSSE observations (thick lines) and two individual observations, VP 337 (Aug, 1994) and 530 (Sep, 1996), representing the a higher and lower luminosity state, respectively. The smooth curves represent the best exponentially cutoff power law fit to each data set. The data sets have been rebinned for display. Error bars are 1σ statistical + systematics. Upper limits are 2σ.

roughly approximates thermal Comptonization spectra (e.g., Zdziarski et al. [54]). The best fit PL-exp spectrum to the average of all OSSE data has an index, $\Gamma = 1.6^{+0.2}_{-0.3}$, and cutoff e-folding energy, $E_c = 96^{+28}_{-19}$ keV.

By fixing the power law index at the average value of 1.6, fits of the PL-exp model to the individual observations provides a possible test for any variation in cutoff energy with source luminosity. As discussed in Johnson et al. [15], there is slight evidence that the cutoff energy, E_c, is correlated with luminosity. This possible positive correlation of flux – cutoff energy and particularly the softer, low-luminosity spectrum seen in Sept, 1996, (VP 530 in Figure 3) relative to the brighter Aug, 1994, spectrum (VP 337 in the figure) is in contrast to the observations in X-rays which indicate a softening of the spectrum as the source brightens (Yaqoob & Warwick [53]). Figure 3 shows the OSSE data from two of the observations (minimum and maximum observed flux) and the total of all OSSE data. The solid lines are the best fit PL-exp models to each data set.

The OSSE spectra of NGC 4151 can also be fitted directly by models of thermal Comptonization. We point out that the model of Sunyaev & Titarchuk [47], used in Table 3, is valid only for nonrelativistic, optically-thick, plasmas, conditions that are not satisfied in NGC 4151. A much more accurate model is provided by Poutanen & Svensson [38]. It yields $kT = 62^{+13}_{-11}$ keV and $\tau = 2.0^{+0.9}_{-0.5}$ at a low $\chi^2 = 35/50$ d.o.f. [15]. The plasma geometry is assumed to be a hemisphere (with τ measured along the radius) located in the vicinity of a cold medium subtending a solid angle of $0.4 \times 2\pi$ (as obtained by Zdziarski

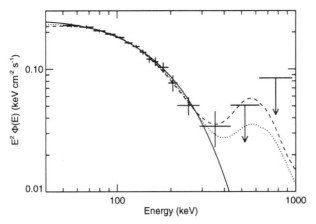

FIGURE 4. The average OSSE spectrum of NGC 4151. The solid curve is the fit by thermal Comptonization and Compton reflection. The dotted curve gives the best fit of the hybrid, thermal/nonthermal, model, and the dotted curve gives that model with the the maximum nonthermal fraction, 15%, allowed by the data. Error bars are 1σ statistical + systematics. Upper limits are 2σ.

et al. [57]), which gives rise to a Compton-reflection component [19,24]. This model is shown by the solid curve in Figure 4.

Thermal plasmas do not give rise to detectable $e^{\pm}$ annihilation features (Maciołek-Niedźwiecki et al. [22]), and indeed no such features are apparent in the data. On the other hand, nonthermal models for the high energy emission from AGN can produce a significant number of $e^{\pm}$ pairs by photon-photon interactions. These pairs lose most of their energy by scattering, rather than escaping, and form a thermal component of the electron/pair distribution. Annihilation of these pairs would then produce a broad feature near 511 keV. A 30 keV thermal pair plasma produces an annihilation line with a width of ~ 150 keV. The OSSE data have been tested for evidence of such emission features near 511 keV. The annihilation emission was modeled as either simple gaussian lines or a proper annihilation spectrum [46] and was fitted along with the broad band continuum described above. There was no statistically significant evidence for 511-keV emission from any of the OSSE observations or from the average spectrum of all data. For the average spectrum, the best fit flux in the annihilation feature, using the model of Svensson et al. [46] and assuming a 30 keV plasma temperature, is $(0.4^{+4.0}_{-0.4}) \times 10^{-5}$ photons cm^{-2}s^{-1}. The luminosity in the annihilation radiation corresponds to $0.2^{+2.1}_{-0.2}\%$ of the total X-ray and γ-ray luminosity (assuming the best-fit e-folded power law model). Similar results were obtained for the simple gaussian line models. The 99% confidence upper limit to the flux is 6×10^{-5} photons cm^{-2}s^{-1}. The limits on annihilation radiation vary with the assumed temperature of the pair plasma because of the resulting variation in width of the emission feature. The 99% confidence limit is 3×10^{-5} for a narrow line, increasing to 8×10^{-5} (which is still only 4.5% of the total X-ray and γ-ray luminosity of the source) for a 60 keV plasma or ~ 250 keV width.

A specific model giving rise to an observable pair-annihilation feature is the hybrid model (Zdziarski et al. [59]), in which a fraction of electrons in the source is accelerated to relativistic, nonthermal, energies and the remaining ones are heated. In the limit of no acceleration, the model corresponds to thermal Comptonization considered above. The maximum fraction of nonthermal acceleration allowed by the average spectrum of NGC 4151 is 15%. The corresponding spectrum is shown in Figure 4. This fraction is about an order of magnitude more than the maximum relative luminosity in pair annihilation. This reflects the maximum efficiency of conversion of nonthermal energy into pair rest mass of about 0.1 (e.g., [20]).

Summarizing, the OSSE data on NGC 4151 (which represent the best available data set on radio-quiet Seyferts) imply that nonthermal processes play at most a minor role in the spectral formation. Instead, thermal Comptonization explains adequately most or all of the intrinsic high-energy spectrum.

3.2 Average Seyfert Spectra

The OSSE spectra from all but the brightest Seyferts are of low statistical precision and cannot be used to place significant constraints on source models. Consequently, Seyfert-class average spectra were formed by summing spectra into one of three classes: 1) radio-quiet Seyfert 1s, 2) radio-loud Seyfert 1s, and 3) Seyfert 2s. (The second class is also classified as Broad-Line Radio Galaxies.) Finally, the average spectrum of all Seyfert classes was formed. These class average spectra can be used to provide more significant comparisions of spectral characteristics among the classes. Subsets of these class averages –those sources observed by both *Ginga* and OSSE (Zdziarski et al. [56]) and by both *EXOSAT* and OSSE (Gondek et al. [10]) – have been used to identify X-ray/γ-ray characteristics of these three classes (see below).

Data from OSSE observations with positive detections (defined as $\geq 2\sigma$) were combined for each Seyfert type. The most intense observations ($> 10\sigma$ detections) were excluded so that single sources would not significantly bias the result. This limit removed NGC 4151, NGC 4388 and IC 4329A from their type's average. Each Seyfert type consists of the following sources:

Radio Loud Sy 1s: 3C 111, 3C 120, 3C 390.3

Radio Quiet Sy 1s: ESO 141-55, MCG +8-11-11, MCG −2-58-22, MCG −6-30-15, MRK 279, MRK 509, NGC 3227, NGC 3783, NGC 5548, NGC 6814, NGC 7213, NGC 7469

Seyfert 2s: MCG -5-23-16, MRK 3, NGC 1275, NGC 4507, NGC 5506, NGC 7172, NGC 7582

TABLE 4. Best fit model parameters for total of OSSE data by Seyfert type and for the average of all OSSE Seyfert data.

Class	Model Function[a]	Γ or τ	kT or E_c(keV)	χ^2_ν (d.o.f.)
Radio Loud Sy 1s	Exponential		46 ± 6	1.4 (10)
	Powerlaw	2.2 ± 0.2		1.4 (10)
	Powerlaw*Exponential	1.5 ± 0.8	165 ± 197	1.4 (9)
	Sunyaev-Titarchuk	1.6 ± 1.1	80 ± 64	1.5 (9)
	Thermal Bremsstrahlung		91 ± 19	1.3 (10)
Radio Quiet Sy 1s	Exponential		39 ± 3	1.7 (12)
	Powerlaw	2.5 ± 0.1		1.3 (12)
	Powerlaw*Exponential	1.7 ± 0.4	118 ± 56	0.8 (11)
	Sunyaev-Titarchuk	2.1 ± 0.5	47 ± 10	0.7 (11)
	Thermal Bremsstrahlung		67 ± 5	0.9 (12)
Seyfert 2s	Exponential		51 ± 4	2.0 (13)
	Powerlaw	2.3 ± 0.1		2.1 (13)
	Powerlaw*Exponential	1.3 ± 0.4	123 ± 60	1.2 (12)
	Sunyaev-Titarchuk	2.3 ± 0.5	53 ± 10	1.0 (12)
	Thermal Bremsstrahlung		93 ± 9	1.2 (13)
Seyfert Average	Exponential		44 ± 2	2.2 (15)
	Powerlaw	2.4 ± 0.1		2.0 (15)
	Powerlaw*Exponential	1.5 ± 0.3	120 ± 37	0.9 (14)
	Sunyaev-Titarchuk	2.2 ± 0.3	50 ± 6	0.7 (14)
	Thermal Bremsstrahlung		77 ± 4	1.0 (15)

[a] Model parameter uncertainties are determined by the 68% confidence interval for joint variation of one or two interesting parameters, depending on the model

Fitting between the energy range of 0.05–0.50 MeV, Table 4 provides the values of the parameters for various models analyzed. The uncertainties in this table are 68% confidence limits for one or two interesting parameters, depending on the model. For both the radio quiet Sy 1s and the radio quiet Sy 2s, a power-law model does not provide a good fit (reduced $\chi^2_\nu = 1.95$ and $\chi^2_\nu = 2.27$, respectively for 11 degrees of freedom). Better fits were achieved using the thermal Comptonization model of Sunyaev & Titarchuk [47] or even a thermal bremsstrahlung model, with cut-off energies ranging from ~ 40 keV to ~ 100 keV. The exponentially truncated power-law model also provided comparably good fits to all three subclasses of Seyferts and resulted in higher cut-off energies between 70–170 keV.

The average spectra of radio-quiet Seyfert 1s and 2s (but including IC 4329A and NGC 4388, respectively) have also been fitted by the thermal-Componization model of Poutanen & Svensson [38] (Zdziarski et al. [58]), see Figure 5. The fitted model also includes Compton reflection and it is

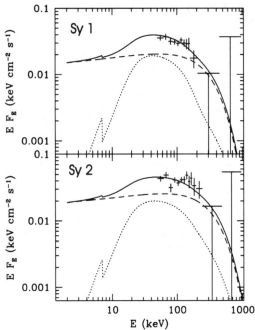

FIGURE 5. The average spectra of radio-quiet Seyferts detected by OSSE. The top panel is average Seyfert 1 (excluding NGC 4151) and the bottom panel is Seyfert 2 average. The dashed curves represent the best fit thermal Comptonization spectra of Poutanen & Svensson [38]. The dotted curves represent the reflected component. The solid curves give the sum. Note that the model spectra do not show effects of X-ray absorption, which is important below a few keV in Sy 1s, and below ~ 50 keV or less in Sy 2s.

constrained in the X-ray range to reproduce the average intrinsic spectrum from *Ginga* [56,10,33]. The plasma parameters obtained for Seyfert 1s and 2s are $\tau = 1.3^{+0.4}_{-0.2}$, $kT = 100^{+10}_{-30}$ keV, and $\tau = 1.2^{+0.6}_{-1.0}$, $kT = 110^{+150}_{-40}$ keV, respectively.

Note that the plasma parameters obtained for Seyfert 1s and 2s are very similar to each other, which is in accord with the AGN unified model (e.g., [1]). The parameters are also relatively close to those obtained for NGC 4151 (see above), confirming that NGC 4151 is indeed intrinsically close to average Seyferts.

Radio-loud Seyferts (Broad-Line Radio Galaxies) may indicate the link between Seyfert AGN – the X/γ-ray emissions of which are presumably accretion disk related – and the Blazar class of AGNs with jet or beamed non-thermal emissions extending to the GeV range. 3C 120 has a jet and displays superluminal motion [51] and, in the OSSE band, may exhibit spectral characteristics of beamed emission which is scattered out of the beam or is viewed from outside the beam opening angle. In fact, the OSSE data provide tentative evidence that radio-loud Seyferts differ somewhat from their radio-quiet counterparts. One difference is that the power-law model provides an acceptable fit, as well as models with a high-energy cutoff (see Table 4). Also, all of the models that provide a cut-off energy as a free parameter produced higher cut-off energies for the radio loud Sy 1s (~ 80 keV to ~ 170 keV) than for either of the two radio quiet Seyfert subclasses. This is also the case for a fit with the Comptonization model of [38], which yields $kT = 160^{+70}_{-50}$ keV, which is more than $kT \sim 100$ keV for radio-quiet Seyferts (see above). The optical depth in the fit is $\tau = 1.1^{+0.4}_{-0.3}$ (Woźniak et al. [52]). These results may indicate that radio-loud Seyferts do have substantial spectral components above ~ 200 keV, in contrast to the case of radio-quiet AGN.

3.3 Centaurus A

The Radio Galaxy Centaurus A is the brightest radio-loud AGN in the OSSE sky. It is a Seyfert-2 like galaxy although the exact classification is difficult due to the strong optical extinction. Due to its brightness in the OSSE band, it may be studied in much more detail than other radio-loud Seyferts, and thus we consider it separately in this review. However, the soft γ-ray spectral properties of radio-loud Seyferts appear very similar to those of Cen A.

Cen A exhibits extended radio lobes, X-ray jets, diffuse X-ray emission, an active nucleus observed in the soft X-ray band [9,7] and in the radio band [39], and a dust lane bisecting the galaxy that obscures the UV and visible nucleus. It is likely that Cen A is a member of the blazar class that is observed from outside the opening angle of the beamed radiation [43] and, as such, provides a unique oportunity to address the role of an accretion disk and the geometry of beamed emissions by detailed modelling of the emission environment and the resultant broad band X-ray/γ-ray spectrum and variability.

BATSE has monitored Cen A since launch and has observed variability by a factor of ~ 5 in the hard X-ray band (see Steinle et al. [44]). It has been observed 8 times by OSSE and 15 times by COMPTEL and EGRET. During the *CGRO* mission, Cen A has been observed to be in a low to intermediate luminosity state in the X-ray/γ-ray band relative to historical observations [16,5]. The BATSE long-term light curve [40,44] shows a declining trend over the *CGRO* mission; the average intensity in 1996 is roughly a factor of 3 less than a similar average in 1991. OSSE observed variability by a factors of 2–3 on long timescales and one 25% drop in flux within 12 hours [17]. The OSSE spectra are well described by broken power law spectra with a break at ~ 150 keV. The power law number index below the break is $\Gamma \sim 1.7$ and the change in index, $\Delta\Gamma$, varies from ~ 0.25 in the lowest intensity spectra to ~ 0.7 in the highest intensity states. As shown by Steinle et al. [44], the combined OSSE, COMPTEL and EGRET data on Cen A require another spectral break at higher energies, the magnitude of which also depends on the intensity of the source. The power law joining the COMPTEL and EGRET observations has index, $\Gamma = 2.6$–3.3, with the softer spectrum describing the more luminous state observed in October of 1991.

The spectrum of Cen A observed by *CGRO* in 1991, shown in Figure 6, extends beyond 100 MeV, which is the first such case among non-blazar AGNs. This spectrum certainly cannot be produced by thermal Comptonization, and the presence of nonthermal processes is implied. On the other hand, this spectrum, extending to such high energy, cannot be due to Compton scattering of jet emission towards our line of sight by some circum-jet material, as proposed by Skibo et al. [43]. We point out that the spectrum of Cen A is rather similar to the average spectrum of radio-loud Seyfert 1s. If this is indeed the case, nonthermal processes have to operate also in those objects, in contrast to radio-quiet Seyferts.

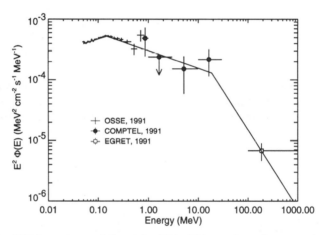

FIGURE 6. *CGRO* spectrum of Cen A in 1991. This spectrum is representative of the intermediate state as identified by Steinle et al. [44], which has been measured in only one *CGRO* observation so far.

4. X-Ray/Gamma-Ray Spectra of Seyferts

Any physical model that attempts to describe the *CGRO*, particularly OSSE, spectra of Seyferts has to confront the broad-band X-ray/γ-ray emissions. In particular, *Ginga* X-ray observations, covering the energy range of ~ 2–30 keV, had a major impact on our understanding of these sources. The early X-ray data (see, e.g., Rothschild et al. [41]) indicated that spectra of Seyferts 1s are (roughly) just simple power laws, with the average photon index of $\Gamma \simeq 1.7$ modestly absorbed at low energies; Seyfert 2s differ primarily in that they show much greater photoelectric absorption (presumably due to the material located on a parsec scale from the nucleus), affecting the X-ray flux at lower energies. On the other hand, the *Ginga* data (see Pounds et al. [37]), showed not only that an Fe Kα line is present in the data, but also that the spectra show a hardening towards high energies at $\gtrsim 10$ keV. This was successfully interpreted as due to an additional component arising from Compton reflection of the primary X-ray continuum, presumably from the matter that is accreting onto the black hole, most likely in a disk-like geometry [19,24]. The Fe Kα line is due to fluorescence from this accreting matter; indeed, in a material at roughly Solar abundances, the combination of the relative content, and the fluorescent yield is such that the Fe Kα line should be the strongest observable spectral feature. This nice picture, at least in a general sense, still holds today. However, *Ginga* had essentially no sensitivity above 30 keV, at the high end of the Compton reflection component, and this is where OSSE made the most important contributions.

The nuclear emission from AGN is variable in all observable energy ranges, and to get meaningful results, one has to observe them simultaneously in other wavebands, requiring concurrent observations with multiple satellites. Below, we discuss some OSSE observations simultaneous with observations in X-rays. An alternative approach is to collect average spectra for one or more objects, with the hope that the effects of variability will average out in multiple observations.

4.1 Radio-Quiet Seyfert 1s

The X-ray bright Seyfert 1 IC 4329A has been observed simultaneously by *ROSAT* and OSSE [23,54]. This data set has been supplemented by earlier *Ginga* observations, normalized to match the *ROSAT* flux at 2 keV. This spectrum has been shown to be quite representative of radio-quiet Seyfert 1s. It can be described well by a primary X-ray power law, modified due to the presence of the circum-nuclear matter. Specifically, at the lowest end, we see a modest amount of absorption due to neutral matter, presumably associated with the host galaxy. Even the *ROSAT* data alone – with their modest energy resolution – show that there is an isolated edge in the data at ~ 0.7 keV. The large depth of the edge indicates that the spectrum is absorbed by material that is partially ionized, such that all the elements lighter than oxygen are completely stripped, and thus essentially transparent to the emergent radiation. This, seen in the soft X-ray spectra of many other AGN (e.g., [18]), is often referred to as a "warm absorber".

Above ~ 2 keV, we observe the primary continuum. In X-rays, it is a power law, with an index of ~ 1.9. At the rest energy of ~ 6.4 keV, we observe the fluorescent iron line. What is remarkable is that this line is broadened, most

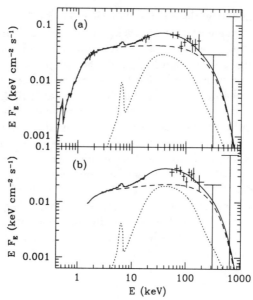

FIGURE 7. Example spectra of radio-quiet Seyfert 1s. (a) IC 4329A as measured by *ROSAT/Ginga*/OSSE, and (b) the average spectrum of 5 radio-quiet Seyfert 1s detected by both OSSE and *Ginga* (from Zdziarski et al. [58]). The dotted curves show the Compton reflection component; the dashed curves show fits with the thermal Comptonization model of [38].

likely by a combination of the gravitational and relativistic effects; this inference from the *Ginga* data was confirmed by an *ASCA* observation (Mushotzky et al. [31]). Perhaps the best example is the shape of the Fe K line observed in another Seyfert, MCG −6-30-15 (Tanaka et al. [49]), showing a characteristic two-pronged shape. This is expected when the line is produced in a disk-like configuration of accreting matter, and the disk has some inclination; we then see the blue-shifted photons emitted by the matter approaching us, and redshifted photons from the matter receding from us. The redshift of the line and the spread of the two prongs indicate that the region where the line is produced must experience strong gravity, and thus it has been interpreted as a good evidence of the presence of a black hole in AGN [49], although this is probably not a unique interpretation of the line shape.

Beyond the Fe Kα line, there is a general hardening of the spectrum, with a slight depression due to an absorption edge from Fe. This is due to the material encountered by the photons reflected from the disk; the continuum photons are reflected roughly at one Thomson depth, and as they emerge from the reflector, they encounter photoelectric absorption, most prominently visible as the Fe K pseudo-edge. Finally, the reflected component peaks at

~ 30 keV, and, at energies $\gtrsim 100$ keV, it drops in intensity. This is due to both Compton recoil and the Klein Nishina effects. The primary spectrum itself exhibits a cutoff beyond ~ 150 keV (Figure 7). When this spectrum is fitted with the thermal Comptonization model of [38], the obtained parameters are $kT = 100^{+10}_{-30}$ keV, $\tau = 1.3^{+0.4}_{-0.2}$ (Zdziarski et al. [58]), very similar to the results of fitting the average spectra of all Seyfert 1s and 2s observed by OSSE (see above).

With the paucity of the simultaneous X-ray and γ-ray data for Seyfert 1s, the best approach available to confirm if this observational picture holds for Seyfert 1s in general, was to add together the X-ray data from *Ginga* or *EXOSAT* with the OSSE data. As it is shown in Figure 7 [58], such an average spectrum is quite similar to that obtained for IC 4329A. The thermal Comptonization fit yields $kT = 110^{+150}_{-40}$ keV, $\tau = 1.2^{+0.6}_{-1.0}$ [58].

This general picture also holds for the recent data for IC 4329A obtained simultaneously by the *Rossi X-ray Timing Explorer* (*RXTE*) and OSSE. The RXTE data are still being analyzed, since there are many systematic effects that have to be carefully taken into account; in general, the results are quite similar to those of [23,54].

4.2 Radio-Quiet Seyfert 2s

How different are the high energy spectra of Seyfert 2s? Generally, unlike Seyfert 1s, where the optical/UV emission line widths are $\gtrsim 2000$ km/s, these objects show only narrow emission lines ($\lesssim 1000$ km/s). Seminal observations of NGC 1068 by Antonucci & Miller [2] revealed that when observed in polarized light, this Seyfert 2 also shows broad emission lines. The model proposed by those authors suggests that Seyfert 2s are essentially the same as Seyfert 1s, but the difference is essentially entirely due to the differences in their orientation with respect to the observer. The central source consists of a nuclear region, which is a relatively thin accretion disk, but at the pc-scale distances, the central region is surrounded by a geometrically thick molecular torus. Along the axis, there is a free-electron scattering medium. If viewed along the axis, we see all the ingredients of the nuclear region, meaning the unabsorbed primary continuum, broad emission lines, and the reflection component. When viewed close to the plane of the disk and the torus, the broad emission lines are shrouded by the torus, and these lines can be viewed only with the help of the scattering medium, which allows the lines to be seen when viewed in polarized light.

This picture has implications for the observable high energy spectra of Seyferts, and in fact, the X-ray/γ-ray observations indicate that, to the first order, this scenario is correct. The intervening torus indeed shows up as photoelectric absorption, essentially absorbing all soft X-ray photons. The amount of the hard X-ray radiation depends on the column of the absorber;

if the medium is only partially Compton thick, we see the X-rays above some energy, depending on the column. In many objects, the column is $\sim 10^{22}$–10^{23} cm^{-2}, where with typical gas-to-dust ratios it is large enough to absorb the broad lines. However, at $\gtrsim 2$ keV, most of the photons can escape, as the absorber is optically thin to both the photoelectric absorption of Fe and to Compton scattering. For these cases, the primary hard X-ray/γ-ray continuum is relatively unaffected, and the similarity of the OSSE spectra of radio-quiet Seyfert 1s and 2s (see above) supports this scenario.

In the case of NGC 1068, the Compton depth is too large to see the primary continuum. A very nice intermediate case is NGC 4945 (see Figure 8, Done et al. [8]): the column is a few times 10^{24} cm^{-2}, such that the Compton depth is only about 3. An interesting aspect of a source with the Compton thickness of a few is that the emergent hard X-ray spectrum depends on the solid angle subtended by the absorber. The main difference is that with a large absorber solid angle, more photons get directed back towards our line of sight. However, these photons are scattered more times (and, at each scattering, they lose a fraction of their energy before they escape the absorber), so the spectrum is generally more peaked towards low energies.

In summary, the X-ray/γ-ray data imply that the Seyfert 1/Seyfert 2 unification scheme holds in general with regard to the high energy emission from these objects.

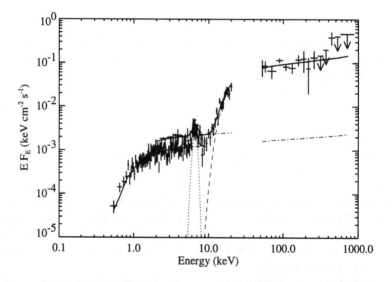

FIGURE 8. The broad band spectrum of NGC 4945 using *Ginga*, *ASCA* and OSSE data from Done et al. [8]. The solid line is the total spectrum; the dashed line above 8 keV is the direct (absorbed) spectrum. The dot-dashed line is the scattered spectrum.

4.3 Radio-loud Seyferts

Radio-loud Seyferts (radio galaxies) appear to be the least understood object among those reviewed here. They do show broad emission lines, but in contrast to other Seyfert galaxies, they often have bright compact radio cores and show superluminal expansion, so the natural question that arises here is: are their spectra blazar-like or Seyfert-like? Recent work by Woźniak et al. [52] assembled much of the available archival X-ray and γ-ray data for those objects, and concluded that the high energy spectra of these objects are clearly different from blazars and somewhat different from ordinary, radio-quiet Seyferts. Despite the fact that they are relatively bright X-ray and soft γ-ray emitters, they show no detectable MeV emission (with an exception of Cen A, see Fig. 6). We thus probably do not observe the direct emission from the relativistic jet.

The X-ray spectra of radio-loud Seyfert 1s show relatively strong Fe Kα lines, so at first, one would think that they are just like radio-quiet Seyferts; blazars generally show no Fe Kα lines, and it is believed that the line emission is "swamped" by the emission of the jet. But radio-loud Seyfert 1s show only weak Compton reflection [52], so a cold, dense accretion disk is either absent or subtends a small solid angle as seen from the source of the primary continuum. Nonetheless, we still have to account for the strong Fe Kα line. A possible scenario [52] is that the material responsible for the iron line has a column density of $\sim 10^{23}$ cm^{-2}, enough to produce the line, but optically thin to Compton reflection. That medium is likely to be a parsec-scale torus surrounding the nucleus.

Comparison of X-ray spectra of radio-loud Seyferts with the OSSE spectra shows that the power laws extrapolated from X-rays have to steepen above ~ 100 keV, but it is not necessary to have an exponential cutoff at ~ 200 keV, and instead, a power law, breaking at ~ 100 keV to another power law, is consistent with the data [52]. In general the spectra of these objects are reminiscent of Seyfert-2–type radio galaxy Cen A, where the spectrum probably extends all the way to GeV energies (Fig. 6). The nature of that emission is probably nonthermal, but its origin remains unknown. Future observations with *INTEGRAL* should help in resolving this issue.

5. Theoretical Interpretation

In the eighties, the most popular model of the intrinsic emission of Seyferts involved nonthermal acceleration of particles to relativistic energies and a subsequent pair cascade and nonthermal Compton scattering (e.g., [48,20]). That model, together with a Compton reflection component, was shown to explain the *Ginga* spectra of Seyferts in Zdziarski et al. [55]. The model predicted a break in the intrinsic spectrum at several hundred keV and a prominent $e^{\pm}$ pair annihilation feature around 511 keV.

However, the OSSE data combined with available X-ray data have not confirmed those predictions. They show that the intrinsic spectra of Seyferts break at energies $\gtrsim 100$ keV and no annihilation features have been detected (e.g., [54,57,15,10]), as discussed in sections above. This rules out the non-thermal pair model in its original version, and it very strongly constrains its modifications [15,10,54].

On the other hand, thermal Comptonization of some soft seed photons (most likely UV photons from an accretion disk) has provided satisfactory fits to the available hard X-ray and soft γ-ray data on Seyferts (see above).

As recently obtained, the typical parameters of the Comptonizing plasma in radio-quiet Seyferts are $kT \sim 100$ keV, $\tau \sim 1$ [57,15,58] (see sections above). In addition, there has to be an optically-thick cold medium subtending a solid angle of $\sim 2\pi$ in order to explain the presence of the Fe Kα lines and Compton reflection in the spectra. The simplest possibility is that the cold medium is a cold accretion disk. Recent *ASCA* results on the shape and variability of the Fe Kα lines indicate that the disk typically extends all the way to the minimum stable orbit (e.g., Iwasawa et al. [14], Nandra et al. [32]).

Thus, the most likely (and most popular) picture is of some active regions above the surface of an accretion disk, or so-called patchy corona (Haardt et al. [11], Stern et al. [45]). The active regions dissipate most of the accretion power. The mechanism of the dissipation remains very poorly understood but it might be related to some kind of magnetic reconnection. The active regions cannot cover the disk completely, as inferred from energy balance arguments [11,45]. Furthermore, the dissipation is likely to take place at some height above the disk. Apart from reducing cooling of the active regions, it allows most of Fe Kα line photons to escape unscattered by the hot plasma, as required by the large typical line fluxes observed from Seyfert 1s (e.g., [32]).

Interestingly, the parameters of the hot plasma are very close to those obtained in the hot inner disk model of Shapiro et al. [42]. In that model, the gravitational energy is transferred to hot ions, which in turn transfer its energy to electrons by Coulomb interactions, and the electrons Compton-upscatter some soft seed photons. The energy balance keeps electrons at $kT \sim 100$ keV [42]. The similarity of the parameters fitted to the OSSE data on Seyferts to those of the hot disk model may indicate the importance of ion heating and the Coulomb process in Seyferts. However, the original model of [42] has to be modified here to include the presence of the cold inner disk. On the other hand, some Seyferts, e.g., NGC 4151, show only weak Compton reflection, and therefore those objects may posses hot inner disks similar to those of [42] with Compton reflection taking place from a cold outer disk.

The hot plasma present in Seyferts may consist either of electrons or of $e^{\pm}$ pairs. The presence of pairs affects on the shape of the spectrum around 511 keV and the ratio of the luminosity in X-rays and γ-rays to the size of the hot source. The former is only poorly constrained by OSSE while the latter is only very crudely constrained by the time variability of Seyferts (e.g., [30]). Thus, no firm conclusion can be achieved yet (see [58] for more discussion). As mentioned earlier, a thermal pair plasma exhibits no pair annihilation features [22], so constraints on the pair annihilation flux cannot be used to constrain the pair abundance.

One case in which the presence of pairs is constrained somewhat more tightly is that of NGC 4151 [57,15]. The quality of the OSSE spectrum of this object is sufficient to rule out the presence of pairs under the assumption that the plasma is purely thermal. However, an admixture of nonthermal processes, allowed by the data (see above), can lead to efficient pair production in the source and pair dominance.

6. Summary

Our main conclusions are:

- The OSSE spectra of Seyferts and radio galaxies are steeper than the corresponding X-ray spectra, which requires the general presence of a break between the OSSE and *Ginga* energy ranges. The average OSSE power law index is $\Gamma \sim 2.4$.

- However, the power-law model provides only a crude description of the the OSSE average spectra of radio-quiet Seyfert 1s and 2s, and the spectrum of NGC 4151. A high-energy break or cutoff in the model around ~ 100 keV is statistically required by the data. Instead of the above two breaks, a gradual steepening of the Seyfert spectra above ~ 30 keV is also consistent with the data.

- The best *physical* model fitting the OSSE data for radio-quiet Seyferts is mildly-relativistic thermal Comptonization (including a Compton-reflection component, required by the X-ray data). The obtained plasma parameters are $kT \simeq 100$ keV, $\tau \simeq 1$.

- There is no statistically-significant difference between the OSSE spectra of Seyfert 1s and 2s, in accord with the unified model of AGN.

- The OSSE data are compatible with the the geometry of typical radio-quiet Seyferts with hot active regions above the surface of a cold accretion disk.

- The spectra of radio-loud objects appear to be harder than those of radio-quiet ones; in particular, the spectrum of Cen A shows a high-energy tail extending to the MeV–GeV range. The latter requires the presence of nonthermal processes in those sources, in contrast to radio-quiet Seyferts.

REFERENCES

1. Antonucci, R. R. J., 1993, ARA&A, 31, 473
2. Antonucci, R. R. J., & Miller J. S., 1985, ApJ, 297, 621
3. Bassani, L. 1985, in Active Galactic Nuclei, ed. J. E. Dyson, p. 252
4. Bassani L., Malaguti G., Paciesas W. S., Palumbo G. G. C., & Zhang S. N. 1996, A&AS, 120, 559
5. Bond, I. A., et al. 1996, A&A, 307, 708
6. Dermer, C. D., & Gehrels, N. 1995, ApJ, 447, 103
7. Döbereiner, S. et al. 1996, ApJ, 470, L15
8. Done, C., Madejski, G. M, & Smith, D., 1996, ApJ, 463, L63
9. Feigelson, E. D., Schreier, E. J., Delvaille, J. P., Giacconi, R., Grindlay, J. E., & Lightman, A. P. 1981, ApJ, 251, 31
10. Gondek, D., Zdziarski, A. A., Johnson, W. N., George, I. M., McNaron-Brown, K., Magdziarz, P., Smith, D., & Gruber D. E. 1996, MNRAS, 282, 646
11. Haardt, F., Maraschi, L., & Ghisellini, G., 1994, ApJ, 432, L95
12. Hartman, R. C., Collmar, W. E., von Montigny, C., & Dermer, C.D. 1997, in Proceedings of the Fourth Compton Symposium, AIP (these proceedings)
13. Hartman, R. C., et al. 1997, the Third EGRET Catalog, to be submitted to ApJS
14. Iwasawa, K., et al., 1996, MNRAS, 282, 1038
15. Johnson, W. N., McNaron-Brown, K., Kurfess, J. D., Zdziarski, A. A., Magdziarz, P., & Gehrels, N., 1997, ApJ, 482, 173
16. Jourdain, E., et al. 1993, ApJ, 412, 586
17. Kinzer, R. L., Dermer, C. D., Johnson, W. N., Kurfess, J. D., Strickman, M. S., Grove, J. E., Kroeger, R. A., Grabelsky, D. A., Purcell, W. R., Ulmer, M. P, Jung, G. V., & McNaron-Brown, K., 1995, ApJ, 449, 105
18. Krolik, J. H., & Kriss, G. A., 1995, ApJ, 447, 512
19. Lightman A. P., & White T. R., 1988, ApJ, 335, 57
20. Lightman, A. P. & Zdziarski, A. A. 1987, ApJ, 319, 643
21. Lin, Y. C., et al. 1993, ApJ, 416, L53
22. Maciołek-Niedźwiecki, A., Zdziarski A. A., & Coppi, P. S., 1995, MNRAS, 276, 273
23. Madejski, G. M., et al., 1995, ApJ, 438, 672
24. Magdziarz, P., & Zdziarski, A. A., 1995, MNRAS, 273, 837
25. Maisack, M., Collmar, W., Barr, P., Bloeman, H., Hermsen, W., Lichti, G. G., McConnell, M., Schonfelder, V., Stacy, J. G., Steinle, H., et al., 1995, A&A, 298, 400
26. Maisack, M., et al., 1993, ApJ, 407, L61
27. Maisack M., Mannheim K., & Collmar W., 1997, A&A, 319, 397
28. Makino, F., et al. 1988, in Physics of Neutron Stars and Black Holes, ed. Y. Tanaka, p. 357
29. Malizia, A., Bassani, L., Malaguti, G., Paciesas, W.S., Palumbo, G.G.C., Zhang, S.N. 1997, in: Winkler C., Courvousier T., Durouchoux P. (eds.) ESA SP-382, The Transparent Universe, p. 439

30. Mushotzky, R. F., Done, C., & Pounds, K. A., 1993, ARA&A, 31, 717
31. Mushotzky, R. F., Fabian, A. C., Iwasawa, K., Kunieda, H., Matsuoka, M., Nandra, K., & Tanaka, Y., 1995, MNRAS, 272, L9
32. Nandra, K., George, I. M., Mushotzky, R. F., Turner, T. J., & Yaqoob, T., 1997, ApJ, 477, 602
33. Nandra, K. & Pounds, K. A., 1994, MNRAS, 268, 405
34. Nolan P. L., et al., 1996, ApJ, 459, 100
35. Parsons, A., et al. 1997, ApJ, submitted
36. Piccinotti G., et al. 1982, ApJ, 253, 485
37. Pounds, K. A., Nandra, K., Stewart, G. C., George, I. M., & Fabian, A.C., 1990. Nat, 344, 132
38. Poutanen, J., & Svensson, R., 1996, ApJ, 470, 249
39. Preston et al. 1983, ApJ, 266, L93
40. Robinson, C. et al. 1997, in: Winkler C., Courvousier T., Durouchoux P. (eds.) ESA SP-382, The Transparent Universe, p. 249
41. Rothschild, R. E., Mushotzky, R. F., Baity, W. A., Gruber, D. E., Matteson, J. L., & Peterson, L. E. 1983, ApJ, 269, 423
42. Shapiro, S. L., Lightman, A. P., & Eardley, D. M., 1976, ApJ, 204, 187
43. Skibo, J.G., Dermer, C.D., & Kinzer, R.L., 1994, ApJ, 426, L23
44. Steinle, H. et al. 1997, in Proceedings of the Fourth Compton Symposium, AIP (these proceedings)
45. Stern, B. E., Poutanen, J., Svensson, R., Sikora, M., & Begelman, M. C., 1995, ApJ, 449, L13
46. Svensson R., Larsson S., Poutanen J., A&AS, 120, C587
47. Sunyaev, R. A., & Titarchuk, L. G., 1980, A&A, 86,121
48. Svensson, R., 1987, MNRAS, 227, 403
49. Tanaka, Y., et al., 1995, Nature, 375, 659
50. Turner, T. J., & Pounds, K. A., 1989, MNRAS, 240, 833
51. Walker, R. C., Benson, J. M., & Unwin, S. C., 1988, in IAU Symp., 129, 29
52. Woźniak, P. R., Zdziarski, A. A., Smith D., Madejski G. M., & Johnson W. N., 1997, MNRAS, in press
53. Yaqoob T., & Warwick R. S., 1991, MNRAS, 248, 773
54. Zdziarski, A. A., Fabian, A. C., Nandra K., Celotti A., Rees, M. J., Done C., Coppi, P. S., & Madejski, G. M. 1994, MNRAS, 269, 55L
55. Zdziarski, A. A., Ghisellini, G., George, I. M., Svensson, R., Fabian, A. C., & Done, C., 1990, ApJ, 363, L1
56. Zdziarski A. A., Johnson W. N., Done C., Smith D., & McNaron-Brown K., 1995, ApJ, 438, L63
57. Zdziarski A. A., Johnson W. N., & Magdziarz P., 1996, MNRAS, 283, 193
58. Zdziarski A. A., Johnson W. N., Poutanen J., Magdziarz P., Gierliński M., 1997, in Winkler C., Courvousier T., Durouchoux P. (eds.) ESA SP-382, The Transparent Universe, p. 373
59. Zdziarski A. A., Lightman, A. P., & Maciołek-Niedźwiecki, A., 1993, ApJ, 414, L93
60. Zhang, S. N., 1997, private communication.

Gamma-Ray Blazars

R. C. Hartman*, W. Collmar†, C. von Montigny§, and C. D. Dermer‡

*NASA/Goddard Space Flight Center, Code 661, Greenbelt, MD 20771 USA
†Max-Planck-Institut für Extraterrestriche Physik, Postfach 1603,
85740 Garching, Germany
§Landessternwarte, 69117 Heidelberg, Germany
‡Naval Research Laboratory, Code 7653, Washington, DC 20375-5352 USA

Abstract. Roughly 60 blazars have been identified with > 100 MeV EGRET γ-ray sources, and show inferred isotropic luminosities as large as 3×10^{49} ergs s^{-1}. The two most remarkable characteristics about the EGRET observations are that the γ-ray luminosity often dominates the bolometric power, and that variability on time-scales of one day or less is seen from several sources. Gamma-ray blazars have been detected with redshifts ranging from $\cong 0.03$ to $\cong 2.3$. The distribution of redshifts is similar to that of flat spectrum radio quasars. The distribution of > 100 MeV photon spectral indices for strong detections is well described by a Gaussian function peaking near 2.16 with a standard deviation of 0.31. In some cases, flare spectra at > 100 MeV energies are harder than spectra in quiescence. Blazar spectra at hard X-ray and soft γ-ray energies have harder spectra and, from analyses of contemporaneous data sets, the change in spectral index clearly exceeds 0.5 for 3C 273 and marginally exceeds 0.5 for PKS 0528+134. An index change of 0.5 is predicted by incomplete Compton cooling models. COMPTEL observations reveal a separate class of 2 identified blazars and 3 unidentified candidates with narrow peaks at MeV energies.

The combined luminosity and variability data strengthen arguments for relativistic beaming or bulk outflow from the active nucleus. Gamma-ray transparency arguments are based on the pair-production attenuation of γ rays in a stationary region defined by the variability time-scale, and many violations are indicated. The Elliot-Shapiro relation compares the minimum black hole mass implied by the Eddington limit with the maximum black hole mass implied by the variability time-scale related to the black hole Schwarzschild radius; weak violations have been observed for PKS 0528+134 and PKS 1622-297.

Most models adopt a beaming paradigm. In models considering leptons as the directly accelerated and radiating particles, Compton scattering is favored to make the γ-ray emission, though opinions differ on whether the soft photons are primarily internal synchrotron photons or accretion disk photons which enter the jet either directly or after scattering off clouds. Hadronic models involve secondary production or photomeson production followed by pair cascades, and predict associated neutrino production.

1. INTRODUCTION

The EGRET detection of roughly 60 blazars in high-energy (> 100 MeV) γ-rays has provided renewed interest and dramatic impetus in the study of these remarkable objects. Although high-energy γ rays from 3C 273 had been discovered by COS-B about 20 years earlier [74], and predictions of additional γ-ray detections had been made [5,45] in the interim, the discovery that the γ-ray emission can provide a major part of the inferred isotropic luminosity and, at least in flaring states, can dominate the luminosity in all other bands by more than a factor of 10, was clearly unexpected. In addition, the dramatic and often short-time-scale variations in blazar γ-ray emission seen by EGRET contrasts with the absence of detected variation in the 3C 273 observations by COS-B [6].

This review describes briefly (Section 2) the discovery of the γ-ray blazar phenomenon and summarizes the present knowledge on detections from EGRET, COMPTEL, and OSSE. Section 3 discusses the intrinsic properties of the objects, including the comparison of the γ-ray emission with that in other observing bands. Sections 4 and 5 address, respectively, the implications for the beaming hypothesis, and the various models which have been proposed for generating the γ radiation. Broader implications, including unification scenarios and particle acceleration requirements, are summarized in Section 6.

2. DETECTION OF GAMMA-RAY BLAZARS

2a. EGRET Discovery of Extragalactic Sources of Gamma Rays > 100 MeV

Although 3C 279 was the first extragalactic object recognized as a γ-ray source in the EGRET data [31], it was not the first one detectable in an EGRET observation. Prior to the formal beginning of observations with the Compton Observatory, several weeks were spent in checkout of the instruments. Because the Galactic anticenter contains the Crab pulsar, known from SAS-2 and COS-B observations to be a strong γ-ray source up to at least 1 GeV, as well as the then-enigmatic γ-ray source known as Geminga, much of the instrument checkout was carried out with the spacecraft pointed in the anticenter region. During those 12 days, the distant ($z \sim 2.0$, determined later) radio-loud quasar PKS 0528+134 was constantly within the EGRET field of view. Because of the combination of limited angular resolution, the proximity of the much brighter Crab source, and lack (at that time) of experience with analysis of the EGRET data, 0528+134 was recognized only later [41,37], despite the fact that it was rather bright during that initial observation.

The 3C 279 discovery in data from the latter half of June 1991 was something of a surprise, primarily because 3C 273 was expected to be the brightest object in that region of the sky, based on the COS-B result. 3C 279 was much brighter than 3C 273 had been in COS-B observations. The observation in June 1991 was actually a Target of Opportunity for SN 1991T, but it was quite fortuitous because of the bright flare in 3C 279 [43]. The emission from 3C 273 was not immediately recognized in the June 1991 data, not only because of the much brighter 3C 279, but because 3C 273 was substantially dimmer than during the COS-B observations [78].

The initial identification of 3C 279 as the bright γ-ray source in June 1991 was made via a NED search of the position error box. Although all EGRET position error boxes contain many astronomical objects, both galactic and extragalactic, the source was sufficiently bright (and its error box therefore sufficiently small) that 3C 279 was easily the most likely candidate identification. Its classification as an optically violently variable (OVV) quasar was noted, as was the radio-through-X-ray flare about three years earlier [66,42,48]. Learning that 3C 279 was experiencing an even more dramatic flare in late 1990 and early 1991 served to solidify the EGRET identification.

As additional medium-to-high latitude observations were completed and analyzed, it became apparent that essentially every one contained at least one fairly bright source, several of which were position-compatible with other OVV quasars (e.g., PKS 0208-512 and 4C 38.41 = 1633+382). A notable exception was the 28 June - 12 July 1991 observation of the region around Seyfert NGC 4151, which contained no immediately obvious sources. However, the similarities between OVV quasars and BL Lac objects suggested the possibility of emission from Markarian 421. A careful analysis indicated a source that was position-compatible with Markarian 421, at a significance well over 4-σ. Although the error box was larger in this case because of the weakness of the source, Markarian 421 was the most "dramatic" object in the error box. Several other similar experiences led to the conclusion that the object-type *blazar* seemed to cover many of the EGRET high-latitude sources, despite the detection of some high-latitude sources for which no obvious candidate was apparent.

The term *blazar* is not well-defined. It was originally used informally to include OVV quasars and BL Lac objects, which have a number of similar properties. Many authors now include high-polarization quasars (HPQ) under the blazar umbrella, and Fugmann [28] concluded that at least 2/3, and maybe all, of the flat-spectrum radio sources are blazars. The EGRET Team eventually adopted the flat-spectrum fraction of the Kühr et al. [44] 1-Jy catalog as a starting point for attempting to identify new high latitude sources. Some of the EGRET sources were initially identified with flat-spectrum radio sources, which have later been shown to have other blazar properties.

In addition to flat radio spectra, the properties that blazars have in common are strong and rapid variability in both optical and radio bands, and strong optical polarization (especially the BL Lac objects. Some of the EGRET-detected blazars have been shown to demonstrate superluminal motion of components resolved with VLBI; it may be that the majority have this property, but VLBI observations are not yet available for more than half of them. The BL Lac objects differ from the quasars in having generally stronger polarization and weaker optical lines. Indeed, some BL Lac objects have no redshift determination because they have no identified lines above their optical continuum. On the other hand, variations in the relative strength of the lines and continuum have been noted in some BL Lacs, so determination of those redshifts may be possible at some time in the future.

BL Lac objects are subdivided into two main categories, commonly designated radio-selected BL Lac objects (RBLs) and X-ray selected BL Lac objects (XBLs). For the RBLs, the multiwavelength νF_ν spectral energy distributions (SEDs) peak in the IR-optical bands, and their SEDs are generally similar to those of OVV and HP quasars. The SEDs of the XBLs peak in the soft X-rays. The XBLs tend to have lower luminosities than RBLs except in X-rays, where the luminosities are similar [68]. The only two extragalactic objects which have been strongly detected in TeV γ rays, Mrk 421 [64,47] and Mrk 501 [65], are XBLs.

More recently, the idea that EGRET is detecting flat-spectrum radio sources has become more quantitative. Figure 1 [77] demonstrates that there are no

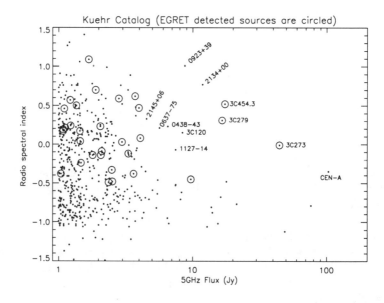

FIGURE 1. Energy spectral indices of > 1 Jy radio sources and EGRET-detected blazars.

EGRET detections of bright radio-emitting objects with radio spectral indices α softer than -0.5.[1] A similar determination has been made quantitatively [56].

Temporal variations have been detected in the fluxes of EGRET-detected blazars on time-scales from a year or more down to well under 1 day. Long-term variations are addressed by von Montigny et al. (1995) and Mukherjee et al. (1997) [77,62]. The shortest time-scale variation detected for a blazar with EGRET is for PKS 1622-297 [57]; its doubling time appears to have been well under 8 hours (The study of such short-time-scale variations is always limited by the small numbers of photons detected in the short time intervals). This object also shows the largest range of γ-ray flux, ranging over a factor of $\gtrsim$ 80. Other sources that have shown variation time-scales on a few day time-scale or less are 3C 279 [43], 3C 454.3 [32,80], 4C 38.41 [55], PKS 1406-076 [79] and PKS 0528+134 [37,61]. Figure 2 shows the long-term and flaring light curves of 3C 279 in the EGRET energy range.

---

[1] Here α is defined by $S(\nu) \propto \nu^\alpha$.

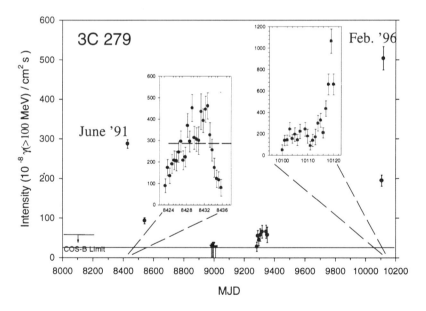

FIGURE 2. Time history of fluxes, including flaring events, of 3C279.

2b. COMPTEL Detections of Blazars

The COMPTEL instrument aboard CGRO detects γ rays in the energy range 0.75 - 30 MeV, covering the central part of the CGRO energy range. This energy range was largely unexplored before the launch of CGRO because of observing difficulties (e.g., a minimum in the cross section for γ-ray interactions, and high backgrounds of γ-ray lines and electron bremsstrahlung). Prior to COMPTEL, no quasar was known to emit at MeV energies. It was anticipated that COMPTEL should detect 3C 273 because it had been successfully observed in the hard X-ray range [4] and at energies above 50 MeV by COS-B [74]. The shape of the hard X-ray and the COS-B spectra suggested two interesting features for the MeV region: first, the maximum energy release across the whole electromagnetic spectrum could occur here; second, there should be a transition between a "hard" X-ray spectrum with $\alpha > -1$ and a softer medium energy γ-ray spectrum with $\alpha < -1$ [4]. Apart from these expectations, there was the prospect of detecting something unexpected, which is usually the case when new hardware with improved sensitivity comes into operation.

Shortly after the launch of CGRO, EGRET detected several blazar-type AGNs at energies above 100 MeV. This stimulated blazar searches in the COMPTEL data, which soon led to the discovery of 3C 273 and 3C 279 at MeV-energies as well [35]. Extrapolating the observed EGRET spectra throughout the COMPTEL energy band revealed PKS 0528+134 as the prime candidate for COMPTEL. A subsequent "close look" immediately revealed evidence for this blazar as well [18]. At the time of writing, COMPTEL has

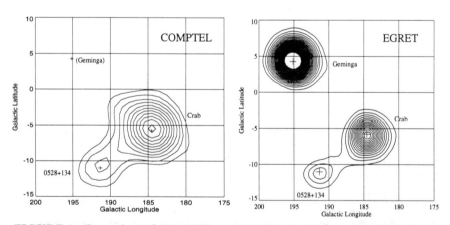

FIGURE 3. Comparison of COMPTEL and EGRET significance maps at the Galactic anticenter. The absence of the Geminga pulsar in the COMPTEL map is due to its hard spectrum, which causes it to fall below the COMPTEL detection threshold despite its prominence in the EGRET data.

detected 8 out of the ~ 60 EGRET blazars. They are marked in Table 1.

In most cases, the COMPTEL detection significance of these sources in individual COMPTEL energy bands is low, between 3 and 4-σ. Comparing detections and non-detections in different observations through the CGRO mission, all of them seem to be time-variable on time-scales from months to years. Because sensitivity-limited statistics determine the shortest variabilty time-scale in the COMPTEL energy range, blazar variabilty at shorter time-scales is likely. The shortest variability time-scale was found just recently from 3C 279 during the Cycle 5 Virgo observations [21]. During VP 511.5 the 10-30 MeV flux was four times larger than 10 days before during VP 511, indicating behavior similar to that observed by EGRET above 100 MeV.

Generally the COMPTEL results on blazars match or even exceed the expectations before launch. Several sources, including 3C 273, have been clearly detected, and the MeV range was found to be a transition region where spectral breaks occur in their spectra. A large fraction of them have the maximal energy released at MeV energies. All eight COMPTEL blazar detections are from the quasar subclass. No BL Lac object has been positively detected yet, which is consistent with the EGRET trend that BL Lac objects have, on average, harder spectra, thereby "falling" underneath the COMPTEL sensitivity.

2c. OSSE Detections of Blazars

By the end of 1994 May, OSSE observed 17 blazar AGNs in the 50 keV – 10 MeV range and has detected 7 [58]. Six of the 7 are EGRET-detected blazars, including PKS 2155-304 which was detected by EGRET [76] following the OSSE detection. Blazars are detected most significantly with OSSE at energies between 50 and 150 keV. With only a few exceptions (e.g. 3C 273), the OSSE measurements are not simultaneous with EGRET and COMPTEL. OSSE blazar detections are indicated in Table 1. Time variability is observed for all of them on scales from days to years. The variability time-scales in the 50 - 150 keV range are as short as 3 days for 3C 273 [59] and PKS 0528+134 [58].

The 50 - 150 keV blazar spectra are well fit by simple power laws with photon spectral indices s ranging from -1.0±0.6 for CTA 102 to -1.7±0.1 for 3C 273 [39] and -2.1±0.4 for 3C 279 [58]. If the OSSE spectra are compared to EGRET/COMPTEL observations, spectral breaks are required for all sources with the exception of 3C 279. The values of spectral index changes between the OSSE range and COMPTEL and EGRET ranges are between 0.3 and 1.7. The energies of the spectral breaks are found between ~ 1 and ~ 20 MeV. Break values significantly larger than 0.5 are found for PKS 0528+134 and for CTA 102, in conflict with simple models of incomplete Compton cooling of relativistic electrons by external photons, which predict a value not larger than 0.5. In such models [22,71], a power-law injection electron distribution leads to a broken power-law Compton-scattered photon spectrum if lower-energy electrons fail to cool through Thomson scattering before leaving a region of high photon density (see Section 5a). The latest OSSE blazar results are presented in these proceedings [40], and a report of a strong OSSE detection of the XBL Mrk 501 has been accepted for publication [16].

3. PROPERTIES OF GAMMA-RAY BLAZARS

3a. Spectra and Spectral Energy Distributions

Although EGRET spectra of blazars typically cover at least two decades in energy, essentially all of them are adequately described by power laws. This is due in part to the large errors on the individual flux points in the upper and lower portions of the energy range. At the upper end, the large errors are due to the common problem of small numbers of counts for a steeply falling counts spectrum. At the lowest energies, the uncertainties are primarily systematic, stemming from poorly known sensitivity calibration. For those few blazars detected by COMPTEL, the poorly determined fluxes at the lowest EGRET energies are less important, since the general shape of the spectrum is fairly well constrained there.

The average blazar spectrum has a photon power-law index of about -2.1 in the EGRET band. Figure 4 shows the spectral index distribution of EGRET blazars with well-defined indices. The BL Lac objects detected by EGRET tend to be less variable in the EGRET band than are the EGRET-detected quasars [62]. There are also indications [60,62] that BL Lac objects have slightly harder spectra in the EGRET energy range than do the EGRET-detected quasars. This result is not statistically compelling, but persists in several independent analyses. Another trend that is not statistically strong, but could be important, is an apparent tendency for spectral hardening at higher flux levels. However, there is at least one clear example of no such hardening, namely the large February 1996 flare of 3C 279 [80]. The statistical limitation in such a study is that the spectral index is not sufficiently well determined in its low-flux state that its difference from the high-state index is significant. The recent long observation of 3C 279 in a very low state may help to address this issue.

Because the COMPTEL detections are in most cases relatively weak and occur over a small portion of the COMPTEL bandwidth, our knowledge on the MeV spectra of blazars is incomplete. Nevertheless, some trends are apparent. For the stronger sources detected with COMPTEL, only 3C 273 has a soft ($s < -2$) spectrum, whereas the others tend to have harder ($s > -2$) spectra as implied by their detections mainly above 3 MeV. Comparing COMPTEL and EGRET spectral measurements, spectral breaks are required. This shows that the COMPTEL energy range represents a transition region for blazar

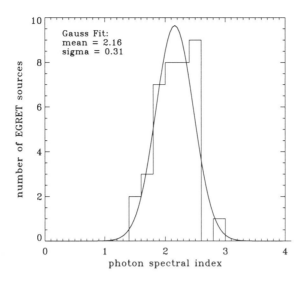

FIGURE 4. Spectral index distribution of EGRET blazars.

spectra, as had been expected prior to the launch of CGRO. The most elaborate COMPTEL spectral analysis has been carried out for PKS 0528+134 [20], and reveals evidence for a spectral change between hard and soft spectral states correlated with γ-ray activity observed by EGRET above 100 MeV. During > 100 MeV γ-ray flares, a hard MeV spectrum is observed, while a soft spectrum is derived for the sum of all other analysed observations. Such a change would be consistent with the emergence of an additional spectral component mainly at energies above 3 MeV. For all high states, the luminosity has its maximum at MeV energies, with a maximum of 4×10^{49} erg s^{-1} between 3 and 30 MeV during a γ-ray flare in March 1993.

Comparison of γ-ray emission with that in other observing bands seems to be most illuminating in the form of SED plots of νF_ν or νL_ν as a function of frequency. Such figures show the relative amounts of energy detected in equal logarithmic frequency bands. Contemporaneous multiwavelength spectra of 3C 279 [53,70] illustrates the point made in Section 1 concerning the power output in γ rays, which can dominate the bolometric luminosity. Based on two epochs about 1.5 years apart, the data indicate an amplitude variation which increases with frequency. However, if spectra from January and February 1996 [80] are added, the picture becomes more complex in that the amplitude variation no longer appears to have a simple frequency dependence.

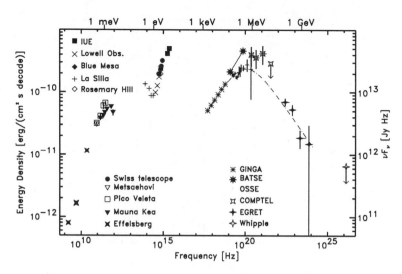

FIGURE 5. Quasi-simultaneous broad-band spectrum of the quasar 3C 273 [46].

Figure 5 [46] shows a quasi-simultaneous multifrequency spectrum of the quasar 3C 273. The spectrum shows (probably) four maxima indicating different emission components and processes, which are interpreted as synchrotron emission from relativistic electrons in the radio- and far-IR band, thermal emission from a dust torus in the IR and from an accretion disk in the UV ("blue bump"), and inverse Compton radiation generated by relativistic electrons and soft photons at X- and gamma-rays. The spectrum shows that the high-energy emission is a significant part of the bolometric luminosity.

Figure 6 [20] shows a non-simultaneous multifrequency spectrum of PKS 0528+134. The different data are selected to be as close as possible to the contemporaneous COMPTEL-EGRET observations in April 1991. This spectrum shows two typical features of several flat spectrum radio quasars in gamma-ray high states: 1) a spectral turnover at MeV energies from a harder spectrum in the hard X-ray range to a softer one at high-energy gamma-rays (> 100 MeV), and 2) the radiated power per natural logarithmic frequency interval peaks at MeV energies.

COMPTEL has also provided evidence for the unexpected class of "MeV blazars" [11,12], which are exceptionally bright in the 1-10 MeV range compared to their fluxes measured simultaneously with EGRET at > 30 MeV energies. Two prime objects, GRO J0516-609 (which has PKS 0506-612 as the probable counterpart) and PKS 0208-512 have been found to exhibit this behavior on occasions. In addition, three unidentified objects have been reported [19,38,81] which display similar spectra, and might belong to the same source class. In contrast to "normal" blazars, where the COMPTEL measurements require spectral softening between the COMPTEL and EGRET

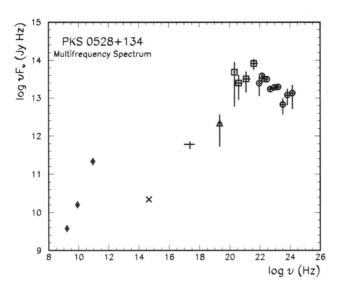

FIGURE 6. Broad-band spectrum of PKS 0528+134 during a gamma-ray high state in April 1991. Only the COMPTEL ("squares") and EGRET ("cicles") data are contemporaneous. The other data are selected to be as close as possible to CGRO observation. For references to the data points, see [20].

energy ranges, spectral "bumps" are found in the spectra of MeV blazars. A simple extrapolation of the simultaneous EGRET spectrum passes well below the COMPTEL points. A power law fit for GRO J0516-609 through the COMPTEL and EGRET measurements between 10 and 100 MeV would give a photon spectral index of ~ -3 [11], which represents one of the steepest blazar spectra reported.

The strong MeV excesses in MeV blazars are time variable. GRO J0516-609 showed variability only during the second and third observations of CGRO Phase I (sky survey). Contemporaneously, a weak (3-σ) EGRET source was found, suggesting that the blazar-type quasar PKS 0506-612 was the counterpart. During the CGRO Phases II and III, neither COMPTEL nor EGRET found any hint for high-energy emission. Recently, a marginal OSSE detection of soft γ-ray emission from PKS 0506-612 has been reported (W. N. Johnson, private communication), supporting it as the counterpart. PKS 0208-512, the second identified MeV blazar, is a well known EGRET blazar soure. It is detected by EGRET in each analyzed CGRO phase, having a variable spectral shape [13]. Evidence for MeV emission larger than predicted by the extrapolation of the EGRET spectrum was observed by COMPTEL during Phase I and II, but not during Phase III [13], again indicating the time-variable nature of the phenomenon.

3b. Redshift Distribution of Gamma-Ray Blazars

Figure 7 shows the redshift distribution of γ-ray blazars detected by EGRET, using the data from Table 1. As can be seen, this distribution is similar to that of flat spectrum radio quasars, implying that the two populations are closely related.

4. GAMMA-RAY TESTS FOR BEAMING

The combination of large inferred γ-ray luminosities L_γ and short observed γ-ray flux variability time-scales implies that the blazar emission region is very compact, as defined by the compactness parameter $\ell \equiv L_\gamma \sigma_T/(4\pi m_e c^3 r)$. Roughly speaking, a factor-of-2 flux variation on an observed time-scale δt_{obs} limits the size r of a stationary isotropically-emitting region to be $r \lesssim c \delta t_{obs}/(1+z)$ by simple light-travel-time arguments. From the blazar observations reviewed above, one finds that very compact emission sites are

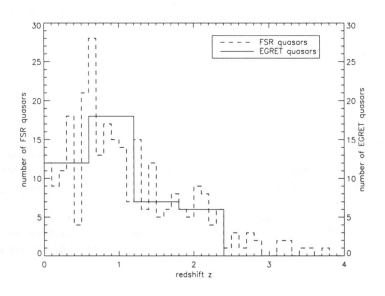

FIGURE 7. Redshift distribution of EGRET blazars.

implied in several cases, leading to arguments that contradict the original assumption of a stationary emitting region, as described in this section.

Blazar research at γ-ray energies therefore faces a similar problem as met thirty years ago in the Compton catastrophe [36], where the small spatial dimensions inferred from the radio and optical variability meant that if this emission was due to incoherent synchrotron radiation, then the radiating electrons would cool catastrophically by Compton scattering their own synchrotron radiation. The most widely-favored resolution to this paradox, which also overcomes the problems posed below, is to relax the assumption of a stationary emission region with isotropic particle distributions. Either the radiating particles are beamed [82], or the emitting plasma is in bulk relativistic motion [7]. The hypothesis of bulk plasma ejection accords with observations of jets in radio sources, which apparently result from plasma being expelled from a central nucleus [3]. It also offers an explanation for apparent superluminal motion as a consequence of relativistically moving radiating plasma viewed nearly along its direction of motion. The relativistic boosting of the radiation poses severe difficulties, however, in cataloguing the number density and luminosity function of blazars. In a flux-limited survey, the Doppler enhancement of the emission from relativistically moving sources which are aligned nearly along the line-of-sight means that these sources can be detected at much greater distances than sources observed at large angles to the beaming axis. Hence the detected blazar population is stongly dependent on orientation effects and source evolution with redshift [17].

4a. Gamma-Ray Transparency Arguments

Gamma-ray transparency arguments [51,55,9] provide an important test for beaming in blazars. These arguments proceed as follows: Assume that the source is at rest in the cosmological frame so that the observed γ-ray variability time-scale δt_{obs} implies a maximum emission-region size-scale from the light-travel-time arguments just mentioned. If X-rays observed from blazars are produced in the same region as the γ rays, then the pair production optical depth $\tau_{\gamma\gamma}$ for γ rays above 100 MeV greatly exceeds unity in several sources [55,51,58], so that beaming is implied. The underlying assumption in these studies is that γ-rays in the EGRET energy range ($E \gtrsim 100$ MeV) are produced cospatially with keV X-rays. This is because X-rays with energies of $\sim 2\,(m_e c^2)^2/E_1$ are most effective at attenuating γ-rays with energy E_1 due to peaking of the $\gamma\gamma \to e^+e^-$ cross section at energies above the pair production threshold. Thus 100 MeV and 1 TeV photons are most strongly attenuated by ~ 5 keV photons and 0.5 eV photons, respectively. Other studies have shown [1] that γ rays do not need to be produced at large distances from the central source if the X-rays are assumed to be produced in an accretion disk, and that the angle-dependent $\gamma\gamma$ opacity provides intrinsic collimation of the γ rays.

It was pointed out [24] that observations at soft γ-ray energies in the OSSE and COMPTEL ranges overcome the need to assume the cospatial origin of X-rays and γ-rays, because soft γ-rays are strongly attenuated by other photons with similar energies. Analysis [58,59] of OSSE observations of blazars provides evidence that the optical depth $\tau_{\gamma\gamma}$ to $\gamma\gamma$ pair production exceeds unity for MeV photons in PKS 0528+134, 3C 454.3, CTA 102, and 3C 273. Thus the detection of MeV photons contradicts the assumption of a stationary, isotropic emitting region, so that models of blazars must employ beamed particle distributions or bulk relativistic plasma motion. Expressions for the optical depth to $\gamma\gamma$ pair production are derived in [26].

4b. Elliot-Shapiro Relation at Gamma-Ray Energies

Another test for beaming from variability and flux measurements is the Elliot-Shapiro relation [27]. If the source luminosity is assumed to be isotropically emitted and Eddington-limited, then the Eddington limit yields a lower limit on the black hole mass $M \equiv 10^8 M_\odot M_8$, assuming that the inferred isotropic luminosity is less than the Eddington luminosity. We can write this as $M_8 \gtrsim 10^2 L_{48}$, where $10^{48} L_{48}$ ergs s^{-1} is the (assumed) isotropic source luminosity in the Thomson regime. The light-travel-time arguments for the maximum size-scale R, assumed to be roughly the Schwarzschild radius, implies $M_8 \lesssim 10^{-2} \delta t_{\rm obs}(s)/(1+z)$. This yields an upper and lower bound upon M, so that observations showing that $L_{48}/\delta t_{\rm obs}({\rm days}) \gtrsim 1/(1+z)$ provide evidence in favor of beaming. Modifications taking into account Klein-Nishina effects on the Compton cross section and Eddington luminosity are essential when interpreting results from γ-ray telescopes [24,63].

Implied minimum black hole masses of blazars, under the assumptions of isotropic radiation and Eddington-limited accretion, give values of $M \gtrsim 8 \times 10^{11} M_\odot$ for PKS 1622-297 [57] from EGRET observations, and $M \gtrsim 7 \times 10^9 M_\odot$ and $M \gtrsim 7.5 \times 10^8 M_\odot$ for PKS 0528+134 from OSSE [58] and COMPTEL [21] observations. Violations of the Elliot-Shapiro relation are claimed for these two blazars, though the OSSE result for PKS 0528+134 is based upon 2.5-σ evidence for factor-of-two variability during one viewing period and represents only a marginal violation of the relation.

5. MODELS FOR THE GAMMA-RAY EMISSION

The favored scenario [7,8] to explain the blazar radio-through-optical continua is that we are viewing nearly along the axis of a relativistically outflowing plasma jet which has been ejected from an accreting supermassive black hole. This broadband radiation is thought to be produced by nonthermal electron synchrotron radiation in outflowing plasma blobs. The high-energy blazar continuum emission appears to constitute a distinct second component in the broadband spectral energy distribution of blazars [53]. Two classes of models have been proposed for the blazar γ radiation, where either leptons or hadrons are the primary accelerated particles, which then radiate directly or through the production of secondary particles which in turn emit photons.

5a. Leptonic Models

In leptonic models, it is argued that the γ radiation emitted by blazars is produced when the nonthermal synchrotron-emitting electrons Compton-scatter ambient soft photons. The ambient soft photons could be the synchrotron photons produced in the jet [51,54,14], photons radiated by an external accretion disk which are intercepted by the jet [22,23], or disk radiation which has been scattered by surrounding clouds of gas and dust before traversing the jet [71,9]. Jet radiation which has been rescattered by the broad-line-region clouds could also provide a source of external soft photons [30]. Probably all processes contribute at some level. An inhomogeneous model involving nonthermal electron injection followed by a $\gamma\gamma$ pair cascade has also been proposed [10]. It has been argued that due to the different energy densities of the external radiation field, the Compton scattering of external soft photons might produce the bulk of the blazar jet radiation in flat-spectrum radio quasars, whereas the synchrotron self-Compton process might dominate in BL Lac sources. Different beaming factors produced by synchrotron radiation and the Compton scattering of photons produced external to the jet [25] can potentially explain the relative number of radio and γ-ray blazars detected in flux-limited surveys. The unification of different classes of radio-emitting active galactic nuclei (AGNs) on the basis of orientation effects has attracted considerable attention [75], and the *CGRO* results have provided important clues for the unification scenario [69].

5b. Hadronic Models

Models have also been proposed in which accelerated hadronic particles carry the bulk of the energy. They interact with ambient particles through secondary production [2] or with ambient photons [49,50] through photo-meson or photo-pair processes to produce $e^{\pm}$. The pairs initiate a cascade through Compton and synchrotron processes to produce a power-law photon spectrum. Electrons and positrons formed though charged pion decay will be accompanied by energetic neutrino production [73,49]. The detection of a strong neutrino flux from blazar jets would definitively identify hadrons as the primary radiating particles, and a diffuse neutrino background from the superposition of blazar jets has been predicted at levels which are within reach of the next generation of neutrino telescopes [29].

5c. Interpretation of MeV Blazars

The MeV enhancement observed with COMPTEL in the "MeV blazars" (see Section 3b) cannot be simply explained by the Compton-scattering models, since they predict power-law spectra. To account for the bump, Doppler-boosted $e^{\pm}$ annihilation radiation emitted by a relativistically moving electron-positron cloud in a jet has been proposed [67,34]. In [67], the nuclei of blazars eject two plasma fluids consisting of a mildly relativistic e^--proton jet which accounts for the jet formation, hot spots and extended radio lobes, and a relativistic $e^{\pm}$ beam with a bulk Lorentz-factor between 3 and 10 which produces the γ-ray emission. Applying their model to the γ spectrum of PKS 0506-612, they derive a plasma density of roughly $n_{\pm} = 4.4 \times 10^8$ cm^{-3}. Recently, Skibo et al. [72] explain the γ-ray bump from MeV blazars by the same general scenario of Doppler-boosted annihilation radiation, though they propose that the MeV-blazar phenomenon is an orientation effect. The beaming pattern of Doppler-boosted thermal annihilation radiation is much broader than the beaming pattern of IC emission from the scattering of external photons by jet electrons. Gamma emission consisting of these two components would naturally lead to the MeV blazar class of sources when viewed at moderate angles ($\sim 15°$) with respect to the axis to the outflowing jet. At these angles, the annihilation radiation is still observable, while the flux from the scattering of external photons by jet electrons is suppressed at these angles.

6. SUMMARY

The *CGRO* results have dramatically changed our picture of the spectral energy distribution and radiation processes which operate in the jets of AGNs. Gamma-ray observations probe the physics of the inner jet in the region between the central engine and the parsec-scale radio jets which can be imaged directly with VLBI. The EGRET results show that particles are accelerated to > GeV energies by processes in the inner jet that produce roughly power-law photon spectra on time-scales as short as a day or less. This suggests that a Fermi-type mechanism is involved in the particle acceleration, but the dominant particle energization mechanism remains an open question, and the question of whether hadrons or leptons carry the bulk of the accelerated energy which emerges as γ rays could be conclusively answered with neutrino detection of blazar jets. Correlated multiwavelength campaigns have been organized to measure contemporaneous blazar SEDs and to examine correlated variability at different frequencies. The central importance of γ-ray observations for understanding blazar properties has been established by the EGRET observations, and will hopefully continue with future *CGRO* observations and successor γ-ray observatories such as *INTEGRAL* and *GLAST*.

TABLE 1. Blazars Detected at Gamma-Ray Energies

Source ID	Max Flux[c] 10^{-6} cm^{-2} s^{-1} ($E > 100$ MeV)	Min Flux[c]	Photon Spectral Index	z	FS/OV/type/ HP/SL/O/C*	Other Names
0130-171[a]	0.12±0.04	similar[g]		1.022	y . Q	PKS
0202+149	0.23±0.06	< 0.06		1.202	y . Q y n . .	4C+15.05,PKS
0208-512	1.34±0.25	0.16±0.09	1.7±0.1	1.003	y . Q y . . y	PKS
0219+428	0.25±0.06	0.12±0.04		0.444	y . R y . . .	3C 66A
0234+285[a,b]	0.31±0.12	0.10±0.04		1.213	y . R y y . .	4C+28.07,PKS
0235+164	0.65±0.09	0.12±0.04	2.0±0.2	0.94	y . R y ? . .	OD 160,PKS
0336-019	1.19±0.22	< 0.11	2.4±0.3	0.852	y y Q y y . .	CTA 26,OE-63
0420-014	0.50±0.10	0.09±0.05	1.9±0.3	0.92	y . Q y y . .	OA 129,PKS
0440-003	0.86±0.12	<0.06	1.8±0.2	0.844	y y Q y . . .	NRAO 190
0446+112	1.10±0.19	0.06±0.03	1.8±0.3	1.207	y	PKS
0458-020	0.32±0.10	0.10±0.03	2.5±0.4	2.286	y y Q y . . .	PKS,4C-02.19
0521-365	0.21±0.06	similar	2.2±0.4	0.055	y . R y . . .	
0528+134	3.51±0.37	0.32±0.14	2.6±0.1	2.06	y y Q . y y y	OG 147,PKS
0537-441	0.91±0.15	0.18±0.09	2.0±0.2	0.894	y . R y . . .	PKS
0716+714	0.46±0.11	0.09±0.05	1.9±0.2	>0.3	y . R y . . .	
0735+178	0.29±0.10	<0.17	2.5±0.5	0.424	y . R y . . .	OI 158
0804+499[a,h]	0.15±0.06	similar	2.7±0.3	1.43	y . Q y . . .	OJ+508
0805-077[a,h]	0.40±0.13	0.21±0.08	2.4±0.6	1.837	y . Q	PKS
0827+243	0.68±0.15	0.24±0.06	2.2±0.4	2.046	y y Q n . . .	
0829+046	0.17±0.05	similar	2.5±0.5	0.18	y . B y . . .	OJ 49
0836+710	0.33±0.09	< 0.07	2.4±0.2	2.17	y . Q n y . .	4C+71.07
0850-122[b,f]	0.44±0.12	< 0.17		0.566	y . Q	
0917+449	0.33±0.10	0.14±0.04	1.9±0.2	2.18	y . ?	
0954+556	0.47±0.16	0.07±0.03	1.7±0.3	0.909	y . Q y . . .	4C55.17,OK 591
0954+658	0.15±0.03	< 0.04	1.7±0.2	0.368	y . R y y . .	
1101+384	0.27±0.06	0.09±0.04	1.7±0.2	0.031	y . X y y . .	Mrk 421
1156+295	1.63±0.41	< 0.05	2.0±0.5	0.729	y y Q y y . .	4C+29.45
1219+285	0.54±0.14	0.11±0.04	1.9±0.4	0.102	y . B y y . .	ON 231,W Comae
1222+216	0.48±0.15	< 0.10	1.9±0.4	0.435	y . Q . ? y y	4C 21.35
1226+023	0.48±0.12	< 0.09	2.4±0.1	0.158	y y Q n y y y	3C 273
1229-021	0.16±0.04	0.09±0.04	2.5±0.2	1.045	y . Q n . . .	4C -02.55
1253-055	2.67±0.11	0.12±0.07	1.9±0.1	0.538	y y Q y y y y	3C 279
1313-333[a]	0.25±0.06	< 0.14	1.8±0.3	1.21	y y Q	OP-322,PKS
1322-428[a]	0.39±0.14	0.14±0.05	2.9±0.4	0.0007	. . . .	Cen A?
1324+224[b]	0.66±0.22	< 0.07			y . ?	
1331+170[a]	0.11±0.03	similar		2.084	y . Q	OP 151
1406-076	0.98±0. 9	< 0.10	2.0±0.1	1.494	y . Q n . . .	OQ-010,PKS
1424-418	0.57±0.16	< 0.14	2.6±0.4	1.522	y . Q y . . .	PKS
1510-089	0.49±0.18	< 0.28	2.6±0.4	0.361	y y Q y . . .	OR-017,PKS
1517+656	no EGRET	detection			. . X . y .	
1604+159	0.42±0.12	0.12±0.05	2.0±0.5	0.357	y . R y . . .	4C +15.54
1606+106	0.63±0.13	< 0.14	2.2±0.3	1.23	y . Q n y . .	4C +10.45
1611+343	0.69±0.15	< 0.10	2.0±0.2	1.40	y y Q n y . .	OS 319

TABLE 1. Blazars Detected at Gamma-Ray Energies (cont.)

Source ID	Max Flux[c] 10^{-6} cm^{-2} s^{-1} ($E > 100$ MeV)	Min Flux[c]	Photon Spectral Index	z	FS/OV/type/ HP/SL/O/C*	Other Names
1622-253	0.66±0.15	< 0.13	2.3±0.2	0.786	y . Q	PKS
1622-297	3.22±0.34	0.15±0.06	1.9±0.2	0.815	y . Q . . . y	PKS
1633+382	0.99±0.09	0.32±0.10	1.9±0.1	1.81	y . Q . y . .	4C +38.41
1730-130	1.04±0.35	0.18±0.07	2.4±0.3	0.902	y y Q n . . .	NRAO 530
1739+522	0.41±0.10	0.15±0.06	2.2±0.4	1.38	y . Q y . . .	4C +51.37
1741-038[d]	0.36±0.10	< 0.16		1.054	y y Q y . . .	OT-68,PKS
1759-396[a,b]	1.46±0.49	< 0.16			. . . .	
1830-210[a]	0.98±0.25	0.18±0.09	2.7±0.3		y . Q	PKS
1908-201	0.32±0.10	< 0.14	2.5±0.2		y	
1933-400	0.94±0.31	< 0.09	2.4±0.2	0.966	y	
2005-489[e]				0.071	y . X y . . .	PKS
2022-077	0.74±0.13	< 0.16	1.5±0.2		y . Q	NRAO 629
2052-474	0.26±0.07	< 0.09	2.4±0.4	1.489	y . Q n . . .	PKS
2155-304	0.30±0.08	< 0.12	1.7±0.3	0.672	y . R y . . .	PKS
2200+420	0.40±0.12	< 0.18	2.2±0.5	0.069	y . X y y . .	BL Lacertae
2209+236	0.14±0.04	similar	2.8±0.5	1.489	y . Q	
2230+114	0.52±0.15	< 0.15	2.6±0.2	1.037	y y Q y . y y	CTA 102
2251+158	1.24±0.19	0.26±0.14	2.2±0.1	0.859	y y Q y y y y	3C 454.3
2356+196	0.26±0.09	0.13±0.06		1.066	y . Q n . . .	OZ 193,PKS

*FS: Flat radio Spectrum/OV: Optically Violently Variable/ type: Q for quasar; X for XBL; R for RBL)/HP: High optical polarization/ SL: Superluminal source/ O: OSSE detected/ C: COMPTEL detected.
[a]Identification questionable in 2nd catalog or supplement.
[b]Identification proposed in [54].
[c]Highest/lowest flux in an individual viewing period, from [31].
[d]Not in 2nd EGRET catalog because significance is just below cutoff.
[e]Detected above 1 GeV at > 4 σ, but at < 4 σ for $E > 100$ MeV.
[f]Identification confirmed in [15].
[g]max. and min. fluxes do not differ significantly
[h]Identification probably ruled out in [31]

REFERENCES

1. Becker, P. A., and Kafatos, M. 1995, *ApJ* **453**, 83
2. Bednarek, W. 1993, *ApJ* **402**, L29
3. Begelman, M. C., Blandford, R. D., and Rees, M. J. 1984, *RMP* **56**, 255
4. Bezler, M., et al., 1984, *A&A* **136**, 351
5. Bignami, G. F., Fichtel, C. E., Hartman, R. C., and Thompson, D. J. 1979, *ApJ* **232**, 649
6. Bignami, G. F., et al., 1981, *A&A* **93**, 71
7. Blandford, R. D., & Rees, M. J. 1978, in *Pittsburgh Conf. on BL Lac Objects*, ed. A. M. Wolfe (Pittsburgh: Univ. Pittsburgh Press), p. 328
8. Blandford, R. D., and Königl, A. 1979, *ApJ* **232**, 34
9. Blandford, R. D. 1993, in *Compton Gamma-Ray Observatory*, ed. M. Friedlander, N. Gehrels and D. J. Macomb (New York: AIP), 533
10. Blandford, R. D., & Levinson, A. 1995, *ApJ* **441**, 79
11. Bloemen, H., et al., 1995, *A&A* **293**, L1
12. Blom, J. J., et al., 1995, *A&A* **298**, L1
13. Blom, J. J., et al., 1996, *A&AS* **120**, 507
14. Bloom, S. D., and Marscher, A. P. 1996, *ApJ* **461**, 657
15. Bloom, S. D., et al., 1997, *ApJ*, in press
16. Catanese, M., et al., 1997, *ApJ*, in press
17. Chiang, J., et al., 1995, *ApJ* **452**, 156
18. Collmar, W., et al., 1993, in *Compton Gamma-Ray Observatory*, ed. M. Friedlander, N. Gehrels and D. J. Macomb (New York: AIP), 483
19. Collmar, W. 1996, in *Heidelberg Workshop on Gamma-Ray Emitting AGN*, ed. J. G. Kirk, M. Camendzind, C. von Montigny, and S. Wagner, (Heidelberg: MPI für Kernphysik), 9
20. Collmar, W., et al., 1997, *A&A*, in press
21. Collmar, W., et al., 1997, these proceedings
22. Dermer, C. D., Schlickeiser, R., and Mastichiadis, A. 1992, *A&A* **256**, L27.
23. Dermer, C. D., and Schlickeiser, R. 1993, *ApJ* **416**, 484
24. Dermer, C. D., and Gehrels, N. 1995, *ApJ* **447**, 103; (e) 1996, *ApJ* **456**, 412
25. Dermer, C. D., Sturner, S. J., and Schlickeiser, R. 1997, *ApJS* **109**,103
26. Dermer, C. D., in Proc. 2nd INTEGRAL Workshop, St. Malo France. 16-20 Sept. 1996, ESA SP-382, p. 405
27. Elliot, J. L., and Shapiro, S. L. 1974, *ApJ* **192**, L3
28. Fugmann, W. 1988, *A&A* **205**, 86
29. Gaisser, T. K., Halzen, F., and Stanev, T. 1995, *Phys. Reports* **258(3)**, 173
30. Ghisellini, G., and Madau, P. 1996, *MNRAS* **280**, 67
31. Hartman, R. C., et al., 1992, *ApJ* **385**, L1
32. Hartman, R. C., et al., 1993, *ApJ* **407**, L41
33. Hartman, R. C., et al., 1997, in preparation
34. Henri, G., Pelletier, G., and Roland, J. 1993, *ApJ* **404**, L41
35. Hermsen, W., et al., 1993, *A&AS* **97**, 97
36. Hoyle, F., Burbidge, G., and Sargent, W. 1966, *Nature* **209**, 751

37. Hunter, S. D., et al., 1993, *ApJ* **409**, 134
38. Iyudin, A., et al., 1996, *A&A* **311**, L21
39. Johnson, W. N., et al., 1997, *ApJ* **445**, 182
40. Johnson, W. N., et al., 1997, these proceedings
41. Kanbach, G. et al., 1992, IAU Circ. 6431, 17 January 1992
42. Kidger, M. R., and Allan, P. M. 1988, IAU Circ. No. 4595
43. Kniffen, D. A., et al., 1993, *ApJ* **411**, 133
44. Kühr, H., Witzel, A., Pauliny-Toth, I. I. K., and Nauber, U. 1981, *A&AS* **45**, 367
45. Königl, A. 1981, *ApJ* **243**, 700
46. Lichti, G. G., et al., 1995, *A&A*, **298**, 711
47. Macomb, D. J., et al., 1995, *ApJ* **449**, L99
48. Makino, F., and Ohashi, T. 1989, IAU Circ. No. 4736
49. Mannheim, K., and Biermann, P. L. 1992, *A&A* **253**, L21
50. Mannheim, K. 1993, *A&A* **269**, 67
51. Maraschi, L., Ghisellini, G., and Celotti, A. 1992, *ApJ* **397**, L5
52. Maraschi, L. et al., 1994, *ApJ* **435**, L91
53. Maraschi, L., Ghisellini, G., and Celotti, A. 1994, in IAU Symp. 159, *Multiwavelength Continuum Emission of AGN*, ed. T. J. L. Courvoisier and A. Blecha (Dordrecht: Kluwer), 233
54. Marscher, A. P., and Bloom, S. D. 1994, in *The Second Compton Symposium*, ed. C. E. Fichtel, N. Gehrels and J. P. Norris (New York: AIP), 573
55. Mattox, J. R., et al., 1993, *ApJ* **410**, 609
56. Mattox, J. R., et al., 1996, *ApJ* **461**, 396
57. Mattox, J. R., et al., 1997, *ApJ* **476**, 692
58. McNaron-Brown, K., et al., 1995, *ApJ* **451**, 575
59. McNaron-Brown, K., et al., 1997, *ApJ* **474**, L85
60. Mücke, A., et al., 1996, *A&AS* **120**, 541
61. Mukherjee, R., et al., 1996, *ApJ* **470**, 831
62. Mukherjee, R., et al., 1997, *ApJ*, to be published
63. Pohl, M. et al., 1995, *A&A* **303**, 383
64. Punch, M., et al., 1992, *Nature* **358**, 477
65. Quinn, J., et al., 1996, *ApJ* **456**, L83
66. Robson, E. I., Smith, M. G., Aycock, J., and Walter, D. M. 1988, IAU Circ. No. 4556
67. Roland, J., and Hermsen, W. 1995, *A&A* **297**, L9
68. Sambruna, R. M., Maraschi, L., & Urry, C. M. 1996, *ApJ* **463**, 444
69. Schlickeiser, R., and Dermer, C. D. 1995, *A&A* **300** L29
70. Shrader, C. R., and Wehrle, A. E. 1997, these proceedings
71. Sikora, M., Begelman, M. C., and Rees, M. J. 1994, *ApJ* **421**, 153
72. Skibo, J. G., Dermer, C. D., and Schlickeiser, R. 1997, *ApJ* **483**, 56
73. Stecker, F. W., Done, C., Salamon, M. H., and Sommers, P. 1991, *PRL* **66**, 2697; (e) 1992, **69**, 2738
74. Swanenburg, B. N., et al., 1978, *Nature* **275**, 298
75. Urry, C. M., and Padovani, P. 1995, *PASP* **107**, 803

76. Vestrand, W. T., et al., 1995, *ApJ* **454**, L93
77. von Montigny, C., et al., 1995, *ApJ* **440**, 525
78. von Montigny, C., et al., 1993, *A&AS* **97**, 101
79. Wagner, S. J., et al., 1995, *ApJ* **454**, L97
80. Wehrle, A., et al., 1997, *ApJ*, submitted
81. Williams, O. R., et al., 1995, *A&A* **297**, L21
82. Woltjer, L. 1966, *ApJ* **146**, 597

Multiwavelength Campaigns

C.R. Shrader* and A.E. Wehrle[†]

*Laboratory for High–Energy Astrophysics, NASA Goddard Space Flight Center
[†]Infrared Processing and Analysis Center, Jet Propulsion Laboratory and California Institute of Technology

Abstract. The first six years of the Compton Gamma Ray Observatory mission has led to significant advances in the study of active galactic nuclei, as shown in various contributions to these proceedings. However, the given the inherent nature of AGN – broadband, non–thermal emission which exhibits variability at all wavelengths over many time scales – empirical insight into the multi-epoch, multi–wavelength behavior is essential before the underlying physics can begin to be deciphered. This is particularly true for the blazar subclass of AGN detected by EGRET, but it also pertains to the lower–luminosity radio–quiet AGN detected with OSSE. Fortunately, this situation was recognized early on and a number of coordinated multiwavelength studies have been successfully carried out both as carefully preplanned monitoring campaigns and as targets of opportunity. Also fortunate has been the overlap with the first 6 years of the CGRO mission of several space–based observatories covering portions of the X-ray, ultraviolet, and infrared spectrum inaccessible from the ground, accompanied by recent advances in ground–based gamma–ray astronomy data acquisition and analysis techniques. We review accounts of multiwavelength studies involving CGRO from the published literature and from new material presented at this conference, noting some of the key results as well as some remaining ambiguities.

INTRODUCTION

Studies of Active Galactic Nuclei since their discovery in the 1960s have led to the basic classifications of radio quiet AGN, and radio loud AGN. In each case a massive black hole fed by accretion is widely believed to comprise the "central engine", but complications arise from several effects. The primary radiation can be reprocessed to higher energies, for example by inverse Compton scattering of synchrotron radiation, or to lower energies, such as by scattering and thermal radiation by dust. Broadly, two classes of gamma–ray emitting AGN have emerged, defined by their redshift and luminosity distributions [1]. These are the Seyferts with $z \sim 0.1$ and with 50–150 keV luminosities of order $10^{41} - 10^{44} ergs/s$ and radio loud quasars or "blazars" with redshifts as high as ~ 2.5 and luminosities as high as $10^{49} ergs/s$ (times a beaming factor). Jets are widely believed to dominate the radiation we detect from blazars. Collimated outflows are present in some Seyferts. Although the jet production mechanism in AGN is unclear, it is becoming widely accepted that they are associated with accretion disks in a variety of Galactic objects where jets are found. For example, there is very direct evidence for such an association [2] thus if a common physics underlies the AGN jet production it is reasonable to speculate on an analogous disk–jet connection. For each AGN subclass, analogous unifying schemes involving viewing geometry or the orientation of an obscuring dust torus, beaming angle orientations in the case of radio loud objects have been proposed and have become generally accepted, although it does not appear that they can account for all of the differences observed.

The Compton Gamma–Ray Observatory (CGRO) has greatly enhanced our current understanding of several AGN subclasses. However, CGRO or any single observatory covering a limited portion of the 25 decades over which AGN emit cannot form an unambiguous picture. Furthermore, in view of their often rapid, often high–amplitude variations at all wavelengths, coordinated observing campaigns offer the only means of establishing a complete empirical understanding. These are difficult to organize, whether they are prescheduled or promptly invoked target–of–opportunity (ToO) campaigns, and they tend to be "expensive" in terms of observing time and the level of individual efforts involved in planning and execution. Fortunately, however, the importance of these campaigns was recognized early in the mission, and a substantial number of successful efforts have been completed.

In this review, we will attempt to highlight some of the contributions to AGN multiwavelength campaigns involving CGRO. It is in this sense that this article is intended to be complimentary to other reviews in this volume [3,4]. Both classes of gamma–ray AGN will be included, although we place a greater emphasis on the blazar subclass.

REVIEW OF SOME INDIVIDUAL CAMPAIGNS

1. Seyfert Galaxies and Radio Quiet Quasars

High–energy gamma rays from the lower–luminosity, generally (but not strictly) radio–quiet Seyfert galaxies and quasars have not been detected with CGRO. The band from hard X–rays to soft gamma–rays is important for distinguishing between thermal and non–thermal scenarios, and in constraining the parameters of those models. Making this distinction requires > 100 keV observations with higher signal-to-noise than is currently possible.

However, the model fitting required to make such a characterization involves a joint fit with the soft X–ray continuum. The Compton reflection component, for example, is indicated by the presence of a fluorescent Fe Kα line at 6.4 keV and the shape of the hard X–ray continuum. As with all AGN subclasses, the sources are highly variable – on time scales as short as days and possibly hours – thus characterization of the continuum energy distribution requires contemporaneous measurements.

NGC 4151: This is the brightest Seyfert galaxy in the OSSE bandpass and it has been the most extensively observed. Its spectrum exhibits a sharp softening at about 100 keV. Caution must be taken however in generalizing this result to the broader class of Seyfert 1 galaxies NGC 4151 has some other properties which suggest it may be rather unique. It has been classified as a "Seyfert 1.5", and its X-ray spectral index in the 2–20 keV region is variable and can deviate significantly from the "canonical" value of $\alpha \simeq 0.7$. Campaigns utilizing Ginga and OSSE during Phase 1 of the CGRO mission were carried out, however, the observations were separated in time by about 1 month [5]. An extensive multiwavelength campaign which included CGRO OSSE as well as ROSAT and ASCA observations was conducted in December of 1993 [6,7]. Low amplitude variability (15%) was seen in the 50–100 keV range, accompanied by larger amplitude variations (30%) in soft X-rays, but without any significant correlation. The difference between the soft and hard X-ray light curves may indicates a steepening of the continuum as the X- ray intensity increases. The broad–band X-ray continuum was found to have two break energies at about 4 keV and 90 keV. The combined X- and gamma-ray continuum is fitted by a thermal Comptonization model [8] with $kT = 49$ keV and $\alpha = 0.64$ [7]. The hard spectrum, relative to other Seyferts, implies that the hot Comptonized plasma is photon starved, such as might be the case for a patchy corona above the surface of an accretion disk. Also, since this source is sufficiently bright, it has been monitored using BATSE in earth–occultation mode, thus we have by far the most complete long–term hard–X-ray flux history of any AGN [9].

IC 4329A: This is the second brightest Seyfert in the OSSE bandpass, and as such it has also been the subject of extensive study as well. Combined Ginga and OSSE spectra have been constructed and simultaneous observations with ROSAT and OSSE have been carried out [10]. The combined ROSAT–OSSE spectrum is well approximated by an absorbed powerlaw with photon index $\Gamma \sim 1$ plus a strong Compton reflection component. The most notable feature is the apparent lack of the ~ 100 keV turnover seen in NGC 4151. The data seem to only weakly constrain the presence of a high–energy cutoff. A similar analysis by the same authors using non–simultaneous Ginga plus OSSE data suggest that the cutoff energy is greater than 250 keV, and it is noted in that case the X-ray intensity at 2 keV is consistent with that measured by ROSAT. Additional analysis of the IC 4349A broad–band X-ray spectrum by either optically thin thermal Comptonization, with a scattering medium characterized by $kT \sim 100$ keV, or alternatively by non–thermal models [5]. In the non–thermal case, the lack of a detectable 511-keV annihilation line suggests that electron injection occurs at low energies, with $\gamma \lesssim 10$. More recently, this object has been observed simultaneously with RXTE and OSSE.

Analysis of the resulting data is presently ongoing [4].

NGC 4945: This Seyfert 2 galaxy was detected by OSSE during Viewing Period 403 (1994.9), and is apparently the brightest object of its class at 100 keV. A composite X-ray spectrum constructed from (non–simultaneous) ASCA, Ginga, and OSSE measurements indicates heavily absorbed emission below 10 keV, with $N_H \sim 10^{24} cm^{-2}$ [11]. At higher energies however, it is a very bright source with a flux $\sim 10^{-10} ergs/cm^2/s$ in the 100–500 keV band. The intrinsic X- to gamma–ray spectral (energy) index is $\alpha \sim 0.9$ which is consistent with Seyfert 1 spectra and thus supports the unified model for low–luminosity AGN. This further supported by the measurement of a Fe Kα line at 6.4 keV, consistent with reflection from cold material. These results prompted a multiwavelength campaign during Cycle 5 utilizing RXTE/ASCA and CGRO, the results of which are forthcoming [12].

MCG 8-11-11: Simultaneous ASCA and OSSE observations of this object have recently been carried out [13]. The soft X–ray spectrum is described by a hard powerlaw, photon index $\Gamma \sim 1.7$, with iron line emission at 6.4 keV. The OSSE data have a softer powerlaw, $\Gamma \simeq 3$ with an apparent cutoff at 300 keV and a strong Compton reflection component.

3C 120: This "radio–loud" Seyfert 1 galaxy was recently observed simultaneously with ASCA and OSSE [14]. The OSSE spectrum was found to be consistent with an extrapolation of the continuum model fitted to the ASCA data. An intense iron line is apparently seen, but the line strength resulting from the fitting is anticorrelated with the strength of a presumed hard (reflection or jet) component. The OSSE measurements could not resolve this issue due to signal–to–noise limitations – the source was in a lower intensity state at the time of this multiwavelength campaign than during past observations [15]. Grandi et al [14] conclude that a Seyfert–like emission mechanism with line and reflection components is most likely present. However, jet mechanism cannot be definitively ruled out with the present data set. Future, higher signal-to-noise OSSE plus ASCA observations could plausibly resolve this issue.

2. Blazars

Blazars are characterized by strong, rapid variability over all wavelengths at which they have been studied. Specific correlations between wavelength bands are not well established, particularly at high–energies. Subsequent to the discovery of the 1991 gamma–ray flare in 3C 279 a number of attempts to orchestrate multi–wavelength campaigns have been made. One of the first targets was, not surprisingly, 3C 279 itself during Phase 2 of the mission, approximately 3 weeks of EGRET, COMPTEL and OSSE time were allocated to such a campaign [16]. In addition to pre–orchestrated campaigns, attempts have been made at ad hoc observing campaigns, i.e. targets of opportunity

(ToOs), triggered by gamma–ray or non–gamma–ray blazar activity.

Also, a number of observers, ground–based optical, millimeter and radio have developed observing programs incorporating known gamma– ray blazars into their monitoring programs – often increasing the density of coverage in time bracketing the CGRO observations. In particular, the onset of flaring activity at millimeter wavelengths [17] and intra–day photometric variability [18,19] have been suggested to be associated with blazar intensities observed in the gamma rays.

In addition to broad–band continuum and variability studies, follow up work incorporating VLBI imaging has revealed important information on the relationship between gamma–ray "alarming" and subsequent jet propagation. Three blazars for example have exhibited new superluminal blobs which were ejected more or less simultaneously with strong gamma–ray detections; 3C 279 [20], NRAO 190 [21], PKS 0528+134 [22]. Also see Piner and Kingham [23] who find a possible connection between the enhanced gamma–ray activity in 1611+343 and the ejection of new VLBI components. The probability of chance association is difficult to estimate ex post facto. Some specific campaigns are described here – also refer to Table 1.

3C 279: This is the first gamma–ray quasar detected with EGRET and has been the subject of the largest number of coordinated multiwavelength campaigns. The July 1991 flare event [24] was of course a surprise to everyone, however, given the nature of 3C 279 it was already included in a number of monitoring campaigns at optical and radio wavelengths. Thus, Hartman et al [25] were able to assemble, in a post–facto manner, a multiwavelength spectrum and light curves. They were able to identify a low–amplitude photometric (R–band) flare which peaked within a few days of the gamma–ray flare and obtain Synchrotron Self Compton (SSC) and External Compton Scattering (ECS) spectral fits to the data. In December of 1992 and extending into January of 1993, the EGRET and COMPTEL FoV was centered approximately on 3C 279, in hopes of catching a similar event. Accompanying observations using ROSAT, IUE, as well as optical, near–IR, millimeter and radio are described in [16]. In this case however, the source was relatively faint. Consequently, only a few conclusions regarding low–state behavior could be made: most notably the ratio of the gamma–ray luminosity to the bolometric luminosity decreased by a factor of order 10 from outburst to quiescence. In early 1996, 3 weeks of observing time was devoted to the 3C 279/3C 273 field. In this case, dramatic activity in both gamma rays and X–rays was recorded [20,26] (Figure 1; Figure 2). The tight correlation between the X-ray and gamma–ray light curves imposes significant constraints on models, as the X–ray photon density establishes the depth of the gamma–ray photosphere. Specifically, by requiring that $\tau_{\gamma\gamma} < 1$ Wehrle et al [20] show that the required beaming factor is $\delta > 6.3$ for 1–GeV photons. A more recent campaign was carried out in 1996/97 with an 8–week of CGRO Z–Axis. Again, as

Figure 1 – Radio-to-γ-ray energy distribution of 3C 279 in pre-flare (open dots) and flaring state (filled dots) in January-February 1996. The UV, optical and near-IR data have been corrected for Galactic extinction. The slope of the ASCA spectrum ($\alpha_\nu = 0.7$) has been normalized at the RXTE point closest in time. The EGRET best-fit power-law spectra referring to the 16-30 January (pre-flare) and 4-6 February (flare) periods are shown, normalized at 0.4 GeV. For comparison, the SEDs in June 1991 (stars) [25] and in Dec 1992 - Jan 93 (crosses) are also shown [16]. Errors have been shown only when they are bigger than the symbol size.

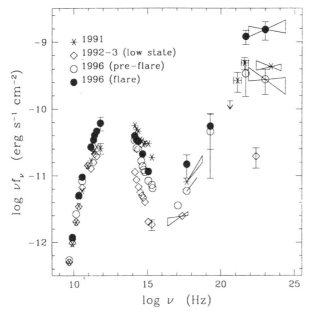

Figure 2 Combined X-ray and gamma-ray light curve for a high-amplitude flare in 3C 279 during early 1996. The combined light curves impose constraints on the beaming factor for gamma-rays as described in the text.

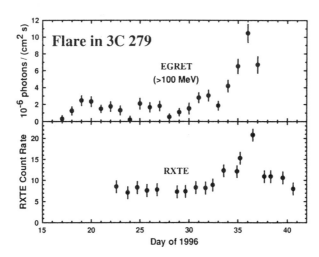

333

in 1992/93 nature was not so cooperative, and activity was relatively minimal.

PKS 0528+108: This blazar has been detected consistently with EGRET throughout the mission. Its fortuitous location in the Crab/Geminga anticenter region has led to coverage in nearly 30 separate viewing periods. Multiwavelength campaigns have been conducted [27], however, it is a difficult source for the optical band and inaccessible in the UV due to its low galactic latitude; A_V is estimated to be $\sim 3-5$. Simultaneous ASCA and EGRET observations during a low state reveal a single X– to gamma–ray SSC component breaking at about 45 MeV [28]. The gamma–ray intensity at that time was the faintest thus far detected and the spectral index was quite steep ($\Gamma \simeq 2.7$). Those authors conclude, based on their analysis of the radio to gamma–ray spectral energy distribution, that although SSC models cannot be ruled out, external Compton scenarios seem to be favored for this object.

PKS 1622–297: This object was the subject of a ToO campaign triggered by gamma–ray variability in EGRET off– axis coverage. It flared to $17 \pm 3 \times 10^6 ph/cm^2/s$, placing it for a brief period among the most luminous quasars measured with EGRET; $L = 2.9 \times 10^{49} ergs/s$ if isotropic [29]. A seven–fold increase in about 7 hours was reported. The magnitude of this flare argues strongly for beaming. Without beaming, the large flux change over such a short period of time is inconsistent with the Elliot–Shapiro condition if it is assumed that the accretion powers the gamma rays. A minimum Doppler factor of $\delta \sim 8$ is suggested based on the inferred pair–production opacity. Photometric monitoring began after the EGRET detection, indicating that the source was about three times brighter than its "quiescent" state. The source was detected in the ultraviolet where some variability was seen, and follow–up VLBA observations were made – this can be combined with later–epoch VLBA studies to monitor propagation of jet components which may be associated with the gamma–ray event.

Mkn 421: This low redshift BL Lac object, despite multiple detections with EGRET, would be considered unremarkable on the basis if its MeV–GeV properties alone – however, it is one of several extragalactic sources detected in the TeV gamma–ray domain. Macomb et al [30] presented results from the Whipple Observatory air Cherenkov experiment during which the flux above 250 GeV increased by nearly an order of magnitude over a 2–day period during 1994. Contemporaneous observations by ASCA showed the X–ray flux to be in a very high state. In addition the first ever simultaneous or nearly simultaneous observations at GeV gamma–ray, UV, IR, mm, and radio energies for this nearest BL Lac object were obtained. While the GeV gamma–ray flux increased slightly, there is little evidence for variability comparable to that seen at TeV and X–ray energies. Other wavelengths show even less variability. Taken collectively, these multiwavelength variability studies provide important constraints on the emission mechanisms. During another TeV flare

Figure 3 Multiepoch, multiwavelength spectral energy distribution for Mkn 421 (provided by Jim Buckley of the Whipple collaboration)). This prototypical XBL is generally a weak, slowly variable EGRET source, but is bright and dramatically variable in the TeV band.

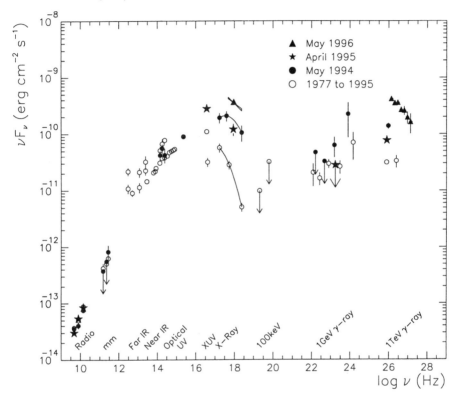

of this source, about one year later, it was shown that the 2– 10 keV X–ray light curve obtained with ASCA exhibits a similar morphology and amplitude to the TeV light curve (at least on the decline side of the flare), arguing strongly that a common electron population is responsible for X–rays and TeV gamma–rays [31]. Recently, even more complete coverage spanning TeV to radio has been presented [32,33] (Figure 3).

Stecker de Jager and Salomon [34] have suggested that the "TeV blazars", among which Mkn 421 is the prototype, are likely to be associated with the X–ray selected BL Lac subpopulation rather than among the radio selected BL Lacs (RBLs). Both TeV blazars currently known are XBLs, whereas 12 of the 14 BL Lacs detected by EGRET are RBLs. Basic physical arguments by those authors involving synchrotron radiation and Compton scattering suggest why this is the case. From the relative variability amplitudes seen in Mkn 421 in the synchrotron and Compton components, they argue that the Thompson – Klein–Nishina transition is constrained to 10 eV, from which a Lorentz factor for the scattering electrons of $\gamma \sim 10^{4.5}$ is derived.

Mkn 501: This low–redshift ($z = 0.033$) BL Lac object is another TeV blazar unlike Mkn 421 however, it has not been detected with EGRET. In 1997 it underwent a major TeV flare which prompted a CGRO ToO declaration. Although no detection was made with EGRET, a strong detection was obtained with OSSE up to 200 keV, and a detection was made with BATSE as well. The photon spectral index was $\Gamma \simeq 2.1$, indicating that this emission was the high–energy tail of the Synchrotron component which normally does not extend to such high energies, nor does it normally encompass such a large fraction of the luminosity [35]. Evidence of correlated variability is seen in the Whipple, OSSE and RXTE/ ASM bands. Short term optical correlations are not evident, but the source seems to have been in a higher photometric state than the previous month by $\sim 10\%$. The mean power output of Mkn 501 appears in these data to peak in the 2–100 keV range. If this is interpreted as an extension of the Synchrotron component, it has the broadest high–energy extension of any such component seen to date.

CTA 102: This gamma–ray blazar has been detected at multiple epochs with EGRET [36] and possibly with COMPTEL [37] – a source of 10–30 MeV emission defines an error box that contains both CTA 102 and 3C 545.3. The MeV emission is likely due to a contribution from both quasars. This suggests that the power–law spectrum measured by EGRET above 50 MeV flattens at lower energies [37]. It has been the subject of coordinated X– and gamma–ray campaigns designed to test the SSC and other emission models, and has been the target of VLBI campaigns as well. It exhibits low–frequency variability and behavior suggestive of expansion with a high apparent transverse velocity [38].

PKS 1406–076: This blazar was detected with high significance by EGRET during a 4-day optical flare [39] (Figure 4). There were enough gamma-ray photons to derive a light curve down to ∼daily time scales. The combined optical and high-energy gamma-ray light curves clearly indicated that the optical preceded the gamma-rays by ∼ 1-day. This supports the external Compton scenarios and would argue against hadronic acceleration models, e.g. [40]. In contrast to flares in 3C 279, the amplitudes of the gamma ray and optical flares over "base" levels were approximately the same, a factor of 3. However, one should regard the overall picture as ambiguous. Other cases are less clear as to which component peaks earlier; as pointed out at this conference by Wagner [19], variability on even shorter time scales occurs in the optical, and undersampling, or too short a timebase of gamma-ray data, could lead to erroneous conclusions.

PKS 0420-014: This flat spectrum radio quasar has been identified as a variable gamma-ray source in the 100 MeV to 5 GeV band. Wagner et al [41] have performed extensive photometric monitoring of this source finding a number of pronounced flares. The highest gamma-ray flux was recorded during the occurrence of the most pronounced optical flux seen up to that time. Additionally, VLBI measurements clearly reveal superluminal knots propagating along a curved trajectory, with both the core component and the knot exhibiting variable intensity. A model involving knots of enhanced particle density propagating along a helical trajectory in a rotation jet provide a reasonable fit to the historical photometric light curve for this source [41].

PKS 2155-304: This BL Lac object has been extensively studied in many wavebands except gamma-rays, revealing rapid variability (down to ∼hour time scales) and correlations between X-ray and UV bands. It is however, only weakly detected at high energies [42]. Those authors measure a spectral index of $\Gamma = 1.7 \pm 0.2$, suggesting that it may be an exceptionally hard gamma-ray source, which combined with its low redshift, make it a viable source for possible TeV detection. A multiwavelength campaign was attempted during Cycle 5. Results are still forthcoming [56], but the source was not found to be in a bright gamma-ray state.

OJ 287: This BL Lac object has been the focus of intensive study at optical and radio wavelengths, and it has been suggested that its outburst history reflects a stable periodicity of about 12 years [43]. Although it resembles the prototypical BL Lac detected in high-energy gamma-rays, it was not detected in the Phase-1 (1991.3-1992.9) all-sky survey, nor in Phase 3 (1993.7-1994.8). Late in 1994 however, it underwent a major flare in the visual band, and was observed as an EGRET target of opportunity [44]. A marginal detection was made, strengthening the optical-gamma-ray correlation and extending this idea into the BL Lac domain.

Figure 4. 1406-076: Comparison of differential light curves in the optical and gamma-ray regimes (from [39]). The gamma-ray data are shown with a grouping of 10 counts each (filled squares). The optical data has an arbitrary scaling (open squares).

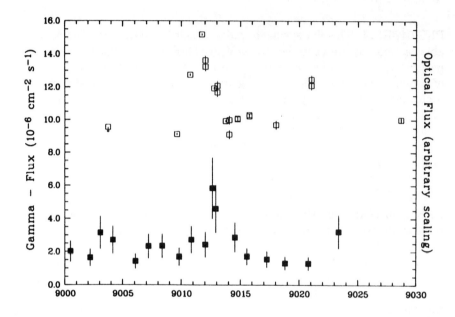

Table 1 Summary of multiwavelength coverage for various blazar AGN campaigns involving CGRO. This is not intended to be an exhaustive list, but reflects some highlights as discussed in the text.

Source ID	VP	TeV	EGRET	COMPTEL	OSSE	RXTE	Ginga	ASCA	ROSAT	IUE	HST	Optical	Infrared	ISO	mm	Radio	VLBI	References
3C 279	204,205,206,5 11, 511.5		x		x	x	x	x	x	x	x	x	x	x	x	x		16,20,24,25,25,26
PKS 0528+134	213,337		x	x	x			x	x			x			x	x	x	27,28,54
PKS 1622-297	423.5		x	x	x							x			x		x	29
Mkn 421	326,418	x	x					x		x		x	x		x	x	x	30,31,32,33,34
Mkn 501	617	x			x	x												35
CTA 102	19,26,28,507		x	x	x												x	36,37,38
PKS 2155-304	404,513		x	x		x		x	x	x	x						x	42,55
PKS 1406-076	205,206,207		x									x					x	41
3C 273	3,204,205,206		x	x	x					x		x	x		x	x	x	53
NRAO 190	337		x				x					x			x	x	x	21,57

339

WHAT HAVE WE LEARNED?

The multiwavelength campaigns conducted during the first six years of the CGRO mission, have contributed significantly to AGN research, but many issues remain unresolved. For example, both SSC and ECS models are still viable for most EGRET blazars, however, the parameter space needs to be stretched to its reasonable limits to match multiwavelength data of the best-studied blazars (3C 279, PKS 0528+134). The question of thermal versus non-thermal emission models in radio quiet low-luminosity AGN still remains. Here we will attempt to summarize some of what has been learned.

The relative infrequency at which EGRET detected major gamma ray flares could imply a low duty cycle, i.e., sources are likely in a high state less than 10-20% of the time; see, for example, the Second EGRET Catalog [36,45]. Optical light curves of blazars show that some have several (2-20) flares per year while radio light curves show flares occurring once a year or less often. Comparison of optical light curves, typically sampled on daily timescales, to infrequently sampled EGRET data (observed at most twice per year) cannot unambiguously identify which, if any, optical flares are associated with gamma-ray detections. At least one carefully monitored blazar, PKS 1406-076 [39] showed that an optical flare occurred a day or two before a gamma-ray detection; at least six other well- monitored blazars show no such unique association. The radio data, on the other hand, shows an excellent correlation between mm flares and gamma ray detections in the sense that gamma ray detections are much more likely to occur when the blazars are in the rising part of a mm flare, rather than in the declining or between-flares states. Valtaoja and Teräsranta [17] conclude "that gamma-rays are produced by the synchrotron self-Compton mechanism in the shocked regions of the jet, with both the relativistic electrons and the seed photons originating in the same shock far away from the core and the accretion disk." Counter examples exist however, as pointed out by Bower et al [57] based on their extensive radio-submillimeter monitoring of NRAO 530.

Gamma ray detections tend to occur when a blazar is in a relatively active state, e.g. when the source is undergoing a radio flare [17]. At first, this appeared to be a selection effect because EGRET sources are usually identified by finding the brightest, i.e., most active, flat-spectrum radio source in the error ellipse. However, as more data become available, a clearer picture has begun to emerge which supports the connection between non-gamma-ray activity and the probability of an EGRET detection.

Evidence is mounting that the emission from blazars is relativistically beamed throughout the entire observed electromagnetic spectrum. The amount of beaming depends on the doppler factor and orientation to line of sight of the jet, both of which may vary. The simultaneity of the x-ray and gamma-ray flare in 3C 279 in February 1996 [20] leads us to believe that both flares are produced in the same region of the jet, probably a shocked area less than one light–day (modulo a beaming factor) in extent. This opens up the possibility of estimating the gamma–ray duty cycle of a single object by monitoring its X– ray flux for a long period, say, of daily observations for one year.

VLBI monitoring of three blazars showed that new superluminal blobs were ejected more or less simultaneously with strong gamma–ray detections (3C 279; [20], NRAO 190, [57], PKS 0528+134 [22]. The probability of chance association is difficult to estimate ex post facto, especially since the evolution timescales between radio and gamma– ray are evidently very different. The question of time–lags between these bands therefore remains open. The moving shock which eventually manifests in a new superluminal blob might have produced a gamma ray flare when it was near the core.

The hard X–ray – soft gamma–ray spectra of Seyfert galactic nuclei do not reveal any evidence of annihilation line emission. There seems to be no real differences between Seyfert 1 and 2 in this range, thus the observations support the unification scenarios, e.g. [4]. The Compton reflection model is generally supported by the combined X– and gamma–ray measurements. The spectra are generally softer above 100 keV than had been anticipated prior to the launch of CGRO – this impacts not only emission models but the diffuse X–ray background problem.

FUTURE PROSPECTS

About 60 blazar AGN have been detected with EGRET during the first six years of the CGRO mission. Due to the limited lifetime of that instrument, this number is not likely to grow significantly, however it is a remarkable legacy of the CGRO mission. A proposed successor to EGRET, the Gamma Ray Large area Space Telescope (GLAST) [46] will offer several orders of magnitude improvement in sensitivity, one order of magnitude improvement in source positioning accuracy, broader energy coverage (extending to 100 GeV) and a wider field of view. In some proposed design schemes, GLAST would include an optical telescope, coaligned with the main detector axis, which would allow us to unambiguously resolve the nature of any optical–gamma–ray flux variability correlations. It is expected that the number of cataloged gamma–ray AGN would likely increased to $\sim 10^3$ leading to a much improved statistical picture of blazar properties and the diffuse gamma–ray background problem. The wide field of view will lead to the discovery of variability or transient phenomena, and thus facilitate target–of–opportunity type multiwavelength studies. Furthermore, with accompanying advances in ground–based gamma–ray observing techniques, the gap between EGRET and current ground–based technology will be effectively bridged. This could potentially impact our understanding of the cosmological IR–UV photon density, which in turn impacts our understanding of the evolution of the universe out to quasar distances.

A major goal for future instruments is to cover, at energies from hard X–rays through MeV gamma–rays, the flux–density history of a statistically significant set of blazars, with regular sampling at intervals of order one day. This kind of unbroken coverage is essential for distinguishing, for instance, a 'high state' from a 'flare'. A definitive test of internal versus external models for the seed photons for gamma–ray emission is broad–band variability – the models predict very different evolution of the broad–band spectrum [47]. Present data are not quite sufficient for this purpose. In addition, observed flares in many blazars are probably a superposition of more than one outburst, such as shock models for radio flares in 3C 279 [49]. Complete time sampling is necessary to verify that we are actually seeing discrete events in the high–energy light curves.

The future of low–energy gamma–ray astronomy, notably the forthcoming INTEGRAL mission [50,51] (also see [52]), offer great promise for the study of both gamma–ray AGN subclasses. A 10-fold increase over OSSE in the 400 keV spectral region could unambiguously resolve the question of thermal – non–thermal emission by detecting the high–energy continuum component and possibly 511–keV line emission associated with the non–thermal scenarios. For blazars, more sensitive low–energy gamma–ray instrumentation will provide more stringent tests of beaming from gamma–gamma transparency and through the Elliot–Shapiro relationship. Unification models for blazars and radio galaxies may be tested through detection of the putative hard (McV) tails of radio galaxies resulting from scattering [53].

REFERENCES

1. Dermer, C.D. & Gehrels, N., 1995, ApJ, 447 103.
2. Livio, M., et al, 1997.
3. Hartman, R.C., Collmar, W., von Montigny, C., & Dermer, C.D., 1997, these proceedings.
4. Johnson, W.N., Zdziarski, A., & Madejski, G., 1997, these proceedings.
5. Zdziarski, A. et al. 1994, proc. Second Compton Symposium, ed. Fichtel, Gehrels & Norris, AIP CP–304.
6. Edelson, R.A., et al, 1993, ApJ, 470, 364.
7. Warwick, R.S., et al, 1996, ApJ, 470, 349.
8. Titarchuk, L., & Mastichadis, A., 1994, ApJ433, 33.
9. Parsons, A., 1997, these proceedings.
10. Madejski, G.M., et al, 1995, ApJ 438, 672.
11. Done, C., Madejski, G.M., & Smith, D.A., 1996, ApJ, 463, L63.
12. Madejski, G., 1997, (private comm).
13. Grandi, P., et al 1997, (these proceedings).
14. Grandi, P., Sambruno, R.M., Maraschi, L., Matt, G. Urry, C.M., & Mushotzky, R.F., 1997, preprint.

15. Johnson, W.N., et al, 1994, proc. "Second Compton Symposium", ed. Fichtel, Gehrels & Norris, AIP CP-304.
16. Maraschi, L., et al, 1994, ApJ, 435, L91.
17. Valtaoja, E., & Teräsrantra, H., 1996, A&ASup, 120, 491.
18. Wagner, S.J., 1996, A&ASup, 120, 495.
19. Wagner, S.J., 1997, these proceedings.
20. Wehrle, A.E., et al, 1997, in preparation.
21. Marchenko, S.G., et al, 1997, these proceedings.
22. Pohl, M., Reich, W., Schlickeiser, R., Reich, P., & Ungerechts, H., 1996, A&ASup, 120, 529.
23. Piner, B.G., & Kingham, K., 1997, ApJ, 479, 684.
24. Hartman, R.C., et al, 1992, ApJ, 385, L1.
25. Hartman, R.C., et al, 1996, ApJ, 461, 698.
26. Wehrle, A.E., et al, 1997, (these proceedings).
27. Mukerhjee, R., et al, 1996, ApJ, 470, 831.
28. Sambruno, R.M., et al, 1996, ApJ, 474, 639.
29. Mattox, J.R., et al, 1997, ApJ, 476, 692.
30. Macomb, D.J., et al. 1995, ApJ, 449, 99.
31. Takahashi, T., et al, 1996, ApJ, 470, 89.
32. Buckley, J., et al, 1996, ApJ, 472, L9.
33. Schubnell, M., et al 1997, these proceedings.
34. Stecker, F.W., de Jager, O.C., and Salomon, M.H., 1996, ApJ, 473, L75.
35. Catanese, M., et al 1997, these proceedings
36. Thompson, D.J., et al, 1995, ApJSup, 102, 259.
37. Blom, J.J., et al, 1995, A&A, 295, 330.
38. Rantakyroe, F.T., Baath, L.B., Dallacasa, D. Jones, D.L., & Wehrle, A.E., 1996, A&A, 310, 66.
39. Wagner, S.J., et al, 1995a, ApJ, 454, L97.
40. Mannheim, K., 1993, Phys. Rev. D, 48, 2408.
41. Wagner, S.J., et al, 1995b, A&A, 298, 688.
42. Vestrand, W.T., Stacy, J.G., & Sreekumar, P., 1995, ApJ, 454, L93.
43. Valtoen, M., Lehto, H., Kokkonen, K., & Mikkola, S. proc. "Blazar Continuum Variability", ed. R.H. Miller, J.R. Webb & J.R. Noble, ASP CF-110.
44. Shrader, C.R., Webb, J.R., & Hartman, R.C., 1996, A&ASup, 120, 599.
45. Thompson, D.J., et al, 1996, ApJSup, 107, 227.
46. Gehrels, N., 1997 these proceedings
47. Marscher, A.P., & Travis, J.P.,, 1996, A&A, 120, 537.
48. Hughes, P.A., & Duncan, G.C., Bull. AAS., 184, 41.14.
49. Hughes, P.A., Aller, H.D. and Aller, M.F. 1990, in "Parsec-scale radio jets" ed. J.A. Zensus and T.J. Pearson, Cambridge Univ. Press, Cambridge.
50. Ubertini, 1997, these proceedings
51. Teegarden, B., 1997, these proceedings
52. Kurfess J., 1997, these proceedings.
53. Dermer, C.D., 1996, presented at "Low/Medium Energy Gamma-Ray Astrophysics Workshop".
54. Lichti, G.G., et al, 1995, A&A, 298, 711.
55. McNaron-Brown, K., et al. 1995, ApJ, 451, 575.
56. Vestrand, T., et al, 1997, these proceedings.
57. Bower, G., Backer, D.C., Wright, M., Aller, H.D., Aller, M.F., 1997, ApJ, 484, 118.
58. McGlynn, T.A., et al, 1997, ApJ, 481, 625.

The Extragalactic Diffuse Gamma-Ray Emission

P. Sreekumar*[1], F.W. Stecker* and S.C. Kappadath**

*NASA/Goddard Space Flight Center, Greenbelt, MD 20771
**University of New Hampshire, Durham, NH 03824

Abstract. The all-sky surveys in γ-rays by the imaging Compton telescope (COMPTEL) and the Energetic Gamma Ray Experiment Telescope (EGRET) on board the Compton Gamma Ray Observatory for the first time allows detailed studies of the extragalactic diffuse emission at γ-ray energies greater 1 MeV. A preliminary analysis of COMPTEL data indicates a significant decrease in the level of the derived cosmic diffuse emission from previous estimates in the 1–30 MeV range, with no evidence for an MeV-excess, at least not at the levels claimed previously. The 1–30 MeV flux measurements are compatible with power-law extrapolation from lower and higher energies. These new results indicate that the possible contributions to the extragalactic emission from processes that could explain the MeV-excess, such as matter-antimatter annihilation, is significantly reduced. At high energies (> 30 MeV), the extragalactic emission is well described by a power law photon spectrum with an index of $-(2.10\pm0.03)$ in the 30 MeV to 100 GeV energy range. No large scale spatial anisotropy or changes in the energy spectrum are observed in the deduced extragalactic emission. The most likely explanation for the origin of this extragalactic γ-ray emission above 10 MeV, is that it arises primarily from unresolved γ-ray-emitting blazars. The consistency of the average γ-ray blazar spectrum with the derived extragalactic diffuse spectrum strongly argues in favor of such an origin. The extension of the power law spectrum to 100 GeV implies the average emission from γ-ray blazars extends to 100 GeV.

INTRODUCTION

The extragalactic diffuse emission at γ-ray energies has interesting cosmological implications since the bulk of these photons suffer little or no attenuation during their propagation from the site of origin [55]. The first all-sky survey in low and medium energy γ-rays (1 MeV–30 MeV) has been carried out by COMPTEL and at higher energies (>30 MeV) by EGRET on board the

---

[1] USRA Research Scientist

Compton Observatory satellite (CGRO). Improved sensitivity, low instrumental background and a large field of view of these instruments have resulted in significantly improved measurements of the extragalactic γ-ray emission and our understanding of its origin. The measurement of the extragalactic γ-ray emission is made difficult by the very low intensity of the expected emission and the lack of a spatial or temporal signature to separate the cosmic signal from other radiation. The measured diffuse emission along any line-of-sight, could be composed of a Galactic component arising from cosmic-ray interactions with the local interstellar gas and radiation, an instrumental background, and an extragalactic component. In the 1 to 10 MeV band, diffuse studies have been traditionally complicated by difficulties in fully accounting for the instrumental background. At higher energies (E>10 MeV), the results are subject to difficulties in accurately accounting for the Galactic diffuse emission.

Before the launch of CGRO, the γ-ray spectrometer flown aboard three Apollo flights [68] and numerous balloon-borne experiments [72] [48], showed the presence of a 'bump' -like feature in the few MeV range that was in excess of the extrapolated hard X-ray continuum. Although the measured intensities varied widely among the numerous experiments, most showed some level of an MeV-excess. It was recognized as early as 1972, that cosmic-ray induced radioactivity is the most dominant source of background in the MeV energy range [19]. Recent results from the COMPTEL [28] [29] and the Solar Maximum Mission (SMM) [70] experiments indicate no evidence for an MeV-bump, at least not at the levels previously reported.

At higher energies (>35 MeV), the SAS-2 satellite [16] provided the first clear evidence for the existence of an extragalactic γ-ray component. This emission, seen as an excess over the the strong Galactic diffuse radiation, was uncorrelated with the column density of matter and was therefore interpreted as being extragalactic in origin [17]. The recent EGRET results [52] extend the high energy measurement to an unprecedented ∼100 GeV. The emission above 30 MeV is well represented by a single power-law of index –2.1 and shows no significant departure from isotropy.

The origin of the extragalactic γ-ray emission has proved to be an elusive goal for theorists over the years. Prior to CGRO, even the question as to whether the radiation is from a truly diffuse process or is, in fact, the superposition of radiation from a large number of extragalactic sources has been difficult to answer. At MeV energies, theoretical efforts to explain the emission, were constrained by the need to explain the MeV excess. The absence of a source class at these energies, whose spectrum displayed this characteristic signature, made this particularly difficult. At higher energies, the SAS-2 and COS-B experiments together, detected one extragalactic source, 3C273, suggesting active galactic nuclei (AGN) as a viable source class that could contribute to the diffuse background [3] [30]. More recently, the detection of >50 γ-ray blazars by EGRET and their spectral properties, have been used to improve theoretical calculations of the diffuse γ-ray emission from blazars.

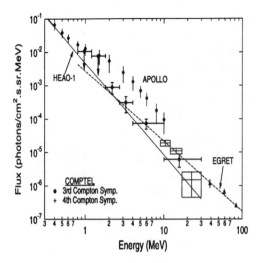

FIGURE 1. COMPTEL results (Kappadath et al. 1996, 1997): The HEAO-1 data are from Kinzer et al. (1997) and the Apollo measurements are from Trombka et al. (1977)

Here we review the current observational and theoretical understanding of the diffuse extragalactic γ-ray emission above 1 MeV. Recent analysis results from the instruments on the Compton Gamma Ray Observatory are summarized. For a more detailed discussion on the COMPTEL and EGRET results, see Kappadath et al. (1996, 1997) [28] [29] and Sreekumar et al. (1997) [52] respectively. Finally, the implications of these new findings on the origin of the extragalactic diffuse emission, are discussed.

RECENT RESULTS FROM CGRO

COMPTEL results (1–30 MeV)

The COMPTEL diffuse emission spectrum is constructed from high-latitude observations, by first subtracting the instrument background and then attributing the residual flux to the extragalactic diffuse radiation. The instrumental background is in general composed of 'prompt' and 'long-lived' components.

The prompt background is instantaneously produced by proton and neutron interactions in the spacecraft. Hence the prompt background refers to the component that modulates with the instantaneous local cosmic-ray flux. The prompt background is seen to vary linearly with the veto-scalar-rates (charge-particle shield rates). Assuming that zero veto-scalar-rate corresponds to zero cosmic ray flux, a linearly extrapolation is used to compute the prompt

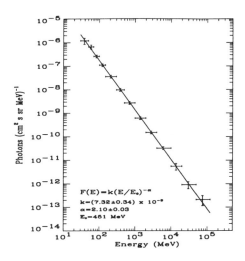

FIGURE 2. EGRET results: The extragalactic diffuse emission spectrum >30 MeV (Sreekumar et al. 1997)

background contribution.

The long-lived background events are due to de-excitation photons from activated radioactive isotopes with long half-lives ($\tau_{1/2} > 30$ sec). Their decay rate is not directly related to the instantaneous cosmic-ray flux because of the long half-lives. The long-lived background isotopes are identified by their characteristic decay lines in the individual detector spectra. Monte Carlo simulation of the isotope decay are used to determine the COMPTEL detector response. The measured energy spectrum is used to determine the absolute contribution of each of the long-lived background isotopes.

The diffuse flux measured by COMPTEL, refers to the total γ-ray flux in the field-of-view (~ 1.5 sr) derived from high-latitude observations. It is important to point out that it includes flux contribution from the Galactic diffuse emission and γ-ray point sources in the field of view. It is important to point out that the COMPTEL results below ~ 9 MeV are still preliminary. The 2–9 MeV flux is significantly lower than pre-COMPTEL measurements. They show no evidence for a MeV-excess, at least not at the levels reported previously. Only upper-limits are claimed below ~ 2 MeV. Recent improvements in the 9–30 MeV spectrum [29], are shown in figure 1. The 9–30 MeV flux is consistent with the previous measurements (mostly upper-limits) and also compatible with the extrapolation of the EGRET spectrum [52] to lower energies.

In examining the isotropic nature of the diffuse radiation, a simple comparison shows that the measured 9–30 MeV spectrum from the Virgo and South Galactic Pole regions are consistent with each other [29].

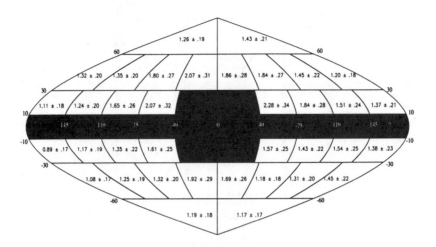

FIGURE 3. The distribution of extragalactic flux (E>100 MeV). The shaded region containing the Galactic plane is excluded due to extreme dominance of the Galactic emission and the region centered on the Galactic Center and extending towards ±30° in latitude is excluded due to difficulties in modeling all of the Galactic emission (Sreekumar et al. 1997).

EGRET results (30 MeV – 100 GeV)

In the EGRET energy range, the primary source of error in estimating the extragalactic emission arises from uncertainties in the Galactic diffuse emission model. In order to derive the extragalactic emission without being sensitive to the Galactic model used, the following approach is adopted for the EGRET data. The observed emission ($I_{observed}$) is assumed to be made up of a Galactic ($I_{galactic}$) and an extragalactic component ($I_{extragalactic}$).

$$I_{observed}(l,b,E) = I_{extragalactic} + B \times I_{galactic}(l,b,E)$$

The slope, 'B' of a straight line fit to a plot of observed emission versus the Galactic model gives an independent measure for the normalization of the input Galactic model calculation. The primary processes that produce the observed Galactic diffuse γ-rays are: cosmic-ray nucleons interacting with nucleons in the interstellar gas, bremsstrahlung by cosmic-ray electron, and inverse Compton interaction of cosmic-ray electrons with ambient low-energy interstellar photons [57]. The Galactic diffuse emission falls of rapidly at higher latitudes, making high-latitude observations, ideally suited to study the extragalactic emission. The possible contribution to the Galactic diffuse emission from unresolved point sources such as pulsars, is uncertain with estimates ranging from a few percent to almost 100% depending on the choice of many model parameters such as the birth properties of pulsars [1]. The evidence for a pion 'bump' (from neutral pion decay) in the Galactic diffuse

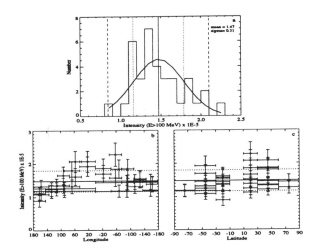

FIGURE 4. The all-sky distribution of integral flux (E>100 MeV) in units of photons $(cm^2\text{-s-sr})^{-1}$ (Sreekumar et al. 1997). (a) Gaussian fit to the flux histogram showing the mean (solid), 1σ (dotted) and 2σ (dashed) confidence intervals; (b)&(c) Intensity values plotted over longitude and latitude respectively. The longitude distribution shows slightly larger intensity values within $\pm 60°$

spectrum [25] can be used to set an upper limit for the contribution from unresolved sources at <50%.

Preliminary results on the extragalactic spectrum above 30 MeV were reported by Kniffen et al. (1996) [34], using the Galactic diffuse model of Hunter et al. (1997) [25] and its high-latitude extension discussed by Sreekumar et al. (1997) [52]. Earlier, Osborne, Wolfendale and Zhang (1994) [44] had shown that the spectrum derived from EGRET is well represented by a power-law of index (2.11 ± 0.05). Even though this was within errors of the previous best estimate of $(2.35^{+0.4}_{-0.3})$ from the SAS-2 experiment [67], the EGRET measurements clearly demonstrated the existence of a well defined, harder power-law spectrum. Independent analysis by Chen, Dwyer and Kaaret (1996) [6] also yielded a spectral index of (2.15 ± 0.06) and an integral flux above 100 MeV of $(1.24\pm0.06)\times 10^{-5}$ photons $cm^{-2}s^{-1}$ sr^{-1} for E>100 MeV. More recently, Sreekumar et al. (1997) [52] using $\sim$ the first 4 years of EGRET observations, carried out a more detailed and careful analysis of the intensity and spectral shape in different regions covering the full sky and for the first time extended the spectrum up to an unprecedented 100 GeV. The differential photon spectrum of the extragalactic emission averaged over the sky is well fit

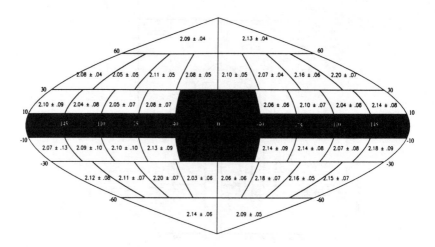

FIGURE 5. The distribution of spectral indices assuming a single power-law fit to the extragalactic spectrum above 30 MeV (Sreekumar et al. 1997).

by a power law with an index of $-(2.10\pm0.03)$. The spectrum was determined using data from 30 MeV to 10 GeV; however as shown in Figure 2, the differential photon flux in the 10 to 20 GeV, 20 to 50 GeV, and 50 to 120 GeV energy intervals are also consistent with extrapolation of the single power law spectrum. The integrated flux from 30 to 100 MeV is $(4.26\pm0.14)\times10^{-5}$ photons cm^{-2}s^{-1} sr^{-1} and that above 100 MeV is $(1.45\pm0.05)\times10^{-5}$ photons cm^{-2}s^{-1} sr^{-1}.

With the availability of high quality data from CGRO, one can for the first time, address the isotropic nature of the extragalactic emission. The derived integral fluxes in 36 independent regions of the sky are shown in Figure 3, with the values ranging from a low of $(0.89\pm0.17)\times10^{-5}$ photons cm^{-2}s^{-1} sr^{-1} to a high of $(2.28\pm0.34)\times10^{-5}$ photons cm^{-2}s^{-1} sr^{-1} [52]. The shaded area in the figure represents the region where the Galactic diffuse emission is dominant and hence not included in the analysis. Figure 4a shows a histogram of integral flux values, overlayed by a Gaussian fit assuming equal measurement errors. The Gaussian fit yields a mean flux above 100 MeV of 1.47×10^{-5} photons cm^{-2}s^{-1} sr^{-1} with a standard deviation of 0.33×10^{-5} photons cm^{-2}s^{-1} sr^{-1} consistent with the all-sky calculation. To further examine the spatial distribution, Figure 4b and 4c shows the same data plotted against Galactic longitude and latitude respectively. No significant deviation from uniformity is observed in the latitude distribution; however as a function of longitude, there appears to be an enhancement in the derived intensities towards l~20°, primarily due to a few regions on the boundary of the excluded Galactic center region. This could arise from unaccounted extended

Galactic diffuse emission. If one excludes the inner regions of the Galaxy ($|r| < 60°$; r=angle made with the direction of the Galactic center), the mean flux is 1.36×10^{-5} photons cm^{-2}s^{-1} sr^{-1} above 100 MeV. Thus the measured extragalactic flux is consistent within errors to a uniform sky distribution in regions outside the inner Galaxy.

Figure 5 shows the distribution of the derived spectral indices from fits to data from 36 independent regions of the sky. The power law indices vary from $-(2.04\pm0.08)$ to $-(2.20\pm0.07)$ and show no systematic deviations from the value of $-(2.10\pm0.03)$, derived from the all-sky analysis.

ORIGIN OF THE EXTRAGALACTIC γ-RAY EMISSION

A large number of possible origins for the extragalactic diffuse γ-ray emission have been proposed over the years (e.g. [56] [18] [20]). The key to understanding the origin of the extragalactic emission may lie in the realization

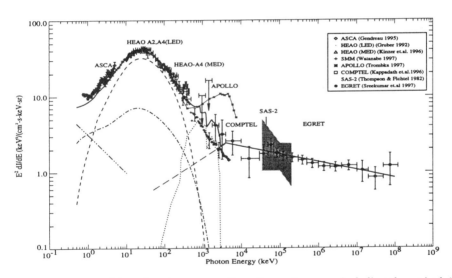

FIGURE 6. Multiwavelength spectrum from X-rays to γ-rays including the revised 1σ upper limits from the Apollo experiment (Trombka 1997). The estimated contribution from Seyfert 1 (dot-dashed), and Seyfert 2 (dashed) are from the model of Zdziarski(1996); steep-spectrum quasar contribution (dot-dot-dashed) is taken from Chen et al. (1997); Type Ia supernovae (dot) is from The, Leising and Clayton (1993). the average blazar spectrum breaks around 4 MeV (McNaron-Brown et al. 1995) to a power law with an index of ~ -1.7. The thick solid line indicates the sum of all the components.

that it may be composed of a number of different components having different origins and that these components in turn may dominate the observed emission only in specific energy ranges (Figure 6). For example, the putative matter-antimatter component would only be important in the energy range between ~ 1 MeV and ~ 100 MeV. The unresolved source models must also be broken up into different source classes and energy ranges. We discuss separately the two distinct possibilities, a truly diffuse origin or the superposition of unresolved sources.

Diffuse origin

The physics of the truly diffuse physical processes which might be involved have been discussed previously [53] [55] [56]. Among those that could potentially contribute to the emission, particularly between 1 and 100 MeV, the most significant could be matter-antimatter annihilation in a globally baryon-symmetric cosmology based on grand unified theories and early universe physics [54] [58]. In the light of recent COMPTEL and SMM data, it appears more likely that the contribution from such a process (if any) is significantly lower than previously reported. Around ~ 1 MeV, it has been suggested that a significant fraction of the emission could be made up of γ-ray production in type Ia and type II supernovae primarily via lines from the decay of ^{56}Ni $\rightarrow ^{56}$Co $\rightarrow ^{56}$Fe and from ^{26}Al, ^{44}Ti, and ^{60}Co [66] [70]. However, uncertainties associated with the existing observational data in the 1–30 MeV makes it difficult to estimate the contributions from the different proposed processes.

Diffuse processes that have been proposed to explain the high-energy emission include , primordial black hole evaporation [46] [24], million solar-mass black holes which collapsed at high redshift ($z\sim 100$) [22], and some exotic source proposals, such as annihilation of supersymmetric particles [50] [47] [27]. All of these theories predict continuum or line contributions that are unobservable above 30 MeV with EGRET.

Unresolved Point Sources

Normal Galaxies

Models based on discrete source contributions have considered a variety of source classes. Normal galaxies might at first appear to be a reasonable possibility for the origin of the diffuse radiation since they are known to emit γ-rays and to do so to very high γ-ray energies [51] [25]. Previous estimates [35] [65] [36] [17] have shown that the intensity above 100 MeV expected from normal galaxies is only about 3% to 10% of what is observed. Further and perhaps even more significant, the energy spectrum of the Galactic diffuse γ-rays is significantly different from the extragalactic diffuse spectrum, being harder at low energies (<1000 MeV) and considerably steeper in the 1 to 50 GeV region [15]. Dar and Shaviv (1995) [12] have suggested that emission arises from cosmic ray interaction with intergalactic gas in groups and clusters of galaxies. Although the authors claim that this proposed explanation leads to a higher intensity level, it is still in marked disagreement with the measured energy spectrum [62]. Finally, cosmic ray electrons and protons that leak out into the intergalactic space could upscatter the CMB to γ-ray energies. Chi and Wolfendale (1989) [10] and Wdowczyk and Wolfendale (1990) [71] have shown these contributions to be not significant.

Active Galaxies: Seyferts

It has been postulated for over two decades by a large number of authors that active galactic nuclei (AGN) might be the source of the extragalactic γ-ray diffuse emission (e.g. [3] [30]). However, prior to the launch of CGRO, only a handful of these objects were claimed to be detected at energies above 1 MeV. With the detection of a large number of AGN with instruments on board CGRO, it was natural to consider that the fainter unresolved sources could collectively make up the observed extragalactic γ-ray emission. Using data from the OSSE experiment, it has been shown that the average Seyfert galaxy spectrum, is characterized by an exponentially falling continuum (e-folding energy $\sim$100 keV) together with a Compton reflection component which contributes mainly in the 10-50 keV band [26] [74]. However, none have been detected above 1 MeV [39] [37]. The generally accepted reason for this is that the photon fields within AGN are so intense that pair production prevents the higher energy γ-radiation from escaping [64] [73] [13]. Thus, although Seyfert galaxies may provide a substantial contribution to the X-ray background, and may also provide the dominant contribution to the high energy neutrino background [59] [60], these objects are not expected to be important sources of high energy γ-radiation. Another sub-class of AGNs, the 'MeV-blazars', have been shown to exhibit peak power at MeV energies [4] [5]. Comastri, Girolamo and Setti [11] have discussed the possible contribution to the diffuse emission from these objects. Since at present, only $\sim$1% of the γ-ray blazars show an MeV excess, observationally, it suggests that MeV-blazars may not be a significant contributor to the extragalactic diffuse emission.

Active Galaxies: Blazars

While radio-quiet AGN do not qualify as significant medium or high energy γ-ray sources, the EGRET observations showed the presence of a class of AGN, characterized by strong time variability at many wavelengths including γ-rays, flat radio spectrum and often exhibit strong polarization and (or) superluminal motion. These are the BL Lac objects and the flat spectrum radio quasars (FSRQ) which together have been classified as 'blazars'. The EGRET team has now reported the detection of over 50 blazars [41] [43]. Of these six have been detected at COMPTEL energies as well. It is believed that most of these objects generally have jets beaming their emission toward us. It is natural to assume that, since the jets are optically thin to high energy γ-rays and since beaming in the jets can produce a large enhancement in the apparent γ-ray luminosity relative to unbeamed components, therefore, the γ-ray emission from these objects is probably beamed and originates in the jets. This hypothesis is supported by the rapid γ-ray time variability observed in many blazars (for eg. 3C279 [33]).

McNaron-Brown et al. (1995) [40] examined the multiwavelength spectra of the six blazars detected by OSSE, COMPTEL and EGRET. Using this data set, one can derive an 'average' blazar spectrum characterized by a broken power-law with a break at ∼4 MeV ([52]). The derived average differential photon spectral index below 4 MeV is determined to be about –1.7 and above 4 MeV to be equal to the average spectral index of EGRET detected blazars (∼–2.1). However below 10 MeV, the blazar contribution alone is not sufficient to explain the observed extragalactic emission. Thus, the true origin of the emission in this energy range is not yet fully understood. We note that while the EGRET spectrum represents the extragalactic diffuse radiation, the COMPTEL spectrum refers to the total γ-ray flux from high-latitude observations (including contribution from the Galactic diffuse and γ-ray point sources in the FOV of ∼1.5 sr).

One of the more important pieces of evidence in favor of the blazar origin of the high-energy portion of the diffuse spectrum is the spectrum itself. Both the spectrum reported here and the average spectrum of blazars may be well represented by a power law in photon energy. The spectral index determined here for the diffuse radiation is $-(2.10\pm0.03)$, and the average spectral index of the observed blazars is $-(2.15\pm0.04)$ [43] These two numbers are clearly in good agreement. A standard cosmological integration of a power law in energy yields the same functional form and slope. Considering the new, well determined γ-ray spectrum, this argues strongly that the bulk of the observed extragalactic γ-ray emission can be explained as ordinating from unresolved blazars.

In order to estimate the intensity of the diffuse radiation from blazars, knowledge of the evolution function is needed, as well as the intensity distribution of the blazars. Two approaches have been utilized to determine the γ-ray evolution. One way is to assume that the evolution is similar to that at other wavelengths. The other alternative is to deduce the evolution from the γ-ray data itself and hence have a solution that depends only on the γ-ray results. The advantage of the former is that, if the assumption of a common evolution is correct, an estimate with less uncertainty is obtained. The positive aspect of the latter is that there is no assumption of this kind, but the uncertainty in the calculated results is relatively large because of the small γ-ray blazar sample.

Several authors have estimated the contribution from blazars using the first approach described above [45] [61] [49] [15] [63] [31] where one accepts the proposition that the evolution function determined from the radio data may be applied to the γ-ray-emitting blazars. Furthermore, the recent work of Mukherjee et al. (1997) [43] shows that there is general agreement within uncertainties between the radio and high-energy γ-ray redshift distributions of both types of blazars, i.e. flat-spectrum radio quasars and BL Lacs. However, a word of caution seems appropriate, since it should be pointed out that the degree and nature of the correlation between the radio and γ-ray emission is still being debated [45] [42] [38]. Most of these calculations show that all or most of the observed mission can be explained as originating from unresolved blazars. Stecker and Salamon (1996) [63] instead of assuming a mean γ-ray spectral index, used the distribution of the observed spectral indices in their calculation. This introduces a curvature in the spectrum; the steep spectrum sources contributing more at lower energies (<500 MeV) and the flat-spectrum sources dominating the emission at higher energies. This is consistent with the curvature in the spectrum reported by Osborne, Wolfendale and Zhang (1994) [44]. However recent instrumental corrections to the 2-4 GeV energy range [14] and the extension of the spectrum to 100 GeV, have weakened any evidence for such a curvature in the spectrum.

Chiang et al. (1995) [8] used the second approach by using the γ-ray blazar data to deduce the evolution function. They used the V/Vmax approach in the context of pure luminosity evolution to show that there was indeed evolution of the high-energy γ-ray emitting blazars and found that the implied evolution is similar to that seen at other wavelengths. However, recent work of Chiang and Mukherjee (1997) [9] argue that an improved calculation of the lower end of the de-evolved luminosity function indicates that only $\sim$25% of the observed emission is made up of unresolved blazars. As stated before, the limited sample of detected γ-ray blazars results in larger uncertainty in the above calculation. Furthermore, Stecker and Salamon (1996) [63] and Kazanas and Perlman (1997) [31] concluded that EGRET has preferentially detected those blazars that were in 'flaring' states. Thus there is a clear need for a much improved evaluation of the true γ-ray luminosity function. The expected detection of a large number of sources using a future more sensitive γ-ray instrument such as GLAST, could make this possible.

If the hypothesis that the general diffuse radiation is the sum of the emission of blazars is accepted, there is an interesting corollary. The spectrum of the measured extragalactic emission implies the average energy spectra of blazars extend to at least 50 GeV and maybe up to 100 GeV without a significant change in slope. Most of the measured spectra of individual blazars only extend to several GeV and none extend above 10 GeV, simply because the intensity is too weak to have a significant number of photons to measure. Intergalactic absorption does not have much effect at this energy except for blazars at relatively large redshift, and, in any case would steepen the spectrum at high energies. Hence, the continuation of the single power law diffuse spectrum up to 100 GeV strongly suggests that the source spectrum also continues without a major change in spectral slope to at least 100 GeV. This conclusion, in turn, implies that the spectrum of the parent relativistic particles in blazars that produce the γ-rays remains hard to even higher energies.

SUMMARY

CGRO observations have lead to a significant advancement in our understanding of the extragalactic γ-ray emission. The recent COMPTEL measurements and SMM results have shown that a significant part of the MeV-excess previously reported around 1–10 MeV, is due to instrumental background events. The new measurements in the 1–30 MeV range are compatible with power-law extrapolations from lower and higher energies. The COMPTEL results on the 9–30 MeV flux represents the first significant detection in this energy range. Above 30 MeV, the EGRET observations have extended the high-energy measurement to an unprecedented $\sim$100 GeV. The 30 MeV to 100 GeV spectrum is well described by a single power-law with spectral index of -2.1. No large scale spatial anisotropy or changes in the energy spectrum is observed in the deduced extragalactic spectrum above 30 MeV. The bulk of the extragalactic emission above 10 MeV appears to arise from unresolved blazars and is supported by the consistency in shape between the two spectra. However, below 10 MeV, the exact nature of the emission is not well understood, partly due to the large uncertainties in the measured diffuse spectrum in this energy range. The average blazar spectrum suggest that only about 50% of the measured emission in the 1–10 MeV range, could arise from blazars. Current observational limits do not provide tight constraints on contributions from additional source classes, or from other truly diffuse processes, making this an important area of investigation for the next generation γ-ray experiments.

ACKNOWLEDGMENTS

The authors wish to thank James Ryan (UNH) for helpful discussions.

REFERENCES

1. Bailes, M., & Kniffen, D. A., *ApJ* **391**, 659 (1992).
2. Bertsch, D. L., et al. *ApJ* **416**, 587 (1993).

3. Bignami, G. F., et al. *ApJ* **232**, 649 (1979).
4. Bloemen, H., et al. *A&A* **293**, L1 (1995).
5. Blom, J. J., et al. *A&A* **298**, L33 (1995).
6. Chen, A., Dwyer, J., & Kaaret, P., *ApJ* **463**, 169 (1996).
7. Chen, L. W., Fabian, A. C., & Gendreau, K. C., *MNRAS* **285**, 449 (1997).
8. Chiang, J., Fichtel, C. E., von Montigny, C., Nolan, P. L., & Petrosian, V., 1995, *ApJ* **452**, 156 (1995).
9. Chiang, J.& Mukherjee, R., *ApJ* (submitted) (1997).
10. Chi, X., & Wolfendale, A. W., *J. Phys. G: Nucl. Phys.* **15**, 1509 (1989).
11. Comastri, A., Girolamo, T.Di., & Setti, G., *A&AS* **120**, 627 (1996).
12. Dar, A., & Shaviv, N., *PRL* **75**, 3052 (1995).
13. Done, C. and Fabian, A.C. *MNRAS* **240**, 81 (1989).
14. Esposito, J. A., et al. (in preparation) (1997).
15. Erlykin, A. D., & Wolfendale, A. W., *J. Phys. G.* **21**, 1149 (1995).
16. Fichtel, C. E., et al. *ApJ* **198**, 163 (1975).
17. Fichtel, C. E., Simpson, G. A., and Thompson, D. J., *ApJ* **222**, 833 (1978).
18. Fichtel, C. E., & Trombka, J. I., *Gamma Ray Astrophysics, New Insights into the Universe* NASA SP-453 (Washington: GPO) (1981).
19. Fishman, G., *ApJ* **171**, 163 (1972).
20. Gehrels, N., & Cheung, C., *Current Perspectives in High Energy Astrophysics* NASA Ref. Pub. **1391**, 113 (1996).
21. Gendreau, K. C., *PhD dissertation*, MIT, Cambridge, MA (1995).
22. Gnedin, N. Y., & Ostriker, J. P., *ApJ* **400**, 1 (1992).
23. Gruber, D. E. *The X-Ray Background*, Cambridge Univ. Press, Cambridge, eds. X. Barcons, & A. C. Fabian, 44 (1992).
24. Hawking, S. W. *Scientific American* **236**, 34 (1977).
25. Hunter, S. D., et al. *ApJ* **481**, 205 (1997).
26. Johnson, W.N., et al. 2nd *Compton Symposium*, AIP Conf. Proc. **304**, 515 (1994).
27. Kamionkowski, M., *The Gamma Ray Sky with Compton GRO and SIGMA*, eds. M. Signore, P. Salati, G. Vedrenne, Kluwer, Dordrecht, 113 (1995).
28. Kappadath, S. C., et al. *A&AS* **120**, 619 (1996).
29. Kappadath, S. C., et al. *these proceedings*
30. Kazanas, D., & Protheroe, J. P., *Nature* **302**, 228 (1983).
31. Kazanas, D., & Perlman, E., *ApJ* **476**, 7 (1997).
32. Kinzer, R. L., et al. *ApJ* **475**, 361 (1997).
33. Kniffen, D.A., et al. *ApJ* **411**, 133 (1993).
34. Kniffen, D.A., et al. *A&AS* **120**, 615 (1996).
35. Kraushaar, W. L., et al. it ApJ **177**, 341 (1972).
36. Lichti, G. G., Bignami, G. F., & Paul, J. A., *Astrophys. Space Sci.* **56**, 403 (1978).
37. Lin, Y.C., et al. *ApJ* **416**, L53 (1993).
38. Mattox, J. R., Schachter, J., Molnar, L., Hartman, R. C., & Patnaik, A. R., *ApJ* **481**, 95 (1997).
39. Maisack, M., Mannheim, K., & Collmar, W., *A&A* **298**, 400 (1995).

40. McNaron-Brown, K., et al. *ApJ* **451**, 575 (1995).
41. von Montigny, C., et al. *ApJ* **440**, 525 (1995).
42. Mücke, A., et al. *A&A* **320**, 33 (1997).
43. Mukherjee, R., et al. *ApJ* (in press) (1997).
44. Osborne, J. L., Wolfendale, A. W., & Zhang, L., *J. Phys. G.* **20**, 1089 (1994).
45. Padovani, P., Ghisellini, G., Fabian, A. C., & Celotti, A., *MNRAS* **260**, L21 (1993).
46. Page, D. N., & Hawking, S. W., *ApJ* **206**, 1 (1976).
47. Rudaz, S., & Stecker, F. W., *ApJ* **368**, 40 (1991).
48. Schönfelder, V., et al. *ApJ* **240**, 350 (1980).
49. Setti, G., & Woltjer, L., *ApJS* **92**, 629 (1994).
50. Silk, J., & Srednicki, M., *PRL* **53**, 624 (1984).
51. Sreekumar, P. et al. *ApJ* **400**, L67 (1992).
52. Sreekumar, P. et al. *ApJ*, (submitted), (1997).
53. Stecker, F.W. *Cosmic Gamma Rays*, Mono Book.Co., Baltimore, (1971).
54. Stecker, F. W., Morgan, D. L., Bredekamp, J., *PRL* **27**, 1469 (1971).
55. Stecker, F.W. *Nature Phys. Sci.* **241**, 74 (1973).
56. Stecker, F.W. *Origin of Cosmic Rays*, ed. J.L. Osborne and A. W. Wolfendale, Reidel Pub. Co., Dordrecht, p.267, (1975).
57. Stecker, F. W. *Apj* **212**, 60 (1977).
58. Stecker, F. W. *Nuc.Phys.B* **252**, 25 (1985).
59. Stecker, F.W., Done, C., Salamon, M.H. & Sommers, P., *PRL* **66**, 2697 (1991).
60. Stecker, F.W., Done, C., Salamon, M.H. & Sommers, P., *PRL* **69**, 2738 (1992).
61. Stecker, F. W., Salamon, M. H., & Malkan, M., *ApJ* **410**, L71 (1993).
62. Stecker, F. W., & Salamon, M. H., *PRL* **76**, 3878 (1996).
63. Stecker, F. W., & Salamon, M. H., *ApJ* **464**, 600 (1996).
64. Svensson, R., *Radiation Hydrodynamics in Stars and CompactObjects*, ed. D. Mihalas and K.-H. A. Winkler, Springer-Verlag, Berlin, p. 325 (1986).
65. Strong, A. W., Wolfendale, A. W., and Worrall, D. M., *J. Phys. A: Math Gen.* **9**, 1553 (1976).
66. The, L.-S., Leising, M. D., & Clayton, D. D., *ApJ* **403**, 32 (1993).
67. Thompson, D. J., & Fichtel, C. E., *A&A* **109**, 352 (1982).
68. Trombka, J. I., et al. *ApJ* **212**, 925 (1977).
69. Trombka, J. I., (private communications) (1997).
70. Watanabe, K., et al. *these proceedings* (1997).
71. Wdowczyk, J., & Wolfendale, A.W., *ApJ* **349**, 35 (1990).
72. White, R. S., et al. *ApJ* **218**, 920 (1977).
73. Zdziarski, A.A. *ApJ* **305**, 45 (1986).
74. Zdziarski, A.A., Johnson, N.W., Done, C., Smith, D., &McNaron-Brown, K., *ApJ* **438**, L63 (1995).
75. Zdziarski, A.A., *MNRAS* **281**, L9 (1996).

HIGH ENERGY GAMMA RAY ASTRONOMY

VHE and UHE Gamma-ray Astronomy in the EGRET Era

Trevor C. Weekes[1], Felix Aharonian[2], David J. Fegan[3], and Tadashi Kifune[4]

[1] *Harvard-Smithsonian Center for Astrophysics, Whipple Observatory, P.O.Box 97, Amado, AZ 85645-0097, USA*
[2] *Max-Planck-Institut fur Kernphysik, Heidelberg, Germany*
[3] *Physics Department, University College, Dublin, Ireland*
[4] *Institute for Cosmic Ray Research, Tokyo, Japan*

Abstract. Although the basic techniques used in Very High and Ultra High Energy γ-ray astronomy have been available for more than 30 years, it is only in the past decade (mostly in the EGRET era) that the disciplines have become viable with the production of verifiable results. Well-established steady sources include the Crab Nebula, PSR1706-44 and Vela as well as the AGNs, Markarian 421 and 501. The early successes of EGRET were a catalyst and accelerated the development of advanced atmospheric Čerenkov telescopes. The useful symbiosis that has developed between space and ground-based γ-ray observations is the major subject of this review.

INTRODUCTION

The dramatic advance in γ-ray sensitivity in the 30 MeV to 10 GeV energy range realized by EGRET over its predecessors (SAS-2, COS-B) has been paralleled by advances in sensitivity in the Very High Energy (VHE) energy range from 300 GeV to 10 TeV by ground-based γ-ray telescopes. Although these latter advances have been incremental (and therefore less dramatic than the step increases seen in the launch of a new space telescope) they are, nonetheless, very significant and have turned VHE γ-ray astronomy into an observational science that exceeds the dreams of its early practitioners. Who could have foreseen in 1987 (when there were no convincing TeV sources) that by 1997 atmospheric Čerenkov telescopes (ACTs) would be recording TeV signals from extragalactic sources with 20σ significance in less than an hour of observation [1,2]? Now, in fact these ACTs have now been shown to have even better sensitivity than EGRET for sources with flat spectra [3].

Not only has the quality of TeV observations increased but the successful detection of the Crab Nebula [4] has led to a sharp increase in the *number* of high quality VHE observatories. Conversely the early promise of Ultra High Energy (UHE) astronomy using arrays of particle detectors has not been fulfilled [5]; the UHE range is usually defined as the γ- ray region from 10 TeV to 10 PeV. No sources have been detected despite improvements in sensitivity, and there has been a marked drop in the number of UHE observatories, particularly at the higher energies.

The increased interest in VHE γ-ray astronomy ensures that, in the inevitable interval between the demise of EGRET and the launch of the next generation γ-ray space telescope, there will be ongoing activity in GeV-TeV γ-ray astronomy. Although the number of sources detected is still small, there has been a wealth of new phenomena which indicate that VHE $-gamma$-ray astronomy is not merely an extension of MeV-GeV γ-ray but a viable discipline in its own right. The TeV γ-ray universe is populated by a variety of objects; it is dynamic and it is a challenge for theoretical interpretation.

TECHNIQUES

The earth's atmosphere is as impervious to photons of energy > 300 GeV as it is to photons in the CGRO range (100 kev to 30 GeV). However, at these very high energies, the effects of atmospheric absorption are detectable at ground level (either as a shower of secondary particles from the resulting electromagnetic cascade or as a flash of Čerenkov light from the passage of these particles through the earth's atmosphere). As in the CGRO range, γ-ray observations > 300 GeV are severely limited by the charged cosmic particle flux which gives superficially similar secondary effects and, for a given photon energy, is 10,000 times more numerous. It is not possible to veto out the charged cosmic ray background with an anti-coincidence shield. At first sight it might seem impossible to do γ-ray astronomy with such indirect techniques. However there are small, but significant, differences in the cascades resulting from the impact of a photon and a proton on the upper atmosphere; the electromagnetic cascade retains the original direction of the photon to a high degree, and the spread of secondary particles and Čerenkov photons is so large that a simple detector can have an incredible (by space γ-ray detector standards) collection area ($> 10,000$ m^2).

In practice, the most successful detectors are ACTs which record the images of the Čerenkov light flashes and which can identify the images of electromagnetic cascades from putative sources with 99.7% efficiency. These detectors were originally proposed in 1977 [6]. The technology is not new (typically arrays of PMTs in the focal plane of large optical reflectors with readout through standard fast amplifiers, discriminators and ADCs) but the technique was only fully exploited in the past decade. Compared to high energy space telescopes such as EGRET, imaging ACTs have large collection areas (>50,000m^2), high angular resolution (~0.1°), relatively good energy resolution (~ 20%), but have a small field of view (FOV) (<5°), a low duty cycle (<10%) and an irreducible background of diffuse cosmic electrons. Most of the results reported to date have been in the energy range 300 GeV to 30 TeV.

Above 30 TeV there are enough residual particles in the electromagnetic cascades that they can be detected at high mountain altitudes using arrays of particle detectors (scintillators or water Čerenkov) and fast wavefront timing [5]. These arrays have large collection area (>10,000m^2), moderate angular resolution (~0.5°), reasonable energy resolution (~ 40%), high duty cycles (~100%) and large FOV (~1 sr); however, their ability to discriminate γ-rays from charged cosmic rays is severely limited. Despite the early promise of these experiments which led to a considerable investment in their development, no verifiable detections have been reported yet by particle air shower arrays.

GALACTIC DISCRETE SOURCES

Supernova Remnants: Plerions

TeV γ-ray detections have been reported of the plerions surrounding three of the pulsars detected by EGRET. There is no strong evidence for pulsed emission from the pulsars at TeV energies.

CRAB NEBULA

Although it was reported as a source of TeV γ-rays more than a quarter of a century ago [7], the steady emission of TeV γ-rays from the Crab Nebula only became credible following the development of imaging ACTs [4,8]. Over the energy range from 300 GeV to 50 TeV, γ-ray emission has been confirmed by many groups [9–14] while in the UHE energy range, only upper limits have been reported [15–18]. Observations made over the last decade indicate no evidence for flux variability in the energy range from 400 GeV to more than 10 TeV, making this object a valuable standard candle for calibration purposes.

The spectrum of the Crab Nebula exhibits a remarkably broad dynamic range, spanning the energy range from less than 10^{-4}eV to at least 5x10^{13} eV. Gould [19] postulated that the entire spectrum (then not even known to extend to X-ray energies) could be explained by a Compton-synchrotron model. From the lowest radio frequencies up to sub-GeV energies, the spectrum is dominated by the synchrotron emission of relativistic electrons with energies extending up to 1 PeV; these are now believed to be accelerated by the pulsar wind shock [20]. The GeV-TeV photons arise from inverse Compton scattering of synchrotron photons or other low energy photons (e.g. the microwave background) in the nebula by the relativistic electrons. Detailed models of the GeV-TeV nebula have been described by [21–23].

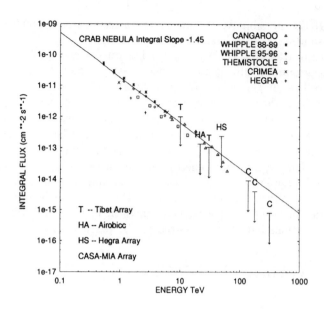

FIGURE 1. Spectrum of the Crab Nebula in VHE and UHE γ-rays.

Over the dynamic range 500 GeV to 5 TeV the spectrum is well represented by a power law of the form $J = (3.2 \pm 0.7).10^{-7} \times (E/1TeV)^{2.49\pm0.06\pm0.05}$ m^{-2}s^{-1}TeV^{-1} [24]. A recent observation by CANGAROO at low elevations provides data in the energy region > 7 TeV, which suggests that a power law spectrum might extend up to 50 TeV [25]. However, the air shower array limits at energies greater than 100 TeV fall below the extrapolation. If this linear power law is extrapolated to lower energies, it passes more than a decade above the upper EGRET point at 5 GeV. A form which is quadratic in Log (E) [26] satisfies both the GeV and TeV data and is consistent with upper limits at energies greater than 100 TeV. This spectrum implies a magnetic field in the nebula of 270μG, close to the equipartition value.

PSR B1706-44.

The CANGAROO group reported the detection of TeV γ-rays with a flux of 8×10^{-12} cm^{-2} s^{-1} [27]. This detection has been confirmed by the Durham group at > 300 GeV [29]. The CANGAROO observations are consistent with a point source (angular size < 0.1°) and corresponds to a luminosity of 3×10^{33} erg s^{-1}, which is $\sim 10^{-3}$ of the spin-down luminosity of the PSR B1706-44 pulsar. There is no evidence for pulsed emission at the 102 ms period of the pulsar. The emission of PSR B1706-44 is pulsed in the radio and high energy γ-ray band of the EGRET detection but unpulsed in the X-ray 0.1-2.4 keV range. The multi-band spectrum thus suggests that the radio and GeV emission originates in the pulsar magnetosphere, and a compact nebula is responsible for the synchrotron X-rays and the inverse Compton TeV γ rays. The pulsar appears to be located at an arc of the shell structure of the SNR; this is different than for the Crab Nebula where there is no shell. The X-ray luminosity, $\sim 1 \times 10^{32}$ erg s^{-1}, is less than the TeV luminosity. If X-rays and VHE γ rays are from common progenitor electrons, the magnetic field in the nebula must be as weak as $\sim 3\mu G$ [28].

VELA.

Evidence of VHE γ rays was found from the Vela pulsar direction by the CANGAROO group [30]. Over four years (1993-96), the flux was $(2.9 \pm 0.5 \pm 0.4) \times 10^{-12}$ cm^{-2}s^{-1} above 2.5 TeV, with a statistical significance of 5.8σ. The luminosity ($\sim 6 \times 10^{32}$ erg s^{-1}) corresponds to $\sim 9 \times 10^{-5}$ of the spin-down luminosity. No pulsed emission was seen. The observed unpulsed VHE emission appears to be displaced from the Vela pulsar position by about 0.13° to the southeast direction from the pulsar (Figure 2). This is not due to a pointing or tracking error, as the telescope tracking is calibrated by observing bright stars in the field of view of the telescope. The spatial size of the emission region is somewhat wider than the point spread function, which has a half width at half maximum of about 0.18°.

It is interesting to note that the position of VHE emission coincides with a bright spot in the soft X-ray distribution observed by ROSAT [31], which can be thought of as the 'birth place' of the Vela pulsar, when the proper motion is traced back 10,000 years (the pulsar's age). By assuming that the X-ray emission is of synchrotron origin, a comparison of X-ray and γ-ray intensities gives an estimate of the magnetic field as low as $4\mu G$ at the birth place [30].

OTHER PLERIONS.

X-ray data from the ASCA and ROSAT satellites suggests that synchrotron nebulae are associated with most of the pulsars with the highest spin-down luminosities and/or at the closest distances [32]. The synchrotron X-rays guarantee that energetic electrons exist in such pulsar nebulae, and encourage the search for VHE γ-rays emitted by the inverse Compton process.

The only plerion detected by EGRET so far is the Crab Nebula. This fact may be explained if the magnetic fields in other nebula which contain the electrons of energy up to ~ 100 TeV are weak but strong in the Crab Nebula. Because the synchrotron and inverse Compton radiation are peaked at the X-ray and TeV bands, respectively, for nebular magnetic fields much weaker than the Crab, the flux in MeV to GeV region is expected to be below the sensitivity of the CGRO detectors. This is similar to the case of the AGNs, Markarian 421 and 501 (see below).

The multi-waveband spectra of the three plerions seen at TeV energies, as well as their spatial structure, suggest significant differences between the three, though the inverse Compton process is considered to be common for all. Spatially extended features will be common in most pulsar nebulae as well as for the ones associated with supernova remnants (SNRs). Current ACTs

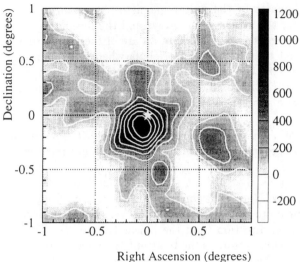

FIGURE 2. A density map of excess counts around the Vela pulsar from CANGAROO data taken between January 1993 and March 1995. The coordinates are right ascension and declination; north is up and east is to the left. The gray scale at the right is in counts degree^{-2}. The mark "star" at the origin of the map indicates the position of the Vela pulsar.

have limited sensitivity to sources with extended emission but those under development will remedy this.

Supernova Remnants: Shell-type

The canonical theory of cosmic ray origins suggests that they emanate in shell-type SNRs. High energy γ-ray observations can indicate which SNRs have a large content of relativistic cosmic ray hadrons, and TeV γ-ray observations, in particular, have the sensitivity and angular resolution to reduce background confusion. EGRET measurements of SNRs are not definitive because the detector has low angular resolution at 100 MeV and measurements are masked by gas clouds. Collisions of cosmic ray nuclei with the interstellar medium result in the production of neutral pions which subsequently decay into γ-rays. These processes result in a secondary γ-ray spectrum which follows the primary cosmic ray spectrum at energies above ∼10 GeV up to ∼1/10 of the maximum proton energy of ∼ 100 TeV.

Calculations [33,34] indicate that the luminosity of nearby SNRs should be sufficient for detection by the most sensitive VHE γ-ray telescopes. If there is a density enhancement from a molecular cloud, current ACTs and EGRET

TABLE 1. SNR Upper Limits

Source	EGRET Flux [36] $\times 10^{-7} cm^{-2} s^{-1}$	Whipple Upper Limit [37,38] $\times 10^{-11} cm^{-2} s^{-1}$
W44	5.0	<3.0
W51	<3.2	<3.6
γ-Cygni	12.6	<2.2
W63	<1.9	<6.4
Tycho		<0.8
IC443	5.0	<2.1

should already be able to detect the γ-ray emission from some objects [35]. For a source spectral index of $\alpha=2.1$, Drury et al. [33] estimate the integral γ-ray flux at earth to be

$$F(> E) \approx 9 \times 10^{-11} \left(\frac{E}{1 \text{ TeV}}\right)^{-1.1} \left(\frac{\theta E_{SN}}{10^{51} \text{erg}}\right) \left(\frac{d}{1 \text{ kpc}}\right)^{-2} \left(\frac{n}{1 \text{ cm}^{-3}}\right) \text{cm}^{-2} \text{s}^{-1} \quad (1)$$

where θ is defined as the fraction of the supernova energy E_{SN} converted into cosmic rays, d is the distance to the SNR, and n is the average density of the ISM around the remnant.

Assuming $\theta \approx 0.15$, a fairly conservative value for the average density of $n \approx 0.2 \text{cm}^{-3}$ and the canonical value $E_{SN} \approx 10^{51}$ ergs, gives a flux $F(> 200 \text{ GeV}) \approx 1.6 \times 10^{-11} (d/1 \text{ kpc})^{-2}$ photons cm^{-2}s^{-1}. This flux lies close to the sensitivity limit of the current generation of TeV telescopes.

For several of the unidentified EGRET sources there is now evidence for an association with a SNR (γ-Cygni, IC443 and W44 [36]). Observations by ground-based experiments do not confirm the expected extension of the γ-ray spectrum [37,38]. Some of these limits are listed in Table 1. They lie a factor of ten below the predicted spectrum, but there is still significant uncertainty in the remnant parameters so that the canonical model is not yet in jeopardy. It is also possible that the EGRET associations with the SNR are incorrect. Although not yet definitive, the upper limits require that the source spectrum for IC443 and γ-Cygni be steeper than $E^{-2.4}$; this would imply that the additional steepening of the cosmic ray spectrum due to propagation effects in the galactic disk should be -0.3 rather than the preferred value of -0.6.

Pulsars

Prior to the development of imaging ACTs the evidence for the detection of VHE and UHE γ-rays sources was based on time-varying features of the event rate, i.e. the detection of a pulsed signal modulated at the spin or orbital period of radio and X-ray pulsars, and/or the observation of a sudden increase of observed events presumably due to an episodic outburst. Several detections of pulsars were reported with marginal significances [39,41,40,42] but remain unconfirmed.

The most convincing detection, a steady pulsed signal from the Crab pulsar, was from the Durham group [43]; the observed flux was $\sim 1 \times 10^{-11}$ cm^{-2}s^{-1} at 1 TeV. However this has not been confirmed by more sensitive observations which show that less than 5% of the total TeV flux is pulsed [11,44]. At UHE energies, the large CASA-MIA experiment finds no statistically significant evidence for pulsed γ-ray emission at the Crab radio period, on an interval of one day or longer, based on analysis of 2.4×10^9 events recorded during the interval March 1990 to October 1995 [18]. The 2σ limit on the pulsed flux for the most significant day of observation is 8.31×10^{-12}cm^{-2}s^{-1}, above 140 TeV. This result does not support earlier claims of pulsed detection by smaller EAS arrays [45].

Evidence for the detection of a pulsed signal was also suggested for the Vela pulsar [46] and for the Geminga pulsar [47,48] (Table 2). However, observations with more sensitive imaging ACTs failed to confirm these early results [49,30]. No TeV pulsed signal has been detected from the CGRO pulsars, PSR B1706-44 [50,27], PSR 1951+32 [51] and PSR B $1055-52$ [52](Table 2). No TeV observations have been reported yet on PSR B0656+14.

The upper limits for the TeV flux from the EGRET γ-ray pulsars are below the fluxes extrapolated from the GeV region of EGRET detection assuming a power law spectrum of constant index. The emission from the pulsar magnetosphere seems to turn off or to fall off steeply in VHE region. It has been have argued [53,54] that the pulsed component will extend to VHE energy in outer gap models.

Unidentified EGRET Galactic Sources

One of the most important EGRET results has been the discovery of 71 unidentified γ-ray sources. These sources may remain the major unsolved mystery left behind by the CGRO. Some of these sources have flat spectra which will permit their detection at higher energies using ACTs. A detection at TeV energies would provide a better source location, the possibility of detecting a pulsar signal, and an opportunity to observe the source over a long period; it should then be possible to identify the source.

The unidentified sources can be subdivided into two groups depending on their distance from the galactic plane. Many of the 39 sources that lie more than 10° from the galactic plane are likely to be extragalactic. There is evidence for another class of source that is not associated with blazars, is not variable, and has a broad galactic distribution. There are also 30 unidenti-

TABLE 2. Flux from Gamma Ray Pulsars

pulsar		EGRET HE flux [55] (10^{-7} cm^{-2} s^{-1})	Group	VHE Flux (10^{-12} cm^{-2} s^{-1})	E_{th} (TeV)
Crab	unpulsed	7.7 ± 0.8	Whipple	8.8 (E/TeV)$^{-1.69}$	0.4
			Crimea	13	1
			ASGAT	27	0.6
			HEGRA	8×10^{-12}(E/TeV)$^{-1.7}$	1
			THEMISTOCLE	2.3 (E/3TeV)$^{-1.4}$	2.3
			CANGAROO	0.76	7
	pulsed	23	Whipple	< 0.2	0.25
			Gulmarg	< 2.5	4
			Tata	9.7 ± 4.3	1
			Durham	7.9 ± 1	1
1706−44	unpulsed	−	CANGAROO	8	1
	pulsed	12	Potchefstroom	< 5.8	2.6
Vela	unpulsed	−	CANGAROO	2.9	2.5
	pulsed	90	CANGAROO	< 0.37	2.5
			SAO/Sydney	< 85	0.3
			Tata	~ 1	5
			Durham	< 50	0.3
			Potchefstroom	< 6	2.3
			Adelaide	< 70	0.8
Geminga	unpulsed	−	Whipple	< 8.9	0.5
	pulsed	37	Durham	30	1
			Tata	21 ± 8	0.8
				4.4 ± 3.5	1.7
			Whipple	< 5	0.5
1055−52	pulsed	2.2	CANGAROO	<0.95	2
1951+32	pulsed	1.6 ± 0.2	Whipple	<5.4	0.3

fied sources within 10° of the galactic plane [55,56]. Geminga-like pulsars are suspected to constitute a considerable portion of these unidentified EGRET sources. An increasing number of the pulsars are found to be associated with synchrotron X-ray nebula, in which the inverse Compton counterpart will be at VHE energies. Detection of an associated nebula would provide a useful means of identifying the pulsar-powered sources.

Only a small fraction of these sources have been observed with ACTs; some upper limits have been published [57]. The next phase of TeV γ-ray astronomy will surely include a systematic survey of all of these sources coupled with a detailed investigation of the galactic plane.

X-ray Binaries and Related Sources

This population of compact galactic sources, in particular its two famous representatives, Cyg X-3 and Her X-1, dominated VHE and UHE γ-ray astronomy in the 80's, and played a crucial role in the renewed interest in ground-based γ-ray observations [1,39,41,42]. Unfortunately, almost all the early reports about the detection of signals from these objects, both in the TeV and PeV domains, were not confirmed by later, more sensitive, observations [58–60]. Although, we cannot exclude a long-term variability of the γ-ray fluxes, many reviewers treat the early claims about the detection of γ-rays from X-ray binaries with some skepticism [40,5,61].

Nevertheless, the jury is still out on the reality of this phenomenon and there are important new results from the X-ray binary, Vela X-1 [62] and the cataclysmic variable AE Aquarii [63]. While the statistical significance of these TeV detections do not match the statistical standards set by imaging ACTs, they do merit further investigation. There is also the possible detection of GeV γ-rays by EGRET from the direction of Cyg X-3 [64] (although without the 4.8 hour modulation that was characteristic of the early detections) and Cen X-3 [65]. Thus it is premature to draw a final conclusion, and it would be wise to monitor these sources with current imaging telescopes in both hemispheres.

Further justification for observations of X-ray binaries by ground-based γ-ray detectors comes from the recently discovered galactic superluminal objects GRS 1915+10 and GRO J1655-40. The relativistic motion of radio components in these hard X-ray transients may be common to many other luminous compact galactic binaries, in particular black hole candidates, as well as sources like Cyg X-3. The possible link between relativistic motion and accretion phenomena in these objects may provide a key insight into the nature of not only these objects, but also the engines of AGNs and quasars.

Diffuse Galactic Gamma Ray Background

Historically, interest in the diffuse γ-ray background appeared in the late fifties in the context of the general problem of the origin of the cosmic rays. Detection of γ-rays above 100 MeV by SAS-2, COS B, and EGRET have already made an essential contribution to the current knowledge of distribution of cosmic rays (CRs) in the Galaxy. In particular, the recent EGRET measurements from the galactic disk show an excess in the GeV γ-ray flux, compared with predictions based on the assumption that the average spectrum of CRs throughout the Galaxy is represented by the local (directly measured) spectrum [66]. The hardening of the γ-ray spectrum can be interpreted (i) as a result of a significant spatial variation of the CR spectrum in the Galaxy [67], or (ii) due to an extra inverse Compton (IC) component of radiation (see e.g. [68]). Both suggestions have an important impact on predictions of diffuse VHE γ-ray fluxes.

π^0 **COMPONENT.** In Fig.3, we present the expected flux of π^0-decay γ-rays (curve 1) in the direction of the inner Galaxy calculated for the local CR spectrum, and assuming the characteristic value of the hydrogen column density $N_H = 10^{22}\,\mathrm{cm}^{-2}$ for latitudes $|b| \leq 10°$. For comparison, we show the average level of diffuse γ-ray flux measured by EGRET from the inner Galaxy region at $|b| \leq 10°$. There is agreement between the calculations and measured flux at 1 GeV. At the same time, the EGRET spectrum at higher energies seems to be harder than the predicted one. Formally, such a discrepancy can be easily overcome by assuming a harder (e.g $\propto E^{-2.5}$) CR spectra in the regions where the *bulk* production of γ-rays takes place (see curve 2 in Fig.3). Do we have a good justification for such an assumption?

The propagation of CRs in the galactic disk on time-scales $\geq 10^7$ yr implies an effective mixture of contributions from individual sources/accelerators of CRs. Therefore, we should not expect a strong gradient of the CR fluxes on 'kpc' scales; such variations are quite possible on smaller scales. Indeed, assuming that CRs with total energy W_p injected by a 'typical' accelerator into the ISM, at the instant t reach a distance R, the mean energy density of CRs in the occupied region is $w_p \simeq 0.5 (W_p/10^{50}\,\mathrm{erg})(R/100\,\mathrm{pc})^{-3}\,\mathrm{eV/cm}^3$. Thus up to 100 pc around the source with $W_p \sim 10^{50}\,\mathrm{erg}$, the energy density of relativistic particles at some stages, depending on the time history of particle injection and character of their propagation in the ISM, may significantly exceed the average level of the 'sea' of galactic CRs, $\sim 1\,\mathrm{eV/cm}^3$. If regions of high density massive clouds exist near the particle accelerators, we may expect enhanced γ-radiation. Speculating now that the main part of the observed diffuse γ-ray background is produced selectively, i.e. it is a result of radiation originating in regions which contain particle acceleration and massive gas clouds, we may explain the hard spectra of the diffuse radiation observed by EGRET above 1 GeV. Indeed, if γ-rays are produced at the interaction of a cloud with relatively fresh (recently accelerated) particles with spectra which have not yet suffered strong modulation (steepening) due to propagation (diffusion) effects in the ISM, the resulting γ-ray spectra should be significantly harder than the typical γ-ray spectra produced by the 'sea' of galactic CRs. Such an assumption agrees with the correlation observed by the EGRET between the CRs and the hydrogen column density [66], which in the galactic plane is contributed essentially by giant molecular clouds (GMCs).

The crucial test for the hypothesis of diffuse GeV background being the superposition of the radiation of GMCs would be the direct detection of individual GMCs. One of the principal parameters which determines the visibility of a GMC in γ-rays is M_5/d_{kpc}^2, where M_5 is the diffuse mass of GMC in units of 10^5 solar masses, and d_{kpc} is the distance to the cloud in kpc. The 'passive' GMCs submerged in the 'sea' of galactic CRs can be detected in γ-rays at the level of the EGRET sensitivity if $M_5/d_{\rm kpc}^2 \gg 1$. Since there are not many GMCs in the galactic disc with $M_5/d_{\rm kpc}^2 \gg 1$, one should expect a limited number of 'passive' GMCs detectable in GeV γ-rays but not in VHE γ-rays. The GMCs located in the proximity of CR accelerators may be visible at GeV energies ($F_\gamma(\geq 1\,{\rm GeV}) \geq 10^{-7}\,{\rm cm^{-2}s^{-1}}$) and TeV ($F_\gamma(\geq 1\,{\rm TeV}) \geq 10^{-12}\,{\rm cm^{-2}s^{-1}}$) γ-rays even for $M_5/d_{\rm kpc}^2 \sim 0.1$ [69].

IC γ-rays. The excess above a few GeV can be explained also by the IC photons. This hypothesis requires a rather bold assumption about the fluxes of CR electrons. This is seen in Fig.3, where the IC γ-ray flux (curve 3) calculated for the locally observed CR electron spectrum [70] is less, by a factor of 10, than the measured γ-ray flux. Thus, in order to explain the EGRET spectrum at $\geq 1\,{\rm GeV}$ by the IC mechanism, one has to accept that the average flux of CR electrons at energies above 50 GeV exceeds by an order of magnitude the locally measured electron flux. In fact, this should not be considered an extraordinary assumption. Due to the severe radiative energy losses of VHE electrons during their diffusion in the ISM, we can see only those particles which are produced within nearest the 100 pc region [70,71]. Formally, this gives significant freedom for speculation about the average *galactic* flux of VHE electrons. Porter & Protheroe [68] emphasized that such an assumption may result in VHE γ-ray fluxes exceeding the contribution of π^0-decay γ-radiation. This possibility is demonstrated in Fig.3 where the IC fluxes of γ-rays (curve 4) are shown for some hypothetical average equilibrium spectrum of galactic electrons with spectral index 3 at energies above 50 GeV, and with a flattening below 50 GeV (in order not to contradict the synchrotron radio measurements).

The predictions of diffuse VHE γ-ray fluxes of the galactic disk are still below the upper limits set both at TeV and PeV energies [60,58,72]. If the sensitive upper limits by CASA/MIA rule out noticeable hardening of the CR proton spectrum and/or of the very high electron fluxes continued well beyond $10^{14}\,{\rm eV}$, the current upper limits at TeV energies are significantly above predictions. New stereoscopic systems of imaging ACTs will be able to probe the predicted range of diffuse fluxes shown in Fig.3.

TEV EMISSION FROM EXTRAGALACTIC OBJECTS.

Active Galactic Nuclei.

The success of EGRET in opening up AGNs, in particular, blazars, to exploration in the MeV-GeV energy range, drew attention to the possibility of studying at TeV energies. Blazars are characterized by highly variable, broad-band continuum emission. The radiation is dominated by emission from the jets viewed at small angles of to the jet axis. BL Lacertae (BL Lac) objects

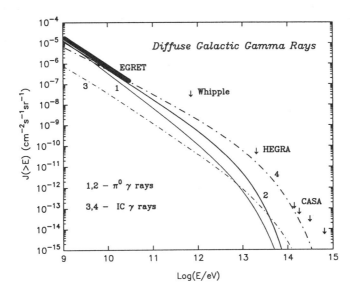

FIGURE 3. Diffuse galactic γ-ray fluxes produced in interactions of cosmic rays with interstellar gas (π^0-decay γ-rays), and starlight and 2.7° MBR photons (IC γ-rays). In calculations it is assumed that the CR protons and electrons have power-law spectra with exponential cutoffs at $E_c = 100$ TeV.

are a sub-class of blazars characterized by weak or absent emission lines; so far they have only been detected at small redshift. The characteristics of such close BL Lac objects may prove important in the overall understanding of AGNs. TeV observations of BL Lacs have a fundamental role to play in determining the *total* spectral energy distribution (SED) and the nature of the emission process which, in some cases, spans in excess of 18 orders of magnitude in frequency. Because the SED of nearby BL Lacs is often very flat in the γ-ray range, measurements made with ACTs have comparable signal-to-noise to those obtained by any instrument on CGRO; this has been one of the most surprising results to come from VHE γ-ray astronomy.

Markarian 421.

With a redshift of z=0.031, Markarian 421 (Mrk 421) is the closest known BL Lac and hence one of the best studied. Although a weak source at EGRET energies [73], Mrk 421 is a strong TeV emitter, with fluxes typically 30-50% that of the Crab Nebula [74]. The relatively strong signal has permitted time variations to be measured with very high resolution [75], correlations to be studied with other wavelengths [76] and the source location to be determined with high accuracy [77] (Figure 4).

An intense multi-wavelength monitoring campaign was undertaken at radio, optical, extreme UV, X-ray, >100 MeV and >300 GeV energies during the interval April 20 to May 17, 1995. The TeV and X-ray light curves showed

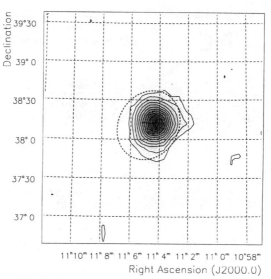

FIGURE 4. Two dimensional plot of the γ-ray emission around Mrk 421 whose position is indicated by the cross. Solid contours correspond to 2σ levels derived from the likelihood ratio statistic. The dashed ellipses give the 95% confidence level position from EGRET [77].

strong correlation at zero relative phase shift with a flux doubling time ~ 1 day and an overall decline with a timescale ~ 1 week (Figure 5(a)). EUVE and optical data also showed correlation with the γ-ray data [77]. Maximum correlation of the X-ray/γ-ray data with the optical and EUVE data occurs when a time lag of one day is allowed for at these latter wavelengths. During this interval the MeV-GeV flux of Mrk 421 was below the detection threshold of EGRET. The multi-wavelength SED for Mrk 421 shows the familiar double peaks seen in many AGNs but shifted to higher energies [78].

Extensive monitoring of Mrk 421 during 1996 resulted in the detection of two dramatic TeV flares in May [2] (Figure 5). The first flare, on May 7, lasted more than 2 hours. During the flare the γ-ray rate increased uniformly from 5 events per minute at the onset of observations to a peak value of 14 events per minute. The average γ-ray rate for Mrk 421 during 1996 (excluding the nights of the large flares) was 0.45 events per minute. Follow-up observations the next day showed the source had returned this level, indicating a decay timescale of <1 day. The second flare on May 15th, though less powerful, was remarkable in terms of the brevity of the emission, with a doubling time of 15 minutes (Figure 5(b)). Contemporary optical observations indicate increased flux and variability during May 1996 [79]. These short variations imply Lorentz factors in excess of 10 and severely constrain source models.

The TeV detection of Mrk 421 has been confirmed by the HEGRA collabo-

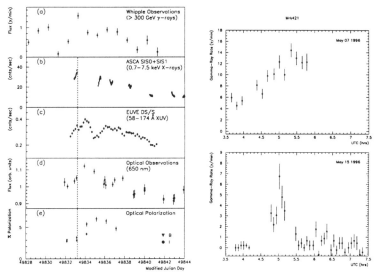

FIGURE 5. (a) Multi-wavelength observations of Mrk 421 in April-May, 1995 [77]. (b) Two nights of observation of Mrk 421 in May, 1996 [2].

ration [80]. Continued TeV observations of Mrk 421 by the Whipple collaboration [77] indicate remarkable source variability, characterized by flickering, on time-scales from minutes to months, with a relatively low (perhaps zero) baseline level of steady emission.

Markarian 501.

Markarian 501 (Mrk 501) is similar in many respects to Mrk421. The TeV detection of this object by the Whipple group [3] based on observations taken between March and July 1995 indicated an average flux level of $8 \times 10^{-12} \mathrm{cm}^{-2} \mathrm{s}^{-1}$ at an energy threshold of 300 GeV, roughly 20% of the Mrk421 flux. EGRET had not detected Mrk501 so this was the first γ-ray source to be discovered by a ground-based telescope. Observations by the HEGRA collaboration [81] substantiate those made by the Whipple collaboration. In 1997 Mrk501 became very active and at times was the brightest TeV source in the sky. Data taken by the Whipple, HEGRA, CAT and Telescope Array groups [82–85] indicate that during the period March to May 1997, Mrk 501 often exceeded the Crab Nebula rate by a factor of five, at energies in excess of 300 GeV (Figure 6(a)).

The SED of Mrk 501 during this epoch was also remarkable [82]. The first of the two peaks is shifted to much higher energies and, if interpreted as a synchrotron distribution, it implies that the electrons have a higher maximum energy than those in any other AGN.

1ES 2344+514.

Evidence for TeV emission from a third AGN, the BL Lac object 1ES

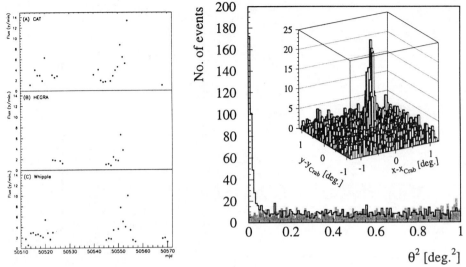

FIGURE 6. (a) Observations of Mrk 501 in March-April 1997 by the Whipple, CAT and HEGRA ACTs (compilation by M.F.Cawley). (b) Distribution of reconstructed directions of showers detected by four telescopes of the HEGRA array in 11.7 hours of observation of the Crab Nebula. The insert shows the two-dimensional distribution.

2344+514, was obtained by the Whipple collaboration in the winter of 1995 [86]. Like Mrk 501, this object was observed as part of a campaign to search for TeV emission from nearby BL Lac objects ($z < 0.2$). Most of the evidence for emission from this object comes from one night of observations when a 6σ excess was detected in 1.5 hours of observations.

If blazars are to be observed at TeV energies, they probably have to be located at redshifts of less than 0.2, otherwise pair production interactions with the intergalactic background light will cause significant absorption of the γ-ray beam [87–89]. To date, three blazars have been detected at TeV energies, Mrk 421 [74], Mrk 501 [3] and 1ES2344 [86], and all three are designated X-ray selected BL Lacs (as distinct from the radio-selected BL Lacs).

As yet there have been no confirmed detections of γ-rays from AGNs at energies > 10 TeV using air shower arrays, despite a number of searches [90–93].

TeV observations will seriously constrain these models by the necessity to explain (a) the production of TeV and higher energy γ-rays; (b) the shape of the full energy spectrum; (c) the correlations in emission between X-rays and TeV γ-rays; (d) the short-term time-variations. There are not sufficient TeV observations yet to confidently select any particular model.

Other Galaxies.

Upper limits on 34 AGNs (15 of which had been detected by EGRET) have been reported by the Whipple Collaboration [95]. Limits on southern hemisphere AGNs have also been reported by the Durham group [94]and by the CANGAROO Collaboration [96]. Conclusions from these null results are limited by the observational evidence that almost all γ-ray AGNs are time variable.

There have been extensive TeV surveys of other classes of extragalactic objects, all of which have produced negative results. Most of these were performed using pre-imaging Čerenkov systems or early vintage imaging systems and are summarized elsewhere [1]. One possible exception is Centaurus A (NGC 5128), a very luminous elliptical galacxy with an AGN-type compact source at its center. This AGN also exhibits both radio and X-ray jets. Three years of observations of this object (1972-1974) by a Smithsonian-University of Sydney group resulted in its detection at a level of statistical significance of 4.6σ [97]. The reported flux was $(\sim 4.4\pm1.0)10^{-11}$ photons-cm^{-2}s^{-1} for E>0.3 TeV. This observation has not been confirmed, although both the Durham group [98] and the JANZOS collaboration [99] have reported upper limits which are consistent with the earlier detection.

It is probably significant that although EGRET and ACTs together have detected more than 60 blazars, they have not detected a single Seyfert, radio-quiet quasar or radio-galaxy. The jet phenomenon and its alignment appears to be central to the issue of detecting γ-rays, at both GeV and TeV energies.

Since nearby starburst galaxies such as M82 and NGC 253 clearly give rise to a large number of supernovae and accelerate cosmic rays to very high energies, they must be considered as reasonable candidates for detection of γ-rays. Only upper limits have been obtained on these objects by EGRET at energies $\sim$ 100 MeV. Neither M31, the next closest galaxy to our own after the LMC and SMC, nor the giant elliptical galaxy M87, have been detected by EGRET. There is still a strong scientific case to be made for long-term observational campaigns on extragalactic objects such as M31, M82, M87 etc. using imaging ACT systems with improved flux sensitivity.

Probing the Extragalactic Diffuse Background

The extension of the spectra of at least two extragalactic sources, Mkn 421 and Mkn 501 well beyond 1 TeV opens a new exciting aspect of γ-ray astronomy – *observational cosmology*. The absorption features in the γ-ray spectra of distant extragalactic sources due to the interaction of primary VHE γ-rays with the diffuse extragalactic background radiation (see e.g. [87,88], and references therein), as well as the secondary pair cascade radiation [100–102], contain unique cosmological information about the intergalactic photon and magnetic fields, and their evolution in time.

Assume that γ-rays are emitted with an initial spectrum $J_0(E)$ from a source at a distance d. An observer looking within a narrow cone centered on the source will see an absorbed spectrum $J(E) = J_0(E)\exp(-\tau)$. The optical depth $\tau(E,d)$ due to γ-γ pair production on isotopically distributed photons of density $n(\epsilon)$ may be presented in a convenient form [103]: $\tau(E,d) = 1.25\,\eta\,d\,\epsilon_s\,\sigma_0\,n(\epsilon_s)$, where $\sigma_0 = 1.7\cdot 10^{-25}\,\mathrm{cm}^2$, $\epsilon_s = 4m_e^2c^4/E$. No deviation of the observed spectrum from the intrinsic (source) spectrum at energy E, e.g. by a factor of ≤ 2 implies $\tau(E) < \ln 2$. For a low-redshift source ($z \ll 1$) this gives an upper limit on the current energy density of the diffuse extragalactic background radiation (DEBRA) at $\epsilon \approx 1\,(E/1\,\mathrm{TeV})^{-1}\,\mathrm{eV}$:

$$u_\epsilon = \epsilon^2 n(\epsilon) < 3.5\cdot 10^{-3}\,(z/0.1)^{-1}\,(E/1\,\mathrm{TeV})^{-1}\,(H_0/100\,\mathrm{km/s\,Mpc})\,\eta^{-1}\,\mathrm{eV/cm}^3$$

Note that formally the shape of the spectrum $n(\epsilon)$ does not enter into this equation. For a power-law spectrum, $n(\epsilon) \propto \epsilon^{-p}$, the correction factor to this approximation, $\eta \sim 1$ for $p = 2 \pm 0.5$. Although the spectrum of DEBRA between 0.1 and 10 eV could be rather irregular in shape, for the purpose of estimates with reasonable ($\leq 50\%$) accuracy, we may adopt $\eta = 1$.

If the lack of evidence for a cutoff in the γ-ray spectra of both Mkn 421 and Mkn 501 ($z \simeq 0.03$) up to 10 TeV [104,83] could be interpreted as an indication for negligible absorption of γ-rays in the DEBRA, one may set a meaningful upper limit on $u_\epsilon \leq 1.1 \cdot 10^{-3}\,\mathrm{eV/cm}^3$ at $\epsilon \approx 0.1\,\mathrm{eV}$. This interesting constraint is below the flux of DEBRA stated by De Jager et al. [105], but still does not contradict a number of other models (see e.g. [89]). Formally, this upper limit may be extrapolated to shorter (optical) wavelengths by assuming an *a priori* power-law background photon spectrum with differential index $p > 2$ (e.g. [105]) and thus we get more valuable (restrictive) information about the DEBRA. However given the strong energy-dependence of the pair production cross-section, this hardly can be justified. The model-independent constraint, for example at 3 eV, comes *only* from the apparent lack of intergalactic absorption feature at $E \sim 300\,\mathrm{GeV}$. In the case of Mrk 421 and Mrk 501, this implies a rather high upper limit $u_\epsilon \leq 4 \times 10^{-2}\,\mathrm{eV/cm}^3$. A deeper probe of the DEBRA at optical wavelengths is contingent only on the discovery of high redshift (≥ 0.1) VHE sources. In this case, extraction of information about the DEBRA at the present epoch requires an additional model assumption about the time history of evolution of DEBRA. Note that the law $(1 + z)^3$, which characterizes, by definition, the evolution of the primordial 2.7 °K blackbody microwave background radiation (MBR), cannot be applied, as is very often done in the literature, to the DEBRA. The evolution of the infrared background is determined, first of all, by the epochs of the galaxy formation, and thus has a much more complicated time history. On the other hand, if 'intergalactic' cutoffs in the spectra of many distant AGNs (at different redshifts) are discovered, the γ-ray observations will be able to address this important cosmological issue as well.

Obviously these interesting possibilities can be successfully utilized only in the case of accurate spectroscopic γ-ray measurements, as well as good understanding of the intrinsic source spectra. The last condition seems to be crucial, especially since the lack of cutoff in γ-ray spectra still does not automatically imply an absence of intergalactic absorption. Interestingly, some DEBRA models predict a *modulation*, rather than *cutoff*, in the energy spectra of nearby AGNs ($d \sim 100\,\text{Mpc}$), at least up to 10 TeV. Such modulation makes steeper the primary spectrum, but still keeps it in the power-law form. From this point of view, the above VHE γ-ray upper limit on the DEBRA density may not be significant.

BURSTS

A feature of γ-ray astronomy has been that as the energy is increased there is an increase in the degree of temporal variation. As seen above, in AGN studies, TeV variations with doubling times as short as 15 minutes have been observed. It would not be unexpected that TeV γ-ray bursts would be observed, either as the tail end of classical γ-ray bursts or as a manifestation of a new phenomenon.

There are a number of exotic suggestions that justify a search for γ-ray bursts at TeV energies. These include emission from the decay of primordial black holes(PBHs) [106] and cosmic strings. Most models predict a final explosion of energy when all possible evaporation channels are available, but the number of degrees of freedom of emission is a highly controversial issue [107]. Upper limits have already been set to the PBH density by atmospheric Čerenkov and air shower array experiments under various assumptions (e.g. [108,109]).

Predictions have been made that (i) non-conducting cosmic strings acquire cusps that are smoothed out by emitting bursts of TeV γ-rays over ill-defined time-scales [110] and that (ii) superconducting strings with a saturated current produce a jet of fermions which decay to TeV γ-rays over a 1 second period [111]. In general these predictions have sufficient free parameters that non-detections are not serious limitations; however the exciting new physics that a positive detection would indicate fully justify the search for new burst phenomena as TeV detectors are developed with improved flux sensitivity.

Since August 1992, the Whipple collaboration has searched for bursts on a one second time-scale, both on-line and in the archival data-base. The number of bursts of 3 or more per 1 second was compared with the expectation value and no significant excess was obtained over the 4 years of the data-base [113].

Although the serendipitous overlap of an atmospheric Čerenkov telescope with its limited field of view with a classical BATSE- detected burst is unlikely, there is some hope of detecting the delayed high energy component seen in some bursts. It is possible to make rapid follow-up observations of BATSE bursts using source positions distributed on the BACODINE network. The effectiveness of these observations is limited by the restricted duty-cycle of ACTs, by the imprecise position locations of the bursts, by the slew speed of the telescope and by its limited field of view. Nonetheless observations by the Whipple Collaboration of 16 BATSE positions, one acquired within two minutes of the reported BATSE burst time, have been reported [112,114]. However in no case did the FOV of the telescope overlap the complete error box in source position uncertainty. No evidence of TeV emission is found and upper limits to the high-energy delayed or extended emission were derived based on assumptions of the source positions.

FUTURE PROSPECTS

It is difficult not to be upbeat about the immediate and long-term future prospects for VHE γ-ray astronomy. Every astronomical discipline has a point in its history when it achieves a threshold of sensitivity, a threshold where the sources first exceed the observational sensitivities. After that the discipline usually develops rapidly. At energies above 300 GeV (and up to energies of at least 10 TeV) that threshold has been achieved with both galactic and extragalactic sources detected. Since only a small section of the sky has been viewed, and at least some of the sources are variable there is the promise of many more sources as the sky is explored with current instrumentation.

But significant progress depends on the development of new instrumentation and new observational techniques. These are already under consideration and, in some cases, construction. Several different approaches will be attempted simultaneously. The reduction in threshold below 100 GeV will be achieved with the conversion to ACTs of existing solar facilities in the USA (STACEE) and France (CELESTE) and the construction of large single reflector imaging ACTs (MAGIC). Complete sky coverage, with good sensitivity to bursts and transients, will come from air shower arrays like the water Čerenkov detector (MILAGRO) in New Mexico and expanded high altitude particle arrays like that in Tibet. Facilities consisting of conventional ACTs that will build on the imaging principle and will operate as state-of-the-art VHE γ-ray observatories into the next millennium include VERITAS, an array of nine 10m telescopes in Arizona, and HESS, an array of 4-16 10m telescopes that may be built in Spain or Namibia. The power of such arrays to increase the angular resolution and to reduce the cosmic ray background has recently been demonstrated by the HEGRA group in observations of the Crab Nebula (Figure 6(b)) [115].

With the death of EGRET there will be a hiatus in GeV γ-ray astronomy until the development of the next generation of space instruments (such as GLAST). Fortunately ground-based γ-ray facilities will continue to operate and will maintain the momentum in high energy γ-ray astronomy. More importantly, they will demonstrate that VHE γ- ray astronomy has become an important component in the study of γ−ray astrophysics.

ACKNOWLEDGEMENTS. The helpful comments of Jim Buckley and Mike Catanese is acknowledged as is the technical assistance of John Quinn and Ann Weekes in the production of the manuscript.

REFERENCES

1. Weekes, T.C. *Physics Reports*, **160** (1 and 2), 1 (1988).
2. Gaidos, J. et al., *Nature* **383**, 319 (1996).
3. Quinn, J. et al., *Astrophys. J. Lett.*, **456**, L63 (1996).
4. Weekes, T.C. et al., *Astrophys. J.*, **342**, 379 (1989).
5. Cronin, J.W., Gibbs, K.G., Weekes, T.C., *Ann.Rev.Nucl.Part.Sci.* **43**, 883 (1993).
6. Weekes, T.C., Turver, K.E., *Proc. 12th ESLAB Symp. (Frascati) ESA SP-124* 279 (1977).
7. Fazio, G.G. et al., *Astrophys. J. Lett.*, **175**, L117 (1972).
8. Vacanti, G., et al., *Astrophys. J.* **377**, 467 (1991).
9. Akerlof, C.W., et al., *Proc. GRO Workshop (Goddard)* 4 (1989).
10. Baillon, P., et al., *Astropart. Phys.* **1**, 341 (1993).
11. Goret, P. et al., *Astron.Astrophys.* **270** 401 (1993).
12. Konopelko, A. et al., *Astropart.Phys.* **4**, 199 (1996).
13. Tanimori, T., et al., *Astrophys. J. Lett.* **429**, L61 (1994).
14. Stepanian, A.A., *Nuclear Physics B (Proc. Suppl.)* **39A**, 207 (1995).
15. Alexandreas, D.E., et al., *Astrophys. J.*, **383**, 653 (1991).
16. Amenomori, M., et al., *Phys. Rev. Lett.*, **69**, 2468 (1992).
17. Cronin, J.W. et al., *Phys. Rev.*, **D45**, 4385 (1992).
18. Borione A. et al., *Astrophys. J.* **481**, 313 (1997).
19. Gould, R.J., *Phys.Rev.Lett.* **15**, 511 (1965).
20. Kennel, C.F., Coroniti, F.V., *Astrophys.J.* **283**, 694 (1984).
21. de Jager, O.C., and Harding, A.K., *Astrophys. J.* **396**, 161 (1992).
22. Atoyan, A.M., and Aharonian, F.A., *Mon.Not.Roy.Ast.Soc.* **278**, 525 (1995).
23. de Jager, O.C. et al., *Astrophys. J.*, **457** 253 (1996).
24. Mohanty, G. et al., *Astropart. Phys.*, submitted (1997).
25. Tanimori, T. et al., *Astrophys. J.* submitted (1997).
26. Hillas, A.M., et al. *Astrophys.J.*, submitted (1997).
27. Kifune, T. et al., *Astrophys. J.* **438**, L91 (1995).
28. Aharonian, F.A., Atoyan, A.M., and Kifune, T., *Mon. Not. Royal Astron. Soc.*, submitted (1997).
29. Turver, K.E., et al., these proceedings, (1997).
30. Yoshikoshi, T. et al., *Astrophys. J.* submitted, (1997).
31. Markwardt, C.B. and Ogelman, H., *Nature* **375**, 40 (1995).
32. Kawai, N. and Tamura, K., *"Pulsars: Problems and Progress" (IAU Colloquium 160)*, Eds. S. Johnston, M.A. Walker and M. Bailes. Publ. ASP Conference Series, **105**, 367 (1996).
33. Drury, L.O'C., Aharonian, F.A., Volk, H.J., *Astron.Astrophys.* **287**, 959 (1994).
34. Naito, T., Takahara, F., *J.Phys.G.: Nucl.Part.Phys.* **20**, 477 (1994).

35. Aharonian, F.A., Drury, L.O'C., Volk, H.J., *Astron.Astrophys.* **287**, 959 (1994).
36. Esposito, J.A., et al., *Astrophys. J.*, **461**, 820 (1996).
37. Buckley, J.H. et al., *Astron.Astrophys.* (in press) (1997)
38. Lessard, R.W. et al., *Proc. 25th ICRC (Durban, South Africa)* (in press) (1997).
39. Fegan, D.J., *Proc. 21st ICRC (Adelaide, Australia)* **11**, 23 (1990).
40. Weekes T.C., *Space Sci. Rev.*, **59**, 315 (1992).
41. Chadwick, P.M. et al., *J.Phys.G.* **16**, 1773 (1990).
42. Fegan, D.J., *High Energy Astrophysics*, Editor: James M. Matthews, published by World Scientific, 107 (1994).
43. Dowthwaite, J.C. et al., *Astrophys.J.* **286**, L35 (1984).
44. Gillanders, G.C., et al., *Proc. 25th ICRC (Durban, South Africa)* (in press) (1997).
45. Gupta, S.K., et al., *Astron. Astrophys.*, **245**, 141 (1991).
46. Bhat, P.N. et al., *Astron.Astrophys.* **81**, L3 (1980).
47. Bowden, C.C.G. et al., *J.Phys.G.* **19**, L17 (1993).
48. Vishwanath, P.R. et al., *Astron.Astrophys.* **267**, L5 (1993).
49. Akerlof, C.W. et al., *Astron.Astrophys.* **274**, L17 (1993).
50. Nel, H.I. et al. *Astrophys.J.* **418**, 836 (1993).
51. Srinivasan, R. et al., *Astrophys. J.*, in press (1997).
52. Susukita, R., *Ph D Thesis, Kyoto University*, unpublished (1997).
53. Cheng,K.S. and Ding, W.K.Y., *Astrophys. J.*, **431**, 224 (1994).
54. Romani, R., *Astrophys.J.* **470**, 469 (1996).
55. Thompson, D.J. et al., *Astrophys.J.* **101**, 259 (1995).
56. Fierro, J., *Ph.D. Dissertation, Stanford University (unpublished)* (1996).
57. Buckley, J. et al., *Proc. 25th ICRC (Durban, South Africa)* (in press) (1997).
58. Reynolds, P.T. et al., *Astrophys.J.*, **404**, 206 (1993).
59. Alexandreas, D.E. et al., *Astrophys.J.* (**405**, 353 (1993).
60. Borione, A. et al., *Phys.Rev. D*, **55**, 1747 (1997).
61. Chardin, G. and Gerbier, G., *Astron.Astrophys.*, **210**, 52 (1989).
62. Raubenheimer B.C. et al., *Astrophys.J.*, **428**, 777 (1994).
63. Chadwick P.M. et al., *Astroparticle Physics*, **4**, 99 (1995).
64. Mori, M. et al., *Astrophys.J.*, **476**, 842 (1997).
65. Vestrand, T.W. and Sreekumar, P., these Proceedings (1997).
66. Hunter, S.D. et al., *Astrophys.J* 481, 205 (1997).
67. Gralewicz, P., Wdowczyk, J., Wolfendale, A.W., *Astron.Astrophys.* **318**, 318 (1997).
68. Porter, T.A., Protheroe, R.J., *J.Phys. G: Nucl. Part. Phys.*, submitted (1997).
69. Aharonian, F.A., Atoyan, A.M., *Astron.Astrophys.* **309** 917 (1996).
70. Nishimura J. et al., *Astrophys.J* **238**, 394 (1980).
71. Atoyan A.M., Aharonian F.A., Völk H.J., *Phys. Rev D* **52**, 3265 (1995).
72. Schmele, D. et al., *Proc. 25th ICRC,(Durban, South Africa)* in press (1997).
73. Lin, Y.C. et al., *Astrophys. J.* **416**, L53 (1993).
74. Punch, M. et al., *Nature* **358**, 477 (1992).

75. Kerrick, A.D. et al., *Astrophys.J.* **438**, L59 (1995).
76. Macomb, D.J. et al. *Astrophys. J.* **449**, L99 (1995); *Astrophys. J.* **459**, L111 (Erratum).
77. Buckley, J.H. et al., *Astrophys. J.* **472**, L9 (1996).
78. Buckley, J.H. et al., *Proc. 25th ICRC (Durban, South Africa)* in press (1997).
79. McEnery, J.E. and Buckley, J.H., *Astrophy. Sp.Sci.*, (in press) (1997).
80. Petry, D. et al., *Astron.Astrophys.* **311**, L13 (1996).
81. Bradbury, S.M. et al., *Astron.Astrophys.* (in press) (1997).
82. Catanese, M. et al., *Astrophys.J.* in press (1997).
83. Aharonian, F.A. et al., *Astron.Astrophys.* (submitted) (1997).
84. Barrau, A., et al., *Proc. 25th ICRC (Durban, South Africa)*, in press (1997).
85. Aiso, S. et al., *Proc. 25th ICRC (Durban, South Africa)*, in press (1997).
86. Catanese, M. et al., *Proc. 25th ICRC (Durban, South Africa)*, in press, (1997).
87. Stecker, F. W., De Jager, O.C., Salamon, M.H., *Astrophys. J.* **390**, L49 (1992).
88. Biller, S., *Astroparticle Physics* **3**, 385 (1995).
89. MacMinn, D., Primack, J. R., *Space Sci. Rev.* **75**, 413 (1996).
90. Alexandreas, D.E. et al., *Astrophys.J.* **418**, 832 (1993).
91. Allen, W.H. et al., *Proc. 23rd ICRC (Calgary)* **1**, 420 (1993).
92. Amenomori, M. et al., *Proc. 23rd ICRC (Calgary)* **1**, 412 (1993).
93. Catanese, M., et al., *Astrophys.J.* **469**, 572 (1996).
94. Bowden, C.C.G. et al., *Proc. 22nd ICRC (Calgary)* **1**, 294 (1993).
95. Kerrick, A.D. et al., *Astrophys. J.*, **452**, 588 (1995).
96. Kifune, T., *Space Science Reviews*, **75**, 31 (1996).
97. Grindlay, J.E. et al., *Astrophys. J.*, **197**, L9 (1975).
98. Carriminana, A. et al. *Astron.Astrophys.*, **228**, 327 (1990).
99. Allen, W.H. et al., *Proc. 22nd ICRC (Dublin, Ireland)* **1**, 344 (1991).
100. Protheroe, R.J., Stanev, T., *Mon.Not.Roy.Ast.Soc.* **264**, 191 (1993).
101. Aharonian F.A., Coppi P.S., Völk H.J., *Astrophys. J.* **423**, L5 (1994).
102. Plaga, R., *Nature* **374**, 430 (1995).
103. Herterich, K., *Nature* **250**, 311 (1974).
104. Krennrich, F. et al., *Astrophys.J.* **481**, 758 (1997).
105. De Jager, O.C., Salamon, M.H., Stecker, F.W., *Nature* **369**, 294 (1994).
106. Page, D. N., & Hawking, S. W., *Astrophys. J.* **206**, 1 (1976).
107. Halzen, F., et al., *Nature* **353**, 807 (1991).
108. Porter, N. A., & Weekes, T. C., *Mon.Not.Roy.Ast.Soc.* **183**, 205 (1978).
109. Alexandreas, D. et al. *Proc. 23rd ICRC (Calgary)* **1**, 428 (1993).
110. MacGibbon, J. H., & Brandenberger, R. H., *Phys.Rev.D* **47**, 2283 (1993).
111. Samura, T., & Kobayakawa, K., *Proc. 23rd ICRC (Calgary)* **1**, 128 (1993).
112. Connaughton, V., et al., *Astrophys. J.* in press (1997).
113. Connaughton, V., et al., *Astroparticle Phys.* (submitted) (1997).
114. Boyle, P.J. et al., *Proc. 25th ICRC (Durban, South Africa)* in press (1997).
115. Aharonian, F.A. et al., these proceedings, (1997).

GAMMA RAY MYSTERIES

Pulsar Counterparts of Gamma-Ray Sources

P.A. Caraveo[1] and G.F. Bignami [2,1]

[1] *Istituto di Fisica Cosmica del CNR, Via Bassini, 15, 20133 Milano, Italy*
[2] *Agenzia Spaziale Italiana, Via di Villa Patrizi 13, Roma, Italy*

Abstract. The EGRET catalogue of unidentified X-ray sources has more objects along the galactic disk than at high galactic latitude, where identifications are comparatively easier. On the other hand, the Egret/GRO mission has already identified several known radio pulsars as gamma-ray sources as well as discovering Geminga's nature as a pulsar. If Geminga is not a unique case, as it is very likely not to be, than other galactic sources could, in fact, be radio quiet isolated neutron stars.
For these, the identification work is extremely difficult and should anyway start from high resolution X-ray/optical data.

INTRODUCTION

Isolated Neutron stars (INSs) are the only galactic objects surely identified as γ-ray emitters. Although not completely understood, as far as the emission mechanism is concerned, their phenomenology is fairly well known.
Combining multiwavelength observations spanning the entire electromagnetic spectrum, we have learned that diversity is the rule amongst γ-ray emitting INSs. With the notable exeption of the Crab, light curves are remarkably different in separate energy ranges (see e.g. Kanbach, 1997, for a review), hinting that alternate beaming geometries are possibly associated to different emission mechanisms. Moreover, the fraction of the rotational energy loss re-emitted in high-energy γ-rays ranges from a fraction of a percent, in the case of the Crab, to the quasi-totality for PSR1055-52 (see e.g. Goldoni et al, 1995). This does not translate directly into luminosity because the rotational energy loss of young pulsars is orders of magnitude larger than that of older objects, so that, in spite of lower efficiency, young pulsars are indeed the brightest γ-ray sources in our Galaxy.
With six successful identifications (counting also PSR 1915+32, which is not included in the 2nd EGRET catalogue since it is detected only as a pulsating

source) amongst the 45 low latitude sources seen by EGRET (Thompson et al, 1995, Thompson et a l, 1996), it is natural to explore the possibility that at least a fraction of the remaining low latitude sources belong to the same class of compact objects. Indeed, many searches for pulsars inside COS-B and EGRET error boxes have been and are being ca rried out: all in all pulsars have received far more attention than any other galactic population thought to be a possible source of high-energy γ-rays.

Why is the pulsar hypothesis so successful amongst γ-ray astronomers?

In a branch o f astronomy hampered by poor angular resolution and low counting rates, a pulsar identification is by far the most unambiguous one. When the light curve obtained folding the γ-ray photons at a known pulsar period is ! statistically compelling, one shou ld not worry about chance superposition nor about lengthy follow-up observations: the source is identified for sure.

This is the appeal of a pulsar identification and this is why pulsar searches have been performed on γ-ray error boxes as soon as COS-B discovered the UGOs (Unidentified Gamma Objects, see Bignami and Hermsen, 1983 for a review). The majority of the searches zeroed in on Geminga, but a fair number of the sources of the 2nd COS-B catalogue were surveyed with no luck. However, one of the COS-B UGOs turned out to be a pulsar which was discovered in a routine survey for southern pulsars ten years after the end of the COS-B mission. It is the case of 2CG342-02, identified in 1992 with PSR 1706-44 (Thompson et al, 1992), a Vela-lik e pulsar at 2 kpc.

COS-B had two, with PSR 1706-44 three, radio pulsars amongst 22 low latitude sources. EGRET has five (six if we count the tentative identification od PSR 0656+14 and seven if we consider also PSR 0540-69, seen by all CGRO instruments but EGRET) amongst 45 sources. In spite of the increased sensitivity of EGRET, the ratio between pulsar identification and total number of sources appears to be constant. Also the lack of results, experienced at the time of COS-B, appears to be unchanged. Dedicated radio searches (Nice and Sayer, 1997), aimed precisely at the search for radio pulsars inside the error boxes of 10 of the brightest EGRET sources, yielded null results, showing that the straightforward radio pulsar identification is not the on ly possible solution to the enigma of the unidentified high-energy γ-ray sources. This is further strengthned by the work of Nel et al. (1997) who investigated 350 known pulsars finding few positional coincidences but no significant timing signature for any of the pulsars in the survey.

A DIFFERENT APPROACH

Indeed, γ-ray astronomy does offer a remarkable example of an Isolated Neutron Star (INS) which behaves as a pulsar as far as X-and-γ astronomy are concerned but has little, if at all, radio emission. As an established representative of the non-radio-loud INSs (see Caraveo, Bignami and Trümper,

1996 for a review), Geminga offers a more elusive template behaviour: prominent in high energy γ-rays, uneventful in X-rays and downright faint in opti cal, with sporadic or no radio emission. Although the latitude distribution of unidentified EGRET sources shows that, in average, they are at least 10 times as distant as Geminga (Mukherjee et al, 1995), the multiwavelength behaviour of this source is har d to beat when one tries to link gamma-ray sources to compact objects in the absence of a radio signal. While one should always keep a totally open mind and be ready to find something new and different, Geminga is the template observers have in mind when planning observing strategies. Unfortunately, in spite of our knowledge of the behaviour of the real Geminga, the study of a Geminga-like source still represents a great challenge to observers since it defies well established techniques. To appreciate such a challenge let us briefly review the many steps that lead to the identification and the understanding of this object with an aim to find the signature to look for.

THE MANY FIRSTS OF GEMINGA

Briefly, the source was discovered in high energy γ-ray by the SAS-2 satellite in 1972 (Fichtel et al, 1975), an X-ray counterpart, suggesting position and distance, has been proposed in 1983 (Bignami et al. 1983) and an optical one, refining the position, in 1987/88 (Bignami et al, 1987, Halpern and Tytler, 1988). However, the breakthrough came with the discovery of the 237 msec periodicity in the ROSAT data (Halpern and Holt, 1992). Finding the same periodicity in the simultaneous high energy γ-ray data of the EGRET instrument (Bertch et al, 1992), as well as in the old archival COS-B (Bignami and Caraveo, 1992) and SAS-2 data (Mattox et al, 1992), yielded the value of the period derivative and thus of the object's energetics. The discovery of the proper motion of the proposed optical counterpart (Bignami et al, 1993) confirmed the optical identification and, thus, provided the absolute positioning of Geminga to within the systematic uncertainty of the Guide Star Catalogue, i.e. 1". Next came the measure of the source parallactic displacement, yielding a precise measure of its distance (Caraveo et al, 1996).

More HST observations, confirming and refining difficult measurements with ground-based instruments, have shown that a broad feature, centered at $\lambda = 5998 Å$ and with a width of 1,300 Å, is superimposed to the Rayleigh-Jeans continuum, as extrapolated from the soft X-rays (Bignami et al, 1996; Bignami, 1997). If interpreted as an ion-cyclotron emission, it implies, for $Z/A=1$, a B field of $3.25 \ 10^{11}$ Gauss as opposed to the value of $1.5 \ 10^{12}$ obtaine d, theoretically, using the Period and Period derivative. This is the first time that the magnetic field of a neutron star is directly measured. Recently, the phenomenology of the source at high energies has been considerably enriched, owing to the very precise positioning of the optical counterpart. The

TABLE 1. GEMINGA(1973-present)

• 1st unidentified γ-ray source
• 1st INS discovered through high-energy emission and identified through its X and γ-rays
• 1st INS identified without the help of radio astronomy
• 1st INS optically identified through its proper motion
• 1st INS the distance of which is measured through its optical parallax
• 1st direct view in optical/UV of the surface/photosphere of a NS
• 1st evidence for an atmosphere surrounding NS crust
• 1st direct measurement of the surface magnetic field of an INS
• 1st INS the timing parameters of which are determined solely by high energy γ-ray data
• 1st optical measurement of absolute position of an INS within 40 mas (This leads to the first measurement of the braking index of a 10^5 y old NS)
• 1st evidence (together with PSR0656+14 and PSR 1055-58) of an INS with a two-component X-ray emission

possibility to link HST data to the Hipparcos reference frame, yielded the position of Geminga to an accuracy of 0.040 arcsec, a value unheard of for the optical position of a pulsar, or of an object this faint (Caraveo et al, 1997). This positional a ccuracy has allowed to phase together data collected over more than 20 years by SAS-2, COS-B and EGRET, unveiling very promising timing residuals (Mattox et al, 1997). The many "firsts" of Geminga are summarized in Table 1.

Quite surprisingly, some of the key parameters of Geminga are now known with an accuracy better than available for the Crab pulsar. This is due in part to the 20 year long chase (see Bignami and Caraveo, 1996 for a review), in part to the remarkable sta bility of this object which rendered possible to phase together such a long time span of γ-ray data.

A STRATEGY FOR THE FUTURE

There is no question that a sizeble fraction of the EGRET UGOs are galactic. Of these. It is reasonable to expect at least several to be radio-quiet INSs, since no further radio pulsars can be identified. Also, no other compact (or star-like) class of ga lactic objects has yet been identified with certainty as a γ-ray emitter. It make sense, therefore, to single out further γ-ray INSs even if the process might be difficult and tiresome, as in the case of Geminga. The potential reward will obviously be a better understanding of the radio-quiet, γ-ray loud INSs as a class. This would yield a precious addition to the general neutron star scanty phenomenology.

Furthermore, a dedicated search process like that required to nail down an INS might well yield, as a bonus, new identifications of γ-ray objects of serendipitous nature.

To plan a strategy for such a search it is as easy as it is difficult to predict, with any confidence, its probability of success. This is to say that the

only obvious way forward is one similar to the "Geminga chase" (Bignami and Caraveo, 1996). At the same time, we know from the start that, although possible Geminga-like, the majority of the UGOs must differ significantly from Geminga itself. The main difference will be in the absolute value of their γ-ray luminosity: objects as (relatively) faint as Geminga, will never be seen at the distances (several kpc) that UGOs must have to show their narrow latitude distribution. This could be d ue to differences in ages, which in term would yield different ratios of the power emitted in thermal versus non-thermal processes, both in optical and X-rays. This will impact on INS visibility.
Nevertherless, until a better one is found, the UGO identification strategy can only be as follows:
1- map the UGO boxes with X-ray imaging devices
2- select those few sources which have a very high F_x/F_v
3- search for possible X-ray pulsations
4- go for optical IDs, using all possible methods, not forgetting proper motion.
With respect to the Geminga chase, carried out in the '80s and early '90s, steps 1 and 4 now benefit of a far deeper penetration in the sky owing to existing (and upcoming) orbiting and ground based telescopes. In the next couple of years, for example, the EPIC instrument on ESA's XMM will be operational as will be (at least) UT1 of ESO's VLT. Their joint usage, if well thought out, will improve by at least three to four magnitudes on the ROSAT/NTT combination, which did much of the work on Geminga. The extremely accurate relative astrometry, possible with the new instruments on HST will detect the much smaller proper motions (and parallaxes) of more distant INSs.
It should be exciting to see, within the next 5 years, how many more Geminga will be seen and if indeed a new different galactic population, other than INSs, is needed to explain EGRET's UGOs.

REFERENCES

Bertsch D.L., et al. 1992, *Nature* 357, 306

Bignami, G.F.,Caraveo,P.A., and Lamb, R.C. 1983, *Ap.J.*, 272, L9

Bignami, G.F. and Hermsen W. *Ann. Rev. Astr. Astrophys.*, 21,67

Bignami,G.F., Caraveo,P.A., Paul, J.A., Salotti,L. and Vigroux, L. 1987, *Ap.J.*, 319, 358

Bignami, G.F., and Caraveo, P.A. 1992, *Nature*, 357, 287

Bignami, G.F., Caraveo, P.A., and Mereghetti, S. 1993, *Nature*, 361,704

Bignami G.F., Caraveo, P.A., Mignani R., Edelstein J. and Bowyer S. 1996, *Ap.J.*, 456,L111

Bignami G.F., and Caraveo P.A. 1996, *Ann. Rev. Astr. Astrophys.*, 34,331

Bignami G.F. 1997 *Adv. Space Research*, in press

Caraveo P.A., Bignami G.F., Mignani R., and Taff L.G. 1996, *Ap.J.*, 461,L91

Caraveo P.A., Bignami G.F., Trümper J.A. 1996, *A & A Review*, 7,209

Caraveo P.A.,et al. 1997, *ESA SP 402*, in press

ESA 1997, *The Hipparcos and Tycho Catalogues* ESA SP 1200
Goldoni P., Musso, C., Caraveo P.A., Bignami ,G.F. 1995 *A & A* 298,535
Halpern, J.H., and Tytler, D. 1988, *Ap.J.*, 330, 201
Halpern, J.P., and Holt, S.S. 1992, *Nature*, 357, 222
Kanbach, G. 1997 *Adv. Space Research*, in press
Fichtel, C.E. et al. 1975 *Ap.J.*, 198,163
Mattox J.R., Halpern J.P. and Caraveo P.A. 1996, *A & A Suppl*, 120C, 77
Mattox J.R., Halpern J.P. and Caraveo P.A. 1997 *Ap.J.* in press
Mattox J.R. et al 1992 *Ap.J.*, 103,638
Mukherjee R. et al 1995 *Ap.J.*, 441, L61
Nel H.I. et al. 1996 *Ap.J.*, 465,898
Nice D.J. and Sayer R.W. 1997 *Ap.J.*, 476,261
Taylor,J., Manchester, R.M. Lyne A.G. 1993, *Ap.J. Suppl.*, 88,529
Thompson D.J. et al, 1992 *Nature* 359,615
Thompson D.J. et al, 1995 *Ap.J. Suppl.* 101,259
Thompson D.J. et al, 1996 *Ap.J. Suppl.* 107,277

On the Nature of the Unidentified EGRET Sources

R. Mukherjee,* I. A. Grenier+ and D. J. Thompson†

*McGill University, Physics, Montreal, H3A 2T8, Canada[1]
+Université Paris 7 & CEA/Saclay/service d'Astrophysique
†NASA/GSFC, Code 661, Greenbelt, MD 20771

Abstract.
Approximately 60% of the high-energy gamma-ray sources in the 2nd EGRET (Energetic Gamma Ray Experiment Telescope) catalog remain unidentified. So far none of these sources has been definitively identified with astronomical objects known at other wavelengths, and this has turned out to be one of the biggest mysteries presented by EGRET. In this article we describe the latitude and longitude distributions of the unidentified sources and discuss their collective properties. The distribution of the unidentified sources is unlike that of the identified classes, like pulsars or blazars, although the unidentified sources do seem to be correlated with our galaxy. In the intermediate Galactic latitudes, between 5° and 30°, there appears to be an excess of unidentified sources that can be associated with the Gould Belt of massive stars and interstellar clouds. Statistical analysis of the EGRET source distribution indicates that many of the plane sources are correlated with OB associations or other tracers of star formation likely to be sites harboring pulsars. Some of the Galactic unidentified sources are strongly correlated with supernova remnants. The association seems strongest for supernova remnants close to molecular clouds, which might be targets for cosmic rays accelerated in the supernova remnants. A few of the high-latitude unidentified sources seen by EGRET may be previously unrecognized blazars or other type of Active Galactic Nuclei (AGN). This article reviews the work done towards understanding the nature of the unidentified sources seen by EGRET.

INTRODUCTION

The EGRET (Energetic Gamma Ray Experiment Telescope) instrument on the Compton Gamma-Ray Observatory (CGRO) has surveyed the gamma-ray sky at energies above 100 MeV. The second EGRET (2EG) catalog [39] of high-energy gamma-ray sources and its supplement (2EGS) [40] list 157

[1)] USRA Research Scientist, NASA/GSFC, Code 610.3

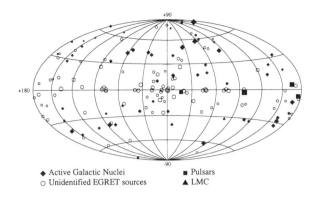

FIGURE 1. Locations of the unidentified 2EG and 2EGS sources in Galactic coordinates. The size of the symbols represents the highest intensity seen for this source by EGRET [40].

sources detected up to the end of Phase 3 (October 1994). Of these, 56 sources at high Galactic latitudes ($|b| > 10°$) and another 40 sources at ($|b| < 10°$) remain unidentified, with no known counterparts at other wavelengths. Figure 1 shows the locations of the unidentified sources, listed in the 2EG and 2EGS catalogs, in Galactic coordinates. The size of the symbol represents the highest intensity seen for each source by EGRET.

The unidentified sources along the Galactic plane have remained a mystery since their discovery with the COS B satellite [37]. Of the 20 or more localized excesses detected by COS B, only four were identified with known objects. Later analysis [22] indicated that several of these excesses could be explained as concentrations of interstellar gas irradiated by cosmic rays, although this still left nearly a dozen sources unidentified. The collective properties of the COS B sources indicated that the sources were largely Galactic [37]. Results from EGRET have shown two of the COS B catalog sources to be pulsars, namely, Geminga [3,23] and 2CG 342−02 (PSR B1706−44) [41].

At high Galactic latitudes, one source is identified with the Large Magellanic Cloud and 51 with blazar-class AGN [30], leaving 56 sources unidentified. Some of these blazar associations are not certain; Mattox et al. [21] find only 42 identifications to have high-confidence. Conversely, some of the unidentified high-latitude EGRET sources are likely to be blazars. In addition to the 42 considered strongest, Mattox et al. [21] note 16 possible associations with bright, flat-spectrum, blazar-like radio sources.

Despite efforts at finding counterparts, however, a large fraction of the high-latitude EGRET sources and the vast majority of the low-latitude sources remain unidentified. The nature of these unidentified sources is one of the outstanding puzzles in astronomy today. This article will review the work that has been done towards characterizing and identifying these sources since the launch of CGRO.

EGRET SOURCE SENSITIVITY

Although EGRET has completed an all-sky survey, not all regions of the sky were observed with the same sensitivity. Two factors enter the calculation of source-detection sensitivity: exposure and background. Figure 2 shows the EGRET exposure above 100 MeV as a function of sky location for the sum of Phases 1, 2, and 3, the data used for the 2EG and 2EGS catalogs. The high and low regions differ by more than a factor of 5. Figure 3 shows the all-sky intensity measured by EGRET in the same energy range. The map is dominated by the bright diffuse emission along the Galactic plane [13]. The gamma-ray source-detection threshold will clearly be higher in regions of limited exposure or high diffuse radiation. Mattox et al. [20] show that the significance of detection S of a source with flux F is related to the exposure E and the background B by $S \sim F\sqrt{E/B}$.

Because the systematic uncertainties in the EGRET analysis are larger in the high-intensity Galactic plane region (due to the highly structured diffuse emission), the catalog adopts two different thresholds: 4 σ for $|b| > 10°$ and 5 σ for $|b| < 10°$. What these different source sensitivities and thresholds mean is that the EGRET catalog cannot be taken as a uniform sampling of the sky. Any suggested correlation with a known source population must take this non-uniformity into account.

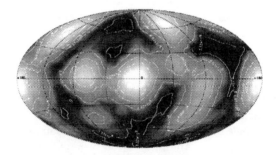

FIGURE 2. EGRET sky exposure in units of cm^2 s (E>100 MeV) for the Compton Gamma-Ray Observatory combined Phase 1 (1991 April - 1992 November), Phase 2 (1992 November - 1993 September), and Phase 3 (1993 September - 1994 October). The contours are 3.33 x 10^8, 6.67 x 10^8, 10.0 x 10^8, 13.3 x 10^8 and 16.7 x 10^8 [40].

HIGH-LATITUDE SOURCES $|b| > 10°$

Figure 4 shows the latitude distribution of the unidentified 2EG and 2EGS sources. To take into account the highly non-uniform exposure, the number of sources observed in a latitude interval was divided by the solid angle and the average exposure. As noted previously by several authors [32,29,12], the unidentified sources show an excess at low Galactic latitudes. Özel & Thompson [32] have compared the source distribution of the high-latitude unidentified sources to the latitude distribution of the 51 Active Galactic Nuclei (AGN) in the 2EG catalog. At the time of their analysis the 2EGS catalog was not available. Özel & Thompson find that the distribution of the AGN is flat, as expected for an isotropic distribution, with an average source density of 1 ± 0.1 (sr $\times 10^8$ cm^2 s)$^{-1}$. This is quite different from the distribution of the unidentified sources. Özel & Thompson note that 26 of the 39 unidentified sources are in the latitude range $10° < |b| < 30°$, compared to only 18 of the 51 AGN in the same range. They have modeled the latitude distribution of the unidentified sources as both a Gaussian plus a constant function, as well as an exponential fall-off from $b = 0°$ plus a constant function. Neither of the two functions was found to fit the data particularly well, owing to the excess of sources in the latitude range $20°$-$30°$ compared to the $10°$-$20°$. However, they did find that the unidentified source distribution is clearly different from the AGN distribution, and demands a large, if not dominant, Galactic component to the unidentified-source population.

Özel & Thompson [32] also compared the $\log N - \log S$ distributions of the high-latitude unidentified 2EG sources with that of the AGN. Figure 5 shows the $\log N - \log S$ distribution of the unidentified sources with a superimposed fit of the form $N \sim S^{-1.4}$, where N is the number of sources with flux greater than S. The function fits both the AGN distribution and the unidentified-source distribution rather well, with a turnover at a flux $\sim 2 \times 10^{-7}$ photon

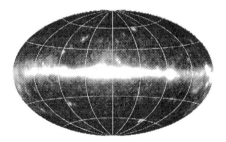

FIGURE 3. Intensity map of the sky seen by EGRET at energies above 100 MeV. The map is in Galactic coordinates, centered on $l = 0°$ [40].

cm^{-2} s^{-1}. Both fits are consistent with the expected $S^{3/2}$ dependence of an isotropic distribution of sources, indicating that the source populations are largely drawn from an isotropic distribution of sources. The only difference between the two distributions is that the highest flux of the unidentified sources is much lower than that of the AGN.

A rather different approach was adopted by Grenier [11,12] to study the medium-latitude 2EG and 2EGS sources. Grenier has taken a statistical approach to find correlations of the 72 unidentified gamma-ray sources having $|b| \geq 2.5°$, with different classes of objects. To take into account the non-uniform exposure and sensitivity of EGRET to different parts of the sky,

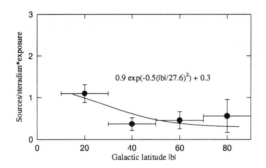

FIGURE 4. Latitude distribution $|b| > 10°$ of the unidentified EGRET sources in the 2EG catalog shown as density of sources per steradian per exposure of 10^8 cm^2 s [32].

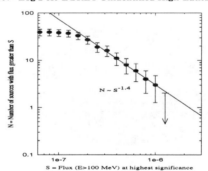

FIGURE 5. Plot of $\log N$ vs $\log S$ for the unidentified EGRET sources with $|b| > 10°$ [32].

Grenier [12] has developed a likelihood test that compares observed and predicted source counts in 5° by 5° bins. To a first approximation, extragalactic sources are isotropically distributed in the sky for any luminosity function. For a Galactic population, however, the contrast in the longitude and latitude profiles strongly depends on their luminosity function. Grenier [12] fit the spatial distribution of the unidentified sources to a linear combination of an isotropic distribution plus various Galactic components. Such components might be a population of sources in a spherical halo of radius 20 kpc, or nearby in the Galactic disc (with a density profile falling off the plane), or spread in the Galaxy with a scale height of 0.4 kpc and radial scale length of 9 kpc, or finally, a population of sources in the Gould Belt. The tests provided strong evidence for the Galactic origin of the majority of the unidentified sources (65 ± 10). Allowing for different luminosity functions for the extragalactic and Galactic sources, the chance probability that the Galactic component is merely due to random fluctuations of an isotropic population range from 10^{-9} to 10^{-5}. Mereghetti & Pellizzoni [26] have reached the same conclusion. Considering the integral distributions in longitude and latitude of all the EGRET sources (unidentified and AGN), they find that ~ 14 AGN may be present in the unidentified sample. In this sample, Grenier [12] has further found a significant correlation between the excess of unidentified sources at medium latitudes, $3° < |b| < 30°$, and the location and extent of the Gould Belt, as mapped either by its OB star or its interstellar gas content. A minimum of 21 ± 6 sources and a maximum of 40 ± 8 sources can be statistically associated with the Belt, depending on their being close to the Belt molecular clouds or being more smoothly distributed inside the Belt.

Özel & Thompson [32], Grenier [12], and Mereghetti & Pellizzoni [26] agree that the medium and high-latitude unidentified sources have a strong Galactic component, in addition to an extragalactic, isotropic contribution.

At medium latitude, neutron stars (and black holes) recently produced by supernovae in the Gould Belt are obvious candidates for the gamma-ray sources associated with the Belt. Using the number of OB stars present today inside the Belt and assuming a constant birth rate and widely different initial mass functions, Grenier [12] finds that 20 to 30 supernovae explosions have occurred per million year in the Gould Belt for the past 2 or 3 million years. Most of the EGRET sources in the Belt appear to be non-variable [24] and therefore excellent pulsar candidates. Even faint pulsars such as Geminga could easily be detected by by EGRET up to the most remote part of the Belt. While pulsar-beam geometry and beaming fraction are poorly known, the current estimates are not inconsistent with the number of sources and supernovae quoted above. So, to date, pulsars are the likeliest candidates for the Gould Belt sources.

The Galactic component of the unidentified source population at $|b| > 10°$ might be correlated with flaring stars such as RS CVn's or winds from massive early-type (O, B, and Wolf-Rayet) stars, according to Özel & Thompson [32].

This is based on the characteristics of the unidentified gamma-ray source population, namely, that they are nearby, relatively low-luminosity, objects that may be time variable. Flare stars and massive early-type stars are known to have non-thermal radio emission, indicating that particle acceleration, a necessary condition for high-energy gamma-ray production, is possible. However, no obvious correlation of these sources was found with the EGRET data.

The isotropic population of the unidentified sources, as discussed earlier, is most likely to consist of gamma-ray loud AGN. A few of the high-latitude unidentified sources seen by EGRET may be previously unrecognized blazars. For example, based on a gamma-ray flare and new radio data, Lundgren et al. [18] indicate that 2EG J0432+2910 is probably a previously unknown blazar. Mattox et al. [21] summarize the possible associations with blazar-like radio sources. Nolan et al. [31] note a possible association of 2EG J1324-4317 with Cen A, the nearest giant radio galaxy, suggesting a second class of gamma-ray AGN.

LOW-LATITUDE SOURCES $|b| < 10°$

Figure 6 (a) shows the longitude distribution of the 40 unidentified 2EG and 2EGS sources within 10° of the Galactic equator. The analysis was done for energy $E > 100$ MeV. It is seen that the distribution of sources is not uniform over longitude, but has a broad peak with a dispersion of $70° \pm 12°$. Although this indicates a concentration of sources towards the central part of the Galaxy, the absence of a strong concentration in longitude within $\sim 30°$ of the Galactic center suggests that the sources are distributed rather uniformly along the disk. The longitude distribution of the unidentified sources implies that they are not very distant objects, and that the typical distance is not even as far as the Galactic center. This puts an upper limit of ~ 6 kpc to the typical distance of the sources [29]. The COS B data showed similar results for sources with fluxes greater than 1.3×10^{-6} photons cm^{-2} s^{-1} [37].

Figure 6 (b) shows the distribution of the latitudes of the unidentified sources with $|b| < 10°$. A fit to a Gaussian distribution, assuming that these sources are a Galactic population centered on $b = 0°$, yields a standard deviation of $1.6° \pm 0.3°$. The *minimum* scale height of a Galactic population that might include these sources can be taken as ~ 40 pc [9]. Using this value, the average distance D of the sources may be estimated from the relation $D = 40/\sin 1.9°$. This implies that the typical distance of the sources is not less than ~ 1.4 kpc. The luminosities of individual unidentified sources seen by EGRET cannot be calculated because their distances are unknown. However, estimates of the average luminosity can be made from the limits on the typical distances of these sources derived above. For isotropic gamma-ray emission the luminosity L of a source is given by $4\pi D^2 <E> F$, where $<E>$ is the mean energy and F is the flux. Assuming an E^{-2} power law, and taking

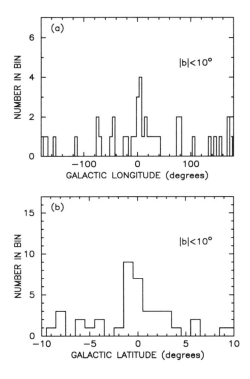

FIGURE 6. Distributions in (a) longitude and (b) latitude of the 40 2EG and 2EGS unidentified sources detected by EGRET at $|b| < 10°$.

a typical flux of 6×10^{-7} photons cm^{-2} s^{-1} for photons with energies greater than 100 MeV [39], and a beaming into 1 sr, the luminosity of these sources lies in the range 7×10^{34} to 1.4×10^{35} ergs s^{-1} [29,15]. The results will be different if a different scale height is used for the population of sources.

Several of the sources concentrated within $\sim 3°$ of the Galactic plane have been positionally associated with observed populations of young objects, e.g., OB associations [14] and supernova remnants (SNRs) [36,6] as well as with hypothetical populations like radio-quiet pulsars [43] or new classes of objects [38]. Kaaret & Cottam [14] have found a statistically significant correlation between the positions of unidentified EGRET sources near the Galactic plane and OB associations. They have found four sources well within OB associations, five sources with high-confidence level position contours overlapping OB association boundaries, and seven sources within 1° of an OB association boundary. All 16 of these sources are within $|b| < 5°$. Figure 7 shows a map of the 2EG sources and the OB associations in Galactic coordinates [14]. The estimate of the chance probability of having 16 or more sources within 1° of an OB association is 6.1×10^{-5}. The main result of Kaaret & Cottam [14] was to establish that the low latitude EGRET sources are Population I objects, i. e. associated with young massive stars or their progeny.

Three-fold associations between a gamma-ray source, OB association, and SNRs were first noticed for COS B sources [27], and have been found again for seven EGRET sources [14,44]. OB associations can be used to trace the population of young pulsars, since there is a strong correlation between pulsar birth sites and OB associations [2]. Pulsars born within an association and having low proper motion will appear near their parent OB association. For e.g., a pulsar with an age of 10^5 years at a distance of 1.5 kpc must have a transverse speed of 250 km s^{-1} to move 1° from its point of origin. Using the known distances to the OB associations to determine the intrinsic luminosity of the unidentified 2EG sources, Kaaret & Cottam [14] find that the distribution of luminosities is consistent with that of the known gamma-ray pulsars. Their results indicate that young pulsars could constitute the majority of the 2EG sources near the Galactic plane.

The idea that young pulsars are potential candidates for the unidentified EGRET sources is also supported by Yadigaroglu & Romani [43,44], based on an outer gap model of pulsars. At an age of $< 10^6$ years, young pulsars have a scale height of ~ 220 pc, comparable with the HI layer, because of

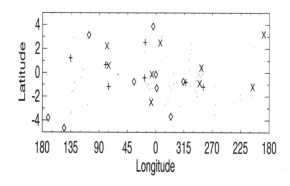

FIGURE 7. Positions of the unidentified EGRET point sources with $|b| < 5°$ in Galactic coordinates. The unidentified 2EG sources within OB associations are indicated by an "X"; those within 1° of an OB association are indicated by a cross; those more than 2° away from an OB association are indicated by a diamond. The positions of OB associations are indicated by dots [14].

their large velocity at birth [19]. A careful estimate using a birth rate of 1/100 year, the beam geometry, and the gamma-ray efficiency of the outer-gap model yields a total of 22 unidentified EGRET sources consistent with this model of young pulsars [43]. The spatial distribution of these pulsars agrees reasonably well with the unidentified source distribution with $|b| < 10°$. In the range $2.5° < |b| < 10°$ these young pulsars may account for a third of the unidentified EGRET sources. Beyond 10° in latitude, however, unless the pulsar scale height at birth is much larger than the 80 pc used by Yadigaroglu & Romani, pulsars in the general disc population are unlikely to explain the Galactic EGRET sources.

While considering the pulsar hypothesis, two important factors regarding the variability and energy spectra of the unidentified sources need to be taken into account. McLaughlin et al. [24] have shown that there is a low-latitude Galactic population of unidentified EGRET sources that are variable, and are, therefore, unlikely to be pulsars. Secondly, Merck et al. [25] and Fierro [8] note that many of the low-latitude sources have rather steep energy spectra, while most of the known pulsars have flat spectra with high-energy cut-offs. The Crab pulsar, which has a steep spectrum, is an exception. Based on multi-wavelength studies, Brazier et al. [4] consider the sources 2EG J0008+7307 (SNR CTA 1) and 2EG J2020+4026 (SNR γ Cyg) to be the best candidates to be radio-quiet gamma-ray pulsars similar to Geminga.

Esposito et al. [6] and Sturner & Dermer [36] have shown that some of the unidentified sources could possibly be SNRs, since there are theoretical reasons to believe that gamma rays may be produced via the interactions of cosmic rays accelerated in supernova shocks [5,33]. Sturner & Dermer have found four superpositions between SNRs and high-confidence, low-latitude sources in the first EGRET catalog [7]. Similarly, Esposito et al. have analyzed the unidentified 2EG sources in the vicinity of 14 radio bright (> 100 Jy at 1 GHz) supernova remnants. They have found a spatial correlation between five SNRs which appear to be interacting with nearby molecular clouds and enhanced gamma-ray emission above the Galactic diffuse emission. Table 1 shows the EGRET measurements of gamma-ray emission from the 14 SNRs, and the 5 EGRET sources that are most likely to be correlated with the SNRs. The two strongest point-like gamma-ray excesses that are coincident with SNRs are 2EG J2020+4026 with γ Cyg and 2EG J0618+2234 with IC 443. The absence of TeV emission from these sources [1,17,34] leaves some doubt about the cosmic ray shock acceleration scenario for the gamma-ray production, however.

EGRET results also suggest other types of Galactic sources. Some examples are:

(1) X-ray binaries: The positional association of 2EG J2033+4112 with Cyg X-3 hints at this to be a gamma-ray source, although no evidence for the characteristic ~ 4.79 hr periodicity observed in X-rays and infrared is seen in

TABLE 1. EGRET measurements of the gamma-ray emission from the supernova remnants in the survey [6].

Name	Flux[a]	EGRET source	EGRET source position (l, b)	σ	E.R.[g]	Notes
W28	55.9 ± 6.6	2EG J1801−2312	6.73, -0.14	8.8	9'	
W44	50.0 ± 8.0	2EG J1857+0118	34.80, -0.76	6.2	27	
W51	< 32.5	-	-	2.9	-	b
CTB 80	< 12.8	-	-	1.4	-	b, c
Cygnus Loop	< 11.6	-	-	2.1	-	b
γ Cyg	126.5 ± 6.9	2EG J2020+4026	78.12, 2.10	21.0	6	
W 63	< 19.4	-	-	2.6	-	b
HB 21	< 22.9	-	-	3.0	-	b
Cas A	< 12.4	-	-	0	-	d
HB 9	< 6.1	-	-	0	-	d
IC 443	50.0 ± 3.9	2EG J0618+2234	189.13, 3.19	15.0	12	
Monocerous	23.0 ± 4.3	2EG J0635+0521	206.17, -0.99	5.9	42	e
Pup A	< 35.6	-	-	3.6	-	b, f
MSH 15-56	< 13.9	-	-	0	-	d

a: Units are 10^{-8} cm^{-2} s^{-1}. The intensities are determined at the location of the TS maximum. The 2σ upper limits are determined at the position of the supernova remnant given in Green 1995 [10].
b: The source significance is between 2.0 σ and 5.0 σ (except for CTB 80 which has significance 1.4 σ).
c: A gamma-ray pulsar has recently been found in the CTB 80 SNR [35].
d: The statistical significance of a source at this position is consistent with zero.
e: The EGRET unidentified source appears extended and lies within the SNR boundary.
f: Pup A lies extremely close to Vela and the quoted upper limit may be high due to contamination from fluctuations in the tail of the Vela point spread function.
g: 95% error contour radius.

the gamma-ray data [28]. Similarly, a transient gamma-ray source coincident with Cen X-3 [42] offers another candidate object.

(2) Flare stars: 2EG J0241+6119 (called 2CG 135+01 in the second COS-B catalog [37]) is associated positionally with the radio flare star GT 0236+610. This unusual object could well be the source of the gamma-rays [16].

(3) Unknown transient sources: As noted by McLaughlin et al. [24], there is a population of low-latitude, time-variable EGRET sources, unlikely to be blazars, pulsars, or SNRs. The most extreme case of such a source is GRO J1838-04 [38], which for a few days in June, 1995, was the second brightest source in the gamma-ray sky. As discussed by Tavani et al. [38], interpretations of this source as an AGN or pulsar are difficult to justify; therefore, this may represent an entirely new class of gamma-ray source.

SUMMARY

The current status of our understanding of the nature of the unidentified EGRET sources is summarized below.

(1) The 2EG and 2EGS catalogs list 56 unidentified sources for $|b| > 10°$ and another 40 sources at low Galactic latitudes ($|b| \leq 10°$). No unambiguous identifications with known astronomical sources are found for any of these sources.

(2) Statistical analysis, latitude distributions, and $\log N - \log S$ plots seem to indicate that the unidentified 2EG and 2EGS sources at high Galactic latitudes ($|b| > 10°$) can be explained largely by a Galactic component together with a small isotropic, extragalactic contribution. Assuming that they are nearby, the implied luminosity of the unidentified sources is in the range $(1-5) \times 10^{32}$ ergs s^{-1} for $E > 100$ MeV [32]. There are no known source populations that can be correlated with these sources, although a correlation with the Gould Belt is found.

(3) The unidentified extragalactic sources could be unrecognized blazars, or other types of AGN not previously seen by EGRET. A correlation with a population of sources of unknown origin cannot be excluded.

(4) The unidentified 2EG and 2EGS sources at low Galactic latitudes have source luminosities in the range 7×10^{34} to 1.4×10^{35} ergs s^{-1} $E > 100$ MeV. Although these sources don't match most of the observed populations, the low-latitude OB associations and SNR associations are the best candidates; there seem to be no corresponding populations at higher latitudes. Under reasonable assumptions, possible source classes that can be used to identify these sources are radio-quiet pulsars, SNRs, X-ray binaries, or transient sources like 2EG J1838−04, which has been seen to flare one time only.

In short, one of the most intriguing questions raised by EGRET has been the nature of its unidentified sources. We hope that the study of the unidentified EGRET sources will continue with further CGRO observations, and perhaps with a future high-energy gamma-ray telescope like GLAST (Gamma-ray Large Area Space Telescope).

R. Mukherjee acknowledges support from NASA Grant NAG5-3696 and would also like to thank Prof. D. Hanna and the High Energy Physics group at McGill University for their hospitality.

REFERENCES

1. Allen, G. E., et al., *24th ICRC* **2**, 443 (1995).
2. Amnuel, P. R., Guseinov, O. H., & Rustamov, Yu. S., *AP&SS* **121**, 1 (1986).
3. Bertsch, D. L., et al., *Nature* **357**, 306 (1992).
4. Brazier, K. T. S., et al., *M. N. R. A. S.* **281**, 1033 (1996).
5. Drury, L., et al., *A&A* **287**, 959 (1994).
6. Esposito, J. A., et al., *ApJ* **461**, 820 (1996).
7. Fichtel, C. E., et al., *ApJS* **94**, 551 (1994).
8. Fierro, J. M., Ph.D thesis, Stanford University, (1995).

9. Giubert, J., Lequeux, J., & Viallefond, F., *A&A* **68**, 1 (1978).
10. Green, D. A., A Catalogue of Galactic Supernova Remnants (1995 July version), Mullard Radio Astronomy Observatory, Cambridge, United Kingdom ("http://www.phy.cam.ac.uk/www/research/ra/SNRs/snrs.intro.html")
11. Grenier, I. A., *Adv. Space Res.* **15**, 573 (1994).
12. Grenier, I. A., *Proceedings 2nd INTEGRAL Workshop, 'The Transparent Universe,'* **ESA SP-382** (1997).
13. Hunter, S. D., et al., *ApJ* **481**, 205 (1997).
14. Kaaret, P., & Cottam, J., *ApJ* **462**, L35 (1996).
15. Kanbach, G., et al., *A&AS* **120**, 461 (1996).
16. Kniffen, D. A., et al., *ApJ* in press (1997).
17. Lessard, R. W., et al., *24th ICRC* **2**, 475 (1995).
18. Lundgren, S. M., et al., *IAU Circular* **6258** (1995).
19. Lyne, A. G., & Lorimer, D. R., *Nature* **369**, 127 (1994).
20. Mattox, J. R., et al., *ApJ* **461**, 396 (1996).
21. Mattox, J. R., et al., *ApJ* **481**, 95 (1997).
22. Mayer-Hasselwander, H. A., & Simpson, G., *Adv. Space Res.* **10**, 89 (1990).
23. Mayer-Hasselwander, H. A., *ApJ* **421**, 276 (1994).
24. McLaughlin, M. A., et al., *ApJ* **473**, 763 (1996).
25. Merck, M., et al., *A&AS* **120**, 465 (1996).
26. Mereghetti, S., & Pellizzoni, A., *Proceedings 2nd INTEGRAL Workshop, 'The Transparent Universe,'* **ESA SP-382** (1997).
27. Montmerle, T., *ApJ* **231**, 95 (1979).
28. Mori, M., et al. *ApJ* **476**, 842 (1997).
29. Mukherjee, R., et al., *ApJ* **441**, L61 (1995).
30. Mukherjee, R., et al., *ApJ* (in press) **490** (1997).
31. Nolan, P., et al., *ApJ* **459**, 100 (1996).
32. Özel, M. E., & Thompson, D. J., *ApJ* **463**, 105 (1996).
33. Pinkau, K., *Phys. Rev. Lett.* **25**, 603 (1970).
34. Prosch, C., et al., *24th ICRC* **2**, 405 (1995).
35. Ramanamurthy, P. V., et al., *ApJ* **447**, L109 (1995).
36. Sturner, S. J., & Dermer, C. D., *A&A* **281**, L17 (1995).
37. Swanenburg, B. N., et al., *ApJ* **243**, L69, (1981).
38. Tavani, M., et al. *ApJ* **479**, L109 (1997).
39. Thompson, D. J., et al., *ApJS* **101**, 259 (1995).
40. Thompson, D. J., et al., *ApJS* **107**, 227 (1996).
41. Thompson, D. J., et al., *Nature* **359**, 615 (1992).
42. Vestrand, W. T., Sreekumar, P., & Mori, M., *ApJ* **483**, L49 (1997).
43. Yadigaroglu, I.,-A., & Romani, R. W., *ApJ* **449**, 211 (1995).
44. Yadigaroglu, I.,-A., & Romani, R. W., *ApJ* **467**, 347 (1997).

A Review of Gamma Ray Bursts

Charles Meegan*, Kevin Hurley†,
Alanna Connors‡, Brenda Dingus¶, Steven Matz**

NASA/Marshall Space Flight Center
†*University of California, Berkeley*
‡*University of New Hampshire*
¶*University of Utah*
**Northwestern University*

Abstract.
Gamma-ray bursts have continued to puzzle astronomers since their discovery thirty years ago. The sources and emission mechanisms are still uncertain. The instruments on the Compton Gamma Ray Observatory, most notably BATSE, have produced a revolution in our understanding of bursts. BATSE found that the burst spatial distribution was isotropic but inhomogeneous, a result inconsistent with any disk population of sources. The currently favored model is one in which the sources lie at cosmological distances. Recent apparent successes in the detection of X-ray and optical counterparts have generated renewed excitement.

1. INTRODUCTION

Gamma ray bursts (GRBs) are perhaps the most puzzling phenomena in modern astrophysics. These brief flashes of non-thermal gamma ray energy, occuring about once per day, continue to defy explanation. The energy is almost exclusively confined to the hard X-ray and gamma ray bands. We do not know what the sources are, what their source of energy is, or how their energy is channeled primarily into gamma radiation.

Before the launch of CGRO, the consensus was that the sources of GRBs were neutron stars located nearby in the galactic disk. We now know from geometrical arguments that this is not the case (see Section 4). The angular and brightness distributions point to a cosmological origin, with the faintest bursts observed by BATSE lying at redshifts of order 1. Alternatively, an extended galactic halo of sources might satisfy the geometrical constraints if the parameters are finely tuned. In either case, the temporal and spectral properties of GRBs present formidable theoretical difficulties.

The lack of counterparts at other wavelengths has traditionally been cited as a reason for the slow progress in understanding bursts. Hence, there is currently great excitement over recent identifications of optical transients associated with GRB970228 and GRB970508. The latter shows absorption features at a redshift of 0.835, which may finally settle the issue of the distance scale in favor of the cosmological hypothesis. These observations are described in Section 6.

Many experiments have contributed to the fund of GRB data. It is fitting, in a Compton Symposium review, to give particular emphasis to the contributions of the instruments on the Compton Gamma Ray Observatory. BATSE was designed specifically for burst observations and has made several important discoveries, including the isotropic distribution that revolutionized the field. OSSE, EGRET, and COMPTEL have provided important spectral, temporal, and location information for events in their fields of view. Data on bursts from BATSE has been made public in a series of burst catalogs [2] [3]. Quite recent data are available from the BATSE World Wide Web site http://www.batse.msfc.nasa.gov.

See [1] for a more complete review of gamma-ray bursts. The proceedings of the Third Huntsville Symposium on Gamma Ray Bursts [4] contain extensive information on the status of GRB research as of late 1995.

2. TEMPORAL STRUCTURE

One of the most striking characteristics of bursts is the wide range of temporal behavior. Bursts can last from less than 0.1 seconds to over 1000 seconds. Their time histories may be relatively smooth are extremely complex and chaotic. Figure 1 shows some examples from the BATSE catalog.

The complex temporal structure makes it difficult to define a burst duration. Commonly used are T_{50} and T_{90}, the time intervals during which 50% and 90% of the background-subtracted counts occur. Figure 2 shows a T_{90} distribution for BATSE bursts that triggered on the 64 ms time scale. The distribution is clearly bimodal with peaks at about 0.1 s and 10 s. The shorter bursts have on average harder energy spectra, but both components are isotropic and inhomogeneous. The significance of this bimodality still eludes us.

OSSE may observe a burst field serendipitously or as the result of an automatic slew using an onboard signal from BATSE [5]. These observations have placed stringent upper limits on gamma ray emission before and after several bursts. In the best cases, the flux immediately before and after a burst must be less than about 2% of the average flux during the event [6] [7].

If burst sources are at cosmological distances, the fainter bursts would be expected to show time dilation. Several studies led by Jay Norris have claimed to see such an effect (e.g. [8]); other studies led by Igor Mitrofanov have not (e.g. [9]). The discrepancy is probably a result of selection effects, since

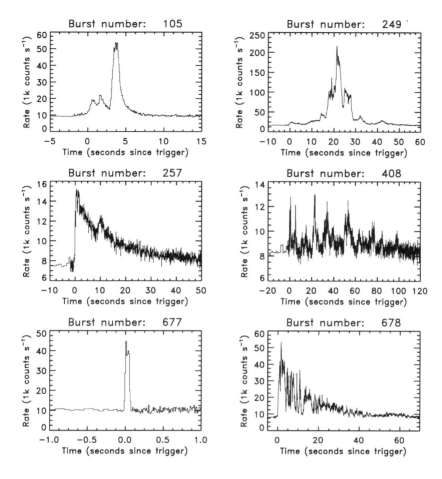

FIGURE 1. A sample of BATSE bursts showing a wide range of temporal behavior

Mitrofanov and Norris agree on the time stretching of a specified set of bursts, but do not agree on which bursts to use for the nearby and the distant samples.

The evidence for burst repetition has been controversial. It is probably fair to say that there is no compelling evidence that bursts repeat, but that upper limits on repetition do not significantly constrain theory. An intriguing possibility of repetition is a series of four BATSE burst triggers with consistent locations [10]. These occurred as two close pairs of triggers two days apart. It is not clear whether these triggers represent 2, 3, or 4 bursts. The chance probability is therefore very uncertain.

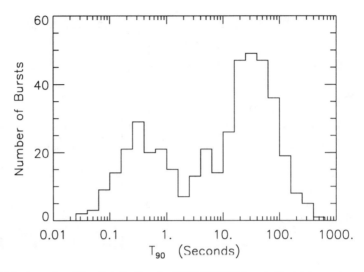

FIGURE 2. Duration (T_{90}) distribution for TBD BATSE bursts that triggered on the 64 ms trigger time scale

3. ENERGY SPECTRUM

The spectral properties of bursts are not nearly as varied as the temporal properties. A typical burst spectrum shows a peak in νF_ν at a few hundred keV. Most burst spectra are well described by a four-parameter Band spectrum [11], consisting of two power laws smoothly joined by an exponential at a break energy. The indices of the power laws are typically 0 to -1.5 below the break energy and -2 to -2.5 above the break energy. Break energies typically range from 100 keV to 1 MeV. COMPTEL has provided statistics on the spectra of bursts above the BATSE energy range. The number spectra have indices ranging from about -1.5 to -3.5 with an average of -2.5 [12]. EGRET observations have extended the energy spectra of bright bursts to above 1 GeV. The spectral index of five events in the 30 MeV to 10 GeV range is about -1.95, with no evidence of a cutoff [13]. The EGRET observations are consistent with all bursts having such power law tails. The most surprising EGRET observation was the discovery of very high energy emission, including one photon at 18 Gev 90 minutes after a burst [14].

Pendleton et al. [15] have identified two distinct types of spectra in bursts. An HE (high energy) spectrum has significant emission extending above 300 keV, while an NHE (no high energy) spectrum exhibits a cutoff that significantly depletes the flux above 300 keV. A single burst may have both HE and NHE pulses, with the HE pulses generally more intense. Bursts that have no HE pulses are thus expected to be intrinsically weaker and therefore closer.

This is supported by the intensity distribution of the NHE-only bursts, which has a −3/2 slope over most of its range.

Mallozzi et al. [16] found that there is an anticorrelation between burst intensity and peak energy that is consistent with cosmological redshift.

4. GLOBAL PROPERTIES

The angular distribution and the intensity distribution of gamma-ray bursts are crucial because they constrain the spatial distribution by model-independent geometrical arguments. Figure 3 shows the angular distribution, in galactic coordinates, of 1635 bursts observed with BATSE. The galactic dipole and quadrupole moments, corrected for anisotropy of the sky exposure, are -0.014 ± 0.014 and 0.004 ± 0.007, consistent with isotropy.

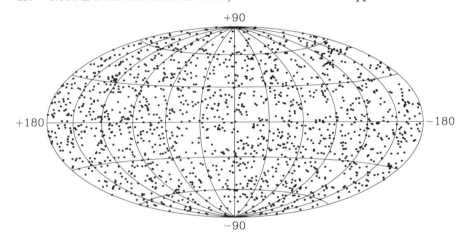

FIGURE 3. The angular distribution in galactic coordinates of 1635 BATSE bursts

The intensity distribution is shown in Figure 4. The integral number of bursts is plotted against the 1.024 s averaged peak flux, and is corrected for detection efficiency, but not for atmospheric scattering. The dashed line represents the −3/2 power law expected for a spatially homogeneous distribution. The deviation of the data from this line may arise from redshift effects (cosmological models) or from a decrease in the source density with distance (extended galactic halo models).

5. COUNTERPART SEARCHES

Much effort has been devoted to trying to find burst counterparts at other wavelengths. Possible counterparts are classified according to the time since

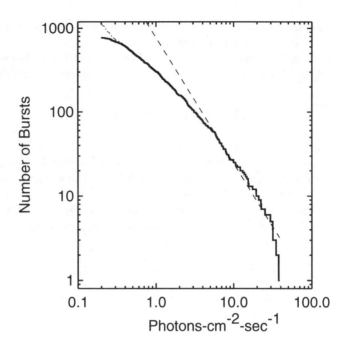

FIGURE 4. The integral intensity distribution for BATSE bursts that triggered on the 1024 ms time scale.

the gamma-ray event. Flaring counterparts are those that occur simultaneously with the gamma rays, fading counterparts would be detectable for some limited time after the event, and quiescent counterparts would be continuously detectable. The searches for counterparts have ranged across the electromagnetic spectrum. The successes have been few, but could be extremely significant. In particular, X-ray and optical counterparts have been associated with GRB970228 and GRB970508, described in the next two sections. In this section, we briefly summarize the current status of counterpart searches. See [17] for a more thorough review.

The search for flaring counterparts requires either a serendipitous burst in the field of view of an observation, or a fast, automated system for responding to burst locations in a few seconds. The latter is the goal of BACODINE, the BATSE COordinate DIstribution NEtwork [18]. The BACODINE system intercepts the real-time GRO data stream at GSFC, finds burst triggers in the BATSE data, computes directions to the bursts, and distributes these data electronically to any user who can make observations of the field. In the best circumstances, an automated optical telescope system, such as the Livermore Optical Transient Imaging Survey (LOTIS), can begin making observations in about 10 seconds, while the gamma-ray event may still be in progress.

The search for quiescent counterparts to well-localized bursts has produced many upper limits and three possible X-ray sources. X-ray sources were found in the error boxes of GRB920501 [19], GRB781119 [20] and GRB960720 [21]. The probability that the X-ray sources are unrelated to the GRB are in the 10^{-3} range. Upper limits on quiescent emission from GRBs are shown in Figure 5.

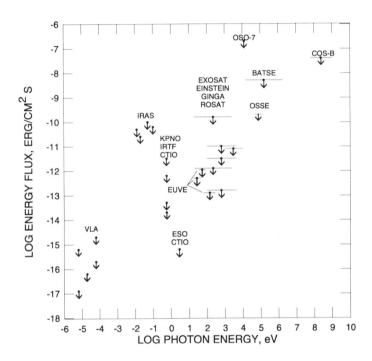

FIGURE 5. Upper limits to quiescent counterparts to GRBs across the EM spectrum

There has been a long-standing controversy as to whether there are suitable candidate galaxies in the error boxes of GRBs - the "no host galaxy" problem. Schaefer [22] has argued that upper limits to the brightness of galaxies in the error boxes of 26 GRBs indicate that bursts cannot be coming from normal galaxies, ruling out models incorporating merging neutrons stars. In a recent examination of four small GRB error boxes using HST, Schaefer *et al.* [23] determined that the maximum absolute B magnitude of the host galaxies is −15.5 to −17.4. On the other hand, Larson *et al.* [24] find appropriate host galaxies in the IR in the error boxes of 6 bright bursts. Vrba *et al.* [25] searched 7 IPN error boxes to V = 24 and found host galaxy candidates.

The gist of the no-host galaxy problem hinges on what is to be considered a suitable candidate galaxy. A small fraction of the brightest galaxies are responsible for most of the luminosity, and presumably most of the baryonic mass, in the Universe. It is therefore natural to expect that a burst, if cosmological, would most likely arise in a galaxy with high luminosity. This would not be true, however, if the bursts are not tracers of the mass. For example, if bursts come from galactic centers (e.g. the Compton attenuation model of Brainerd [26]), they might occur more frequently in low luminosity galaxies, which are far more numerous. There is no problem finding such galaxies; at B=22.5 to 24 there are 17200 galaxies per square degree and the median redshift is ~ 0.46 [27].

6. GRB970228

GRB 970228 provided the first convincing case for a fading counterpart to a GRB. This burst was discovered by BeppoSAX [28] and also seen by TGRS [29] and Ulysses [30]. The burst triggered the BeppoSax Burst Monitor and was detected by the Wide Field Camera. Within eight hours, BeppoSax was reoriented to observe with the narrow field instruments. A fading X-ray source was detected [31]. OSSE observations of the field [32] only 33 minutes after the burst found no detectable flux above 40 keV. Optical observations of the field [33] revealed a source at $R \sim 21$ on Feb. 28 that faded to $R > 23.3$ by March 4. Figure 6 shows the images on these two days. An extended source was later found at the location of the transient. Subsequent observations using the Hubble Space Telescope [34] found the point source still visible and ~ 0.5 arc seconds from the center of the extended source. At the time of the preparation of this review, there are controversial and conflicting reports on the nature of the transient and the extended emission.

7. GRB970508

Quick response by the BeppoSAX team [35] resulted in the identification of X-ray and optical counterparts for GRB970508. Bond [36] found a brightening optical source within the 3 arcminute error box over the next day. Subsequent observations by a number of groups established a peak at about R=19.7 two days after the GRB, followed by a decay of about 2 magnitudes over the next 10 days. The most noteworthy observation of this transient was the discovery by Metzger et al. [37] of absorption lines at a redshift z=0.835. This represents a minimum distance to the source of the optical emission and corresponds to an absolute peak intensity that is about 100 times greater than a type Ia supernova. Frail and Kulkarni [38] found a flaring radio source coincident with the optical emission. VLBA observations [39] confirmed the radio source and placed a limit of 0.3 milliarcseconds on the size of the source.

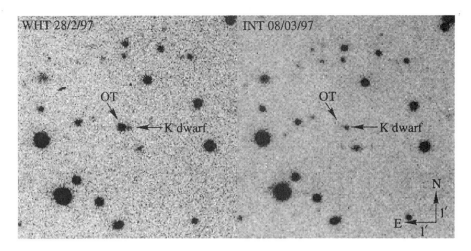

FIGURE 6. Optical transient associated with GRB970228

At the time of this review, optical and radio observations are continuing, and the community is struggling to digest all of the implications of the myriad observational data. Before GRB970508 can be considered the "smoking gun" that confirms the cosmological nature of gamma-ray bursts, it must be demonstracted that the optical variable is not an unrelated source, such as a BL Lac object.

8. MODELS

The BATSE location and intensity data strongly constrain models. The combination of isotropy and inhomogeneity is not compatible with a flattened disk distribution, regardless of the dimensions or sampling distance. These data indicate that we are close to the center of an approximately spherical distribution of burst sources. This conclusion is based on geometric arguments alone, and excludes most distance scales.

The current top contender for the distance scale is the cosmological hypothesis, with z of order unity for the faintest observed bursts. Here the apparent inhomogeneity is due to expansion of the Universe, and the isotropy arises naturally. Assuming isotropic emission, the burst energy required is about 10^{52} ergs, and the required rate is about once per 10^6 years per galaxy.

The only reasonable alternative to the cosmological hypothesis is to have the sources in an extended galactic halo (EGH) whose size is much larger than the distance to the Galactic Center (to make the dipole moment negligible), but not so large that bursts from M31 would be detectable. The favored candidates for source objects are high velocity neutron stars. A number of parameters of the model must be adjusted to reproduce the directional and intensity distributions. Specifically, there must be a delayed turn-on to allow the neutron stars to get far from the disk before bursting, and bound neutron stars must be prevented from bursting. In the EGH model, the faintest bursts are at distances of 100-300 kpc, and emit about 10^{43} ergs.

A successful model of GRBs must satisfy the contradictory requirements of short time scale, large energy release, and non-thermal gamma-ray spectrum. Current attention has focused on "fireballs" - relativistic pair plasmas that convert kinetic energy to gamma radiation in shocks [40] [41]. A difficult constraint is that there must be very low contamination by baryons, which would prevent the expansion from becoming relativistic. Suggested energy sources include coalescing neutron stars [42], collapse of magnetic white dwarfs [43], and "failed supernovae" [44]. See Hartmann [45] for review of theoretical models of gamma-ray bursts.

9. FUTURE WORK

The discovery of optical counterparts to GRB970228 and GRB970508 has re-energized efforts to obtain burst counterparts. BeppoSax will provide additional fast and accurate burst locations. The BATSE team has instituted a rapid alert system to obtain better locations for coordinates distributed by BACODINE. Upgrades to LOTIS will eventually yield simultaneous observations to magnitude 17. Having finally achieved some success, we can anticipate a greater commitment of telescope time to the search for optical counterparts. In 1999, HETE-2 will join the battle to find the sources of gamma-ray bursts.

There is reason to be optimistic. We are now starting to accumulate X-ray, optical, and radio data on GRBs, and more such cases will surely follow. This wealth of new information should provide the clues to unraveling the GRB puzzle. On the other hand, the resolution of the distance scale is only one small hurdle. Profound theoretical difficulties remain. It has always been easier to show that gamma-ray bursts are impossible than it is to explain them.

REFERENCES

1. Fishman, G. and Meegan, C., Ann. Rev. Astron. Astrophys., **33**, 415 (1995)
2. Fishman G. et al., *ApJS*, **92**, 229 (1994)
3. Meegan, C. et al., *ApJS*, **106**, 65 (1994)
4. Kouveliotou, C., Briggs, M., and Fishman, G., *Gamma-Ray Bursts: 3rd Huntsville Symposium* AIP Conference Proceedings, 384 (1996)
5. Matz, S. et al., *Astrophys.&Space Sci.* **231**, 127 (1995)
6. Matz, S. et al., *Astrophys.&Space Sci.* **231**, 123 (1995)
7. Matz, S. et al., *Gamma-Ray Bursts: 3rd Huntsville Symposium*, AIP Conference Proceedings 384, 846 (1996)

8. Norris, J. et al., *ApJ*, **424**, 540 (1994)
9. Mitrofanov, I. et al., *ApJ*, **459**, 570 (1996)
10. Connaughton, V. et al., *Proceedings of 18th Texas Symposium on Relativistic Astrophysics* A. Olinto: ed., Annals of the New York Academy of Sciences (in press) (1997)
11. Band, D. et al., *ApJ*, **413**, 281 (1993)
12. Kippen, M. et al., *Gamma-Ray Bursts: 3rd Huntsville Symposium*, AIP Conference Proceedings 384, 197 (1996)
13. Dingus, B., *Astrophys.& Space Sci.* **231**, 187 (1995)
14. Hurley, K. et al., *Nature*, **372**, 652 (1994)
15. Pendleton G. et al., *ApJ*, **489**, in press (1997)
16. Mallozzi, R. et al., *ApJ*, **454**, 597 (1995)
17. Vrba F. *Gamma-Ray Bursts: 3rd Huntsville Symposium*, AIP Conference Proceedings 384, 565 (1996)
18. Barthelmy. S. et al., *Gamma-Ray Bursts: 3rd Huntsville Symposium*, AIP Conference Proceedings 384, 580 (1996)
19. Hurley K. et al., *ApJ*, **464**, 342 (1996)
20. Hurley K. et al., *ApJ*, **469**, L105 (1996)
21. Piro L. et al., IAU Circular 6480 (1996)
22. Schaefer B., *Gamma-Ray Bursts: Second Workshop*, AIP Conference Proceedings 307, 382 (1994)
23. Schaefer B., Cline, T., Hurley K. and Laros, J., preprint astro-ph/9704278 (1997)
24. Larson, S., McLean, I., and Becklin, E., *ApJ*, **460**, L95 (1996)
25. Vrba F. et al., *ApJ*, **446**, 115 (1995)
26. Brainerd, J. J., *ApJ*, **428**, 21 (1994)
27. Glazebrook, K. et al., *MNRAS* **273**, 157 (1995)
28. Costa E. et al., IAU Circular 6572 (1997)
29. Palmer D. et al., IAU Circular 6577 (1997)
30. Hurley K. et al., IAU Circular 6578 (1997)
31. Costa E. et al., IAU Circular 6576 (1997)
32. Matz S., Grove, J. and Share, G., IAU Circular 6578 (1997)
33. van Paradijs J. et al., *Nature*, **386**, 686 (1997)
34. Sahu K. et al., *Nature*, **387**, 476 (1997)
35. Costa E. et al., IAU Circular 6649 (1997)
36. Bond, H., IAU Circular 6654 (1997)
37. Metzger, M. et al., IAU Circular 6655 (1997)
38. Frail D. & Kulkarni S., IAU Circular 6662 (1997)
39. Taylor, G., Beasley, A., Frail, D. and Kulkarni, S., IAU Circular 6670 (1997)
40. Meszaros, P. and Rees, M., *ApJ*, **405**, 278, 1993
41. Sari, R. and Piran, T., *MNRAS*, **287**, 110 (1997)
42. Narayan, R., Paczynski, B. and Piran, T., *ApJ*, **395**, L85 (1992)
43. Usov, V., *Nature*, **357**, 472 (1992)
44. Woosley, S., *ApJ*, **405**, 273 (1993)
45. Hartmann, D., *Astron. Astrophys. Suppl. Ser*, **120**, 31 (1966)

Gamma-Ray Line Transients

MICHAEL J. HARRIS

*USRA/GVSP, Code 661, NASA/Goddard Spaceflight Center,
Greenbelt, MD 20771*

Abstract.
Early interpretations of the variability of the narrow 0.511 MeV annihilation line from the Galactic center (GC) were confused by the unrecognized presence of a very extended component (mid 70s — mid 80s). Transient behavior was suggested by comparisons of measurements made by instruments with comparable fields of view, and by detection of other lines which could be interpreted as broadened and Doppler-shifted versions of the line. A new generation of instruments (late 80s — early 90s) provided more compelling evidence, by improved imaging capability (identifying a known transient high-energy candidate source, 1E 1740.7-2942), and by improved sensitivity. Additional broadened, shifted, and Compton-scattered derivatives of the annihilation line were detected from this and other sources. At the same time, theoretical developments suggested exciting new physics which might arise in the accretion regimes around black holes, of which these features would be important diagnostics.

In the GRO era the existence of such transient lines has been brought into question by the failure of long-term monitoring to detect them. A seemingly critical observation of the GC by three different instruments simultaneously, in which transient emission was detected only by one, provides the strongest evidence that the earlier results were in error. Recent developments in accretion disk theory have weakened the theoretical case for the lines. I draw conclusions concerning the nature of scientific inquiry and public outreach efforts in science.

I INTRODUCTION: EARLY DISCOVERIES OF THE 1970s

Transient events pose considerable problems in a young field such as γ-ray astronomy, in which the measurements are subject to poorly understood systematic errors. If two instruments disagree significantly when observing the same source, this may be due to the variability of the source in between measurements. Alternatively, the systematic errors in the two instruments may be responsible. Given the very poor spatial resolution of most γ-ray instruments, there is a third possibility — the two may have been looking at different sources.

These difficulties go back to the very beginning of cosmic γ-ray line astrophysics, when a series of balloon-borne experiments by Rice University teams[1-4] detected a line from the Galactic center (GC) region near the expected 0.511 MeV line from e^-e^+ annihilation. Discrepancies between the line energies and fluxes measured in the four Rice flights could be explained by differences in the aperture exposed to a diffuse source, or by time variability; Haymes et al.[4] were unable to decide the question.

The fourth Rice experiment contributed a further example to the pathology of γ-ray line astrophysics — "phantom" detections, whereby lines detected by one experiment can never be recovered by later experiments with adequate sensitivity. In such cases it can never be clear whether a systematic errors are to blame, or whether a genuine line transient has been discovered, with a low duty cycle — which of course becomes ever lower with each successive failure to recover the line. In this case[4], the reported lines were broad 4.4 and 6.1 MeV features towards the GC, such as are expected from cosmic-ray interactions with the ISM (specifically, from cosmic ray ^{12}C and ^{16}O nuclei impacting on protons[5]); these lines, whose fluxes were each $\simeq 4 \times 10^{-3}$ photon cm^{-2} s^{-1} rad^{-1}, have not been seen since by more sensitive instruments[6].

Perhaps the most famous "phantom" from this era was the Jacobson transient[7]. This event, detected by a balloon-borne Ge spectrometer from JPL, consisted solely of four narrow lines (0.413, 1.790, 2.219 and 5.946 MeV) lasting for 20 min. These energies are significant; they represent the 0.511 MeV annihilation line (redshifted), redshifted and unshifted versions of the prompt line from neutron capture ^{1}H(n,γ)^{2}H, and a redshifted capture line from ^{56}Fe(n,γ)^{57}Fe. In all cases the redshift is consistent with the value $z = 0.285$ expected at the surface of a neutron star. If accretion onto a neutron star produced copious positrons and neutrons, their reactions with accreted hydrogen and the original ^{56}Fe on the star's surface, and with hydrogen much further out in an accretion disk or secondary star, can explain all the lines[8]. The appeal of this scenario has kept interest in the Jacobson transient alive, despite the long-term failure to detect another[9,65,66].

On the edge of the field of view (FOV) containing the Jacobson transient lay the Crab pulsar, which itself contributed reports of transient lines almost from the beginning. We shall say little of these, since they have been reviewed recently by Owens[10]. The early reports of these lines were around energies 75 and 400 keV, the latter of which was interpreted as a red-shifted 0.511 MeV line from a neutron star surface. Owens[10] even discerned a long-term increase of a few percent in the lines' energies, corresponding to an outward movement of the emitting region.

The main line of development in the unfolding history of γ-ray line transients was in direction of black hole candidates, not neutron stars (see next section). However the stories of the Jacobson transient and the Crab lines illustrate a tendency to improve the chances of detecting a line by allowing for "plausible" Doppler shifts, usually redshifts attributed to black hole or neutron star gravity.

II THE MAKING OF A SYNTHESIS: THE 1980s

The steady diffuse level of Galactic emission of a narrow line at 0.511 MeV is not the subject of this review. However, a brief account of how it was recognized must be given, since it has been extremely important in studies of line transients, as the benchmark against which time variability has been measured. Following the Rice University experiments (§1), Leventhal, McCallum & Stang[11] inaugurated a series of flights of the Bell-Sandia high-resolution Ge detector. Their results fixed the energy of the line at 511 keV, at a flux level consistent with the Rice measurements given a diffuse source; they also provided evidence for an enhanced continuum immediately below the line due to annihilation via positronium (Ps) formation. As the 1980s proceeded, it became clear that there was a general tendency for the line flux measured by any given instrument to increase with aperture — a sure sign of an extended source. The exceptions from this trend became candidates for variability of one or more sources.

The good energy resolution of the Bell-Sandia experiment also paid off in the discovery of another transient line which had an obvious relationship to the 0.511 MeV line. This was a line at 170 keV, which was naturally explained by Compton scattering of the annihilation line through 180°, i.e. from material lying beyond the annihilation line source in the observer's line of sight[11]. An accretion disk or torus would be a natural medium of this kind.

All the measurements descibed above were made from balloons. In the late 1970s the first satellite experiments began in this energy range, offering the possibility of frequent monitoring of suspected sources over long periods. The most important of these was HEAO-3, whose Ge spectrometer observed the Galactic center region twice, in fall 1979 and in spring 1980. It detected a striking fall-off in 0.511 MeV line flux (3.5σ significance) between the two observations[12]. This finding was rapidly confirmed by Leventhal and co-workers with reflights of the Bell-Sandia instrument[13,14]; indeed, their data (taken with a smaller FOV, and thus less contaminated by the diffuse 0.511 MeV line) were consistent with zero flux in both flights. It looked as though a strong, point-like source had suddenly turned off at the beginning of 1980.

Another satellite experiment which played a part in the Galactic center 0.511 MeV story was *SMM*, launched in spring 1980. The FOV of this instrument was so large ($\sim 130°$) that its view of the 0.511 MeV line was dominated by the entire diffuse Galactic glow. However its long lifetime in orbit (1980–1989) made it valuable for variability studies. During the 1980s it detected no variation on year-to-year time-scales in the Galactic 0.511 MeV line[15]. If the variable 0.511 MeV source existed, it was clearly quiescent for years at a time.

At the end of the 1980s the source appeared to turn on again. Following on from the Bell-Sandia Ge instrument was the much larger GRIS, whose first flights in 1988 detected the 0.511 MeV line from the GC once again[16]. The flux level was close to 10^{-3} photon cm^{-2} s^{-1}; very soon afterwards (1988 November) the French-Italian FIGARO II measured a similar point source flux after correcting for the diffuse 0.511 MeV emission in their broad FOV[17]. Further confirmation came from another balloon experiment, HEXAGONE, early in 1989[60]. HEXAGONE also detected the 170 keV Compton back-scattered line[61], which had been observed on a previous occasion when the 0.511 MeV flux was high[11].

The picture had begun to emerge of a source whose variations were slow, spending months or years in a high state and becoming quiescent for years at a time. Its flux in outburst was up to 2×10^{-3} photon cm^{-2} s^{-1}. It was the most convincing example yet found of a γ-ray line transient. Although there were sceptics who pointed out that the variability could be explained away by changing the corrections made for the diffuse 0.511 MeV line[15], the changes seen between pairs of observations by the same instruments with the same pointings (i.e. Bell-Sandia[11,13,14] and HEAO-3[12]) should not have suffered from this problem.

A How Other Transients Fitted into the Picture

The contributions made by HEAO-3 were not confined to the 0.511 MeV line, nor to the GC. The black hole candidate Cyg X-1 was observed repeatedly, and great progress was made in understanding the relation between its soft and hard X-ray and γ-ray emissions[18,19]. Discrete emission levels — "high", "low" and "superlow" — were identified, with the "superlow" state having a much harder spectrum than the others. The hardening was accompanied by a striking and significant ($> 3\sigma$) feature of FWHM ~ 1 MeV at an energy close to 1 MeV.

The discoverers[18] offered two possible explanations for this feature, which has become known as the "MeV bump". Both involve an increase in the rate of accretion which creates a very hot plasma in the inner accretion disk. In the more modest proposal this plasma attains a temperature ~ 400 keV, and soft photons from the outer disk are Comptonized by the hot electrons; the resulting spectrum is Wien-like[20]. However radiation in equilibrium with a 400 keV plasma will have a substantial "tail" of photons above the 1.022 MeV threshold for e^-e^+ pair creation. Recent theoretical work showed that a stable situation was possible where positrons could replace protons as the dominant contributor to the charge balance in the plasma[21]; and that the annihilations occurring in this "pair-dominated" plasma would give rise to a line that was strongly blueshifted[22], by $\sim kT = 400$ keV to around 1 MeV. Being thermally broadened, this line at ~ 1 MeV would attain the width of the observed MeV bump. As we shall see, this annihilation-line interpretation became favored, because it could be related to other line transients in an overall unifying scheme.

This HEAO-3 observation of the "MeV bump" prompted a reconsideration of a number of earlier "phantom" observations of fugitive MeV emissions. For example, Perotti et al.[23] had detected a broad bump around 1 MeV in the spectrum of NGC 4151 with the balloon-borne Compton telescope MISO; although a later flight of the same instrument detected only a rather weak continuum at this energy, this continuum was hard (extending up to 7 MeV[24]). (The MISO instrument, incidentally, corroborated the HEAO-3 measurement of Cyg X-1 on a later flight[42]). Another Compton telescope (MPI-Garching) had detected a hard continuum up to ≥ 10 MeV from Cen A in 1982[25], although since no flux was detected around 1 MeV, this might equally well be interpreted as a broad feature centered at ~ 5 MeV. The interest in relating these observations to those of Cyg X-1 is obvious, in view of the long-held belief that AGN are powered by accretion onto black holes.

Reanalyses of the HEAO-3 data provided successive links strengthening the chain connecting e^-e^+ annihilation, black holes, and galactic nuclei. Riegler et al. followed up their work on the 0.511 MeV line with an analysis of the GC spectrum as a whole[26]. They found considerable spectral variability above a few hundred keV. In particular, the HEAO-3 fall 1979 GC observation, in which the 0.511 MeV line was strong (§2.1), showed a marked excess of emission around 1 MeV, which was absent in the later spring 1980 period. A hint of this feature had been seen earlier by HEAO-1 with very poor statistics[27]. This HEAO-3 GC MeV bump was remarkably similar in shape and intensity to that seen in Cyg X-1. However Riegler et al. interpreted it in terms of a thermal shape, thus obscuring the connection with e^-e^+ annihilation.

Exactly complementary to this discovery, Ling & Wheaton found evidence of a narrow 0.511 MeV line in a reanalysis of the HEAO-3 Cyg X-1 data[28]. In this case the signal was weak (1.9σ), and the authors expressed caution about the result. However there were now two sources whose properties appeared to confirm each other. The agreement between the spectra became even clearer when the GC Ps continuum was subtracted off; the case was made that it did not share the variability of the line, and must be entirely due to the diffuse Galactic source[29].

B The Debut of 1E1740.7-2942

Since Cyg X-1 is a long-standing black hole candidate, it was natural to assume that the unknown point source in the GC was one also. The identification of this source became a priority in high-energy astrophysics; none of the observations mentioned previously could locate it to better than several degrees. Two lines of attack on the problem developed. In the first, historical data on the known GC hard X-ray sources (emitting at somewhat lower energies, but in known locations) were examined, to see which of them had been most active during the time before 1980 when the annihilation-line source was strong. This led to the candidate sources GX 1+4, GX 5-1, A 1742-294 and 1E 1740.7-2942[29,30] on comparing images from the 1970s and 1980s; a strong case was made for GX 1+4[31], but this approach was by its nature doomed to be inconclusive.

The other approach was the development of γ-ray imaging instruments which could detect the line and image its source. Compton telescopes were not effective at such relatively low energies, so the coded-mask technique had to be applied. The first such attempt, the balloon-borne GRIP[32] in 1988, immediately picked out 1E 1740.7-2942 as a bright, hard source with a spectrum up to at least 200 keV which resembled that of Cyg X-1 (in X-ray high state[19]). However GRIP failed to detect the 0.511 MeV line, which, on the basis of GRIS and other balloon measurements[33], was then believed to be in outburst again (§2.2).

Long-term observing of the source by an imaging instrument was soon undertaken by a new satellite platform, the coded-mask *SIGMA* NaI detector on the Russo-French GRANAT mission. The resemblance between the 1E1740.7-2942 and Cyg X-1 spectra was quickly confirmed[34] in 9 separate $\sim$1-day observations (1990 March through October). During the tenth observation (1990 October 13–14) the source was detected in an outburst lasting at least 17 hours[35]. The continuous spectrum of this outburst was much like the standard Cyg X-1–like continuum seen earlier, but superimposed upon it was a strong, broad line centered near 400 keV, having FWHM$\simeq$ 200 keV and flux $\sim 10^{-2}$ photon cm^{-2} s^{-1}. The outburst could not have lasted much longer than 17 hr, since it was followed a few hours later by another GC observation, in which 1E 1740.7-2942 had the standard spectrum.

This seemed to confirm 1E 1740.7-2942 as the required point source of transient annihilation radiation — if the 400 keV feature was interpreted as a strongly broadened and redshifted annihilation line, which became the favored point of view (*SIGMA* did not see any significant narrow line at 511 keV). The similarity to Cyg X-1 suggested that 1E 1740.7-2942 should be a black hole candidate, though a significant difference from Cyg X-1 was soon discovered by ground-based mm observations: 1E 1740.7-2942 is surrounded by a dense molecular cloud[36]. Whereas Cyg X-1 accretes from an orbiting companion star, 1E 1740.7-2942 may accrete directly from the cloud. It was also revealed to be a nonthermal radio source[37], and perhaps its most black hole-like feature is a pair of jets, like those seen in many AGNs which are supermassive black hole candidates, detected by radio interferometry[38].

Both 1E 1740.7-2942 and Cyg X-1 continued to produce transient emissions, which tended to confirm their status as known, variable line sources.

Broad lines very similar to the 1990 October 13–14 line from 1E 1740.7-2942 were seen by *SIGMA* on two subsequent occasions, in 1991 October[39,40] and on 1992 September 19–20[40,41]. Both lines were only ~40% of the strength of the 1990 event, but the 1991 transient lasted for almost 3 weeks (October 1–19). Earlier (1984), a balloon-borne coded-mask experiment (DGT) had detected substantial MeV emission from Cyg X-1, in a form even harder than the MeV bump. The DGT spectrum peaked at about 5 MeV[43]; it was interpreted as pair-plasma annihilation feature, similar to the MeV bump, but at a very much higher temperature $kT \sim 3.5$ MeV. Weak evidence for similar bumps, corresponding to a variety of temperatures, was disinterred from earlier balloon observations[44].

Finally, in addition to the "established" sources 1E 1740.7-2942 and Cyg X-1, three other sources of broad line emission below 0.511 MeV were proposed. One was X-ray Nova Mus 1991, in which *SIGMA* detected a line at 480 keV during a period of 13 hours about 12 days after discovery[45,46]. The line was much narrower than the previous ones (FWHM 55 keV), and might have lasted up to 12 days beyond the observed time. Like 1E 1740.7-2942 and Cyg X-1, Nova Mus is a black hole candidate, according to observations of its companion. The second possible new source was discovered in a reanalysis of HEAO-1 data from the 1970s by Briggs and co-workers[47]. This line appeared in only one of three 6-month GC observations; its properties were very similar to those which *SIGMA* mesured from 1E 1740.7-2942, i.e. it was redshifted and very broad. The source is highly uncertain, due to HEAO-1's poor spatial resolution; the discoverers favored a low-mass X-ray binary, 1H 1822-371. If true, this would not fit the emerging pattern whereby annihilation-related line transients came only from black hole candidates. The third possible source was detected by the balloon-borne coded-mask scintillator detector EXITE in 1989, which detected a line around 100 keV from a previously unknown source, EXS 1737.9-2952[54]; this was interpreted as a *doubly* back-scattered 0.511 MeV annihilation line at 102 keV. The nature of EXS 1737.9-2952 is not known.

C A Synthesis of Diverse Line Transients

The observations which I have emphasized in the previous sections can be made to fall into a very elegant pattern, which I refer to as "the synthesis". As well as elegance, the synthesis involved distinctly novel physics, since the key concept was that of the pair-dominated plasma, which may be called a new, hitherto unobserved state of matter.

The concept of a pair-dominated plasma developed gradually during the 1970s and early 1980s (see eg. refs. 21, 48–50). Ramaty & Mészáros[22] identified its observational signature (a broad blueshifted annihilation line corresponding to the observed MeV bump), and Liang & Dermer[51–53] set up a detailed model for individual sources, first Cyg X-1 and then the GC. In this model, a central black hole is surrounded by an accretion disk, which contributes an underlying hard continuum from Comptonization of soft disk photons by a warm plasma (~ 50 keV). This is the object's spectrum when it is not in outburst. The outburst occurs when there arises, in the innermost region around the black hole, a plasma sphere at a temperature of a few hundred keV. Under these conditions an equilibrium developes between e^-e^+ pair production and annihilation into 0.511 MeV photons; there are two possible equilibria, one a "conventional" plasma in which the e^+ density remains small and the charge balance is dominated by protons, and the other in which the e^+ dominate. This novel pair-dominated plasma corresponds to a state of high compactness and optical depth. The photons and pairs are in thermal equilibrium (unlike, say, the non-thermal "pair cascade" models which have been invoked to explain AGN spectra[55]); the basic physical processes shaping the output spectrum are self-bremsstrahlung and pair annihilation providing the source spectrum, which is modified by self-Comptonization. The computed output spectrum, when added to the underlying disk spectrum, agrees with the transient spectra measured by HEAO-3.

A number of minor flaws were soon noted in this model; for example, there must be a rigorous separation between the accretion disk photons and the plasma, otherwise the hard X-ray spectrum is completely distorted (Liang[56] introduced a "transition region" with its own properties and spectrum to effect this). Also, the solution with two equilibria (e^+-rich and proton-rich) is strictly valid only if there is no escape of e^-e^+ from the plasma. This conflicted with one of the most appealing features of the model, its potential role as a source of Galactic positrons.

The observable variability of the narrow GC 0.511 MeV line (§2.1) suggested that the e^+ involved do not come from radioactive decays, which typically vary on unobservably long time-scales. The characteristic variable flux of $\sim 10^{-3}$ photon cm^{-2} s^{-1} requires an annihilation rate of 1.3×10^{43} s^{-1}. The pair plasma model provided a natural source for these positrons, which could vary on almost any required time-scale. However, allowance would have to be made for the slowing down of the relativistic e^+ coming out of such a source, since the line width precluded annihilation in a warm medium[57]. This requirement in turn constrained the density of the ambient medium to a rather high value $> 10^5$ atoms cm^{-3}, in order to perform the slowing down quickly enough.

The identification of 1E 1740.7-2942 as a black hole candidate, emitting e^-e^+ jets like a radio galaxy, and surrounded by a high-density molecular cloud (§2.3), made this element of the synthesis seem very plausible. The subsequent discovery by *SIGMA* of the broad, redshifted (400 keV) transient brought the synthesis to its highest point. In its fullest form it appears in ref. 58. This model specified that the pair plasma occurs for time-scales ~ 1 day, recurring on a time-scale $\sim$months. Some of the e^+ produced in the pair plasma escape and encounter the inner edge of the accretion disk, where annihilation takes place. The annihilation line produced at this location is gravitationally redshifted to 400 keV, and thermally broadened by the disk temperature. The time-scale for this is $\ll$ 1 day, so its variability is governed by the e^+ production time-scale. Other escaping positrons leave the system entirely and diffuse into the molecular cloud surrounding 1E 1740.7-2942, slowing down continually until they annihilate with the cold ambient material; the line produced is narrow, and at 0.511 MeV. The time-scale for the slowing down and annihilation is $\sim$ 1 year. Hence the variability of the narrow 0.511 MeV line would be expected to be much slower than those of the other annihilation lines. The fluxes in the 400 keV, 511 keV and 1 MeV lines are consistent with each other[59]. This model had great predictive power; given an observation of either the MeV bump or the broad redshifted line, the temporal behavior of the narrow line could be predicted for more than a year ahead, and so could its flux. It could also be extended to other black hole sources with different masses[53].

The Compton back-scattered annihilation line, coming from reflection from the inner accretion disk, must occur on the shorter ($\sim$ 1 d) time-scale. The observations require redshifting of both the annihilation region and the scattering region[62], which is plausible given the existence of a line at 400 keV. It is less clear why the 400 keV line and the back-scattered line are not seen simultaneously, although the original Bell-Sandia spectrum did show an apparent red wing below the 0.511 MeV line[11,62].

Thus in summary the synthesis explained the following transients:
(a) The MeV bump in Cyg X-1 and the GC, which was emitted directly by the pair-dominated plasma around a black hole.
(b) The broad 400 keV feature, which was due to annihilation of e^+ escaping from the pair plasma and encountering the inner edge of the accretion disk.
(c) The 170 keV back-scattered annihilation line, from Compton scattering through 180° off accretion disk material.
(d) The variable narrow 0.511 MeV line source in the GC, which was due to e^+ from the pair plasma escaping from 1E 1740.7-2942 and annihilating in the surrounding molecular cloud.
(e) Variable emission in the form of broad lines and quasi-continua at MeV energies from Cyg X-1 and various AGNs, which represented pair plasma emission at higher temperatures than that of the MeV bump.

III RETREAT FROM THE SYNTHESIS: THE 1990s

A The GRO era

The launch of the *Compton* Gamma Ray Observatory in 1991 April introduced a new capability which had especial importance for the study of transients — the ability to perform long-term monitoring. Except for *SMM*, previous satellites had had short lifetimes. The new mission was designed for a much longer lifetime; one of its four instruments (BATSE) was intended specifically as a sky monitor, and another (OSSE) could be independently targeted, making frequent observations possible for any given source. In practice, given the energies of most of the transient lines, OSSE was of more use for "known" line sources, while BATSE was better suited to monitoring continuum emissions at unknown locations. The whole spacecraft could be maneuvered to point to any variable source which was declared a "target of opportunity".

Of the monitoring work done by *SMM* during 1980–1989, we have already noticed Share et al.'s demonstration of the constancy of the narrow 0.511 MeV line[15]. The other line transient searches carried out by *SMM* had similar negative results. The MeV bump was searched for from several sources from which MeV emission had been reported[63], and the broad 400 keV line was searched for from the GC[64]. Due to the relatively long periods over which *SMM* viewed a given source uninterruptedly, the negative results could be expressed as an upper limit on the duty cycle with which the expected transient occurred. These upper limits were no more than a few percent in almost all cases — only the longer ($\sim$ 3-week) broad 400 keV transient at the level seen by *SIGMA* from 1E 1740.7-2942 might have occurred more often during 1980–1989.

One of GRO's main tasks was the resolution of this somewhat unsatisfactory situation. The GC, 1E 1740.7-2942 and Cyg X-1 have been regular OSSE targets throughout the mission, while background-subtraction techniques have been developed for BATSE which enable the kind of monitoring which was done with *SMM*. The results of these monitoring exercises, periodically updated, have been uniformly negative.

B The Narrow 0.511 MeV Line from the GC

The OSSE aperture at energy 0.5 MeV is a rhombus of dimensions $3.8° \times 11°$. This is well-suited for the mapping of the Galactic diffuse emission line, since it is intermediate between the wide-FOV instruments such as HEAO-3 (which detect the line in full strength, but cannot resolve it spatially) and the high-resolution imaging capability of *SIGMA* (which can detect point sources but not the diffuse emission). In terms of the trade-off between spatial resolution and signal strength, the aperture is broad enough for the diffuse emission at the GC to be detected during a single 2-week viewing period (VP), but not so broad that the diffuse signal swamps any variable point-source signal which may be superimposed. The instrument is therefore also well-suited for the detection of variable line emission on $\sim$week time-scales.

Between mid-1991 and late 1995 OSSE observed the GC during 28 VPs in 122 separate pointings. The average interval between observations was 50 d, and no interval was longer than 10 months. With this level of coverage, the activity of variable point sources of the 0.511 MeV line was for the first time adequately monitored at finer time-scales than the year-to-year monitoring by *SMM*[15]. Purcell and co-workers have used these data to search for point sources and for variability, measured against the background of a much larger set of observations of the GC as a whole[67,68]. From the larger set they derived a model of the diffuse emission, which was subtracted from their measurements in FOVs at the GC. No evidence was found for a point source above the diffuse emission, although one cannot be ruled out, given OSSE's moderate spatial resolution. Neither was any evidence found for variability above or below the diffuse flux, and this result was more constraining. Of the 28 GC observations, 25 had good or moderate exposure to 1E 1740.7-2942[69]; the typical 3σ upper limit on the variable flux from this source was 2.6×10^{-4} photon cm^{-2} s^{-1}. This is smaller than the fluxes measured during the outbursts in the 1970s and late 1980s by a factor 4–8 (§2.1). According to the synthesis, this narrow 0.511 MeV line ought to vary on time-scales ~ 1 yr (§2.3); the OSSE observations showed no variation on any time scale up to 4 yr (1991–1995). Nor did they show higher fluxes at the beginning of the period, which might be expected if annihilation in the molecular cloud had begun following an outburst in 1988–1989 (§2.1), and had then decayed from a flux level $\sim 10^{-3}$ photon cm^{-2} s^{-1} according to the scheme of ref. 58.

Monitoring of the GC by OSSE continues. It has been joined by another instrument, the occulted Ge detector TGRS on the WIND spacecraft, observing the GC continually since 1994 November. The effective FOV of this instrument is $\sim 10° \times 50°$, so that its signal is dominated by the diffuse emission. However, this has shown no variability from month to month, down to a 3σ upper limit of 10^{-3} photon cm^{-2} s^{-1} (ref. 70).

Even before the results of OSSE's monitoring were published, the actual status of this key element of the synthesis had come into question. The original HEAO-3 GC data in which the 0.511 MeV line varied (§2) were reanalyzed, and it was found that the discrepancy between the 1979 fall and 1980 spring line intensities was much smaller than previously thought[78]. The new results ($1.25 \pm 0.18 \times 10^{-3}$ photon cm^{-2} s^{-1} in fall 1979, $0.99 \pm 0.18 \times 10^{-3}$ photon cm^{-2} s^{-1} in spring 1980) were consistent with no variability.

C The Broad 400 keV Line from 1E 1740.7-2942: A Confrontation

As noted above (§2.2), the *SIGMA* instrument has been observing the GC at intervals since 1990. The observations have not been optimized for monitoring, especially since exhaustion of the gas supply in 1995 restricted maneuverability. However, the monitoring capabilities of OSSE (with which *SIGMA* tried wherever possible to coordinate) and BATSE have enabled an independent evaluation of one of the transients described in §2.2, that of 1992 September

19–20.

During VP 40, OSSE pointed at the GC from September 17–October 8. From this observation Jung et al.[71] extracted a spectrum simultaneous with the transient, subtracted from it the diffuse Galactic spectrum and the contributions of other point sources, and compared it with *SIGMA*'s transient spectrum as OSSE would have seen it. Compared with the *SIGMA* flux measurement of $4.3^{+2.7}_{-1.5} \times 10^{-3}$ photon cm^{-2} s^{-1}, OSSE saw no line, down to a 3σ upper limit in the range 1.0–2.4 $\times 10^{-3}$ photon cm^{-2} s^{-1} (depending on the exact line energy and width).

Whereas BATSE could in principle have monitored both of the *SIGMA* GC transients which occurred since GRO launch (1991 October and 1992 September 19), the difficulty of obtaining backgrounds separated from the source by > 14 d (i.e. the typical GRO pointing period) has led attention to focus on the shorter 1992 September 19 event. Smith et al.[72] confirmed the negative OSSE result for this event, obtaining a 3σ upper limit of 1.8×10^{-3} photon cm^{-2} s^{-1} for any transient simultaneous with the *SIGMA* report. They continued monitoring 1E 1740.7-2942 for 665 d during 1991–1993, seeing no events on 1 d time-scales down to such typical sensitivities[73]. They conclude that these events cannot occur at both the flux level and the duty cycle seen by *SIGMA*, the mutual probability being 7×10^{-10}. Attempts to confirm or refute the 1991 October event have been inconclusive[74].

These two independent refutations of the *SIGMA* report are probably the most serious attacks on the observational foundations of the synthesis (§2.3.). No resolution of the disagreement has yet been found.

D The Back-Scattered Annihilation Line from 1E 1740.7-2942

Smith et al.[75] undertook the monitoring of the line at 170 keV, attributed to Compton back-scattering of the 0.511 MeV line, in OSSE data. They analyzed data from 180 d of OSSE observations of 1E 1740.7-2942 between mid-1991 and mid-1993 on a day-to-day timescale. The results of the search were negative. The typical 3σ upper limits for each day were 6.8×10^{-4} photon cm^{-2} s^{-1}, compared with the earlier observations of fluxes close to 1.3×10^{-3} photon cm^{-2} s^{-1} (§2). It was concluded that 1E 1740.7-2942 is not a source of random outbursts in this line on 1 d time-scales, with 99.3% confidence; the possibilities of non-random variability or of variability on longer time-scales were rejected less strongly.

E The 0.511 MeV Line and MeV Bump in Cygnus X-1

Cygnus X-1 has also been well-observed by OSSE, since it has frequently been a target of opportunity. Phlips et al.[77] were able to use 122 d of data from 17 VPs during 1991–1995 to monitor its spectrum, in which they searched specifically for broad and narrow lines around 0.511 MeV, for the MeV bump, and for the kind of very high energy line seen by DGT (§2.2), on a day-to-day time-scale.

Even on a 1 d time-scale, all the OSSE measurements of the MeV bump fell far below the level of the line seen by HEAO-3, the typical 3σ upper limits $\sim 3 \times 10^{-3}$ photon cm^{-2} s^{-1} being a factor of 5 smaller. The correlation which HEAO-3 found between the X-ray "superlow" state and the MeV bump was not confirmed either. The 3σ upper limits on a narrow 0.511 MeV line from day to day were typically $\sim 1 \times 10^{-3}$ photon cm^{-2} s^{-1}.

F Conclusions

Line by line and source by source, careful monitoring by the GRO instruments has failed to reveal evidence for any of the features which were predicted by the synthesis to be time-variable. Some of the strongest apparent observational supports of the synthesis have been controverted by simultaneous measurements with equal or greater sensitivity (§3.3). The synthesis has been seriously weakened, even if it is not yet dead.

While the synthesis was a remarkable intellectual achievement, its weakness lay in the experimental measurements, not in the theory — inconsistent measurements were interpreted as evidence of variability rather than as systematic errors. This is one aspect of a larger problem in γ-ray line astronomy — in searching for a specific line, "degrees of freedom" are multiplied until any feature can be so interpreted. Compared to the reference line, the observed line may be red- or blueshifted, narrower or broader; the time-scale during which it appears, and the interval between recurrences, are also free. In the case of the GC 0.511 MeV line, the relative contribution of the non-varying diffuse line was in the past another degree of freedom.

Monitoring of the sources over the long term is the only real cure for this problem; monitoring by CGRO appears to have cast doubt on almost all the line transients on which the synthesis was based (§3). The best remaining candidate for a genuine γ-ray line transient is now the broad 0.480 MeV line seen in Nova Mus 1991 by *SIGMA* (§2.2). The reason for this is simply that X-ray novae are rather rare, and so have not been subject to the kind of monitoring that the GC and Cyg X-1 have. Since Nova Mus, about half a dozen black hole X-ray novae have occurred[79], of which four were monitored by BATSE with negative results[73].

In summary, it appears that, with the obvious exceptions of solar flares and lines from supernova nucleosynthesis, *there are no well-established sources of transient γ-ray line emission in astrophysics.*

G Epilog: The Crab in the Nineties

In the case of lines from the Crab Nebula, there is an additional degree of freedom in γ-ray line searches — the line may occur only at certain pulse phases. This was the case with a line at 440 keV seen by the balloon experiment FIGARO II on two separate flights[80,81] during the interpulse and second peak phases. Unfortunately the Crab has not been a frequent OSSE target during the GRO mission; OSSE's pulsar mode is well suited to phase-resolved spectroscopy. Ulmer et al.[88] searched for several reported lines during two VP in 1991 and 1992, with negative results. These upper limits were inconsistent with most previous reports, but not with the FIGARO II measurement.

Attempts at monitoring the Crab over longer periods have been made by *SMM*[82] and BATSE[73], but these lacked the temporal resolution necessary to resolve the pulse phase. However, BATSE's monitoring capability became important in a different context when *SIGMA* reported a transient γ-ray line at 536 keV about one year after GRO launch[83]. Coverage of the Crab by BATSE on this date (1992 March 10-11)[72] placed a 3σ upper limit of 1.7×10^{-3} photon cm^{-2} s^{-1} on the line flux at 536 keV, compared to the *SIGMA* measurement of $5.1 \pm 1.7 \times 10^{-3}$ photon cm^{-2} s^{-1}. Thus, the reported lines from the Crab appear to be just as elusive as the various lines from the GC and Cyg X-1 which went into the synthesis.

H Epilog: Possible Annihilation-Line Transients from EGRET

Nuclear γ-ray lines do not occur at energies above a few MeV, but other classes might be found at higher energies within the EGRET range. It is not clear that any of them would be time-variable. Lines from the annihilation of hypothetical dark-matter particles (e.g. neutralinos[84]) are outside the scope of this review, and are not likely to be variable. There is a well-known broad line in the Galactic diffuse emission at a few hundred MeV[85], arising from the decay of π^0 produced in cosmic-ray proton collisions, but this is not likely to vary either, unless particle acceleration in stellar winds contributes a point-like component to the diffuse emission[86]. It is possible in principle that bulk collisions of matter and antimatter could produce a time-variable $p\bar{p}$ annihilation line around 70 MeV due to π^0 decay. There are, however, no known *natural* sources of antiprotons in bulk[87], and so far as is known no searches have been made for this type of transient.

IV SOME IMPLICATIONS

A For the scientist as scientist

The synthesis described in §2 is a classic example of a universal human tendency — the tendency to integrate imperfect information into logical patterns. In sensory terms, this tendency forms the basis of many well-known optical illusions. In science it appears in a more abstract and logical form.

It would be easy to make the synthesis an example of a scientific failure, by which brilliant minds deluded themselves into seeing a pattern which was not there. I believe such a judgement would be mistaken. At the time when the synthesis was being built, great plans were afoot for exploring the new γ-ray spectral window — including the planning for GRO. These plans had to be guided by some overall scientific vision; if it had not been the synthesis, it would have been something else. I believe that the overall success of GRO justifies the synthesis, even if it turns out factually to have been erroneous. "In war it is better to do the wrong thing than to do nothing"[89], and the same applies to big-science enterprises like space research.

B For the scientist as citizen

Efforts at outreach by scientists to the general public usually focus on the material benefits, but it is questionable whether most scientific research produces benefits on any time-scale which the ordinary citizen can perceive. A more important contribution could be made by emphasizing the conceptual apparatus of science — our (relative) objectivity, skepticism and especially our use of mathematical reasoning at every step in an argument.

The human susceptibility to pattern-recognition where no pattern exists (§4.1) is fraught with great dangers, about which the scientist ought to warn the public. The history of conspiracy theories in American conservative politics is a dramatic example[90]. However there are more insidious examples. Many prosecutions are based on superficially attractive ("circumstantial") patterns of events, in the absence of a witness or a confession. The scientist should publicly warn of the dangers of deception, and act accordingly when performing jury service.

I believe that the scientist is morally obliged to take this stand even in unpopular cases involving race or religion. I think some of us have not met this obligation. I regret to say that during the recent O.J. Simpson trial every single one of my scientific acquaintances supported the prosecution[91]. Similarly, although scientists have frequently denounced paranoid cults and sects, they have generally ignored the potential for this sort of thinking among the more mainstream religions. If we are going to contribute our conceptual expertise to society, then we are going to have to face the fact that the thought processes of science and religion are totally incompatible.

REFERENCES

[1] R.C. Haymes et al., ApJ, 157, 1455 (1969).
[2] W.N. Johnson, F.R. Harnden, & R.C. Haymes, ApJ, 172, L1 (1972).
[3] W.N. Johnson & R.C. Haymes, ApJ, 184, 103 (1973).
[4] R.C. Haymes et al., ApJ, 201, 593 (1975).
[5] R. Ramaty, B. Kozlovsky & R.E. Lingenfelter, ApJS, 40, 487 (1979).
[6] M.J. Harris, G.H. Share & D.C. Messina, ApJ, 448, 157 (1995) and refs.
[7] A.S. Jacobson et al., *Gamma-Ray Spectroscopy in Astrophysics* (NASA T.M. 79619), 228 (1978).
[8] R.E. Lingenfelter, J.C. Higdon, & R. Ramaty, *Gamma-Ray Spectroscopy in Astrophysics* (NASA T.M. 79619), 252 (1978).
[9] G.H. Share et al., A&AS, 97, 341 (1993).
[10] A. Owens, *Gamma Ray Line Astrophysics*, ed. P. Durouchoux & N. Prantzos (New York: AIP), 341 (1991).
[11] M. Leventhal, C.J. MacCallum, & F.R. Stang, ApJ, 225, L11 (1978).
[12] G.R. Riegler et al., ApJ, 248, L13 (1981).
[13] M. Leventhal et al., ApJ, 260, L1 (1982).
[14] M. Leventhal et al., ApJ, 302, 459 (1978).
[15] G.H. Share et al., ApJ, 358, L45 (1990).
[16] M. Leventhal et al., Nature, 339, 36 (1989).
[17] M. Niel et al., ApJ, 356, L21 (1990).
[18] J.C. Ling et al., ApJ, 321, L117 (1987).
[19] J.C. Ling, *Nuclear Spectroscopy of Astrophysical Sources*, ed. N. Gehrels & G.H. Share (New York: AIP), 315 (1988).
[20] R.A. Sunyaev & L.G. Titarchuk, A&A, 86, 121 (1980).
[21] R. Svensson, MNRAS, 209, 175 (1984).
[22] R. Ramaty & P. Mészáros, ApJ, 250, 384 (1981).
[23] F. Perotti et al., Nature, 282, 484 (1979).
[24] F. Perotti et al., ApJ, 247, L63 (1979).
[25] P. von Ballmoos, R. Diehl & V. Schönfelder, ApJ, 312, 134 (1987).
[26] G.R. Riegler et al., ApJ, 294, L13 (1985).
[27] J.L. Matteson, *The Galactic Center*, ed. G.R. Riegler & R.D. Blandford (New York: AIP), 109 (1982).
[28] J.C. Ling & W.A. Wheaton, ApJ, 343, L57 (1989).
[29] R.E. Lingenfelter & R. Ramaty, ApJ, 343, 686 (1989).
[30] G.K. Skinner et al., Nature, 330, 544 (1987).
[31] J.E. McClintock & M. Leventhal, ApJ, 346, 143 (1989).
[32] W.R. Cook et al., ApJ, 372, L75 (1991).
[33] N. Gehrels et al., ApJ, 375, L13 (1991).

[34] R.A. Sunyaev et al., ApJ, 383, L49 (1991).
[35] L. Bouchet et al., ApJ, 383, L45 (1991).
[36] J. Bally & M. Leventhal, Nature, 353, 234 (1991).
[37] D.A. Leahy, MNRAS, 251, 22p (1991).
[38] I.F. Mirabel et al., Nature, 358, 215 (1992).
[39] E. Churazov et al., ApJ, 407, 752 (1993).
[40] M. Gilfanov et al., ApJS, 92, 411 (1994).
[41] B. Cordier et al., A&A, 275, L1.
[42] L. Bassani et al., ApJ, 343, 313 (1989).
[43] M.L. McConnell et al., ApJ, 343, 317 (1989).
[44] A. Owens & M.L. McConnell, Comments Astrophys., 16, 205 (1992).
[45] R.A. Sunyaev et al., ApJ, 389, L75 (1992).
[46] A. Goldwurm et al., ApJ, 389, L79 (1992).
[47] M.S. Briggs et al., ApJ, 442, 638 (1995).
[48] G.S. Bisnovatyi-Kogan, Ya.B. Zeldovich & R.A. Sunyaev, Sov. Astr/AZh, 15, 17 (1971).
[49] E.P. Liang, ApJ, 234, 1105 (1979).
[50] A.P. Lightman, ApJ, 253, 842.
[51] E.P. Liang & C.D. Dermer, ApJ, 325, L39 (1988).
[52] C.D. Dermer & E.P. Liang, *Nuclear Spectroscopy of Astrophysical Sources*, ed. N. Gehrels & G.H. Share (New York: AIP), 326 (1988).
[53] E.P. Liang, ApJ, 367, 470 (1991).
[54] J.E. Grindlay et al., A&AS, 97, 155 (1993).
[55] A.P. Lightman & A.A. Zdziarski, ApJ, 319, 643 (1987).
[56] E.P. Liang, A&A, 227, 247 (1991).
[57] R. Ramaty & R.E. Lingenfelter, *The Galactic Center*, ed. D.C. Backer (New York: AIP), 51 (1987).
[58] R. Ramaty et al., ApJ, 392, L63 (1992).
[59] R. Ramaty, J.G. Skibo & R.E. Lingenfelter, ApJS, 92, 393 (1994).
[60] P. Wallyn et al., ApJ, 403, 621 (1993).
[61] J.L. Matteson et al., *Gamma Ray Line Astrophysics*, ed. P. Durouchoux & N. Prantzos (New York: AIP), 45 (1991).
[62] R.E. Lingenfelter & X.-M. Hua, ApJ, 381, 426 (1991).
[63] M.J. Harris et al., ApJ, 416, 601 (1993).
[64] M.J. Harris, G.H. Share & M.D. Leising, ApJ, 433, 87 (1994).
[65] J.P. Heslin et al., BAAS, 13, 901 (1981).
[66] D.M. Palmer et al., in preparation (1997).
[67] W.R. Purcell et al., *Second Compton Symposium*, ed. C.E. Fichtel, N. Gehrels, & J.P. Norris (New York: AIP), 403 (1994).
[68] W.R. Purcell, INTEGRAL96 Workshop
[69] W.R. Purcell, private communication (1996).
[70] B.J. Teegarden et al., ApJ, 463, L75 (1996), and these proceedings.
[71] G.V. Jung et al., A&A, 295, L23 (1995).
[72] D.M. Smith et al., ApJ, 458, 576 (1996).

[73] D.M. Smith et al., ApJ, 471, 783 (1996).
[74] P. Wallyn et al., A&AS, 120 295.
[75] D.M. Smith et al., ApJ, 443, 117 (1995).
[76] P. Wallyn & P. Durouchoux, ApJ, 414, 178 (1993).
[77] B.F. Phlips et al., ApJ, 465, 907 (1996).
[78] W.A. Mahoney, J.C. Ling, & W.A. Wheaton, ApJS, 92, 387 (1994).
[79] Y. Tanaka & N. Shibazaki, ARA&A, 34, 607 (1996).
[80] E. Massaro et al., ApJ, 376, L11 (1991).
[81] J.F. Olive et al., A&AS, 97, 321 (1993).
[82] M.J. Harris, G.H. Share, & M.D. Leising, ApJ, 420, 649 (1994).
[83] M. Gilfanov et al., ApJS, 92, 411 (1994).
[84] S. Rudaz & F.W. Stecker, ApJ, 368, 406.
[85] S.D. Hunter et al., ApJ, 481, 205 (1997).
[86] R.L White & W. Chen, ApJ, 387, L81 (1992).
[87] M.J. Harris, *Bioastronomy (Lecture Notes in Physics 390)*, ed. J. Heidmann & M.J. Klein (Springer-Verlag: Berlin), 300 (1991).
[88] M.P. Ulmer et al., ApJ, 432, 228 (1994).
[89] A.J.P. Taylor, *Essays in English History*, (Penguin: London) (1976).
[90] R. Hofstadter, *The Paranoid Style in American Politics*, (Harvard UP: Cambridge) (1996).
[91] Dr. C.D. Dermer has requested that he be exempted from this statement.

Constraints from Undetected Gamma-Ray Sources

C.E. Fichtel and P. Sreekumar*

NASA/Goddard Space Flight Center, Greenbelt, MD 20771
* Universities Space Research Association

Abstract. The number of positive gamma-ray source detections by the Compton Gamma-Ray Observatory has been large, and the significance of the results has been substantial, as is described in other reports at this conference. In addition, searches for gamma rays were made that have led to null results, but that nonetheless deserve attention because of their important scientific implications. In the gamma-ray observations there has been as yet no indication of antimatter in the universe other than that produced in cosmic-ray interactions. The upper limit for microsecond bursts has implications for attempts at unified theories of physics. The search for signatures from supersymmetric particles has not revealed any positive evidence. The measurements of the high-energy gamma-ray emission from blazars and its interpretation in terms of the relativistic particles and their origin has been one of the highlights of the EGRET results. However, there are a few flat-spectrum radio quasars that are very strong in the radio regime, but have not been seen in high-energy gamma rays. The derived upper limits from external normal galaxies indicate no evidence for significant deviation from equipartition arguments. A minimum value of $0.4\mu G$ for the intracluster magnetic field is derived from gamma-ray upper limits on the Coma cluster.

I INTRODUCTION

The goal of the Compton Gamma-Ray Observatory was to examine the gamma-ray sky in order to learn about the large transfers of energy in astrophysics and to study the properties of energetic particles in space, as well as to explore this relatively unknown region of the astronomical electromagnetic spectrum. Many positive observations have been made and have contributed substantially to the knowledge of many different facets of astronomy. The gamma-ray emission from neutron-star pulsars and AGN has been well studied and the basic properties of both are now known. The Galactic diffuse emission is well mapped and explained in terms of cosmic-ray interactions. Gamma-ray lines have been seen from sources, as well as in the diffuse radiation. With the exception of a small (but important, as will be discussed here) region in the gamma-ray spectrum around 1 MeV, the extragalactic diffuse radiation has been measured well from the soft X-ray region to 50 GeV, and seems well explained as the sum of AGN emission: Seyferts at low energy and blazars at high energy. Quite interesting results on solar gamma rays have been obtained, speaking to the acceleration and subsequent history of energetic solar particles. Although their origin remains a mystery, there is now an extensive knowledge of the characteristics of gamma-ray bursts. For a summary of these results and a discussion of their significance, see, e.g., the other papers in this conference and the second edition of the book by Fichtel and Trombka (1997).

These very exciting results naturally have tended to overshadow, to some extent, the searches for gamma-ray emission that have led only to upper limits, even though, before the launch, some of these searches were placed high on the list of important areas of study for the Compton Observatory. This paper is intended to be a brief summary of some of the searches that did not lead to the detection of gamma rays, including why it was felt that they were important, and, in so far as possible, what the null result implies.

THE SEARCH FOR ANTIMATTER IN THE UNIVERSE

From the very earliest days of interest in gamma-ray astronomy, it was realized that antimatter could be revealed through the distinct signature of its annihilation, especially the gamma rays from neutral pions produced in the annihilation of a nucleon and antinucleon, and in the 0.511 MeV line of the electron-positron annihilation. In a typical proton-antiproton annihilation there are about five pions produced, about a third of which are, on the average, neutral. If the particles have relatively small kinetic energies, the resulting gamma rays will have energies concentrated around 67.5 MeV. The observed energy can be considerably lower, if the gamma rays had been formed in the past as a result of the cosmological red shift. Thus, the predicted spectrum depends on the specific model. It was, of course, realized that these antiparticles could also be the result of cosmic-ray or other energetic particle interactions and of nucleosynthesis in supernovae, but many of the early theoretical predictions were well in excess of radiation from these processes.

The first important upper limits came well before the Compton Observatory. If there were continuous creation of matter and antimatter throughout the universe at the level necessary to keep the density constant in the steady state theory of the universe, the level of diffuse gamma-ray emission caused by particle-antiparticle annihilation would have been several orders of magnitude in excess of that observed by the early satellites, such as OSO-3 and SAS-2. (See, for example, Fichtel and Kniffen, 1974)

There are alternate matter-antimatter baryon symmetry models that would predict much lower gamma-ray intensities than the one just mentioned. Omnes (1969), for example, considered a big-bang model which, initially, has a very high temperature and density, and proceeds to show that, with baryon symmetry, a separation of matter from anti-matter may result, leading to regions of pure matter and antimatter on a very large scale. Present results suggest that this scale must be at least on the order of super clusters of galaxies, or the interactions at the boundaries would probably lead to a detectable level of gamma rays. The spectrum that would be expected is the sum over the history of interactions, and hence the peak in the radiation occurs at a much lower energy than that mentioned above because of the cosmological red shift. The expected spectrum for interactions back to a z of about 100 has been calculated by Stecker, Morgan and Bredekamp (1971), with a peak of the radiation falling in the MeV range. Although early work suggested the possibility of such a feature in the diffuse radiation, more recent work by Kappadath et al. (1995) and Watanabe (1996) does not. Their work sets upper limits in this energy region well below the earlier reported results; this difference is now understood to be the result from underestimation of a steady background component. For the presently measured combined extragalactic diffuse spectrum from the X-ray region through the high-energy gamma-ray region, see Figure 1, which is taken from Sreekumar et al. (1997).

Turning to the annihilation of electrons and positrons, there is no evidence for a uniform extragalactic 0.511-MeV-line radiation. The limit on the average fraction of antimatter in the universe is, however, less severe than that which can be set by the lack of any evidence for nucleon-nucleon annihilation. See,

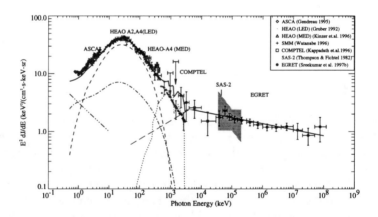

FIGURE 1. Multiwavelength spectrum from X-rays to gamma rays. See Sreekumar et al., 1997, for more details).

e.g., Fichtel and Kniffen (1974). The observed Galactic diffuse 0.511-MeV-line radiation was first seen by Leventhal, McCallum and Stang (1978) and now has been viewed extensively by several telescopes, including OSSE (Purcell et al., 1993 and 1994). The origin of this radiation is generally attributed to the decay of products of various Galactic nucleosynthesis processes, particularly supernovae, and there seems to be reasonable, quantitative agreement with theoretical estimates of the intensity. For a further discussion of this radiation and other half MeV line considerations, see, e.g., Ramaty and Lingenfelter (1994), Ramaty (1996), and Timmes et al. (1996).

There is, therefore, at the present time no evidence from the observations by the Compton Observatory or others of the existence of any radiation that can be attributed to matter-antimatter interactions that cannot be explained by either energetic particle interactions or nucleosynthesis associated with supernovae or other processes. Current upper limits are well below many early predictions that antimatter is created in the universe, either at an early stage or throughout time. There is not much reason to believe that the current upper limits will be reduced significantly in the near future, both because of the low level of expected emission from these sources and because of the ultimate realistic size of gamma-ray instruments.

A POTENTIAL UNIFICATION THEORY AND THE SEARCH FOR MICROSECOND GAMMA-RAY BURSTS

In the process of searching for a unification theory for physics, Hawking (1974) was led to conclude that black holes could emit particles, contrary to the belief at the time that black holes could not emit anything. This was followed by the prediction that mini black holes, whose masses are only a small fraction of the Sun, could have been created in the early universe, even though they cannot be created now because the necessary compression forces do not exist (Page and Hawking, 1976; Hawking, 1977). These mini black holes would gradually decay, emitting electrons, positrons, neutrinos, and photons. The limit on the photon radiation that can be set by the observed level of gamma radiation provides one constraint to this theory, but there is an even more intriguing prediction. As the mini black hole continues to lose its mass, its temperature rises, and it begins to emit particles of higher rest mass, until finally it ejects all of its remaining mass in a very short time. The heavy hadrons emitted in the final release would decay very rapidly, emitting a short burst (about 10^{-7} s) of hard gamma rays with an energy spectrum peaking around 250 MeV and having a total energy of approximately 10^{34} ergs (Page and Hawking, 1976).

The EGRET instrument was designed to have the capability of detecting a burst of gamma rays of the type just described. Following the first detection of a gamma ray in the EGRET detector, there is a time interval of approximately 6×10^{-7} s before the tracking system is triggered and the tracks in the system are recorded. Hence, any high-energy gamma ray converting to an electron pair in that interval, or shortly before it, will be seen. Since this time is very long compared to the burst time, all of the gamma rays from such a burst converting in the chamber will be recorded. As a verification, multiple gamma-ray events generated in the accelerator target on the ground during the calibration of EGRET were seen clearly in the data. Hence, unless the event is vetoed by the anticoincidence system, gamma rays will be recorded. The EGRET data have been systematically searched for events having more than one gamma ray from the same direction. The details of the approach to processing the gamma-ray events as described by Thompson et al. (1993) and Fichtel et al. (1994) have shown that the probability of missing a multiple gamma-ray event is small.

There has been no evidence in the EGRET data for a microsecond burst of high energy gamma rays (Fichtel, et al., 1994; Fichtel et al., 1996). The procedure for calculating the upper limit for the emission is given in the first paper (Fichtel et al., 1994). For the total energy mentioned above and the energy spectrum of the gamma rays used by Page and Hawking (1976), the upper limit calculated is 3×10^{-2} pc^{-3} yr^{-1} (Fichtel et al., 1996). The relevant distance range is 20 to 100 pc. Below 20 pc, there is a high probability that the event would be vetoed and beyond 100 pc the probability of detection becomes small. Note that this distance range is a very relevant one, because it represents the local region of the Galaxy. Page and Hawking (1976) stated that one might expect that the mini black holes might be concentrated in the gravitational potential wells of galaxies.

For completeness, it should be added that in the "Standard Model" the high-energy gamma-ray emission from the black hole is predicted to occur much more slowly than in the "Hagerdon" model, which is the one that was just discussed. (See, e.g., MacGibbon and Webber, 1990; Halzen et al., 1991; references therein.) In the Standard Model alternative, a burst of the order of seconds is expected for a threshold energy of 10 TeV, and the length of the burst is proportional to E_t^{-3}; so a group of gamma rays in the 10^2 to 10^4 MeV range would be so long in time that it would, in effect, not be a burst. Cline and Hong (1992, 1993) have proposed still other alternatives, wherein the bursts would be either a small fraction of a second in duration or even milliseconds in length. Attempts to see bursts of two of these types have been made, using ground-based telescopes that record secondaries produced in the atmosphere. Porter and Weekes (1978), using a large Cerenkov telescope, set an upper limit for the 10^{-7}s 10^2 MeV type bursts similar to the one determined by EGRET. Alexandreas et al. (1993), using an ultra-high-energy extensive air-shower array, have set an upper limit for the proposed 1 s TeV bursts of 8.5×10^5 pc^{-3} yr^{-1}, although this limit is theory dependent. Connaughton et al. (1993) have also set an upper limit for bursts of this type.

In summary, there is no evidence for the existence of high-energy microsecond bursts from mini black holes; however, data from the Compton Observatory have set reasonably severe upper limits.

SEARCHES FOR ANNIHILATION OF SUPERSYMMETRIC PARTICLES AND DECAY OF LONG-LIVED PRIMORDIAL RELICS

It has been realized for some time that, should these particles exist in adequate numbers and have lifetimes in an appropriate range, they might be seen in the diffuse radiation. Given the current experimental situation associated with the diffuse background, the relevant question is, what upper limits can be set. The answer to this question is not simple, because, in each case, it depends on many assumptions, including the nature of the universe and the characteristics of unknown particles. Nonetheless, some general statements can be made.

Consider first the WIMP (Weakly Interactive Massive Particle). If the WIMP is stable or has a very long lifetime, it might account for the dark matter thought to exist throughout the universe. Under the right conditions, the annihilation of the WIMP (possibly in the Galactic halo) could lead to a distinct, observable intensity of high-energy gamma rays in the GeV to TeV range (Rudaz and Stecker, 1991). There has been no observation of gamma-ray lines in this energy range. The telescopes, including EGRET, do not have sufficient energy resolution and sensitivity to detect this emission.

A more general consideration of long-lived primordial relics has been undertaken by Kribs and Rothstein (1996). They use data from COMPTEL and EGRET (see Figure 1), to conclude that relics with the critical density decaying predominantly through radiative (hadronic) channels are excluded for lifetimes in the range from 10^{12} s to 10^{22} (10^{29})s. The upper bound on the excluded lifetime assumes the worst case. The reader is referred to their paper for the specific bounds for particular masses.

THE FLAT-SPECTRUM RADIO QUASARS THAT EGRET DID NOT SEE

This subject is somewhat different from the other important searches discussed previously, in that, although it is concerned with radiation that was not seen, other objects believed to be of a similar nature were seen. The concern here is then why some were seen, when others were not.

Approximately sixty Flat-Spectrum Radio Quasars (FSRQs) and BL Lacs have been seen by EGRET in high-energy gamma rays (e.g., von Montigny et al., 1995a; Mukherjee et al., 1997; Hartman et al., 1997; and Fichtel and Trombka, 1997; references therein). They belong to the blazar class of flat-spectrum radio sources, which are characterized by beamed jets directed towards earth. In general, the ones that are observed are those that are most intense in the radio-frequency range. However, there are a few (six) notable ones that are very intense and have flat spectra in the radio range, but are not seen in the high-energy gamma-ray range, even though there were reasonably good observations by EGRET (see particularly von Montigny et al., 1995b; and Hartman et al., 1997.).

It is not known for certain why these six have not been seen in high-energy gamma rays, but three reasons have been suggested (See, e.g., von Montigny et al., 1995a.). The first, and probably the most likely, is that the flux from these source was in a low state at the time of the EGRET observation(s). The EGRET observations have shown that the high-energy gamma-ray flux from blazars is very time-variable (See, e.g., Mukherjee et al., 1997; Hartman et al., 1997.). Time variability is a very common characteristic of blazars in other wavelengths as well, although the gamma-ray region seems to have the greatest variation. Thus, it may be that the six blazars in question were simply not in a high gamma-ray state when observed. A second explanation that has been suggested involves the nature of the beam. One thought is that the gamma-ray beam may be narrower. For example, if the gamma-ray emission occurs in the inner region of the jet reasonably close to the core, as now appears quite likely, the bulk Lorentz factor may be much larger and the beam of relativistic particles may be much narrower in that region. Alternately, the beam may be bent and the gamma rays come from a different region than the bulk of the radio emission. Third, it may be that the blazars unseen by EGRET are just quiet in the gamma-ray range for some reason, although this seems to be the least likely explanation at the moment.

NORMAL GALAXIES AND CLUSTERS OF GALAXIES

Only one normal galaxy has been seen (Sreekumar et al., 1992), the Large Magellanic Cloud (LMC), and its intensity was at the level expected on the basis of quasi-stable dynamic balance and cosmic-ray interactions with matter and photons (Fichtel et al., 1991). No galactic clusters have been seen. The questions to be considered here are: Should other normal galaxies have been seen and should clusters of galaxies been seen? The most relevant aspect of the consideration is the intensity in the high-energy gamma-ray region, since it is in this energy range that the expected intensities are closest to what might be observable with the Compton Gamma-Ray Observatory.

Magellanic Clouds

Consider first the Small Magellanic Cloud (SMC). Sreekumar and Fichtel (1991) calculated the expected intensity for three different cases: a disrupted state, a quasi-equilibrium state such as appears to exist for the LMC, and a scenario of a metagalactic cosmic-ray density common to our galaxy, the LMC and the SMC. The results of their calculation for the flux above 100 MeV are: 2.4×10^{-7} photons cm^{-2} s^{-1} for the metagalactic cosmic-ray case, 1.2×10^{-7} photons cm^{-2} s^{-1} for the quasi-equilibrium state, and $(2 \text{ to } 3) \times 10^{-8}$ cm^{-2} s^{-1} for the disrupted state. Sreekumar and Fichtel also showed that, based on the synchrotron data, the expected magnetic field strength, and the assumption of the same electron-to-proton ratio in the cosmic rays as in our galaxy, the cosmic-ray density in the SMC would be small, well below that in our galaxy, if the cosmic rays are Galactic and not metagalactic. This conclusion is consistent with the SMC being in a state of irreversible disintegration and in agreement with the experimental findings of Mathewson, Ford, and Visvanathan (1986, 1988), and the tidal interaction model of Murai and Fujimoto (1980).

The upper limit to the flux from the SMC measured by EGRET was initially determined to be 0.5×10^{-7} cm^{-2} s^{-1} (Sreekumar et al., 1993), and more recent results set an even lower limit of 0.4×10^{-7} cm^{-2} s^{-1}. From the previous discussion, it is clear that this result is only consistent with the SMC being in a disrupted state. This upper limit is also the strongest argument to date for the Galactic origin of cosmic rays.

M31 and M87

No measurable flux of high-energy gamma rays from M31 is seen in the EGRET data in any energy interval. EGRET observations yield a 95% confidence upper limit of 0.8×10^{-7} photons cm^{-2} s^{-1} for energies above 100 MeV, consistent with the predicted flux of 0.2×10^{-7} cm^{-2} s^{-1} above 100 MeV by Özel and Fichtel (1988).

M87, a giant elliptical galaxy located near the Virgo Cluster, is 30 times further away than M31, but the radio luminosity from M87 is 1000 times greater than M31 or the Milky Way. Previous efforts to examine the high-energy emission from M87 using the SAS-2 and COS-B satellites provided large upper limits: 1.0×10^{-6} photons cm^{-2} s^{-1} from SAS-2 above 100 MeV (Fichtel et al., 1975), and an upper limit equivalent to 3×10^{-7} above 100 MeV from COS-B (Pollock et al., 1981).

EGRET observations yield a 2σ upper limit for E>100 MeV emission of 0.4×10^{-7} photons cm^{-2} s^{-1} (Sreekumar et al., 1994). Dermer and Rephaeli (1988) determined the relativistic electron distribution in the central region and halo from radio-continuum data, inferred the proton contribution and predicted the gamma-ray emission from M87. Their predicted flux depends on the assumed value of the magnetic field in the disk and halo. The observed upper limit from EGRET constrains the disk field to be $\geq 7\mu G$ and that in the halo to be $\geq 4\mu G$ if the assumptions made by Dermer and Rephaeli are valid. The lower limit on the halo field is consistent with the values derived by Andernach et al., (1979) and Feigelson et al., (1987) from equipartition arguments. Hence gamma-ray observations provide further evidence for the existence of energy equipartition in M87 at least on a galactic scale. The derived gamma-ray upper limit is also consistent with the gamma-ray flux of 1.5×10^{-8} photons cm^{-2} s^{-1} (including electron bremmstrahlung contribution to the π^0 component derived by Dermer and Rephaeli (1988)) expected from cosmic-ray interactions with the interstellar gas assuming a cosmic-ray density level equal to that observed in our Local Galactic neighborhod.

These results indicate the mean cosmic-ray energy density in M31 and M87 is not significantly different from equipartition arguments.

Galaxy Clusters

One of the interesting questions that can be addressed using gamma-ray observations, is the role of cosmic rays in heating the intracluster medium. Early radio observations of the Coma cluster by Large, Mathewson and Haslam (1961) showed a diffuse, steep spectrum radio halo. Coma appears to have the strongest radio halo amongst those clusters that sustain such an extended radio emission. It is believed that synchrotron emission from relativistic electrons in an intracluster magnetic field gives rise to this radio emission. Currently, the origin of these energetic electrons is still not fully understood. One possible scenario is that the halo emission arises from electrons that leak out of active radio galaxies. However, the spatial extent of the halo is not consistent with a diffusion model (Jaffe, 1977). Another possibility assumes that Coma-type radio halos are powered by secondary electrons generated by cosmic-ray protons interacting with the thermal ions in the intracluster medium (Dennison 1980; Vestrand 1980, 1994). X-ray observations indicate that the intergalactic medium in such clusters is filled with thin, hot gas. Numerous studies (Lea and Holman, 1978; Tucker and Rosner, 1983) have addressed the possible reheating of the intracluster medium by cosmic rays streaming out of the galaxies. Böhringer and Morfill (1988) estimate that an energy input of $\sim 10^{44}$ ergs s^{-1} is required from cosmic rays to balance the cooling of the intracluster gas. Clearly, both the models discussed above require the presence of energetic cosmic rays in the cluster. These cosmic rays could give rise to gamma-ray emission due to their interaction with hot gas in the cluster. Assuming no proton contribution, Vestrand (1994) calculated the gamma-ray flux from electron bremsstrahlung to be

$$F(E_\gamma) = 7.5 \times 10^{-11} B^{-2.3} (E_\gamma/100 MeV)^{-3.6} cm^{-2} s^{-1} MeV^{-1}$$

Assuming a range of field strength B = 0.1 to 1.0 μG, the predicted flux above 100 MeV lies in the range (7.5 to 0.03)$\times 10^{-7}$ photons $cm^{-2}s^{-1}$ respectively. EGRET observations of the Coma cluster showed no detectable flux and the derived 2σ upper limit is 4×10^{-8} photons $cm^{-2}s^{-1}$ for E$\geq$ 100 MeV (Sreekumar et al., 1996). Our upper limit to the gamma-ray flux from Coma translates to a minimum intracluster magnetic-field strength of 0.4 μG. The minimum field strength would be even larger if there is also a finite contribution from cosmic-ray protons. However, there is no simple method to determine such a contribution. This estimate is 4 times larger than the minimum field of 0.1μG obtained by Rephaeli, Ulmer, and Gruber (1994) from the OSSE data utilizing the absence of a hard Compton X-ray component.

REFERENCES

1. Andernach, H., Baker, J.R., von Kap-herr, A., and Wielebinski, R. 1979, A&A, 74, 93.
2. Alexandreas, D.E., 1993, Phys. Rev. Letters, 71 2524.
3. Böhringer, H., and Morfill, G. E., 1988, ApJ, 330, 609.
4. Cline, D.G., and Hong, W., 1992, ApJ, 401, L57.
5. Cline, D.G., and Hong, W., 1993, Gamma Ray Bursts, Second Workshop, AIP Conference Proceedings 307. ed. G.J. Fishman, J.J. Brainerd, K. Hurley, 470.
6. Connaughton, V., et al., 1993, Gamma Ray Bursts, Second Workshop, AIP Conference Proceedings 307, ed. G.J. Fishman, J.J. Brainerd, K. Hurley, 470.
7. Dennison, B., 1980, ApJ, 236, 761.
8. Dermer, C., and Rephaeli, Y. 1988, ApJ, 329, 687.
9. Feigelson, E.D., Wood, P.A.D., Schreier, E.J., Harris, D.E., and Reid, M.J. 1987, ApJ, 312, 101.
10. Fichtel, C.E., and Kniffen, D.A., 1974, "Gamma Ray Astronomy", in High Energy Particles and Quanta in Astrophysics, The MIT Press, ed. Frank B. McDonald and Carl E. Fichtel.
11. Fichtel, C.E., et al., 1975, ApJ, 198, 163.
12. Fichtel, C.E., Özel, M., Stone, J., and Sreekumar, P., 1991, ApJ, 374, 134.
13. Fichtel et al., 1994, ApJ, 434, 557.
14. Fichtel et al., 1996, Third Compton Burst Symposium, Huntsville, AL, October 1995, AIP Conference Proceedings, to be published.
15. Fichtel, C.E., and Trombka, J.I., 1997, Gamma Ray Astrophysics, New Insight into the Universe, second edition, NASA Reference Publication 1386.
16. Halzen, F., Zas, E., MacGibbon, J.H., and Weekes, T.C., 1991, Nature, 353, 807.
17. Hartman, R., et al., 1997, submitted to ApJS.
18. Hawking, S.W., 1974, Nature, 248, 30.
19. Hawking, S.W., 1977, Scientific American, 236, 34.
20. Jaffe, W. J., 1977, ApJ, 212, 1.
21. Kappadath, S. C., et al., 1996, A&AS, 120, 619.

22. Kribs, G.D., and Rothstern, I.Z., 1996, submitted.
23. Large, M.I., Mathewson, D.S., and Haslam, C.G.T., 1961, MNRAS, 123, 113.
24. Lea, S. M., and Holman, G. D., 1978, ApJ, 222, 29.
25. Leventhal, M., McCallum, C.J., Stang, P.D., 1978, ApJ, 225, L11.
26. MacGibbon, J.H., and Webber, B.R., 1990, Phys. Rev., D41, 33052.
27. Mathewson, D.S., Ford, V.L., and Visvanathan, N., 1986, ApJ, 301, 664.
28. Mathewson, D.S., Ford, V.L., and Visvanathan, N., 1988, ApJ, 333, 617.
29. Montigny, C. von, et al., 1995a, ApJ, 440, 525.
30. Montigny, C. von, et al., 1995b, A&A, 299, 680.
31. Mukherjee, R., et al., 1997, submitted to ApJ.
32. Murai, T., and Fujimoto, M., 1980, Publ. Astronomical Society of Japan, 32, 581.
33. Omnes, R., 1969, Phys. Rev. Letters, 23, 38.
34. Özel, M., and Fichtel, C.E., 1988, ApJ, 335, 135.
35. Page, D.N., and Hawking, S.W., 1976, ApJ, 205, 1.
36. Pollock, A., et al., 1981, A&A, 94, 116.
37. Porter, N.A., and Weekes, T.C., 1978, MNRAS, 183, 205.
38. Purcell, W.R., et al., 1993, ApJ, 413, L85.
39. Purcell, W.R., et al., 1994, AIP Conference Proceedings, 280, 70.
40. Ramaty, R., and Lingenfelter, R.E., 1994, in "High Energy Astrophysics, Models and Observations from MeV to EeV", ed. J.M. Matthews, World Scientific Publishing Co., 32.
41. Ramaty, R., 1996, A&A (Third Compton Symposium), to be published.
42. Rephaeli, Y., Ulmer, M., and Gruber, D., 1994, ApJ, 429, 554.
43. Rudaz, S., and Stecker, F.W., 1991, ApJ, 368, 406.
44. Sreekumar, P., and Fichtel, C.E., 1991, A&A, 251, 447.
45. Sreekumar, P., et al., 1992, ApJ, 400, L67.
46. Sreekumar, P., et al., 1993, Phys. Rev. Letters, 70, 127.
47. Sreekumar, P., et al., 1994, ApJ, 426, 105.
48. Sreekumar, P., et al., 1996, ApJ, 464, 628.
49. Sreekumar, P., et al., 1997, submitted to ApJ.
50. Stecker, F.W., Morgan, D.L., Bredekamp, J., 1971, Phys. Rev. Letters, 27, 1469.
51. Timmes, F.X., Woosley, S.E., Hartman, D.H., and Hoffman, R.D., 1996, ApJ, 464, 332.
52. Thompson, D.J., et al., 1993, ApJS, 86, 629.
53. Tucker, W. H., and Rosner, R., 1983, ApJ, 267, 547.
54. Vestrand, W. T., 1980, Ph.D thesis, Univerisity of Maryland, College Park.
55. Vestrand, W. T., 1994, unpublished work.
56. Watanabe, K., 1996, PhD dissertation, Clemson University, Clemson, SC.

HIGH ENERGY PHYSICS AND ASTROPHYSICS

Gamma Ray Implications for the Origin and the Acceleration of Cosmic Rays

R. Schlickeiser[1], M. Pohl[2], R. Ramaty[3], J. G. Skibo[4]

1 Max-Planck-Institut für Radioastronomie, Postfach 2024, 53010 Bonn, Germany;
2 Danish Space Research Institute, Juliane Maries Vej 30, 2100 Copenhagen, Denmark;
3 Laboratory for High Energy Astrophysics, Code 665, NASA, Goddard Space Flight Center, Greenbelt, MD 20771;
4 E. O. Hulburt Center for Space Research, Code 7653, Naval Research Laboratory, Washington, DC 20375-5352

Abstract. We examine the key observations of the Compton Observatory that constrain the origin of cosmic rays, in particular the locations of cosmic ray sources and the physical acceleration mechanisms. The observations of the Magellanic Clouds and measurements of the extragalactic diffuse hard X-ray and gamma-ray background radiation exclude any metagalactic origin of cosmic-ray electrons and cosmic-ray nucleons with energies below 100 GeV. The nature of cosmic-ray sources in our Galaxy is constrained by the observations of galactic supernova remnants, measurements of the diffuse galactic continuum radiation and of gamma-ray lines. The measured energy spectrum of the diffuse galactic continuum radiation provides new insight into the relative acceleration of cosmic-ray electrons and nucleons. We also address the consequences of the gamma-ray line observations from the Orion molecular clouds.

I INTRODUCTION

Continuum and line γ-radiation is produced in interactions of energetic cosmic ray particles with ambient target matter and photon gases. It therefore traces the global spatial distribution of these particles, knowing the distribution of the target fields. As much as continuum radio astronomy has been used to infer the global distribution of cosmic ray electrons, a prime motivation for the development of γ-ray astronomy ([1], [2]) has been to use γ-radiation to deduce the global distribution of cosmic ray nucleons. However, there is no clear-cut division between cosmic ray electrons radiating only at radio and not at γ-ray frequencies. Quite to the contrary, cosmic ray electrons are responsible for a substantial part of the galactic γ-ray emissivities, and this "electron contamination" limits the conclusions on the global distribution of cosmic ray nucleons.

The organisation of this review is as follows: in section II we discuss the *Compton* observatory results on γ-radiation with photon energies above 10 MeV. No line emission occurs in this energy range. As we shall see, these results rule out any metagalactic origin of cosmic ray nucleons with energies below 100 GeV. Gamma radiation below 10 MeV is a mixture of line radiation generated by low-energy cosmic ray nucleons and continuum radiation produced by cosmic ray electrons [3]. The electron generated inverse Compton continuum radiation, discussed in section III, also excludes any metagalactic origin of these particles, and the detected galactic electron bremsstrahlung radiation indicates the abundance of many semirelativistic cosmic ray electrons in the Galaxy. The interpretation of the strong emission from the Orion region as nucleon generated line emission allows conclusions on the cosmic ray source abundances, as is discussed in Section IV.

II γ-RAY EMISSION ABOVE 10 MEV

Since the pioneering work based on SAS-2 data ([4] and references therein) observations of the diffuse γ-ray emission of galaxies have been useful for studies of cosmic rays and the acceleration of energetic particles in general. Reviews of the situation at the end of the COS-B era [5] and early in the CGRO era [6] have been published in the past.

The γ-ray sky above 10 MeV, as we know it today, is a composite of the emission from point sources and diffuse emission. The most prominent feature is the galactic plane, in which interactions between cosmic rays and the thermal gas lead to γ-ray emission by π^0-decay and bremsstrahlung. On top of this diffuse emission we see point sources all over the sky, part of them being pulsars, another part distant AGN, and also a significant fraction of yet unidentified sources. But we also know that the diffuse emission extends out of the galactic plane up to the poles. There appears to be an isotropic emission component which is presumably extragalactic in origin and may be understood as blend of unresolved AGN, but there is also considerable galactic emission. At higher latitudes interactions between cosmic rays and thermal gas still play a role, but inverse-Compton scattering of ambient photons by cosmic ray electrons becomes increasingly important. This emission tells us about the propagation of cosmic rays from their sources to the interaction regions and thus complements the direct particle measurements by balloon and satellite experiments.

A The locality of cosmic rays

There has been a long-standing debate on whether cosmic rays in the GeV energy range are galactic or extragalactic [7]. The electrons have to be galactic since the energy losses by comptonisation of the microwave background prevent propagation from one galaxy to another. In case of protons the situation has been improved significantly by the Compton observatory results. EGRET observations of the Magellanic Clouds have revealed that the γ-ray flux of the Large Magellanic Cloud is weakly inconsistent [8] and that of the Small Magellanic Cloud is strongly inconsistent [9] with cosmic ray protons having uniform density in space. *Therefore the bulk of the locally observed protons at GeV energies must be galactic,* and we have to think about which galactic accelerators are capable of producing cosmic rays with a source power of $\sim 3 \cdot 10^{40}$ erg/sec. A galactic origin of cosmic rays also has impact on dark matter studies [10], since this fact allows a substantial amount of baryonic dark matter to be hidden at 10 kpc or more from the Galaxy without inducing observable γ-ray emission.

There has been a debate also on whether the γ-ray flux of LMC and SMC in relation to their radio emission would allow equipartition between the magnetic field and cosmic rays. It turns out that this equipartition is still possible provided one allows the cosmic ray electron-to-proton ratio to be different from that in the solar vicinity. The γ-ray data of the Magellanic Clouds are best explained when the e/p ratio is higher than locally observed so that basically only the density of cosmic ray nucleons is reduced in LMC and SMC [11]. The results mentioned were derived from the volume-integrated emission of the Magellanic Clouds. A new study including spatial and spectral information is under way.

Neither our neighboring Galaxy M31 nor the nearby starburst galaxies M82 and NGC 253 have been detected by EGRET [12], in accord with detailed model predictions for the flux of these sources being below the EGRET threshold [13,14].

B Confusion and point sources

Any analysis of the galactic diffuse emission can be seriously hampered by unresolved galactic point sources which may have a sky distribution similar to that of gas. Because of six objects being already detected, pulsars are the most likely input from discrete sources. Many authors have addressed this problem on the basis of pulsar emission models [15–20] and consistently estimates the contribution of pulsars to the diffuse γ-ray intensity above 100 MeV integrated over the whole sky to be a few percent. Another strategy is to base the analysis only on the observed properties of the six identified γ-ray pulsars, which allows also an inspection of the spectrum of the unresolved pulsars [21]. It is found that pulsars contribute mostly at γ-ray energies above 1 GeV, and preferrentially exactly in the galactic plane where they can provide more than 20% of the observed emission for a reasonable number of directly observable objects. This also provides limits to the contribution of pulsars to the acceleration of cosmic ray electrons, since the γ-ray emission of pulsars takes a large fraction of their energy loss. The power in escaping particles will be less than the power in their total γ-ray emission, and thus can be only a small fraction of the total power in galactic cosmic ray electrons.

Estimates for the contribution of other (than pulsars) discrete sources are very uncertain due to the lack of a clear identification of γ-ray sources with any known population of galactic objects. It is interesting to see that roughly ten unidentified EGRET sources can be associated with supernova remnants (SNR) or with OB associations, or with both (SNOBs) [22–25]. Obviously these sources may also be radio-quiet pulsars or highly dispersed radio pulsars.

Especially SNR are likely sites of particle acceleration in our Galaxy in view of the energetics. We know by their synchrotron emission that SNR accelerate electrons, in case of the Crab nebula ([26]) and SN1006 ([27,28]) up to TeV energies. However, there still is no clear observational proof to date that the nuclear component of cosmic rays is produced in SNR. Models predict that the detection of SNRs in the EGRET range is difficult but not impossible, and that detection at TeV energies appears more feasible [29]. Prime candidates for detection are SNR overtaking a molecular cloud like e.g. G78.2+2.1 [30], which however often are located in crowded regions of the sky. As a result of the limited spatial resolution, confusion is the major problem for the identification of SNR as emitters of GeV γ-rays . At TeV energies no detection of a SNR has been published yet, and for many promising sources the upper limits are well below the earlier model predictions [31,32]. Very recently the detection of the remnant SN1006 at TeV energies has been announced [33]. The lineless X-ray emission of this SNR [27], when interpreted as synchrotron emission of 100 TeV electrons, implies TeV γ-ray emission at observable flux levels from the comptonisation of the microwave background [34], but with a spectrum different from that of nucleonic γ-rays. The main lesson of this is that, even if other SNR are detected by Čerenkov telescopes, we would not have evidence of shock accelerated nucleons either, since substantial emission of TeV energy electrons may not be unique to SN1006. However, leptonic TeV γ-ray emission is still a superb opportunity to measure the magnetic field strength in supernova remnant shells and to model electron acceleration.

C The γ-ray emissivity

In galactic halos inverse Compton scattering of ambient photons by cosmic ray electrons is an important production process for γ-rays. Due to the large scale height of far-infrared photons and the universal nature of the cosmic

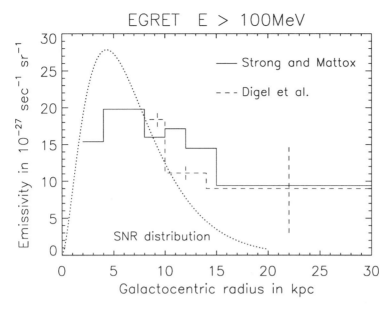

FIGURE 1. A comparison between the γ-ray emissivity gradient (solid histogram) to the distribution of SNR as possible acceleration sites (dotted line). The statistical uncertainties of the gradient are typically below 10%. The obvious discrepancy implies that either SNR are not accelerating the bulk of GeV cosmic rays, or diffusive reacceleration is operative, or galactic cosmic rays are confined on a scale of many kpc's. Please note that locally derived emissivities (dashed histogram) can differ significantly from the global trend.

microwave background radiation, the latitude distribution of this emission is much broader than in the case of the bremsstrahlung and π^0-decay, and may reflect directly the latitude distribution of the radiating cosmic ray electrons [35]. Thus the inverse Compton component can tell us about the life and the propagation history of cosmic ray electrons in the Galaxy.

By comparison of the spatial distribution of γ-rays, unrelated to the interstellar gas, to the brightness temperature distribution in radio surveys, the average intensity of inverse Compton emission at high latitudes is determined to be $(5.0 \pm 0.8) \cdot 10^{-6}$ cm^{-2} sec^{-1} sr^{-1} for $E > 100$ MeV with a photon spectral index of -1.85 ± 0.17 [36].

To investigate the γ-ray emission originating from π^0-decay and bremsstrahlung, we need some prior knowledge of the distribution of interstellar gas in the Galaxy. This includes not only HI but also H$_2$, which is indirectly traced by CO emission lines, and HII, which is traced by Hα and pulsar dispersion measurements. Even in case of the directly observable atomic hydrogen we obtain only line-of-sight integrals, albeit with some kinematical information. Any deconvolution of the velocity shifts into distance is hampered by the line broadening of individual gas clouds and by the proper motion of clouds with respect to the main rotation flow. The distance uncertainty will in general be around 1 kpc.

It thus may be appropriate to use only a few resolution elements and investigate the γ-ray emissivity per H-atom in galactocentric rings, i.e. to assume azimuthal symmetry for the emissivity. This kind of approach has already been successfully applied to the earlier COS-B data. When repeated for the analysis on EGRET data at γ-ray energies above 100 MeV one finds [37] that the γ-ray emissivity per H-atom declines with galactocentric radius, confirming earlier COS-B results. In Fig. 1 we compare this gradient to the distribution of SNR [38] as possible acceleration sites for the cosmic rays. Obviously the cosmic ray gradient derived from the γ-ray data is rather weak compared to the SNR distribution, which implies that, if SNR are the sources of galactic cosmic rays, either diffusive reacceleration is operating, or that the spatial propagation disperses the distribution of cosmic rays from that of their sources by many kpc's with obvious consequences for the halo size in diffusion models, or a galactic variation of the H_2 column density to ^{12}CO intensity ratio [39].

It should be noted that the emissivity gradient does not necessarily hold locally. A comparison of the γ-ray emissivity in the Perseus arm at 3 kpc distance in the outer galaxy to the Cepheus and Polaris flare at 250 pc gives a difference of a factor 1.7±0.2 [40], far exceeding the overall gradient (see Fig.1).

D Spectral information

The observed γ-ray spectrum of emission related to the gas is difficult to understand. It deviates significantly from the expected superposition of π^0-decay and bremsstrahlung [41]. At energies above 1 GeV there is a clear excess of emission related to gas which may be only slightly relaxed by variation of the input proton spectrum to the pion production process [42]. This excess can not be explained as superposition of unresolved sources, since pulsars as the only known candidates with appropriate spectrum fail to provide the required intensity and latitude distribution [21]. Quite curiously, the observed spectrum can be modelled very reasonably as being completely due to cosmic ray electron generated bremsstrahlung and inverse Compton radiation [43] with an electron power law source spectrum with differential number spectral index of -1.7±0.1, and absolute numbers in agreement with radio studies [44].

The excess at higher γ-ray energies may also indicate that the average spectrum of high energy electrons is different from what we observe locally. This would be a natural consequence if electrons were not accelerated isotropically throughout our Galaxy, but in discrete objects like SNR [45]. It is however unclear whether this effect can account for all the excess.

III CONTINUUM GAMMA RAY EMISSION BELOW 10 MEV

The OSSE and COMPTEL instruments have provided us with the first clear picture of the galactic plane in the light of $\simeq 0.1-10$ MeV photons. In this photon band the observed emission is a mixture from continuum radiation processes of cosmic ray electrons and line emission resulting from positron-electron annihilation, radioactive decays and de-excitation of nuclei in the interstellar medium and cosmic ray nuclei. Results in the sense of a clear separation of continuum and line emission at MeV energies are still preliminary.

In the context of cosmic ray origin, the information from the inverse Compton and bremsstrahlung continuum and the cosmic ray de-excitation lines are most conclusive. First and most important, the OSSE and COMPTEL instruments have not detected huge fluxes of inverse Compton scattered microwave background radiation photons, which are to be expected [46] for any metagalactic origin of cosmic ray electrons. In fact, *this non-detection rules out any metagalactic origin of cosmic ray electrons.*

Moreover, the OSSE [49] and COMPTEL [50] instruments provided evidence that the diffuse galactic continuum emission extends down to photon energies below 100 keV, see Fig. 2. The diffuse origin of this radiation has been suggested by correlated SIGMA measurements, to estimate the galactic point source contribution [49], and the analysis of the GINGA measurements of the galactic ridge emission at much lower energies [47,48]. Although one should keep an open mind to alternative explanations as e.g. synchrotron radiation from ultrahigh-energy ($> 10^{15}$ eV) secondary electrons [51], the γ-ray emission in this energy band is most likely electron bremsstrahlung in the interstellar medium, which implies the presence of many low-energy (<10 MeV) cosmic ray electrons in the Galaxy. The power in cosmic ray electrons to produce a given amount of bremsstrahlung is a fixed quantity that depends only on the energy spectrum of the radiating electrons and (weakly) on the ionization state of the interstellar medium. Attributing this power input to injection in cosmic ray electron sources it has been estimated that, intergrated over the whole Galaxy, a source power of about $\sim 4 \times 10^{41}$ erg s^{-1} [52] or even $\sim 10^{43}$ erg s^{-1} [53] in low-energy (<10 MeV) electrons is required, to maintain these electrons against the severe Coulomb and ionization losses. The increase in the later estimate results from the GINGA result that the diffuse bremsstrahlung emission may extend down to photon energies ~ 10 keV, well below the OSSE energy band. Taking the first number as a lower limit, this electron power exceeds the power supplied to the nuclear cosmic ray component by at least an order of magnitude.

Recently, the extension of the bremsstrahlung continuum emission to these low energies has been attributed to the existence of in-situ stochastic electron acceleration by the interstellar plasma turbulence [54], rather than to

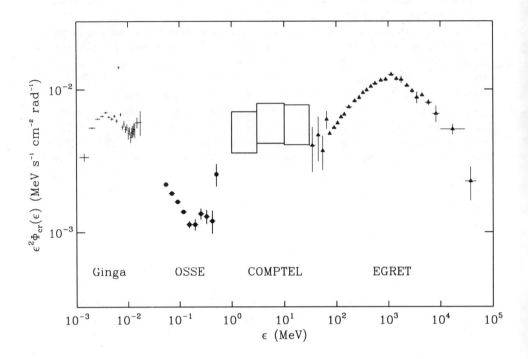

FIGURE 2. The current best estimate of the diffuse high energy continuum emission from the central radian of the Galaxy (Ginga - Yamasaki et al. 1996 ; OSSE - Purcell et al. 96 ; COMPTEL - Strong et al. 1996 ; EGRET - Hunter et al. 1997). The left vertical axis is the measured flux multiplied by ϵ^2.

the existence of a second electron source component. This turbulence with a measured energy density of $\simeq 4 \times 10^{-14}$ erg cm^{-3} [55] is an important additional energy source of cosmic ray particles, and thus weakens considerably the power requirements on the injection of particles in nominal cosmic ray sources. The interplay of interstellar in-situ acceleration and Coulomb and ionization losses generates an equilibrium energy spectrum of low-energy electrons that rises steeply at small electron energies, providing many low-energy electrons below the minimum electron injection energy, as required by the γ-ray observations, This interpretation requires that most of the interstellar plasma turbulence is damped by accelerating cosmic ray electrons on time scales less than 3×10^4 yrs, and that this acceleration process preferentially favours cosmic ray electrons over hadrons. This definitely rules out reacceleration by gyroresonant interactions with incompressive shear Alfven waves since these would preferentially accelerate hadrons over electrons. But transit-time damping of fast magnetosonic waves in a low-beta interstellar medium is a suitable acceleration mechanism [56]. Proving the truly diffuse nature of the galactic continuum emission below 1 MeV would be of utmost importance to pin down the most relevant particle acceleration process.

IV GAMMA RAY LINE EMISSION BELOW 10 MEV FROM THE ORION REGION

Intense γ-ray emission in the 3-7 MeV energy band was observed from the Orion molecular cloud complex with the COMPTEL instrument [57,58]. Since the source of emission appears to be spatially extended, models based on line emission from ^{12}C and ^{16}O deexcitations in a diffuse medium [59] seem to be favoured over scenarios of blueshifted annihilation lines from an active galactic nucleus [60] or nuclear deexcitations in the vicinity of a black hole [61]. A related issue, discussed at this meeting [62,63], is the amount of the expected X-ray emission that accompanies the observed gamma ray lines. ROSAT data was used to place an upper limit on the total X-ray emission from the gamma ray emitting region [62]. The predicted inverse bremsstrahlung and bremsstrahlung from secondary knock-on electrons are not inconsistent with this upper limit. However, the predicted line emission following electron capture onto fast O ions is quite large, and potentially inconsistent with the ROSAT upper limits [63]. But this line emission depends sensitively on the spatial distribution of the nuclear γ-ray deexcitation line emission, which is rather uncertain, and on the metallicity and column density of the interstellar medium towards and in Orion, which determines the level of photoelectric absorption.

The observed spectrum suggests that the the 3-7 MeV γ-ray emission mostly consists of line emission from ^{12}C and ^{16}O deexcitations. Such line emission can only be produced by accelerated particle interactions. The observed spectrum suggests that the emission mostly consists of broad lines, resulting from collisions of accelerated C and O nuclei with ambient H and He, rather than narrow lines due to the collisions of accelerated protons and He nuclei with ambient C and O [64]. Apart from the observed emission in the 3-7 MeV band, the COMPTEL observations revealed only upper limits at the neighbouring γ-ray energy bands [57,58]. In particular, the upper limit on the 1-3 MeV sets tight constraints on the accelerated Ne-Fe abundances relative to those of C and O. A further constraint is set by Orion's γ-ray emission at photon energies above 30 MeV [65] which is consistent with pion production and bremsstrahlung due to the irradiation of the molecular clouds by standard Galactic cosmic rays. As such cosmic rays underproduce the observed line emission by at least three orders of magnitude, the gamma ray line production in Orion must be predominantly a low-energy cosmic ray phenomenon, requiring very large low energy cosmic ray fluxes in Orion. The bulk of the line emission is produced by accelerated particles with energies of at least several tens of MeV/nucleon but not exceeding about 100 MeV/nucleon [64]. The explanations proposed for the suppression of both, the Ne-Fe abundance because of the 1-3 MeV upper limit, and the $p-He$ abundance because of the limits on the narrow lines [64], relative to C and O invoked the injection into the particle accelerator [66,67] of seed particles from a variety of sources: the winds of massive stars in late stages of evolution [59], the ejecta of supernovae from massive star progenitors [67,59,68], and the acceleration of ions resulting from the breakup of interstellar dust [69]. Apparently these nuclear gamma ray lines allow for the first time valuable conclusions on the injection and acceleration processes of low-energy cosmic rays in interstellar environment.

V SUMMARY AND CONCLUSIONS

The instruments on board of the Compton observatory have provided important constraints on the origin and the acceleration of cosmic rays. Most importantly, they verified that cosmic ray particles are of galactic origin: the EGRET observations of the Magellanic Clouds and the OSSE and COMPTEL measurements of the extragalactic diffuse hard X-ray and gamma-ray background radiation exclude any metagalactic origin of cosmic ray nucleons with energies below 100 GeV and cosmic ray electrons, respectively. Cosmic ray nucleons with energies above 100 GeV are not constrained by the Compton observatory results.

More specific properties of the cosmic ray sources in our Galaxy can be derived from the observations of individual galactic supernova remnants, gradient and spectral measurements of the diffuse galactic continuum radiation and of gamma-ray lines. The now available measured energy spectrum of the diffuse galactic continuum radiation over more than six frequency decades provides new insight into the relative contributions from cosmic ray nucleons and cosmic ray electrons, and thus into the relative acceleration of cosmic ray electrons and nucleons. The interpretation of the strong 3-7 MeV emission from the Orion molecular clouds as cosmic ray nucleon generated deexcitation line emission allows conclusions on the cosmic ray source abundances.

REFERENCES

1. Morrison P.: 1958, *Nuovo Cimento*, **7**, 858
2. Ginzburg V. I.: 1973, in *NASA Intern. Symposium and Workshop on Gamma-Ray Astrophysics*, eds. F. W. Stecker & J. I. Trombka, NASA X-641-73-180
3. Ramaty R., Murphy R. J.: 1987, *SSR*, **45**, 213
4. Fichtel C. E., Kniffen D. A.: 1984, *A&A*, **134**, 13
5. Bloemen H.: 1989, *ARA&A*, **27**, 469
6. Strong A. W.: 1995, *SSR*, **76**, 205
7. Brecher K., Burbidge G. R.: 1972, *ApJ*, **174**, 253
8. Sreekumar P. et al.: 1992, *ApJ*, **400**, L67
9. Sreekumar P. et al.: 1993, *PRL*, **70**, 127
10. Salati P. et al.: 1996, *A&A*, **313**, 1
11. Pohl M.: 1993, *A&A*, **279**, L17
12. Sreekumar P. et al.: 1994, *ApJ*, **426**, 105
13. Pohl M.: 1994, *A&A*, **287**, 453
14. Paglione T. A. D., Marscher A. P., Jackson J. M., Bertsch D. L.: 1996, *ApJ*, **460**, 295
15. Harding A. K.: 1981, *ApJ*, **247**, 639
16. Schnepf N. G. et al.: 1992, *1st Compton Symposium*, eds. M. Friedlander, N. Gehrels & D. Macomb, New York, AIP, p. 85
17. Bailes M., Kniffen D. A.: 1992, *ApJ*, **391**, 659
18. Hartman D. H., Brown L. E.: 1993, *2nd Compton Symposium*, eds. C. E. Fichtel, N. Gehrels & J. P. Norris, New York, AIP, p. 101
19. Yadigaroglu I.-A., Romani R. W.: 1995, *ApJ*, **449**, 211
20. Sturner S. J., Dermer C. D.: 1996, *ApJ*, **461**, 872
21. Pohl M. et al.: 1997, *ApJ*, submitted
22. Sturner S. J., Dermer C. D.: 1995, *A&A*, **293**, L17
23. Esposito J. et al.: 1996, *ApJ*, **461**, 820
24. Kaaret P., Cottam J.: 1996, *ApJ*, **462**, L1
25. Yadigaroglu I.-A., Romani R. W.: 1997, *ApJ*, **476**, 347
26. DeJager O. C., et al.: 1996, *ApJ*, **457**, 253
27. Koyama K. et al.: 1995, *Nature*, **378**, 255
28. Reynolds S.: 1996, *ApJ*, **459**, L13
29. Drury L. O'C., Aharonian F. A., Völk H. J.: 1994, *A&A*, **287**, 959
30. Aharonian F. A., Drury L. O'C., Völk H. J.: 1994, *A&A*, **285**, 645
31. Prosch C. et al.: 1995, *24th ICRC*, **2**, 471
32. Lessard R. W. et al.: 1995, *24th ICRC*, **2**, 475
33. Tanimori et al.: 1997, *25th ICRC*, in press

34. Pohl M.: 1996, *A&A*, **307**, L57
35. Schlickeiser R. Thielheim K. O.: 1977, *ApSS*, **47**, 415
36. Chen A. et al.: 1996, *ApJ*, **463**, 169
37. Strong A. W., Mattox J. R.: 1996, *A&A*, **308**, L21
38. Case G., Bhattacharya D.: 1996, *A&AS*, **120**, C437
39. Sodrowski T. J. et al.: 1995, *ApJ*, **452**, 262
40. Digel S. W. et al.: 1996, *ApJ*, **463**, 609
41. Hunter S. et al: 1997, *ApJ*, **481**, 205
42. Mori M.: 1997, *ApJ*, **478**, 225
43. Skibo J. G.: 1997, *this conference*
44. Pohl M., Schlickeiser R.: 1991, *A&A*, **252**, 565
45. Cowsik R., Lee M. A.: 1979, *ApJ*, **228**, 297
46. Fazio G. G., Stecker F. W., Wright J. P. .: 1966, *ApJ*, **144**, 611
47. Yamasaki N. Y. et al.: 1996, *A&AS*, **120**, C393
48. Yamasaki N. Y. et al.: 1997, *ApJ*, **481**, 821
49. Purcell W. R. et al.: 1996, *A&AS*, **120**, C389
50. Strong A. W.: 1996, *A&AS*, **120**, C381
51. Mörsberger U., Schlickeiser R.: 1997, *25th ICRC*, in press
52. Skibo J. G., Ramaty R.: 1993, *A&AS*, **97**, 145
53. Skibo J. G., Ramaty R., Purcell W. R.: 1996, *A&AS*, **120**, C403
54. Schlickeiser R.: 1997, *A&A*, **319**, L5
55. Minter A. H., Spangler S. R.: 1997, *ApJ*, in press
56. Schlickeiser R., Miller J. A.: 1997, *ApJ*, submitted
57. Bloemen H., et al.: 1994, *A&A*, **281**, L5
58. Bloemen H., et al.: 1994, *ApJL*, **475**, L25
59. Ramaty R., Kozlovsky B., Lingenfelter R.: 1996, *ApJ*, **456**, 525
60. Pohl M.: 1996, *A&A*, **120**, C457
61. Miller J. A., Dermer C. D.: 1995, *A&A*, **298**, L13
62. Dogiel V.: 1997, *this conference*
63. Ramaty R., Kozlovsky B., Tatischeff V.: 1997, *this conference*
64. Kozlovsky B., Ramaty R., Lingenfelter R.: 1997, *ApJ*, in press
65. Digel S. W., Hunter S. D., Mukherjee R.: 1995, *ApJ*, **441**, 270
66. Nath B. B., Biermann P.: 1994, *MNRAS*, **270**, L33
67. Bykov A. M., Bloemen H.: 1994, *A&A*, **283**, L1
68. Casse M., Lehoucq R., Vangioni-Flam E.: 1995, *Nat*, **373**, 318
69. Ip W.-H.: 1995, *A&A*, **300**, 283

Spectral Signatures and Physics of Black Hole Accretion Disks

Edison Liang* and Ramesh Narayan[†]

*Rice University, †Harvard University

Abstract. This review summarizes the x-and-gamma-ray signatures of Galactic black hole candidates and highlights some of the accretion disk models relevant to the formation of the observable spectra and spectral variability.

I INTRODUCTION

Broadband x-and-gamma-ray monitoring of Galactic black hole (GBH) candidates by CGRO-BATSE and RXTE, coupled with pointed observations by ground and space-based telescopes, have dramatically expanded our knowledge of GBHs in the last few years [1]. This rapidly increasing volume of spectral and temporal variability data allows us to systematically confront the different models of black hole accretion disks. Such unprecedented diagnostics will eventually help to clarify the physical conditions near the horizon, the origin of accretion disk viscosity, the driving engine of relativistic outflows and jets, the amount of intrinsic spin of the hole, and whether black holes are electron-positron pair factories. At the same time, theoretical advances will help experimentalists to plan the strategies for future space experiments.

We review the spectral states of GBHs in Sec.2 and introduce the currently popular conceptual picture in Sec.3. In Sec.4 we discuss the possible origins of electron-positron pairs and their outflows from accretion disks. In Sec.5 we review models of accretion disk structure and dynamics which motivate the concepts of advection-dominated flows in Sec.6 and other transonic solutions in Sec.7. Finally in Sec.8 we speculate on the origin of nonthermal processes that may be responsible for both the power-law MeV tail and radio emissions.

II GBH SPECTRAL STATES

Fig.1 shows the outburst spectra of 2 GBH X-ray Novae (XRNs) which epitomize the two canonical spectral states of GBHs: (a) the hard Compton-

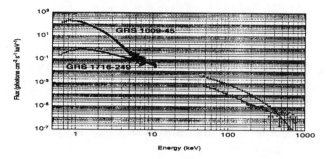

FIGURE 1. Joint ASCA-OSSE spectra of two XRNs during outburst showing the canonical spectral states of GBHs (from Moss [5]).

like [2] state with photon index =1.5+/-0.5 and exponential cutoff above 200 keV; (b) the soft power-law state with photon index > 2.2 above 10 keV and no apparent cutoff out to MeVs. The νF_ν distribution of the hard state shows a clear peak at 100 - 200 keV whereas that of the soft state shows no peak (Fig.2).

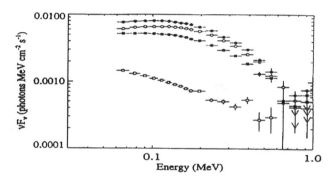

FIGURE 2. OSSE spectra of Cyg X-1 at different intensity states showing the presence of a spectral peak in the hard state and no peak in the soft state, which also has the ultralow intensity (from Phlips et al [6]).

The soft state is often accompanied by the blackbody-like soft x-ray high state below 10 keV. In addition, Comptel data suggest that in the case of Cyg X-1 and GROJ0422, the hard state is accompanied by a power-law tail in the MeV range (Fig.3) [3,4], which may be of nonthermal origin.

It is tempting to associate this MeV tail with the extension of the soft state power-law but we need further confirmation. Cyg X-1 goes into the soft state when its hard x-ray flux becomes ultralow [6]. We speculate that, at least for Cyg X-1, the two spectral states originate from physically distinct regions, and the size and output of each region increase or decrease at the expense of

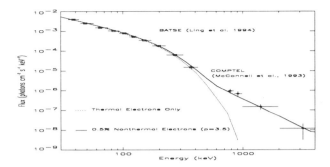

FIGURE 3. Joint BATSE-COMPTEL spectrum of Cyg X-1 showing the MeV power-law tail. Solid curve is best Compton model fit with hybrid thermal plus nonthermal leptons (from Crider et al [41]).

the other.

In addition to the above two spectral states, GBHs may episodically exhibit a "superhard" state with enhanced emissions near or above 500 keV. Various experiments have claimed to observe an "MeV bump" from Cyg X-1 [7], narrow 480 keV feature from Nova Muscae [8], and broad 500 keV features from 1E1740.7 [9] and the Briggs source [10], which are interpreted as pair annihilation signatures [11]. The reality of such features is controversial since they have not been confirmed by other experiments. There may also be alternative interpretations of the narrow feature even if it is real (e.g. nuclear emission from Li, [12]). In this review we will tentatively treat the superhard state as real but of very low duty cycle. Unlike Seyfert galaxies, most GBHs exhibit weak or no Fe-K fluorescence lines [13,14], and the evidence for a Compton reflection component is weaker.

III CONCEPTUAL PICTURE

Fig.4 illustrates the different phases of a GBH accretion disk in the plane of the local electron temperature T_e and vertical column density Σ. Depending on the accretion rate $\dot{M}$, viscosity α, and radius r, a ring in the disk can be anywhere in this phase plane and emit the corresponding local spectrum. For GBHs the optically thick disk blackbody peaks at keVs. The unsaturated Compton regime corresponds to cooling by copious soft photon sources and the Comptonized bremsstrahlung regime corresponds to quenching of the soft photon source. Above a few hundred keV temperature pair production and annihilation dominate cooling and the output is a combination of Comptonized bremsstrahlung and pair annihilation, leading to the appearance of the characteristic "MeV bump". A 511 keV feature will appear if enough pairs escape

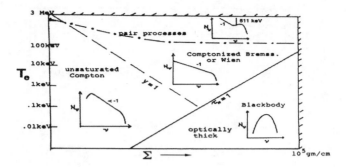

FIGURE 4. Local phase diagram of a black hole accretion disk showing the different regimes and respective output spectra. y is the Komponeet Compton parameter and τ_* is the effective absorption depth.

from the disk and annihilate in the circumstellar or interstellar medium after they cool.

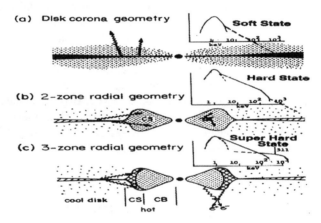

FIGURE 5. Artist conception of the 3 different disk configurations that may be responsible for the 3 spectral states of GBHs. Dotted region corresponds to a very hot or nonthermal tenuous corona.

When we combine the above local outputs with the global disk geometry we end up with the following conceptual picture (Fig.5) for the origin of the different spectral states discussed in Sec.2. In the soft power-law state accompanied by the soft x-ray high state, the SS disk penetrates down to the last stable orbit ($6GM/c^2$ for a nonrotating hole) sandwiched by a corona, which is likely nonthermal or at least ultrahot ($T_e > MeV$) and Thomson thin ($\tau_T < 1$). The soft power law is likely produced by the Comptonization of blackbody disk photons by the corona. In the hard state the SS disk and its corona retreat from the last stable orbit. The innermost disk is then replaced by a hot optically thin torus cooling via unsaturated Comptonization of soft photons, both external and internally generated synchrotron photons. But if the soft photon source is ever quenched, due to a combination of lowered magnetic field and shielding of the external blackbody soft photons, the innermost disk may heat up to relativistic T_e with copious pair production, leading to the superhard state. In the next section we will discuss the consequences of such superheating and possible origins of the relativistic outflows of the microquasar GBHs.

IV PAIRS AND RELATIVISTIC OUTFLOWS

The OSSE narrow 511 keV line survey has recently uncovered a secondary hot spot 7° above the Galactic center [15,16] . While more observations will be needed to clarify the true nature of this diffuse source, it is tantalizing that the centroid of this hot spot coincides with the foreground XRN GROJ1719-24. This raises the question of whether this or other GBHs may be significant pair factories. Independent of the reality of the superhard state of GBHs discussed in Sec.3, radio observers have now discovered 4 microquasars among GBHs, with at least two of them superluminal sources [17-20]. Jet-like features may also be present around other GBHs (e.g. GX339-4). Since it is much easier to accelerate pairs than protons to relativistics velocities it is likely that these relativistic outflows are related to copious pair production episodes. Hence we need to explore the connection between the accretion disk and pair production events.

Nonthermal electrodynamic processes caused by the differential rotation of the disk or spin of the hole can in principle generate large macroscopic electrostatic potentials, leading to particle accelerations and pair production. However, at this point it is unclear what would trigger such events. The overall efficiency of nonthermal processes is also highly uncertain. On the other hand, the physics of thermal accretion disk is better understood. Liang and Li [21,22] proposed that quenching of the soft photon source in the innermost disk would naturally superheat the inner disk to relativistic temperatures and emission of copious gamma-rays. If the gamma-ray compactness ($l = L_\gamma/R/3.7.10^{28}\text{erg.s}^{-1}.\text{cm}^{-1}$) is $\ll 100$, most of the gamma-rays would escape, leading to the observed superhard state of Cyg X-1 [7]. On the other hand if the compactness is > 100 almost all of the gamma-rays will be absorbed via gamma-gamma and gamma-x collisions into pairs. Provided the pairs are driven out by radiation pressure or poynting flux they may escape before reannihilating, leading to a pair outflow. Li and Liang [22] found that to achieve a compactness > 100, a minimum requirement is that the accretion rate > 0.2 Eddington accretion rate. Such a minimun luminosity is also needed to accelerate the pairs to the observed bulk Lorentz factor of $\Gamma = 2.5$ for

GROJ1655 and GRS1915 via radiation pressure. These results are illustrated in Fig.6 and Fig.7. Future coordinated BATSE RXTE and radio observations should help to confirm or disprove such a scenario.

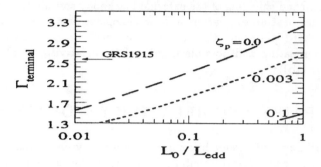

FIGURE 6. Terminal bulk Lorentz factors of the ejecta as a function of luminosity/Eddington luminosity for different proton loadings (from Liang and Li [21]).

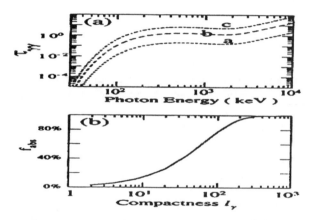

FIGURE 7. (a) γ-γ pair production optical depth as a function of photon energy. Curves a, b, c are for compactness $l=12, 100, 392$ respectively; (b) The fraction of gamma-rays absorbed into pairs as a function of l. Note that it exceeds 80% for $l > 100$ (from Li and Liang [22]).

V ACCRETION DISK STRUCTURE

The structure of the early accretion disks models around GBHs [2] is based on the assumption of small radial drift and purely local radiative cooling. However these assumptions break down (a) close to the horizon, (b) at high accretion rates and (c) when the disk is very hot and optically thin. When radial drift and advection of energy terms are included the disk exhibits new solution branches [27,23]. At fixed radius r, the local disk solution in the $\dot{M}$ - Σ plane bifurcates into two separate branches (Fig.8).

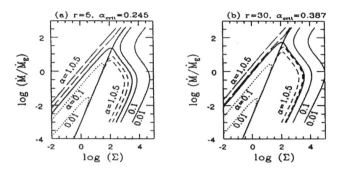

FIGURE 8. Changing topology of the local disk solutions in the $\dot{M}$ - Σ plane as α varies, for two sample disk radii (details see Chen et al [24]).

For α near unity, there is the hot optically thin advection dominated accretion flow (ADAF) [23,24] separated from the inverted-U branch discovered by Liang and Wandel [25] which connects the optically thin SLE solution [26] to the optically thick SS solution [2]. At low α however, the ADAF solution "reconnects" with the inverted-U branch to form the optically thin inverted-V branch and the S-shaped optically thick slim-disk solution [27]. In this case the optically thin ADAF solution exists only below a critical $\dot{M}_{crit}$. Conversely, for a given $\dot{M}$ there is a maximum radius r_{tr} beyond which there is no ADAF solution. For application to the different spectral states the key challenge is how to connect such local solutions into a self-consistent global disk model relevant to the picture of Fig.5. In the next two sections we review some of the recent developments.

VI ADVECTION-DOMINATED ACCRETION FLOWS

Two kinds of advection-dominated accretion flow (ADAF) are known. One solution branch corresponds to high values of $\dot{M}$ where radiation is trapped by the high optical depth of the accreting gas and is advected into the central star [28,29,27]. The other solution branch is present at low values of $\dot{M}$ [23,30–32,24]. See Narayan 1997 [33] for a review. The accreting gas is optically thin and becomes a two-temperature plasma [26,34], with the ions being much hotter than the electrons. Because the gas radiates inefficiently, most of the viscous energy is retained in the gas as thermal energy and advected into the central star. Low-$\dot{M}$ ADAFs are stable [35] and have been used to model a variety of black hole X-ray binaries and active galactic nuclei. Some recent results are discussed here.

The most successful applications have been to black holes in the quiescent state whose luminosities are several orders of magnitude below the Eddington limit. The basic paradigm, developed by Narayan, McClintock & Yi [36] and Narayan, Barret & McClintock [37], is shown in the bottom panel of Fig.9 (taken from Esin, McClintock & Narayan 1997 [38]). According to this model, the accretion flow consists of two zones: an ADAF extending from the black hole horizon to a transition radius r_{tr}, and a thin accretion disk extending from r_{tr} to the outer radius r_{out} of the accretion flow. In the case of Roche-lobe overflow systems like the soft X-ray transients, $r_{tr} \sim 10^4$ (in Schwarzschild units), whereas in wind-fed systems like Cyg X-1, r_{tr} may be as small as $\sim 10^2$ or even smaller [38].

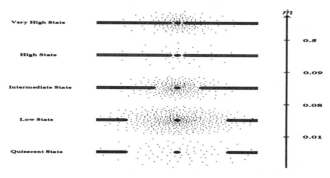

FIGURE 9. Configuration of the accretion flow in different spectral states, shown schematically as a function of the mass accretion rate $\dot{m}$ [38]. The ADAF is indicated by dots and the thin disk by horizontal bars. The proposal for the very high state is speculative.

The switch from the outer thin disk to the inner ADAF is presumed to occur through evaporation of material from the surface of the disk into a corona. The mechanism by which this happens is not fully understood; it could be the result of electron conduction from the corona [39] or turbulent energy transfer from the inner regions of the ADAF [40]. Apart from this uncertainty, however, the ADAF models are well-developed and have achieved self-consistency in the flow dynamics, thermal balance (separately for ions and electrons in the two-temperature ADAF), and radiative transfer [37,38,42]). The optically thin emission from the ADAF consists of several components: synchrotron radiation, bremsstrahlung, and Comptonization of soft synchrotron photons and radiation from the outer disk.

The models assume equipartition between magnetic pressure and gas pressure in the ADAF (i.e. $\beta \equiv P_{gas}/P_{total} = 0.5$) and use a viscosity parameter

$\alpha_{ADAF} \sim 0.25 - 0.3$. These parameters are not usually adjusted. If the other system parameters such as the black hole mass, the inclination and the distance are known, the model essentially has only one fitting parameter, viz. the mass accretion rate $\dot{m}$ (in Eddington units); a relatively unimportant second parameter is the transition radius.

The ADAF model has successfully explained the spectra of three black hole SXTs for which optical and X-ray data are available in quiescence: A0620–00 [36,37], V404 Cyg [37], and GRO J1655–40 [43]. The observations of these quiescent SXTs cannot easily be reconciled with a pure thin accretion disk model, but fit naturally into the ADAF paradigm. According to the model, the accretion flow is radiatively very inefficient; most of the viscous energy is advected into the black hole and only a small fraction ($< 0.1\%$) of the rest mass energy is radiated. This feature allows interesting tests for the presence of event horizons in accreting black holes [36,37,44]). Apart from explaining quiescent spectra, the ADAF model explains the long recurrence times between outbursts in SXTs [45], which is difficult to understand with a pure thin disk model, and also reproduces the remarkable 6 hour delay observed between the optical and X-ray rise in a recent outburst of GRO J1655–40 [46,43].

The ADAF model has also had success with some quiescent galactic nuclei: Sgr A* at the center of our Galaxy [47,48], NGC 4258 [49], M87 [50]. Fabian & Rees [51] and Mahadevan [52] show that most giant elliptical galaxies may harbor large black holes in their nuclei accreting quietly via ADAFs, while Yi [53] argues that ADAFs may be relevant for understanding quasar evolution at low luminosities.

Of interest to the Compton Observatory and this conference are more luminous systems. Black hole X-ray binaries when not in quiescence display a number of distinct spectral states: the famous low state and high state, plus the so-called very high state and intermediate state [54–57,1]. The soft X-ray transient, Nova Muscae, is a particularly good example of a source that has displayed all the spectral states. Fig.10 shows the 2–12 keV and 20–100 keV lightcurves of Nova Muscae during its 1991 outburst (taken from [58]. Following the peak of the outburst, Nova Muscae went through the following stages in sequence: very high state from day 7 to day 60 (a relatively soft spectrum with significant flux in a hard tail), high state from day 60 to day 130 (an ultrasoft spectrum with a very weak hard tail), intermediate state from day 130 until about day 200 (dramatic switch from a soft to a hard spectrum), and low state from day 200 until the end of the observations on day 238 (a very hard spectrum with no evidence of a soft component). Clearly, this is a sequence of decreasing mass accretion rate.

The ADAF model provides a fairly robust and consistent explanation for all the spectral states except the very high state. Fig.10 shows the results of detailed modeling of Nova Muscae by Esin et al [38]. The theoretical lightcurves (thick and thin lines; see the original paper for details) agree well with the

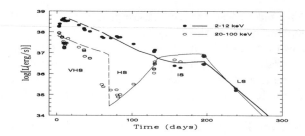

FIGURE 10. Lightcurves of Nova Muscae 1991. Filled and open circles are data from Ebisawa (1994) [58] corresponding to the 2–12 keV and 20–100 keV bands. The heavy and thin lines are model predictions [38] based on the ADAF paradigm shown in Fig.9. The symbols VHS, HS, IS, LS refer respectively to the very high state, high state, intermediate state and low state.

observations. The basic paradigm on which these calculations are based is due to Narayan [36] and is indicated schematically in Fig.9. We discuss the various states in sequence below.

Let us begin with the ADAF model of the quiescent state and increase $\dot{m}$ by stages. (We define $\dot{m} = \dot{M}/\dot{M}_{\rm Edd}$, where $\dot{M}_{\rm Edd} = L_{\rm Edd}/0.1c^2 = 1.39 \times 10^{18}(M/M_\odot)$ g s^{-1}.) For a range of $\dot{m}$ up to a critical value $\dot{m}_{\rm crit} \sim 0.08$, the geometry of the flow remains the same as in the quiescent state. (The value of $\dot{m}_{\rm crit}$ depends on α; the value given here is for $\alpha = 0.25$, cf. Esin et al [38].) Since the radiative efficiency of the ADAF increases rapidly with increasing $\dot{m}$ [31] the luminosity becomes significantly higher than in the quiescent state. In fact, by the time $\dot{m} \sim \dot{m}_{\rm crit}$, the radiative efficiency of the ADAF is almost as high as that of a thin disk. Strictly, ADAFs with $\dot{m} \sim \dot{m}_{\rm crit}$ are not advection-dominated since they radiate as much energy as they advect. Nevertheless, these flows belong to the same solution branch that at lower values of $\dot{m}$ is truly advection-dominated. Therefore, by continuity, it is appropriate to refer to them as ADAFs [38].

Detailed calculations show that for $\dot{m} \lesssim \dot{m}_{\rm crit}$ most of the cooling in the ADAF occurs via Comptonization of synchrotron radiation. The calculated spectra are very hard, with photon indices in the range 1.4−1.9 and a cut off at ~ 100 keV. It is natural to identify these solutions with the low state (Fig.9). The calculated fluxes and spectral indices fit observations of Nova Muscae in the low state (days 200 to 238 in Fig.10) quite well. Fig.11 shows another example, GRO J0422+32, where data obtained in the low state are compared with the spectrum predicted by the ADAF model. For simplicity, we show by the solid line the spectrum of Nova Muscae on day 200 [38] normalized for GRO J0422+32. The dashed line is another low state model for a slightly

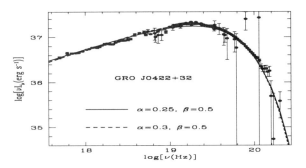

FIGURE 11. The symbols and error bars correspond to spectral data on GRO J0422+32 taken with OSSE, TTM and HEXE (courtesy Eric Grove). The source was in the low state during the observations. The solid line is the predicted spectrum of the ADAF model for $\alpha = 0.25$ [38]. The dashed line corresponds to $\alpha = 0.3$.

different value of α. The ADAF model reproduces the main features in the spectrum, notably the spectral index and the high energy cutoff. The model also predicts that in the low state the electron temperature of the ADAF increases with decreasing luminosity. This correlation has been observed in GRO J0422+32 [59].

When $\dot{m}$ exceeds $\dot{m}_{\text{crit}}$, the density of the ADAF becomes large and the gas begins to cool efficiently. An ADAF solution is then no longer possible. Over a narrow range of $\dot{m}$ between ~ 0.08 and ~ 0.09 the transition radius moves in and the ADAF zone is squeezed down to a progressively smaller size. This stage can be identified with the intermediate state (Fig.9). As r_{tr} decreases, soft radiation from the outer disk penetrates deeper into the ADAF and cools it more efficiently. The effect is particularly strong once r_{tr} falls below about 30, and the spectrum switches quite rapidly from hard to soft. In the case of Nova Muscae, the transition occurred between days 130 and 200, the critical period being around day 130 when the transition radius changed from 3 (the marginally stable orbit) to about 30 Schwarzschild radii.

Fig.12 shows another example, Cyg X-1, which made a low-to-high and a reverse high-to-low transition during 1996. Both transitions were well-observed with BATSE and RXTE (the upper and lower panels in Fig.12). The middle panel shows spectra predicted by the ADAF model. Once again, for simplicity, model spectra corresponding to Nova Muscae (around day 130) are shown. The qualitative agreement with the observations is quite remarkable.

According to the ADAF model, the low-to-high transition corresponds to a nearly constant $\dot{m}$ (between 0.08 and 0.09 for the assumed value of α) and it is the change in r_{tr} that causes the change in the spectral shape. Furthermore, the radiative efficiencies in the hard and soft states are not very different. Therefore, the bolometric luminosity changes by no more than about 50%

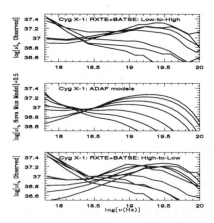

FIGURE 12. The upper and lower panels show spectra of Cyg X-1 observed during a low-to-high and high-to-low transition respectively (courtesy Nan Zhang and Wei Cui). The data were collected with BATSE and RXTE. The middle panel shows predicted spectra of the ADAF model. Model spectra of Nova Muscae are shown (taken from Esin et al [38]) corresponding to the period around day 130 (see Fig.10) when the source went through a high-to-low transition.

during the transition. This has been confirmed in Cyg X-1 [68]. Note also the characteristic pivoting of the model spectra around 10 keV. This too is seen in the data.

Once $\dot{m}$ rises above about 0.09, the transition radius comes down to 3 Schwarzschild radii and the thin disk extends down to the marginally stable orbit (Fig.9). The spectrum now is dominated by a multicolor blackbody from the thin disk. However, there is still a residual ADAF in the form of a corona above the disk which produces a weak hard tail in the spectrum. This is the high state. The cooling of the corona increases with increasing $\dot{m}$; this explains why in Nova Muscae the hard tail is weaker at higher luminosities (days 80 to 130). According to the ADAF model, this should be a generic feature of systems in the high state.

The ADAF model does not have a viable explanation for the very high state.

VII TRANSONIC SOLUTIONS

Motivated by the ADAF results, Luo and Liang [60] recently investigated the global structure of disks including the transonic inner boundary condition near the horizon [61-63]. Instead of using the self-similar approach of the ADAF solutions, they solve the full radial Euler equation with the transonic boundary condition as in the Parker solar wind and Bondi accretion solutions. Instead of specifying detailed cooling laws for the disk, they specify a power law form for the (subKeplerian) angular momentum which allows the disk structure equations to be solved semi-analytically. The results are rather interesting. In general they find an optically thick outer disk gradually transitioning first into an optically thin cooling-dominated advective flow (CDAF) before finally becoming an ADAF very close to the horizon, usually near or inside the sonic radius (Fig.13).

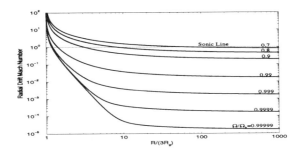

FIGURE 13. Radial drift Mach number profiles of the transonic solution showing that the sonic point moves farther out as the angular momentum is lowered (details see Luo and Liang [60]).

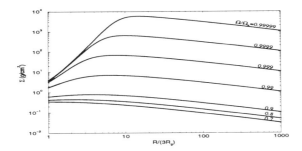

FIGURE 14. Disk vertical column density profiles of the transonic solution showing that the disk is optically thin everywhere at low angular momentum, but as the angular momentum approaches the Keplerian value the outer disk becomes optically thick while the very inner disk stays optically thin (from Luo and Liang [60]).

The global structure of such transonic solutions are very sensitive, not just to $\dot{M}$ and α, but also the angular momentum J at the outer boundary (Fig.14). For J very close to Keplerian, the disk is cool optically thick and physically thin everywhere except inside the sonic radius. But for low J the disk is hot optically thin out to large radii, similar to the ADAF solutions (Fig.14).

Hence such transonic solutions, though somewhat specialized in their angular momentum profiles and cooling laws, illustrate how the optically thick outer disk transitions into an optically thin inner disk gradually due to increasing radial drift. However, despite the increasing drift the solution usually does not become advection-dominated until close to the horizon. Most of the hot optically thin regin is a CDAF, and that is where most of the hard state luminosity originates. The biggest difference between the ADAF scenario of the last section, and the CDAF scenario here, is that the ADAF transition radius r_{tr} between the optically thick and thin zones is insensitive to the outer disk angular momentum, whereas the CDAF transition radius is very sensitive to the outer disk angular momentum J, which implies that the multi-state behavior of GBHs may be triggered by changes in the overall angular momentum of the accretion stream. This is especially attractive for Cyg X-1 which is known to accrete from both the companion wind (with low J) and through the inner Lagrange point (with high J).

Another interesting result of the transonic solution is that the global disk structure is also very sensitive to the viscosity α. As α decreases r_{tr} increases, and the inner disk gets optically thinner while the outer disk gets optically thicker.

VIII NONTHERMAL PROCESSES

Most GBHs show some low level core radio emission which is likely of nonthermal origin. In addition during radio outburst and relativistic ejections the emission has nonthermal spectrum. This coupled with the MeV power-law tail of Cyg X-1 motivates us to serious consider nonthermal processes around GBHs. This is a relatively underexplored field, compared with thermal disk models. Here we briefly list some of the potential sites and origins of nonthermal emissions from GBHs.

1. A nonthermal extended corona around the SS disk, energized by reconnecting magnetic loops anchored in and twisted by the disk, and wave turbulences propagating from the disk [64]. 2. Macroscopic electric fields generated by the disk rotation or the spin of the hole [65]. 3. Shocks and boundary layers in the winds and outflows from the disk and the companion star. 4. Comptonization in a collisionless radial converging flow [66]. 5. A collisonless shock in a radial accretion flow [67].

There have been evidences that the low level core radio emission is often correlated with the hard x-ray flux [17]. A possible picture for such correlated emission is the following. Relativistic nonthermal particles are usually accelerated mainly along magnetic field lines, and so emit little synchrotron radiation. However, whenever they scatter against photons the recoil gives them a small transverse momentum, leading to the emission of synchrotron photons together with the upscattered hard x-ray and gamma-ray. Hence the synchrotron radio flux would be correlated with the hard-x-gamma ray flux. Future coordinated observations in radio and gamma-rays, especially polarization studies, would be needed to confirm or disprove this picture.

IX ACKNOWLEDGEMENTS

EL is supported in part by NASA grant NAG5-3824. RN is supported in part by NASA grant NAG5-2837.

REFERENCES

1. Liang, E., *Physics Report* in press (1997).
2. Shakura, N.I. and Sunyaev, R.A., *Ast. Ap.* **24**, 337 (1973).
3. McConnell, M. et al., *Ap. J.* **424**, 933 (1994).
4. Van Dijk, R. et al., *Ast. Ap.* **296**, L33 (1995).
5. Moss, M., Ph.D. Thesis, Rice University (1997).
6. Phlips, B. et al., *Ap. J.* **465**, 907 (1996).
7. Ling, J. et al., *Ap. J. Lett.* **321**, L117 (1987).
8. Goldwurm, A. et al., *Ap. J.* **389**, L79 (1992).
9. Paul, J. et al., *AIP Conf. Proc.* **232**, p.17, ed. Durouchoux, P. and Pranzos, N. (AIP, NY 1991).
10. Briggs, M. et al., *Ap. J.* **442**, 638 (1995).
11. Chen, W. et al., *Ap. J.* **426**, 586 (1993).
12. Charles, P.A. et al., *Mon. Not. Roy. Ast. Soc.* **245**, 567 (1996).
13. Sheth, S. et al., *Ap. J.* **468**, 755 (1996).
14. Sheth, S. et al., *4th Comp. Symp. Proc.* to appear (1997).
15. Cheng, L. et al., *Ap. J.* **481**, L43 (1997).
16. Purcell, W. et al., *4th Comp. Symp. Proc.* to appear (1997).
17. Mirabel, F. et al., *Nature* **358**, 215 (1992).
18. Mirabel, F. and Rodriquez, L.F. *Nature* **371**, 46 (1997).
19. Rodriguez, L. et al., *Ap. J. Lett* **401**, L15 (1992).
20. Hjellming, R. and Rupen, M., *Nature* **375**, 464 (1995).
21. Liang, E. and Li, H., *Ast. Ap.* **298**, L45 (1995).
22. Li, H. and Liang, E., *Ap. J.* **458**, 514 (1996).
23. Narayan, R. and Yi, I., *Ap. J.* **428**, L13 (1994).
24. Chen, X. et al., *Ap. J.* **443**, L61 (1995).
25. Liang, E. and Wandel, A., *Ap. J.* **376**, 746 (1991).
26. Shapiro, S. et al., *Ap. J.* **204**, 187 (1976).
27. Abramowicz, M.A. et al., *Ap. J.* **332**, 646 (1988).
28. Katz, J., *Ap. J.* **215**, 265 (1977).
29. Begelman, M.C., *Mon. Not. Roy. Ast. Soc.* **184**, 53 (1978).
30. Narayan, R. and Yi, I., *Ap. J.* **444**, 231 (1995).
31. Narayan, R. and Yi, I., *Ap. J.* **452**, 710 (1995).
32. Abramowicz, A. et al., *Ap. J.* **438**, L37 (1995).

33. Narayan, R., *Accretion Phenomena and Related Outflows* eds. Wickramasinghe, D.T. et al. in press (1997).
34. Rees, M.J. et al., *Nature* **295**, 17 (1982).
35. Kato, S. et al., *PASJ* **48**, 67 (1996).
36. Narayan, R. et al., *Ap. J.* **457**, 821 (1996).
37. Narayan, R. et al., *Ap. J.* **482**, 448 (1997).
38. Esin, A.A. et al., *Ap. J.* submitted (1997).
39. Meyer, F. & Meyer-Hofmeister, E., *Ast. Ap.* **288**, 175 (1994).
40. Honma, F., *PASJ* **48**, 77 (1996).
41. Crider, A. et al., contribution in this volume (1997).
42. Nakamura, K.E. et al., *PASJ* in press (1997).
43. Hameury, J.-M. et al., *Ap. J.* in press (1997).
44. Narayan, R. et al., *Ap. J.* **478**, L79 (1997).
45. Lasota, J.P. et al., *Ast. Ap.* **314**, 813 (1996).
46. Orosz, J.A. et al., *Ap. J.* **478**, L83 (1997).
47. Narayan, R., Yi, I. & Mahadevan, R. *Nature* **374**, 623 (1995).
48. Narayan, R. et al., *Ap. J.* submitted (1997).
49. Lasota, J.P. et al., *Ap. J.* **462**, 142 (1996).
50. Reynolds, C.S. et al., *MNRAS* **283**, L111 (1997).
51. Fabian, A. & Rees, M.J., *MNRAS* **277**, L5 (1995).
52. Mahadevan, R., *Ap. J.* **477**, 585 (1997).
53. Yi, I. *Ap. J.* **473**, 645 (1996).
54. van der Klis, M., *Ap. J. Supp.* **92**, 511 (1994).
55. Nowak, M.A., *PASP* **107**, 1207 (1995).
56. Tanaka, Y. & Lewin, W., *X-Ray Binaries* eds. Lewin, W. et al. 126 (Cambridge, UK, 1995).
57. Tanaka, Y. & Shibazaki, N., *ARAA* **34**, 607 (1996).
58. Ebisawa, K. et al., *PASJ* **46**, 375 (1994).
59. Kurfess, J.D., *Ap. J. Supp.* **120**, 5 (1996).
60. Luo, C. & Liang, E., *Ap. J.* submitted (1997).
61. Liang, E. & Thompson, K.A., *Ap. J.* **240**, 271 (1980).
62. Abramowicz, M. et al., *Ap. J.* **242**, 772 (1980).
63. Chakrabarti, S.K., *Theory of Transonic Astrophysics Flows* (World Sci., Singapore, 1990).
64. Li, H. et al., *Ap. J.* **460**, L29 (1996).
65. Blandford, R.D. & Znajck, R.L., *MNRAS* **179**, 433 (1977).
66. Chakrabarti, S.K. & Titarchuk, L.G., *Ap. J.* **455**, 623 (1995).
67. Meszaros, P. & Ostriker, J.P., *Ap. J.* **273**, L59 (1983).
68. Zhang, S.N. et al., *Ap. J.* in press (1997).

Comptonization Processes in Galactic and Extragalactic High Energy Sources

Lev Titarchuk*

*NASA/Goddard Space Flight Center and George Mason University

Abstract. We review the principal radiation mechanisms which can be responsible for the production of high energy emission of Galactic and extragalactic sources. The spectral properties of Comptonization processes are studied. We present a rigorous treatment of the problem in terms of the Boltzmann kinetic equation formalism. We also overview the different exact and approximate analytical and numerical techniques which decouple the photon transport in six dimensions to that in the configuration space and in energy space separately. *CGRO*, *RXTE*, and *SAX* observations of Galactic black hole candidates, neutron star systems, and Seyfert nuclei show that their spectra can be fitted by thermal Comptonization models in the hard state, and two components in the soft state. In the latter case, these components show a blackbody like spectrum with the color temperature which is a fraction of keV and tens of eV for Galactic and extragalactic sources, respectively, and an extended power-law with spectral index between 1 and 2 (in the neutron star systems in their soft state this power-law component is not detected). We give arguments for an explanation of these two types of spectral states in the framework of the thermal and bulk motion Comptonization processes.

1.INTRODUCTION

1.1 General Background

The discovery of high energy (*i.e.* X– and γ-ray) emitting sources in the early 70's, established the concept of accretion disk as the engine of thermalization of the accretion kinetic energy and the site of production of the associated high energy emission. The presence of an accretion disk is thought inevitable since the angular momentum of matter in orbit near a compact (i.e. of a few Scwarzschild radii in size) object is much smaller than that of matter at much large distances where the matter originates. It has been thought that the dynamics of the matter which powers the emission from this class of sources would be dominated by the kinetic energy in the azimuthal direction, having radiated away the energy in the other degrees of freedom, leading to a formation of a thin disk, dissipating its kinetic energy through the action of a viscous agent into the observed radiation. Since the seminal work of Shakura & Sunyaev (1973) the notion of the viscous, thin accretion disk has served as the foundation of any models for accretion powered high energy sources whether galactic or extragalactic. The inability of current (and most likely any instrument in the foreseeable future) to resolve the radiation emitting region, has forced upon us a situation akin to that encountered in the study of spectroscopic binaries, in which the kinematics and dynamics of the system must be inferred (rather than observed) from the resulting high energy spectra associated with this class of objects and the associated variability.

In fact, the discovery of milisecond variability detected by the Rossi X-ray Timing Explorer (RXTE) in the X-ray emission from the low-mass X-binaries (e.g. van der Klis et al. 1997) and the explanation of these results in terms of the rotational spitting of an intrinsic frequency (which is order of the local Keplerian frequency) caused by an accretion disk (Titarchuk & Muslimov 1997) can be considered as the observational evidence of the presence of the disk in these systems.

However, the case with accretion powered objects is considerably more complicated, involving the largely unknown dynamics of accreting matter and the feedback induced by its interaction with the emitted radiation. Within this framework, the detailed study of the spectra (both in steady state and in time varying conditions) resulting under specific dynamical situations takes on special importance, since it can serve as the instrument for uncovering the underlying dynamics and the spatial distribution of the accreting matter. *The main thrust of this review is hence related to modeling the spectra accreting sources in a number of diverse situations and their use in the interpretation of the observations.*

While the qualitative features of the emission (i.e. the frequency at which most of the radiated luminosity is emitted) are consistent with those predicted by the viscous accretion disk models, the observations present us invariably with features which require the presence of additional components in addition to the disk alone, such as hot coronae, additional plasma clouds, warm absorbers, relativistic jets and non-thermal particle populations.

Accretion powered sources cover a very large range in luminosity, from $\sim 10^{37}$ ergs s^{-1} associated with emission from galactic X-ray binary sources to $\sim 10^{48}$ erg s^{-1} appropriate for some of the most luminous quasars. Despite this disparity in scales their high energy spectra are not very different in form. This rather surprising at first sight feature has an explanation in a simple scaling associated with the radiative transfer of sources powered by accretion: If the luminosity of the source is expressed in units of the Eddington luminosity (the natural luminosity for accretion onto a gravitating object) and the length scale in units of the Scwarzschild radius, the Thomson depth τ_T of a flow close to free–fall is of order 1, in the vicinity of the black hole (where the energy release takes place) for objects accreting close to their Eddington luminosity. This value of τ_T is the one demanded by the high energy observations, indicating that the similarity in the spectra of these sources is related to the fact that they are powered by accretion.

1.2 Phenomenology

Observationally, the high energy spectra of accretion powered sources exhibit two distinct types: The first (say type I) is characterized by a hard power-law (energy index $\sim 0.5 - 0.8$) with a high energy cutoff at about 50-200 keV. The second (type II) consists of a steeper power-law (energy index 1.0 - 2.0) without a detected break up to very high (~ 1 MeV) energies, as determined by the OSSE instrument aboard the CGRO (Grove et al. 1997). It is interesting that both types of high energy spectra have been observed from both galactic or extragalactic high energy sources. This is in confirmation of the arguments given above concerning the similarity of such diverse kind of sources vis–á–vis the physics of radiative transfer and the formation of high energy spectra.

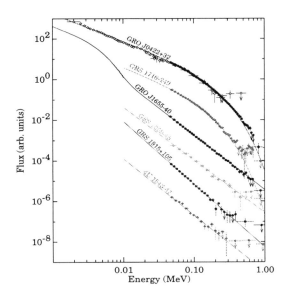

FIGURE 1. Broadband X-ray photon spectra of BHC sources, GRO J0422+32 (Sunyaev et al. 1994) observed on Mir-Kvant module (TTM+HEXE), GRO J1655-40 observed by ASCA and OSSE and four other observed by OSSE (see details in Grove et al. 1997). Spectra have been shifted along the vertical axis for convenience.

In addition to the power-law high energy spectra described above, both galactic and extragalactic sources exhibit thermal–like components believed to be due to thermal emission from the viscous quasi–thermal accretion disk. This component is referred to as the "Big Blue Bump" in the AGN context and in the galactic X–ray binary context as the "Multicolor Black Body Disk". The spectral fits of these quasi–thermal components with spectra appropriate for "standard" (optically thick, geometrically thin) disks are quite good and have established the belief that this is indeed their correct interpretation.

In particular, it has been noticed that, generally, the spectra of galactic sources tend to be of type I when their overall luminosity is low compared to that of Eddington. Interestingly, this appears to be the case independent of the nature of the compact source, i.e. whether it is a neutron star or a black hole. This is a rather surprising result considering the drastically different boundary conditions (both from the dynamic and the radiative transfer point of view) associated with these two types of objects.

Furthermore, sources for which the compact object is thought to be a black hole, their spectra turn into type II when their luminosity increases. This specific correlation has been noticed first in the black hole candidate source Cyg X–1, and more recently in a number of X–ray novae. It is quite remarkable that type I spectra associated with much large emission-region scales than that seen in the high-soft state (type II spectra). The effect reported by Cue et al. (1997) is detected in the process of RXTE observations of Cygnus X-1 of the transition from the hard state to the soft state. This change of the size of the emission region was predicted by Chakrabarti & Titarchuk (1995, hereafter CT95) in terms of Titarchuk, Mastichiadis & Kylafis (1996, hereafter TMK96) model.

The X–ray nova sources are particularly suited for studies of this kind because they cover within the duration of an outburst (~ 100 days) a wide range in luminosity allowing the study of such spectrum - flux correlations in a well defined way. Interestingly, most of these X–ray nova sources have been found to harbor black holes rather than neutron stars. This correlation has been well documented for accreting black hole sources, so that it is considered to be a defining characteristic of black hole candidate sources. In contrast, type II spectra have not yet been observed in sources which are known to harbor neutron stars. It should be added here that these type II spectra are also accompanied at low energies by the soft quasithermal component attributed to emission from a "standard", thin, viscous disk. Finally in the case of neutron stars, when the luminosity approaches that of Eddington, their spectra exhibit only this quasithermal component without any indication for the presence of high energy power law tail.

These systematics are shown clearly in Figure 1 where the spectra of a number of black hole candidate sources (GRS 1915-105, GRO J1655-40, GRO J0422+32 *etc*) are presented.

The hard power laws are apparent in the cases of GRO J0422+32, GRS

1716-249, as well as the steep power law and the multicolor black body disk spectrum in the case of GRO J1655-40 (see also Figure 4).

The simultaneous presence of both a quasithermal *and* a hard X-ray component in AGN and galactic binary spectra, suggest that the standard accretion disk cannot alone account for the observed phenomenology. An additional hot electron (several tens of keV) component is also necessary in the vicinity of the viscous disk. Such components have been postulated in the past and have been assumed to be present in an ad hoc way, usually referred to as coronae. These components, as indicated by the observations are the source of the "hot" radiation observed by inverse comptonizing the "soft" photons of the quasithermal component. They have been generally assumed to lie over these thin, quasithermal, viscous disks. However, the more likely geometry should be slightly different, since the observed spectra indicate that they should intercept only a fraction ($\lesssim 50\%$) of the available viscous disk photons (if this is not the case the "hard" coronal luminosity should be much greater than that of the soft component a fact not generally observed when both components are present). Furthermore, Kazanas, Hua & Titarchuk (1997) give strong arguments for the presence of hot quasi-spherical extended atmosphere surrounding the viscous disk in the hard state (type I spectra) of the galactic sources.

Thus the origin, structure and hydrodynamics of these coronae have not been adequately accounted for.

These facts have recently motivated to study a hybrid dynamic model which can reproduce the requisite hot and cool electron components. Many current models of high-energy radiation from BHC (for example CT95; Narayan 1996; Skibo & Dermer 1995, Cherny, Witt & Zycki 1996) associate the transition from the hard to the soft state with the change of the mass accretion rate $\dot{M}$. But in CT95 it was demonstrated that the transition is regulated by redistribution of mass accretion rates between sub-Keplerian and the Keplerian disk components. In principle, for the same $\dot{M}$ (or the bolometric luminosity) the system can be in soft state or the hard state (compare with observations, Zhang et al. 1997) depending on the disk $\dot{M}_d$ (the soft photon supply).

Interestingly, the dynamics of the bulk motion model (TMK96, see also CT95) can provide a natural account of the steep power law component observed in the type II spectra. Within this model, this component is the result of Comptonization of the Keplerian disk's photons by the *bulk* motion of the electrons which are free-falling onto the black hole in the centrifugal barrier (see section 2). Since this free-falling component is not present for accretion onto neutron stars, one naturally expects the absence of this extended power law component when the accretion takes place on a neutron star. The different spectral behavior between neutron stars and black holes raises the interesting question of whether the nature of the underlying compact object can be determined only by spectral observations in high energy X-ray regime. We believe that this question can in fact, be answered with a combination of theoretical developments and observations.

2. RADIATIVE TRANSFER THEORY

We first summarize the works done so far and show that these works are only partially adequate to explain many, if not most, of the observational results from compact sources. We divide this Section into two sub-Sections: the first one deals with the overview of the main physical processes and the second one deal with the recent developments in the study of the spectral models with the radiative transfer formalism and with the various numerical and analytical results of the radiative transfer theory. Since the spectrum of the hard radiation is going to be presented in terms of upscattering of the soft photons off (centrifugal barrier region or corona) hot electrons the approach to solve our radiative problem has to basically follow the footsteps of pioneering works in the Comptonization problem and in the spherical accretion paradigm. Thus, in this Section, we shall review the solutions of the hard radiation spectral formation.

2.1 The Physical Processes

Three types of radiation processes have been found to be of prime importance for the spectral formation in hot astrophysical plasmas: bremsstrahlung, synchrotron (cyclotron) radiation, and Compton scattering. In sufficiently compact and hot plasmas, these processes are supplemented by pair production and annihilation. It has become clear that a proper interpretation of the transient and quiescent spectra of X-ray and γ-ray sources and a complete understanding of the underlying physical processes is impossible without the detailed self-consistent calculations of photon production and scattering effects along with energy balance (cooling and heating) in the emission region. The electron scattering occurs in a variety of forms (thermal and nonthermal, upscattering and downscattering), and includes such processes as inverse-Compton scattering (in which the photon gains energy from the electron), and Comptonization (repeated scattering). Variations in the scattering geometry give rise to reflection and back-scattering phenomena, because the energy and angular dependences are coupled by Compton interactions. The very important aspect of this problem is related with electron scattering in a moving medium. In the presence of high electron velocity the significant photon energy is seen in the observer (the laboratory) frame. The photon scattering off the energetic electrons are followed by their random walk throughout the moving medium. In case of accretion (bulk motion inward) the photon can be rather trapped by the bulk motion and ultimately disappear (being absorbed) in the inner boundary than escape towards the observer. The whole picture is also complicated by the gravitational redshift and gravitational bending (and distortion) of photon trajectory very close to the central object (black hole or neutron star). Another very important effect in the presence of subrelativistic and relativistic bulk motion is the coupling between bulk and thermal velocities. The standard Maxwellian electron distribution with the appropriate shift due to the bulk velocity cannot be used anymore. Thus the entire problem of spectral formation is formulated as a 6-D problem (3-D for photons and 3-D for electrons) along with the self-consistent calculation of the energy balance. This problem in its general form is beyond the capability of any modern or future computer. Any additional processes included in this formulation (e.g. magnetic field) make the situation with the problem solution even worse and uncertain.

The main breakthrough in the solution of this problem has been the realization that power law spectra are solutions of the *exact, full* kinetic equation (Titarchuk & Lyubarskij 1995, hereafter TL95) in the upscattering regime (*i.e.* when the electron energy is much higher than that of the photon in the electron rest frame). Thus the Green's function of the problem can be presented in separable form where the energy part of the solution is expressed through power laws and the solution is reduced to a 2D-eigenvalue problem (in radius and angle for the spherically symmetric case) for the spatial part of the solution.

2.2 Compton Scattering. Thermal and Bulk Motion

Qualitative description

Comptonization is the problem of energy exchange in the scattering of photons off electrons. The average energy exchange per scattering is determined by the relation between the photon and electron energies. For a thermal electron distribution with temperature κT_e and nonrelativistic electron energies E, $(E, \kappa T_e \ll m_e c^2)$, we have $<\Delta E>/E = (4\kappa T_e - h\nu)/m_e c^2$. When the photon energy E is much less than the mean electron energy κT_e, the photon gains energy due to the Doppler effect, i.e. $<\Delta E>/E = 4\kappa T_e/m_e c^2$. In the opposite limit, $(E \gg \kappa T_e)$, the photon loses its energy because of the recoil effect.

Two important cases are to be distinguished: optically thick and optically thin. In the optically thick (diffusion) regime where there is no preferential direction for the photon propagation in a plasma cloud, the radiation field is almost isotropic. The solution in this case is well known from the Fokker-Planck approximation, in which the radiative transfer equation for the intensity along a certain direction is replaced by the equation for the intensity averaged over all photon directions (ST80, T94). We can confirm this result using the exact solution of the kinetic equation (see TL95, §2c, Eq.22). For a relativistic plasma cloud (when the electron velocity v is very close to the speed of light c, i.e. $v/c \approx 1 - 1/2\gamma^2$), the situation is quite different. In order to gain energy, the primary photons have to undergo a certain number of collisions k in the hot electron gas. A soft photon of energy E_1 attains an energy E_2,

$$E_2 = E_1 \frac{c - v \cos\theta_1}{c - v \cos\theta_2} \qquad (1)$$

when scattering off an electron moving with velocity v. $\theta_{1,2}$ are the angles between the electron velocity and the direction of the photon propagation (before and after scattering respectively). The energy gain is maximal when the electron moves toward the photon, and scatters the photon backward, $\theta_1 = -\pi$, $\theta_2 = 0$. The maximum energy of the scattered photon is greater than the initial energy by the factor $4\gamma^2$.

In a medium with moderate optical depths, the probability of the photon to be scattered k times decreases rapidly with k. Therefore only photons that scattered effectively enough, i.e. $\theta_1 \approx -\pi$, $\theta_2 \approx 0$, contribute to the hard tail of the spectrum because they get the required energy after a minimal number of scatterings. Photons that are scattered ineffectively escape before they reach high energy. *So photons in the hard tail may be considered as scattering only in the backward and forward direction.* This dictates the specific time distribution of the photons emerging from the plasma cloud [Hua & Titarchuk (1995, hereafter HT95)].

Also this produces a specific angular distribution of the emergent radiation as seen by an observer on the surface of the plasma cloud. In a disk geometry the effective propagation of those high energy photons is predominantly in the plane of the disk because photons propagating at a large angle to this plane escape readily. Similarly, in a spherical geometry, photons scattering forward and backward sufficiently large number of times should propagate along the diameter of the sphere. The specific angular distribution determines the dependence of the spectral index α on the electron temperature and optical depth.

As it was shown by Zel'dovich (see the review by Pozdnyakov, Sobol, & Sunyaev 1983 and Ebisawa, Titarchuk & Chakrabarti 1996), the spectral index of the upscattered spectrum can be understood in terms of the scattering probability p and the fractional photon energy gain η per scattering. The intensity I_n of the emerging photons, which have undergone n scatterings in a plasma cloud, is proportional to p^n, i.e., $I_n \propto p^n$ On the other hand, a photon with initial energy E_0 gains energy and ends up with $E = E_0(1+\eta)^n$ after n scatterings. Solving the latter equation for n and substituting into the former equation one obtains that $I_n \propto (E/E_0)^{-\alpha}$, with spectral index $\alpha = -\ln p / \ln(1 + \eta)$. The scattering probability p is determined by the mean number of scatterings $< N_{sc} >$ that an emerging photon underwent in the plasma cloud. Because by definition, the mean number of scatterings $< N_{sc} > = p/(1 - p)$, we can present the above spectral index equation (for $N \gg 1$) as follows:

$$\alpha \approx -\frac{\ln(1 - 1/<N_{sc}>)}{\ln(1+\eta)}. \qquad (2)$$

The exact spectral index formula in the case of pure thermal Comptonization as a function of the plasma cloud temperature and optical depth is presented in TL95 (see below Eq. 9). Now I want to restrict myself in the qualitative estimates of spectral index behaviour as function of the optical depth τ_0 (or the dimensionless mass accretion rate $\dot{m} = \dot{M}/\dot{M}_{Ed}$, and $M_{Ed}c^2 = L_{Ed}$, where L_{Ed} is the Eddigton luminosity). For the case of accretion into black holes, the average number of scatterings $<N_{sc}>$, that the emergent photons (which form the power-law spectrum) have undergone, is proportional τ_0. This is to be contrasted with the isotropic diffusion result $<N_{sc}> \propto \tau_0^2$ (ST80). Thus by using (2) we find α is independent of τ_0 (or $\dot{m}$) if $\dot{m} \gg 1$. The numerical calculation of the spectral index confirms this qualitative result (see Fig. 2).

Now let us investigate the relative importance of the bulk (converging flow) and the thermal motion of the electrons to the mean photon energy change per scattering in the relatively simple case, the nonrelativistic diffusion one.

The mean energy gain per scattering of a photon by thermal Comptonization $<\Delta E_{tc}>$ is proportional to $(v/c)^2$, On the other hand, the mean energy gain $<\Delta E_{cf}>$ in the presence of a converging flow is proportional to v/c (TMK97), i.e. $<\Delta E_{cf}> \approx 4E(\dot{m}^{-1})$ Thus, we have $<\Delta E_{tc}>/<\Delta E_{cf}> \approx \kappa T_e/\dot{m}/m_e c^2 < \delta^{-1}$, where $\delta \equiv 1/\dot{m}\Theta = 51.1 \times T_{10}^{-1}\dot{m}^{-1}$, with $T_{10} \equiv \kappa T_e/(10$ keV$)$ and $\Theta = \kappa T_e/m_e c^2$. Hence, the bulk motion Comptonization dominates the thermal one if $\dot{m}T_{10} < 51$. It is worth noting these estimates are obtained in the nonrelativistic approximation. In the relativistic case there is coupling of the thermal motion with the hydrodynamical bulk motion (Titarchuk & Zannias 1998. hereafter TZ98) which affects the latter inequality (the product $\dot{m}T_{10}$ is much less than 51).

When $E \ll m_e c^2$, the recoil effect is negligible, resulting in a pure power-law spectrum. At high energies the two effects are comparable and the cutoff in the spectrum occurs at $E_c/m_e c^2 \approx 4[\dot{m}^{-1} + (v_b/c)^2/3 + \Theta]$.

Basic equations and their solutions

We now illustrate the formalism behind the computation of the theoretical spectra which result from Comptonization in plasma clouds.

The general relativistic **exact** integro-differential transfer equation (GTE) describing the interaction of low energy photons with a Maxwellian distribution of hot electrons has been analyzed in a Schwarzschild black hole background by Titarchuk & Zannias (1998). The spherically symmetric accretion onto a compact object with rate $\dot{m}$ was considered. Equation for the occupation number was derived [see also the relativistic particle transport theory by Lindquist (1966)]. It is proven in TL95 that the upscattered part of solution can be looked in the factorized form

$$n(\nu, r, \Omega) = F(r, \Omega) z^{-(\alpha+3)}. \tag{3}$$

where $z = h\nu/m_e c^2$, Ω is incoming photon direction. Thus the whole problem is reduced to the space radiative problem for the space part of the solution $F(r, \Omega)$

$$\ell^\alpha F = \sigma_T n_e [-F(r, \Omega) + B(r, \Omega)], \tag{4}$$

$$B(r, \Omega) = \frac{1}{4\pi} \int d\Omega' R(\eta') F(r, \Omega') \tag{5}$$

with the phase function $R(\eta')$ (see TL95, Eq.8 and TZ98, Eq. 17), where η' is the cosine of scattering angle between photon incoming and outgoing directions in the electron rest frame. The differential operator ℓ^α can be found very easily from differential part of GTE (see TZ98, Eq. 22)

In order to derive the dispersion equation for the determination of the spectral indices of the appropriate Green's function we use the method of the expansion of the phase function $R(\eta')$ in series over Legendre polynomials: (see also Sobolev 1975 and TL95).

In Figure 2 the results of the calculations of the spectral indices as a function of mass accretion rates are presented. It is clearly seen for the same electron

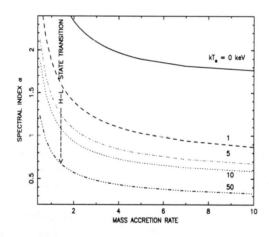

FIGURE 2. Plot of the energy spectral index (photon index minus one) and the (total) mass accretion rate for BH system.

temperature the spectral index is a weak function of mass accretion rates in the wide range of $\dot{m} = 2 - 10$.

The relativistic kinetic formalism presented in TZ98 (Eq.12 there) is the basis of the wide class of the spectral models. Because the limited space of this review does not allow us to embrace the whole variety of the models following from this formalism we point out some of them which are widely used for the spectral data analysis.

i. Among them the simplest one is the upscattering model. In this case the Green's function $I_\nu(x, x_0)$ (the spectrum for the monochromatic low-frequency injection) is presented by the broken power-law with the spectral indices determined as a solution of the appropriate eigenvalue problem

$$I_\nu(x, x_0) = \frac{C_N}{x_0} \left(\frac{x}{x_0}\right)^{\alpha+\zeta} \quad \text{for } x \leq x_0, \tag{6}$$

and

$$I_\nu(x, x_0) = \frac{C_N}{x_0} \left(\frac{x}{x_0}\right)^{-\alpha} \quad \text{for } x \geq x_0. \tag{7}$$

Here $C_N = \alpha(\alpha + \zeta)/(2\alpha + \zeta)$ is the normalization constant, $x = E/\kappa T_e$. The Green's function is normalized in such a way that the photon number is $1/x_0$ (the photon number is conserved in the scattering process).

The above relations correspond to a monochromatic photon injection with energy E_0. In order to get the resulting spectrum F_ν for an arbitrary source spectrum $g(E_0)$ one has to convolve this with the Green's function I_ν(TMK97):

$$F_\nu(E) = \int_0^\infty I(E, E_0) g(E_0) dE_0. \tag{8}$$

We remind the reader that the spectral indices α of the high energy part are taken from Fig. 2. The above convolution (8) is insensitive to the value of the Green's function low-energy spectral index $\alpha + \zeta$, which is always much greater than one. For example, it is $\alpha + 3 + \delta \gg 1$ (TMK97) in the bulk motion case and $\alpha + 3+ \gg 1$ in the static case (Sunyaev & Titarchuk 1980, hereafter ST80).

ii. The spectral model for the static atmosphere is the particular case of the general relativistic model (TZ98) where one considers the plasma cloud as a spherical layer with the boundaries situated quite far away from the horizon. In this case the bulk motion can be neglected and the upscattering is determined by the thermal motion only. In framework of this formulation Titarchuk & Lyubarskij 1995 (and recently Giesler & Kirk 1997 confirm this result) demonstrate explicitly the separation of the problem into its configuration and energy space parts. In this case, the problem reduces to an eigenvalue equation for the source function. They also provide analytic approximations for the computation of the power-law index over a very large range of optical depth and temperature. The spectral index is determined from the solution of the transcendental equation,

$$\alpha = \frac{\beta}{\ln[1 + (\alpha + 3)\Theta/(1 + \Theta) + 4d_0^{1/\alpha}\Theta^2]} \quad (9)$$

where,

$$d_0(\alpha) = \frac{[(\alpha + 3)\alpha + 4]\Gamma(2\alpha + 2)}{(\alpha + 3)(\alpha + 2)^2}. \quad (10)$$

From the above formula it is seen that only two parameters, the plasma temperature κT_e and the dimensionless escape rate, β, control the spectral index value.

iii. When the electrons are nonrelativistic, the fractional energy transfer per scattering is small. The kinetic equation may be expanded to the second order in this small quantity and thus the whole problem is reduced to Fokker-Planck problem. Kompaneetz (1957) was the first who derived the Fokker-Planck equation for the infinite homogeneous medium, which after that was generalized by Blandford & Payne (1981) for the bounded medium including the nonrelativistic bulk motion. The exact solution of the Fokker-Planck problem for the plasma cloud of the finite size was obtained by Sunyaev & Titarchuk (1980). In the wake of this important result, many authors investigated methods of extending the range applicability of the Fokker-Planck approximation. Hua & Titarchuk (1995) found the exact analytical solution of the Fokker-Planck equation in the subrelativistic regime of the photon energy and the plasma temperature. The equation was derived in HT95 by using the results of Cooper (1971), Grebenev & Sunyaev (1987), Prasad et al. (1988), Titarchuk (1994, hereafter T94). The HT95 solution verified by the Monte-Carlo calculations (see Fig. 3 in HT95) demonstrates high accuracy over a wide range of optical depth ($\tau_0 > 1$) and temperature ($\kappa T_e < 150$ keV). It naturally includes ST80 as the particular solution for the nonrelativistic case. For example in the case of low-frequency sources ($x_0 \ll 1$), the Green's function for $x \geq x_0$ is

$$I_\nu(x, x_0) = \frac{\alpha(\alpha + 3)e^{-sx}}{\Gamma(2\alpha + 4)x_0} \left(\frac{x}{x_0}\right)^{-\alpha} \int_0^\infty t^{\alpha+\epsilon-1}(\rho x + t)^{\alpha-\epsilon+3}e^{-t}dt \quad (11)$$

where $\Gamma(z)$ is the gamma function; $s = (1 + \rho + \delta)/2$, $\rho = \sqrt{(1-\delta)^2 + 4a_2}$, $\epsilon = [2(\rho-1)+a_1+2\delta]/\rho$, $a_1 = 7.4\gamma_0\Theta(1-0.42\Theta)$, $a_2 = 13.5\gamma_0\Theta^2(1-1.05\Theta)$, $\delta = -4.6\Theta$ and $\Theta = \kappa T_e/m_e c^2$, $\gamma_0 = \beta/\Theta(1+f_0)$ and $f_0 = 2.5\Theta + 1.875\Theta^2(1-\Theta)$. For a given optical depth τ_0 the values of β are uniquely determined (T94).

The above solution takes into account (in contrast to the ST80) that the Compton scattering cross section drops substantially when the electron temperature and the photon energy go up. In other words, a cloud is more transparent to higher energy photons and less capable in changing the photon energies and directions. Consequently even in plasma clouds with fairly large τ_0 and $kT_e > 100$ keV, most of the photons do not reach the Wien barrier $3kT_e$ and the Wien hard tail $x^3 \exp(-x)$ is more or less suppressed.

iv. When the electron temperature is much less than the average photon energy, the downscattering effect is very important in the formation of the emergent spectra. Because this particular case is outside the scope of this review, we can only refer to publications where the reader will find a discussion of this issue in detail. They are, for instance, Basko et al. (1974), ST80, George & Fabian (1991), CT95.

3. SPECTRAL MODELS AND INTERPRETATION OF THE RECENT HIGH ENERGY OBSERVATIONS

In this section we provide an example of the relevance of the existing models to the recent high energy observations. We demonstrate that the bulk motion model described in detail in the previous section fits nicely BHC X-ray spectra in their soft state.

Figure 3 shows a comparison between the analytical approximation and the numerical solution of the bulk motion problem in terms of Fokker-Plank treatment (TMK96, 97)

The analytical approximation represents the emergent spectrum quite reliably in the whole energy range below the exponential turnover. In this region, two effects, the bulk motion upscattering and the Compton (recoil) downscattering, compete forming the hard tail. Because the analytical representation does not take into account properly the inner boundary condition at energies $E > E_c$ there is a significant deviation of the analytical curve from the numerical solution close to the cut-off. When one takes into consideration the relativistic corrections, the hard tail becomes flatter, since Compton downscattering is less efficient at high energies (Laurent & Titarchuk in preparation).

Here we have assumed the external illumination of the converging flow by the low-energy black body radiation of an accretion disk having a characteristic temperature T_{bb}. Furthermore we have assumed the accretion disk radiation illuminates the bulk motion atmosphere at angle $\theta = 17.5°$ (i.e. the cosine of the angle of incidence $\mu = 0.3$). All spectra are pure power laws in the high energy range from 15 kev up to the exponential turnover, which occurs at energy $E_c \lesssim m_e c^2$. It is shown in TMK97 that the emergent spectrum observed at infinity can be presented by two components. One is a soft (disk) component coming from those input photons which escaped after a few scatterings without any significant energy change and another one is the component made of those photons which underwent significant upscattering. The luminosity of the upscattering component is relatively small compared to that of the soft component. Thus we can describe this bulk motion spectral model in terms of a sum of blackbody like (disk) component and the convolution of some fraction of the soft component with the upscattering Green's function (see Eq.8).

Shrader et al. (1997) constructed composite high-energy spectra for GRO J1655-40 during an outburst in the spring of 1996, covering the 2-200 keV spectral region and fit these data by the bulk motion model. This was accomplished using summed data from the RXTE/PCA and the BATSE/LAD earth-occultation data. The galactic hydrogen column density we used was constrained to within ±25% of the ASCA value of 0.5×10^{21} cm^{-2} (Inoue et al., 1994, [IAUC 6063]) for all fitting. They found that the lowest energy channels ($<$ 4 keV) proved to be problematic in that they contributed significantly to the overall chi-square, although generally did not lead to significant differences in the resulting fit parameters thus, these channels were ignored. They additionally found PCA channels 49 and 51 to be problematic in some cases, presumably a result of inaccuracy in the background estimation, thus we ignored these in this fitting as well. The BATSE 16-channel data were summed over Viewing Period 522.5, which was coincident in time with the RXTE observation for 3 of the Large Area Detectors. The bulk motion model described in sections 2, 3 were imported into the XSPEC software which was used to perform all of the model fitting described here. One of our resulting

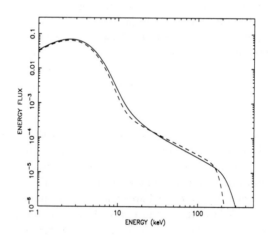

FIGURE 3. Plot of the emergent spectral energy flux versus photon energy for $\dot{m} = 3$, $kT_e = 1$ keV, color blackbody temperature $kT_{bb} = 1$ keV, $\mu = 0.3$. The solid line is a numerical solution of the Fokker-Planck equation with the appropriate boundary conditions. While the dashed line corresponds to the approximate analytical solution.

'fits, and the inferred parameters are shown in Figure 4.

A similar procedure as described above was applied to RXTE/PCA and BATSE data obtained during the November 1996 period of activity in the BHC X-ray nova GRS 1915+105.

For GRO J1655-40 Shrader et al. (1997) obtained a blackbody temperature of 1.1 keV for the soft photon source, a spectral index of 1.6, and an (Log) of the covering factor A (a fraction of the total soft photon flux illuminating the bulk motion inflow atmosphere) of -0.5 (similar results were obtained for GRS 1915+105). An optically thick accretion disk spectrum was adopted for the spectrum of the soft component fully taking into account the electron scattering in the disk atmosphere. The hardening factor (ratio of color temperature to effective temperature) due to the electron scattering is 2 ± 0.3 (Shimura & Takahara 1993). The disk component makes the main contribution to the soft emission at < 8 keV. It is easy to show that the main parameters of the system, the black hole mass and the distance to the source can be retrieved by using the derived parameters, the blackbody temperature, spectral index and the theoretical dependence of the spectral index on the mass accretion rate the plasma temperature (the same as the blackbody temperature). We

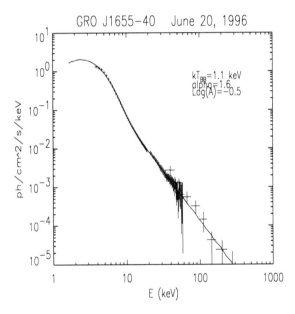

FIGURE 4. RXTE/PCA and BATSE/LAD spectrum for GRO J1655-40 fitted to a model involving spherical accretion onto the central black hole. Radiative diffusion of the soft disk photons, which on average gain energy, produces the hard tail. The free parameters in this case were kT_{bb}, an covering parameter A, and a spectral index parameter α which characterizes the hard (> 20 keV) component.

found the the derived distance (3.8 ± 1.4 kpc) was consistent with previous determination, and the black hole mass we calculate could be reconciled with the measured value of Orosz and Baiyn (1997). It is worthwhile to emphasize that the estimates of the black hole mass and distance obtained by using the free fall model are in fact, the lowest limits of the above quantities. A higher mass accretion rate is required in the rotation case in order to provide the same efficiency for the soft photon upscattering. Thus the increase of the mass accretion rate with respect to the nonrotating case leads to higher values of the blackhole mass and distance estimates.

REFERENCES

1. Basko, M.M., Sunyaev, R.A., Titarchuk, L.G., *A&A* **31**, 249 (1974).
2. Blandford, R. D., Payne, D. G., *MNRAS*, **194** 1033 (1981).
3. Chakrabarti, S. K. & Titarchuk, L. G., *ApJ* **455**, 623 (1995) (CT95).
4. Cherny, B., Witt, H. J. & Zycki, P.T., *Acta Astr.* **46**, 9 (1996).
5. Cue, W., Zhang, S.N., Focke, W., & Swank, J.N., *ApJ.* **484**, 363 (1997).
6. Cooper, G. 1971, *Phys. Rev.* D, 3, **10**, 2312 (1971).
7. Ebisawa, K., Titarchuk, L.G. & Chakrabarti, S, *PASJ*, **48**, 59 (1996).
8. George, I.M., & Fabian, A.C. *MNRAS*, **249**, 352 (1991).
9. Giesler, U.D.J & Kirk, J.G., *A&A* **323**, 259 (1997).
10. Grebenev,S.A. & Sunyaev, R.A., *AZh Pis'ma*, **13**, 1042 (1987).
11. Grove J.E., et al., *ApJ*, in press (1997).
12. Hua, X.-M. & Titarchuk, L. G., *ApJ*, **449**, 188 (1995) (HT95).
13. Kazanas, D., Hua, X.-M. & Titarchuk, L. G., *ApJ*, **480**, 735 (1997).
14. Kompaneetz, A.S. *Sov. Phys. JEPT*, **4**, 730 (1957).
15. Lindquist, R., *Annals of Physics* **37**, 487 (1966).
16. Narayan, R., *ApJ*, **462**, 136 (1996).
17. Orosz, J. & Bailyn, C.D., *ApJ*, **482**, 1086 (1997).
18. Prasad, M.K., Shestakov, A.I., Kershaw, D.S., Zimmerman, G.B., *J. Quant. Spectrosc. Rad. Transf.*, **40**, 29 (1988).
19. Pozdnyakov, L.A., Sobol, I.M., & Sunyaev, R.A., *Ap. Space Phys. Rev.*, **2**, 189 (1983).
20. Shakura, N.I. & Sunyaev, R.A., *A&A*, **24**, 337, (1973) (SS73).
21. Shimura, T., & Takahara, F. *ApJ*, **445**, 780 (1995).
22. Shrader, C., Chen, W. & Titarchuk, L.G., in preparation (1997).
23. Skibo, J. G., & Dermer, C. D., *ApJ*, **455**, L25 (1995).
24. Sobolev, V. V., *Light Scattering in Atmospheres*, Oxford: Pergamon, (1975).
25. Sunyaev, R. A., Borozdin, K.N., et al., *Pis'ma Astron. Zh.* **20**, 890 (1994).
26. Sunyaev, R.A., Titarchuk, L.G., *A&A*, **86**, 121, (1980) (ST80).
27. Titarchuk, L. G. *ApJ*, **434**, 570 (1994) (T94).
28. Titarchuk, L. G. & Lyubarskij, Yu. E., *ApJ*, **450**, 876 (1995) (TL95).
29. Titarchuk, L. G., Mastichiadis, A., Kylafis, N. D., A&AS, **120**, C171 (1996). (TMK96)

30. ———— *ApJ*, **487**, (1997)(TMK97).
31. Titarchuk, L. G., Muslimov, A., A&A, **323**, L5 (1997).
32. Titarchuk, L. G. & Zannias, T. *ApJ*, **493**, (1998) (TZ98).
33. van der Klis, M., Wijnands, R., Horne, K. & Chen, W., *ApJ*, **481**, L97 (1997).
34. Zhang, S. N. et al., *ApJ* in press (1997).

Radiation Processes in Blazars

Marek Sikora*

*Nicolaus Copernicus Astronomical Center, PAN, Bartycka 18, 00-716 Warsaw, Poland

Abstract. We present an overview of the current theoretical models attempting to describe the structure and radiative processes operating in blazars and discuss the observational constraints that these models must confront. Many of these objects are found by *CGRO* to be strong high energy γ-ray sources. Their spectra consist of two broad components: the low-energy component, peaking between the infrared and soft X-ray band, and the high-energy component, peaking in the high energy γ-rays. The overall energetics is often dominated by the latter, by as much as an order of magnitude over the former. Both components show rapid large-amplitude variability, which indicates a very compact emission region.

Most current models for the structure of blazars, to avoid the problems with excessive opacity of γ-rays to pair production, invoke beaming of electromagnetic emission from the radiating matter moving in a jet pointed close to the line of sight towards the observer, in an analogy to the jet-like structure inferred from other observations. The non-thermal shape of the spectrum of the low-energy component as well as its polarization suggest synchrotron emission. For the high-energy component, most current models invoke Comptonization of the lower energy photons, either those internal to the jet, as in synchrotron self-Compton (SSC) models, or external to the jet (UV radiation from the accretion disk or from the emission-line region, or IR radiation from dust), as in External Radiation Compton (ERC) models. Comparison of the energy densities of the external radiation fields with the energy density of the synchrotron radiation field (both as measured in the jet comoving frame) suggests that while the SSC model may be adequate to explain the γ-ray emission for BL Lac objects, the ERC models are probably more applicable for flat spectrum radio-quasars (FSRQ).

1. INTRODUCTION

Blazars are extragalactic objects with core dominated, flat spectrum, variable radio sources. Radio spectra smoothly join the infrared-optical-UV spectra, and in all these bands flux is highly variable and polarized. These properties are shared by BL Lac objects as well as by most flat spectrum radio-quasars (FSRQ), and are successfully interpreted in terms of synchrotron radiation produced in relativistic jets and beamed into our direction [6,3]. This interpretation is strongly supported by direct observations of superluminal motions observed in radio cores in VLBI data [42], and lays the grounds for the unified scheme of radio-loud active galactic nuclei (AGN) [18].

As was recently discovered by *CGRO*, many blazars are strong and variable sources of high energy γ-rays [43]; in a few sources, the spectrum extends up to the TeV energies [32,33]. The γ-ray radiation forms a separate spectral component, with the luminosity peak located in the MeV-TeV range. Variability of GeV/TeV radiation itself provides evidence for relativistic speeds of radiating plasma [29,14], and the lack of high energy γ-rays in AGNs other than blazars proves that γ-rays must be at least as well collimated as synchrotron radiation.

Production of high energy radiation was predicted many years ago by synchrotron-self-Compton (SSC) models [19,27]. In this process, the same electrons that produce synchrotron radiation also upscatter some of synchrotron photons to γ-ray energies. However, as was recently recognized, the SSC process is not necesarily the one which produces most of the γ-rays. The competing process, at least in quasars, may be the Comptonization of external radiation. Models based on this process are called ERC (external radiation Compton) and their original variants have been investigated by Dermer, Schlickeiser, & Mastichiadis [13], Sikora, Begelman, & Rees [37], Blandford & Levinson [4], Ghisellini & Madau [16]. Production of γ-rays is also predicted by so-called hadronic models, where ultrarelativistic electrons/positrons are injected by relativistic protons [25,24,1,10].

Different scenarios of γ-ray production in AGN jets are reviewed in §2. Their predictions are confronted with multiwavelength spectral and variability data in §3, and discussion of what we have learned already and can learn in the nearest future about AGN jet physics from the blazar observations is presented in §4.

2. PRODUCTION OF HIGH ENERGY γ-RAYS

The fact that the spectra of synchrotron components blazars extend up to optical, UV, and even X-ray bands indicates that AGN jets contain highly relativistic electrons/positrons, with Lorentz factors up to $10^4 - 10^6$. Energy losses of such electrons are so rapid that they must be accelerated/injected *in situ*, i.e. at the locations where they radiate. These locations are the sites of the energy dissipation events, which propagate along the jet at moderate ($\Gamma \sim 10$) relativistic speeds. The dissipation events can result from interaction of the jet with external obstacles, annihilation of magnetic fields, and/or collisions of inhomogeneities in a jet [3]. During such events, a part of the dissipated energy is converted to relativistic electrons and protons.

Models based on the assumption that the high energy γ-rays are produced by directly accelerated electrons are discussed in §2.1, while hadronic models for γ-radiation production are reviewed briefly in §2.2.

2.1. ERC process vs. SSC process

Assuming that the momentum distribution of relativistic electrons is isotropic in the comoving frame of dissipative events [12] and that SSC and ERC operate in the Thomson regime, one can compare radiation production in these processes using formulae for electron cooling rates

$$\left(\frac{d\gamma'}{dt'}\right)_{SSC} = \frac{4}{3}\frac{c\sigma_T}{m_e c^2} u'_S \gamma'^2 \tag{1}$$

and ([39])

$$\left(\frac{d\gamma'}{dt'}\right)_{ERC} \simeq \frac{16}{9}\frac{c\sigma_T}{m_e c^2} u'_{ext} \gamma'^2 \tag{2}$$

where γ' is the random Lorentz factor of relativistic electrons,

$$u'_S = L'_S/4\pi a^2 c \tag{3}$$

is the energy density of synchrotron radiation produced by the source with radius a,

$$u'_{ext} \simeq \Gamma^2 \xi L_{UV}/4\pi r^2 c \tag{4}$$

is the energy density of the external radiation field [39], ξ is the fraction of the central radiation isotropized by reprocessing and/or rescattering at a distance scale corresponding with a distance $\sim r$ from the central source, and all primed quantities are as measured in the source comoving frame.

Using formulae (1) - (4) one can find that

$$\frac{L'_{ERC}}{L'_{SSC}} = \frac{(d\gamma'/dt)_{ERC}}{(d\gamma'/dt')_{SSC}} \simeq (\Gamma \theta_j)^2 \frac{\xi L_{UV}}{L'_S} \tag{5}$$

where $\theta_j = a/r$. The above formula can be used to find the ratio of the observed SSC and ERC luminosities, provided that the angular distributions of radiation fields are known. For isotropic (tangled) magnetic field, the synchrotron radiation in the source frame is isotropic, and for jets with $\theta_j \ll 1/\Gamma$, the observer located at θ_{obs} sees $L_S = \mathcal{D}^4 L'_S$, where $\mathcal{D} = 1/(\Gamma(1 - \beta \cos\theta_{obs}))$ is the Doppler factor. As was pointed out by Dermer [11], this is not the case for the ERC process. Compton scattered radiation, as measured in the source comoving frame, can have quasi-isotropic distribution only for $\Gamma \simeq \Gamma_{eq}$, where Γ_{eq} is the Lorentz factor at which the flux of the external radiation field is zero.

As was shown by Sikora et al. [39]

$$\Gamma_{eq} \simeq \left(\frac{3}{16\xi}\right)^{1/4}. \tag{6}$$

At $r \ll 1$ pc, ξ is provided by rescattering of central radiation by hot gas present in coronae or winds around accretion disks. Up to $r \sim 0.1 - 1.0\sqrt{L_{UV}}$ pc, this is dominated by the fraction of the central radiation converted to emission lines by optically thick clouds/filaments, and at larger distances, it is determined by the fraction of the central radiation reemitted by dust. For a typical quasar environment, ξ is expected to be $\sim 0.01 - 0.1$, and, for such values, equation (6) gives $\Gamma_{eq} \leq 2$. This is much less than the Lorentz factors of AGN jets which are typically enclosed in the range $5 < \Gamma < 20$ and often reach values > 10 [42,30,26]. The external radiation is seen by the emitting cloud of plasma moving at $\Gamma > \Gamma_{eq}$ as coming from the front. In such a field, the Compton scattered radiation is produced with a highly anisotropic distribution, with a large deficiency of radiation scattered into the $\theta' > \pi/2$-hemisphere. As measured in the external frame, such radiation is beamed more strongly than the synchrotron radiation and the observed luminosity $L_{ERC} \propto \mathcal{D}^6$. However, for comparable total emitted luminosities, the respective observed luminosities L_{SSC} and L_{ERC}, if averaged over the $1/\Gamma$-cone, are also comparable. Then, for the case $\theta_j \sim 1/\Gamma$, for which the angular distribution of radiation within the $1/\Gamma$-cone is smeared out, the formula

$$L_{ERC}/L_{SSC} \sim L'_{ERC}/L'_{SSC} \sim (\Gamma\theta_j)^2 \frac{\xi L_{UV}}{L_S} \Gamma^4 \qquad (7)$$

can be used for all $\theta_{obs} \leq 1/\Gamma$-observers. Since in quasars $\Gamma \sim 10$ and $\xi L_{UV} \sim 10^{44} - 10^{45}$ erg s^{-1}, while L_S observed in FSRQ is in the range $10^{46} - 10^{47}$ erg s^{-1}, the above formula seems to prove strong domination of the ERC process over the SSC process in quasar jets.

However, it should be noted here that because of the different energy distributions of the ambient radiation fields, the respective high energy components produced by a given population of electrons will not overlap entirely. As a result, a less luminous SSC component produced by far infrared radiation can still be visible in soft/mid X-rays [20], while relativistic electrons injected at a distance ~ 1pc can produce in the MeV range two separate "bumps", one due to Comptonization of external UV radiation and one due to Comptonization of external near-IR radiation.

The situation is less clear in BL Lac objects, where the radiative environment in the central region is not very well known. The lack of strong emission lines and of UV excesses (even during lowest states) suggest that in these objects ξL_{UV} can be very low. Noting also that in BL Lac objects Γ factors are typically smaller than in quasars [30,26], domination of SSC over ERC in these objects is very likely.

One should be warned, however, that for BL Lac objects with the high energy spectra extending up to TeV energies the formula (7) cannot be applied directly; this is because the Klein-Nishina effect reduces the efficiency of the Compton process, and this reduction is different for different energy distributions of the ambient radiation fields.

2.2. Hadronic Models

In all particle acceleration processes, the injection of relativistic electrons/positrons is accompanied by injection of relativistic protons. Their energy can be converted to high energy radiation following such processes as direct synchrotron radiation of protons, proton-photon pair production, photomeson production, and nuclear collisions. The first three processes are known to be very inefficient, and in AGN jets can become important only for proton energies $\geq 10^8 - 10^{10}$ GeV. Only for such high energies can the time scales of the proton energy losses become comparable to or shorter than the propagation time scale of the source in a jet. Energy losses of such energetic protons are dominated by photomeson production, and this process was used by Mannheim and Bierman [25] to model γ-ray production in luminous blazars.

The radiation target for photomeson production is dominated by the near/mid-infrared radiation. In quasars, such radiation is provided by hot dust at distances $\sim 1 - 10$ parsecs from the central source and by synchrotron radiation in a jet, produced by directly accelerated electrons. The main output of the photomeson process are single pions. They take about 30% of the protons' energy and convert it to photons, neutrinos, and through muons, to electrons and positrons. The photons injected by neutral pions are immediately absorbed by soft photons in the pair production process. These pairs and electrons/positrons injected by muons have Lorentz factors $\gamma' \geq 10^{11}$. For such energies, Compton scattering with the ambient radiation field takes place deeply in the Klein-Nishina regime and, therefore, their energy losses are dominated by synchrotron radiation. Most of this radiation is so energetic that it produces two more generations of photons and pairs. The final output of this synchrotron-supported pair cascade is the high energy component, enclosed within or cut off at energies above which photons are absorbed by $\gamma\gamma$-pair production process. This maximum energy can be ~ 30 GeV in FSRQ, as determined by external UV radiation, and ~ 1 TeV in low luminosity BL Lac objects, as determined by infrared radiation of dust [31].

The weakness of the "photomeson" model is that it requires fine tuning in order to avoid situations where the luminosity peak lies below MeV energies. This is because, after 3 pair generations, the location of the peak depends on the 6th power of the maximum proton energy. Also, even if a model is successful in locating the peak of the high energy component, there is still the problem of how to obtain the hard X-ray spectra after three generations of the pair cascade process [40]. To overcome this difficulty, Mannheim [24] proposed that transition from softer γ-ray spectra to harder X-ray spectra results from a break in the pair injection function. This, however, requires the ambient radiation to be transparent to γ-rays up to energies $\sim 10\sqrt{\Gamma/B'}$ TeV, while external UV and near-IR radiation fields are expected to cut the spectrum in quasars at ~ 30 GeV for $r < \sqrt{L_{UV}}$ pc and at ~ 1 TeV for larger distances.

The photomeson scenario was also suggested to explain the production of TeV radiation in low luminosity BL Lac objects [31]. The recent discovery of variability on time scales < 1 hour seems to jeopardize this idea. This is because to get proton energy losses on such short time scales, much higher IR luminosities are required than are observed.

Much less extreme proton energies are required in models based on the assumption that proton energy losses are dominated by collisions with the ambient gas. The final output of these collisions is the same as in the photomeson process, i.e., relativistic electrons/positrons, photons and neutrinos. The process can be efficient only if the column density of the target is $n_H \geq 10^{26}$ cm^{-2}. Bednarek [1] proposed as a target the funnels formed around the black hole by a geometrically thick disk, while Dar and Laor [10] suggested interactions of jet with clouds and/or stellar winds. The shortcoming of such models is that relativistic protons, before colliding with the nuclei, may easily suffer deflections by magnetic fields; this generally results in a lack of collimation of the radiation produced following pp collisions.

3. MULTIWAVELENGTH SPECTRA

3.1. ERC and SSC luminosities vs Synchrotron Luminosity

If both high-energy and low-energy spectral components are produced by the same population of relativistic electrons, and the production of high energy radiation is dominated by the SSC process, then

$$\frac{L_{SSC}}{L_S} \sim \frac{L'_{SSC}}{L'_S} \sim \frac{u'_S}{u'_B} \tag{8}$$

where $u'_B = (B')^2/8\pi$ is the energy density of magnetic field. Equations (3) and (8) give

$$u'_B \simeq \frac{1}{4\pi a^2 c \Gamma^4} \frac{L_S^2}{L_{SSC}}, \tag{9}$$

which, in the case of steady flow with $\theta_j \sim a/r$, determines the flux of magnetic energy

$$L_B \simeq c u'_B \pi a^2 \Gamma^2 = \frac{1}{4\Gamma^2} \frac{L_S^2}{L_{SSC}}. \tag{10}$$

For $\Gamma \sim 10$, $L_S \sim 10^{46} - 10^{47}$ergs^{-1} this gives $L_B \sim 10^{43}$ergs^{-1}, which is about 3 orders of magnitude less than the typical power of the quasar jets [34,8,15].

Thus, the observed high γ-ray luminosities can be explained in terms of the SSC models only if one assumes that the jets are very weakly magnetized.

In the case of ERC models we have

$$\frac{L'_{ERC}}{L'_S} \sim \frac{u'_{ext}}{u'_B}, \tag{11}$$

and provided that L_{ERC} and L_S are luminosities observed at $\theta_{obs} \leq 1/\Gamma$ and that $\theta_j \sim 1/\Gamma$, we can use the scaling $L_{ERC}/L_S \sim L'_{ERC}/L'_S$. With this scaling equations (4) and (11) give

$$u'_B \simeq \Gamma^2 \frac{\xi L_{UV}}{4\pi r^2 c} \frac{L_S}{L_{ERC}}. \tag{12}$$

This gives

$$L_B \simeq c u'_B \pi a^2 \Gamma^2 = \frac{(\theta_j \Gamma)^2}{4} \Gamma^2 \xi L_{UV} \frac{L_S}{L_{ERC}} \tag{13}$$

which is 2-3 orders of magnitude greater than in the case of the SSC model.

Another interesting aspect of comparing ERC and SSC radiation components with the synchrotron component is the angular distribution of these radiation fields. As was shown by Dermer [11] and discussed in §2.1, ERC radiation is much more strongly collimated than the synchrotron and SSC radiation. Since SSC and synchrotron radiation fields have the same angular distribution (they both are produced isotropically in the source comoving frame), the predicted high-energy to low-energy luminosity ratio doesn't depend on θ_{obs}. In contrast, the ERC model predicts this ratio to drop very rapidly with viewing angle outside the $1/\Gamma$-cone. Dermer proposed that this can explain why a significant fraction of FSRQ do not show γ-ray activity, even though they are otherwise recognized as typical blazars on the basis of the low energy component properties.

3.2. Production of X–rays

X-rays in different sub-classes of blazars can have different origins. In most BL Lac objects X-ray spectra are steep ($\alpha \sim 1 - 3$) and variable, and lie on an extrapolation of the UV spectrum. This indicates that X-rays in these objects represent high energy tails of the synchrotron component. In FSRQs, the X-ray spectra are usually very hard ($\alpha \simeq 0.5 - 0.7$), showing weaker variability than in other spectral bands. These spectra are often interpreted as low energy tails of the γ-ray components; however, one cannot exclude the possibility that they are superposed from two or more components. And finally, there are intermediate objects where the soft X-rays are dominated (at least occasionally) by the synchrotron component, while higher energy X-rays belong to the high-energy Compton component [23]. These differences in the X-ray spectra seem to follow the general trend where in less luminous blazar, the peaks of the low-energy (synchrotron) component are located at higher energies [36,20].

The simplest interpretation of the hard X-ray spectra of FSRQs is that, together with the γ-rays, they form a single component produced by the ERC process. In the model, the spectral slope changes from a steeper one in the γ-ray band to a harder one in the X-ray band, as a result of incomplete cooling of electrons below certain energy which is determined by an equality of the ERC cooling time scale and the propagation time scale [37]. The break is located in the $1 - 30$ MeV range if the distance of radiation production is $\sim 0.1 - 3$ pc.

In ERC models, the X-ray spectra imprint the distribution of Lorentz factors of relativistic electrons down to $\gamma' \sim 10$ for the X-rays produced by Comptonization of IR radiation, and even down to $\gamma' \sim 1$ for the X-rays produced by Comptonization of UV radiation. Since for distances > 0.1 pc the low energy electrons cool very inefficiently, to produce the observed X-ray flux by ERC process requires such a large number of electrons that the jet must be strongly pair dominated in order to avoid unreasonable high kinetic energy flux.

Another possibility is that hard X-ray spectra are superposed from partial spectra produced over a wide range of distances and having low-energy cutoffs at energies which increase with distance [38]. In this model, the soft X-rays are produced closely to the black hole, and therefore the production of X-rays can be accomplished by a lower number of electrons, and thus the jet plasma need not be pair-dominated.

Finally, the hard X-ray spectra can be produced by SSC radiation, while production of high energy γ-rays can be dominated by ERC process [20]. In this model, a wide range of n_e/n_p is acceptable.

3.3. Bulk-Compton Radiation

Very interesting constraints on the AGN jets come not only from what we *do* observe, but also from what we *do not* observe. A feature that was predicted – but not confirmed observationally – is radiation produced by cold electrons in a jet. Such electrons, dragged by the protons and/or magnetic fields in the jet, for $\Gamma > \Gamma_{eq}$, should scatter external UV photons and produce a collimated beam of bulk-Compton radiation [2,39]. The predicted observed luminosity is

$$L_{BC} \sim \Gamma^2 n_e \pi a^2 r dE_e/dt \tag{14}$$

where

$$\frac{dE_e}{dt} = \frac{4}{3} c\sigma_T \xi u_{ext} \Gamma^2, \tag{15}$$

$u_{ext} \sim L_{UV}/4\pi r^2 c$, and r is the distance at which this process is most efficient. The bulk-Compton spectrum should have a peak at $h\nu_{BC} \sim \Gamma^2 n_{UV} \sim 1\text{keV}$. Since this is not observed, L_{BC} must be smaller than the luminosity L_{SX} of the nonthermal X-ray spectrum in the soft X-ray band. This gives an upper limit for the Thompson optical thickness in a jet

$$\tau_{j,max} \equiv n_{e,max} a \sigma_T \sim \frac{3}{\Gamma^3(\theta_j \Gamma)} \frac{L_{SX}}{\xi L_{UV}} \tag{16}$$

i.e., ~ 0.03 for $\Gamma \sim 10$, $L_{SX} \sim 10^{46}$ erg s^{-1}, and $\xi L_{UV} \sim 10^{45}$ erg s^{-1}, With this limit, the processes scaled by $(\tau_j)^2$ (like annihilation, bremsstrahlung and Coulomb interactions) are inefficient, and play a negligible role in shaping the spectra of blazars. This is because in such thin plasmas, the time scales of these processes are much longer than the time scale of plasma propagation in a jet [9].

The upper limit for τ_j also gives interesting constraints on the $e^+ e^-$ pair content of a jet. If r_{min} is the radius where τ_j is maximal and if for $r > r_{min}$ the pair flux is conserved, then for a conical jet $\tau_j \propto 1/r$ and, therefore, the bulk-Compton radiation is mostly contributed by the innermost parts of the jet. Assuming that the kinetic energy flux in a jet is dominated by cold protons, we have

$$L_K \simeq n'_p m_p c^3 \pi a^2 \Gamma^2 \tag{17}$$

where n'_p is the number density of protons in the jet comoving frame. Noting that $n_p = n'_p \Gamma$, we have

$$n_e \simeq \frac{n_e}{n_p} \frac{L_K \Gamma}{\pi m_p c^3 a^2 \Gamma^2}. \tag{18}$$

Substituting this into the formula for L_{BC} evaluated for $r = r_{min}$, we obtain

$$L_{BC} \simeq \frac{n_e}{n_p} \frac{r_g}{r_{min}} \frac{L_K}{L_{Edd}} \xi L_{UV} \Gamma^3, \tag{19}$$

where $L_{Edd} = (4\pi m_p c^3/\sigma_T) r_g$. Then, the condition $L_{BC} \leq L_{SX}$ gives that for powerful ($L_K \sim L_{Edd}$) and pair-dominated ($n_e \gg n_p$) jets, overproduction of soft X-rays can be avoided only at very large distances ($10^3 - 10^5 r_g$) from the black hole.

4. SUMMARY

Discovery of strong and variable γ-ray radiation in blazars by CGRO provided an exceptional possibilities to explore and verify the operation of various nonthermal processes in AGN jets, and to study the structure, energetics and matter content of these jets. The main achievements of such studies and future prospects are listed below:

- γ-ray radiation provides independent evidence that blazar radiation is produced by relativistic jets [29,14]. This is because the compactness of the source derived from the observed γ- and X-ray luminosities and variability time scales is so high that if it was intrinsic (true) compactness, all γ-rays would be absorbed by $\gamma\gamma$ pair production process. This implies that the true source compactness must be much lower than the observed one, and this is the case if the observed radiation originates from plasma propagating in our direction at relativistic speed.

- γ-rays can be also absorbed by external radiation fields, and because the compactness of such fields decreases with distance, this gives the minimum distance from which the γ-rays can escape. Of course, since the opacity for $\gamma\gamma$ interactions depends on energy and is higher for more energetic γ-rays, the minimum escape distance is smaller for less energetic γ-rays. Such stratified γ-ray production was suggested by Blandford and Levinson [4] in their inhomogeneous version of ERC model.

- Huge apparent γ-ray luminosities, reaching in some FSRQ $10^{48} - 10^{49}$ erg s^{-1}, provide independent evidence that AGN jets must be very powerful, with $P > (\bar{L}_\gamma/\Gamma^2)/\epsilon_{rad}$, i.e., $\sim 10^{46}$ erg s^{-1} for radiation efficiency $\epsilon_{rad} \sim 0.1$, $\Gamma \sim 10$ and time-averaged observed luminosity $\bar{L}_\gamma \sim 10^{47}$ erg s^{-1}.

- Comparison of the ERC efficiency with the SSC efficiency for typical quasar radiation fields implies that the production of γ-rays in FSRQ should be strongly dominated by Comptonization of external radiation. However, SSC can still contribute visibly to the X-ray band [20].

- The hadronic models have problems explaining hard X-ray spectra in FSRQ and very short variability time scales in TeV BL Lac objects.

- Comparing ERC and SSC spectra with the synchrotron spectra, one can attempt to derive physical parameters of radiating plasma - such as maximum electron energies, magnetic fields, electron injection function, and distance of the source from the black hole [37,17,38,41,28]. However, such analyses must be performed by taking into account that the observed spectra, especially their lower energy parts (in both the synchrotron and the Compton components) may well be superposed by two or more components.

- In the case of FSRQ, simultaneous observations in the γ-ray and X-rays bands must be used to verify various mechanisms of X-ray production. This can help to establish the pair content in AGN jets. In particular, the correlation and absence of a time lag between the γ-ray and the X-ray flares (as seen in Jan, 1996 in 3C279; [44]) may indicate co-spatial production of both, and in the case of X-rays produced by the ERC process, this would indicate a pair-dominated plasma.

- Soft X-ray limits for the bulk-Compton process prove that AGN jets are optically thin. This implies that such processes as bremsstrahlung, annihilation and Coulomb interactions are inefficient in these jets.

- For a given n_e/n_p, the upper limit for the bulk-Compton radiation gives a minimum distance for jet formation. For strongly pair-dominated plasmas this distance is $0.1 - 1.0$ pc. However, jets with energy flux $P \geq 10^{46}$ erg s^{-1} must be powered very near the black hole, by its rotation [7] or by the innermost parts of an accretion disk [5]. These two inferences – the large distances of formation of pair-dominated jets and the very central source of the jet energy – can be reconciled if over the first 3 decades of distance, the jet is strongly dominated by the Poynting flux. This very wide distance range of conversion of the magnetic energy to the bulk kinetic energy can result from radiation drag [22]. This transition process can be accompanied by pair production in shocks and magnetic field reconnection sites [21,35].

ACKNOWLEDGMENTS

I wish to thank Mitch Begelman and Greg Madejski for valuable comments which helped improving the paper. This work has been supported in part by the NASA grant NAG5-4106 and by the Polish KBN grant 2P03D01209.

REFERENCES

1. Bednarek, W. *ApJ* **402**, L29 (1993)
2. Begelman, M. C., and Sikora, M., *ApJ* **322**, 650 (1987)
3. Blandford, R. D., and Königl, A., *ApJ* **232**, 34 (1979)
4. Blandford, R. D., & Levinson, A., *ApJ* **441**, 79 (1995)
5. Blandford, R. D., & Payne D. G. , *MNRAS* **199**, 883 (1982)
6. Blandford, R. D., & Rees, M. J., in *Pittsburgh Conference on BL Lac Objects*, ed. A. N. Wolfe (Pittsburgh University Press), 328 (1978)
7. Blandford, R. D., & Znajek, R. L., *MNRAS* **179**, 433 (1977)
8. Celotti, A. and Fabian, A. C., *MNRAS* **264**, 228 (1993)
9. Coppi, P., and Blandford, R., *MNRAS* **245**, 453 (1990)
10. Dar, A. and Laor, A., *Apj* **478**, L5 (1997)
11. Dermer, C. D., *ApJ* **446**, L63 (1995)
12. Dermer, C. D., and Schlickeiser, R., *ApJ* **416**, 458 (1993)
13. Dermer, C. D., Schlickeiser, R., and Mastichiadis, A., *A&A* **256** L27 (1992)
14. Dondi, L., and Ghisellini, G., *MNRAS* **273**, 583 (1995)
15. Falce, H., Malkan, M. A., and Biermann, P. L., *A&A* **298**, 375 (1995)
16. Ghisellini, G., and Madau, P., *MNRAS* **280**, 67 (1996)
17. Ghisellini, G., Maraschi, L., and Dondi, L., *A&AS* **120**, C503 (1996)
18. Ghisellini, G., Padovani, P., Celotti, A., and Maraschi, L., *ApJ* **407**, 65 (1993)
19. Königl, A., *ApJ* **243**, 700 (1981)
20. Kubo, H., PhD Thesis, University of Tokyo (1997)
21. Levinson, A., *MNRAS* **278**, 1018 (1996)
22. Li, Z., Begelman, M. C., and Chiueh, T., *ApJ* **384**, 567 (1982)

23. Madejski, G., et al., *ApJ* **459**, 156 (1996)
24. Mannheim, K., *A&A* **269**, 67 (1993)
25. Mannheim, K. and Biermann, P. L., *A&A* **253**, L21 (1992)
26. Maraschi, L., and Rovetti, F., *ApJ* **436**, 79 (1994)
27. Marscher, A., and Gear, W. , *ApJ* **198**, 114 (1985)
28. Mastichiadis, A. and Kirk, J. G., *A&A* **320**, 19 (1997)
29. Mattox, J. R., et al., *ApJ* **410**, 609 (1993)
30. Padovani, P., and Urry, C. M. , it ApJ **387**, 449 (1992)
31. Protheroe, P. and Biermann, P. L., *Astroparticle Physics* in press
32. Punch, M. et al. *Nature* **358**, 477 (1992)
33. Quinn, J. et al. *ApJ* **456**, L83 (1996)
34. Rawlings, S. and Saunders, R., *Nature* **349**, 138 (1991)
35. Romanova, M. M., and Lovelace, R. V. E., *ApJ* **475**, 97 (1997)
36. Sambruna, R., Maraschi, l., and Urry, C. M., *ApJ* **463**, 444 (1996)
37. Sikora, M., Begelman, M. C., and Rees, M. J., *ApJ* **421**, 153 (1994)
38. Sikora, M., Madejski, G., Moderski, R., and Poutanen, J., *ApJ* in press
39. Sikora, M., Sol, H., Begelman, M. C., and Madejski, G. M., *MNRAS* **280**, 781 (1996)
40. Svensson, R., *MNRAS* **227**, 403 (1987)
41. Takahashi, T., et al. *ApJ* **470**, L89 (1996)
42. Vermeulen, R. C., and Cohen, M. H., *ApJ* **430**, 467 (1994)
43. von Montigny, C., et al., *ApJ* **440**, 525 (1995)
44. Wehrle, A., et al. *ApJ* submitted

COMPTON INSTRUMENTS AND HISTORY

Overview of the Compton Observatory Instruments

J.D. Kurfess[1], D.L. Bertsch[2], G.J. Fishman[3], and V. Schönfelder[4]

[1] *Naval Research Laboratory*
Washington, DC 20375
[2] *NASA/Goddard Space Flight Center*
Greenbelt, MD 20771
[3] *NASA/Marshall Space Flight Center*
Huntsville, AL 35812
[4] *Max-Planck-Institut fur Extraterrestrische Physik*
D-85740 Garching, Germany

Abstract. The Compton Gamma Ray Observatory and its complement of four scientific instruments has completed over six years of outstanding on-orbit operation. In this paper we summarize the capabilities of the four CGRO experiments. The enhanced capabilities that have been achieved since the beginning of the mission are discussed. Recently, the spacecraft has been re-boosted to a higher orbit that could extend the mission lifetime for another ten years. This will enable continued monitoring of the hard X-ray sky by BATSE, continued mapping of the Galaxy across the gamma ray spectrum, the capability to observe rare events such as extragalactic supernovae, and solar flare observations during the next solar maximum by all instruments. The instrumental capabilities for an extended mission are discussed.

INTRODUCTION

The Compton Observatory has completed over six years of very successful operation. Launched on April 5, 1991, the mission has achieved many of its objectives, and provided several important and unanticipated discoveries. Major accomplishments include the BATSE determination of the isotropy of gamma ray bursts, EGRET's discovery of the blazar class of gamma ray sources, OSSE observations of softening spectra of Seyfert galaxies, BATSE monitoring of galactic X-ray sources and transients, COMPTEL maps of ^{26}Al in the Galaxy and first-ever all-sky maps of the MeV sky, OSSE discovery of an annihilation cloud above the central region of the Galaxy, and COMPTEL and OSSE observations of line gamma ray emission from SN1987A, Cas-A, and Orion.

The CGRO instruments have performed exceptionally well. Descriptions of the instruments and their performance characteristics have been published elsewhere (Johnson et al. 1993, Thompson et al. 1993, Schönfelder et al. 1993, Fishman et al. 1989). In this paper, we briefly describe the current status of the instruments and point out recent capabilities that have been developed that will augment the scientific output from an extended mission. The only instrument consumable is the spark chamber gas on EGRET. Operating on a reduced duty cycle and with a limited FOV has enabled the EGRET team to extend the instrument lifetime by several years.

The spacecraft also has performed exceptionally well. The failure of the tape recorders during the first year of the mission has resulted in the loss of some data. This has necessitated the reliance on real-time data through TDRSS. The loss of data has been minimized with the positioning of a TDRSS spare satellite and ground station to cover the $\sim$ 20% zone of exclusion between the Atlantic and Pacific satellites, and the use of on-board storage of selected data by OSSE and BATSE. The real-time acquisition of nearly all data has also been used to advantage with the development of the BACODINE system (Barthelmy et al. 1994) that enables fast response by ground-based observatories to the occurrence of gamma ray bursts.

Recently, the observatory has been re-boosted to an altitude of about 515 km. The mission lifetime from this orbit is likely to exceed ten additional years, and it appears likely the unique capabilities of CGRO will be available for many years to come.

BATSE

In its seventh year of continuous operation on CGRO, the Burst and Transient Source Experiment (BATSE) is operating extremely well, returning a wealth of scientific data in many diverse areas of astrophysics, solar physics and even geophysics. New capabilities have been developed by the BATSE operations and analysis team since launch. The flight software is virtually unchanged, although modifications have been made to store critical gamma-ray burst data for later recovery when realtime data transmissions are not available. The early loss of the CGRO tape recorders has been largely overcome through the use of extensive realtime telemetry coverage, by which over 80% of the data are returned. In addition, lower rate data from the BATSE large area detectors (LADs) supplement these data, primarily for use in Earth occultation analyses.

The primary scientific objective of BATSE continues to be the study of gamma-ray bursts (GRBs). This experiment has been the major source of spectral and temporal data over a wide energy range where most of the energy of gamma-ray bursts is emitted. The BATSE observations showing the isotropy and inhomogeneity of large numbers of GRBs have made local models of GRBs virtually untenable, based upon geometrical arguments alone. Galactic halo models are severely constrained and highly contrived in order to be compatible with the observed global distribution of nearly two thousand bursts.

For many years it has been thought that progress in gamma-ray burst research would come only after identification of counterpart objects. The BACODINE system (Barthelmy et al. 1996) was developed with this goal: to provide an automated, rapid notification of a GRB location to other observers, with a typical accuracy of about 8°. That system has been in operation since 1994, providing near real-time burst location messages to about thirty sites, worldwide. Although many observations have been made with this system, no counterparts had been found.

A major breakthrough in the field was made in 1997 by the team that operates the Italian-Dutch satellite, BeppoSAX. Launched in 1996, this satellite has the unique capability of obtaining rapid and precise gamma-ray burst locations. This has resulted in the identification of X-ray, optical, and radio afterglow observations in several gamma-ray bursts. These discoveries are having a major impact on the field and new models to account for the afterglow emission are already being developed.

Inspired by the BeppoSAX discoveries, a high-priority, effort was undertaken to improve the accuracy of the BATSE-derived locations in a rapid manner in order to enable other spacecraft and wide-field, ground-based facilities to observe GRB afterglows in other wavelength regions. This new rapid response system requires manual intervention and processing of BATSE data to achieve the best location accuracy. The BACODINE system is used to acquire these data and to distribute the derived locations, while personnel in Huntsville are on call to perform the required burst location processing. The typical location accuracy of this new system is about 2° in radius. This considerably improved location allows other spacecraft, such as RXTE and Beppo-SAX, a reasonable small sky region in which to scan for counterpart X-ray afterglow sources.

The other primary function of BATSE is to serve as an all-sky monitor for the Compton Observatory. In addition to this CGRO function, it has been serving the entire high-energy astrophysics community with an unprecedented, nearly-continuous monitor of the sky above 20 keV. Strong transients can usually be detected within one or two days of their occurrence by the Earth occultation technique. Since the typical risetime of most of these events has been at least a few days, the initial flare is usually detected well in advance of the peak flux of the transient event.

Daily occultation source flux reports are made and distributed within the BATSE occultation group. These, in turn, are combined into semi-weekly BATSE Bright Source reports which are distributed to all operational high-energy astronomy spacecraft centers. These data are also available on the Web. This service function will be useful for source target planning for future missions such as AXAF, XMM, Spectrum-X and INTEGRAL. In addition to these bright source reports, the INTEGRAL mission plans to use all-sky BATSE hard X-ray maps in several broad energy bands in their background model.

The all-sky, continuous observations with BATSE are particularly useful for those sources which are short (days-to-weeks) transients or have a low-duty cycle. Many of these sources would likely be missed by narrower-field, survey instruments. Figure 1 shows the active phases of most of the transient sources that have been continuously monitored by BATSE.

About 110 sources of hard X-ray and gamma-ray emission are continuously monitored and about ten to fifteen of these are detected daily. Bright, new sources may be found by detection of occultation steps which occur at times inconsistent with those of sources which are being monitored. Weaker sources at known locations may be monitored by multi-day summations of occultation data. Increasingly, BATSE hard X-ray data are being used in long-term and multi-wavelength observations of many high-energy objects.

Occultation transform images are also used extensively to find strong new sources and to resolve ambiguities in crowded galactic regions. A new capability, developed in collaboration with researchers at the Harvard-Smithsonian Center for Astrophysics (Barret et al. 1996) allows processing of these occultation images in a systematic way to search for occurrences of new and/or transient sources. Initially, this processing will concentrate on the Galactic plane, but it may evolve into an all-sky imaging and detection system.

As the next solar active period begins, BATSE will again provide hard X-ray solar flare data to the solar community for coordinated observations. Over five thousand impulsive solar hard X-ray flares were cataloged by BATSE in the last solar cycle.

COMPTEL

COMPTEL covers the middle gamma-ray energy range of the four instruments from about 1 to 30 MeV. It is the first gamma-ray telescope that has performed a complete survey of the sky at MeV energies. Before the launch of the Observatory, the MeV range was nearly unexplored – only very few celestial objects had been seen. COMPTEL has now flung open this new window to astronomy.

With its large field-of-view of about 1 steradian different sources within this field can be resolved if they are separated by more than about 3 to 5 degrees. Its energy resolution of 5% to 10% FWHM is an important feature for gamma-ray line investigations. COMPTEL has an unprecedented sensitivity: within a 2-week observation period, it can detect sources that are about ten times

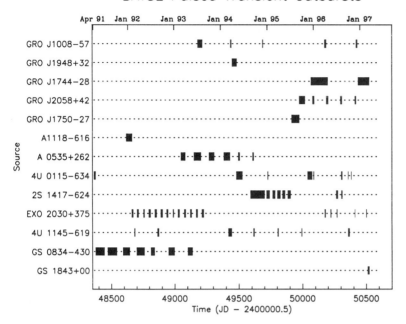
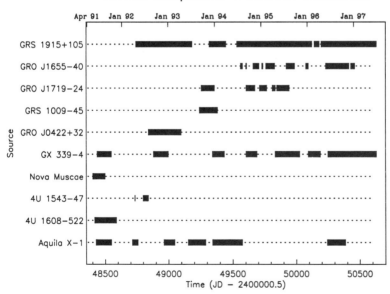

FIGURE 1. Timelines for BATSE observations of transient sources.

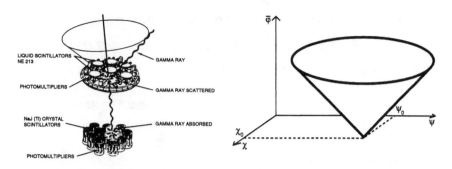

FIGURE 2. Principle of measurement of COMPTEL. Left side: schematic view of COMPTEL; right side: schematic diagram of the COMPTEL spatial response in the dataspace spanned by scatter direction angles (χ, ψ) and scatter angle $\bar{\varphi}$. The cone apex is the source direction. The blurring due to measurement imperfections widens the cone as a whole and the energy thresholds cut the tip of the cone at the bottom.

weaker than the Crab. A comprehensive description of the capabilities and characteristics of the instrument was provided by Schönfelder et al. (1993).

A schematic view of COMPTEL is shown in Figure 2 (left). A gamma-ray is detected by a Compton collision in the upper detector (consisting of 7 modules of liquid scintillator NE 213) and a subsequent interaction in a lower detector, consisting of 14 modules of NaI (Tl). The arrival direction of a detected gamma ray is known to lie on a circle on the sky. The center of each circle is the direction of the scattered gamma ray, and the radius of the circle is determined by the energy losses in E1 and E2 in both interactions. The detected photons are binned in a 3-dimensional (3-D) data space, consisting of the scatter direction (denoted by the arbitrary angular coordinates χ and ψ) and by the scatter angle $\bar{\varphi}$ (derived from the measured energy losses in both interactions; see Fig. 2 right). Each detected photon is represented by a single point in the 3-D data space. The signature of a point source with celestial coordinates (χ_o, ψ_o) is a cone of 90° opening angle with its axis parallel to the $\bar{\varphi}$-axis.

The apex of the cone is at (χ_o, ψ_o). The blurring due to measurement imperfections widens the cone patterns, and the energy thresholds truncate the cone at its apex. Imaging with COMPTEL involves recognizing the cone patterns in the 3-D dataspace.

In some ways, COMPTEL is similar to an optical camera: the upper detector, analogous to the camera's lens, directs the light into the second detector, comparable to the film, in which the photon is absorbed. The main difference is that the photons are not focused as in the case of the camera.

Two main techniques of imaging are applied: one is a maximum-entropy method that generates model independent images (Strong et al. 1992) and the other one is a maximum likelihood method that is used to determine the

TABLE 1. COMPTEL Event selections

ΔE$_\gamma$ [MeV]	E1 [keV]	E2 [keV]	ToF [channel]	PSD [channel]	Horizon angle [degree]	$\bar{\varphi}$-range [degree]
0.75 – 1	> 70	> 650	115 – 130	0 – 110	> 5	4 – 50
1 – 3	> 70	> 650	115 – 130	0 – 110	> 5	4 – 50
3 – 10	> 70	> 650	115 – 130	0 – 110	> 5	4 – 50
10 – 30 (standard)	> 70	> 650	115 – 130	0 – 110	> 5	4 – 36
10 – 30 (optimized*)	> 70	> 650	113 – 126	65 – 85	> 0	4 – 36

statistical significance, flux and position uncertainty of a source (de Boer et al. 1992).

The sensitivity of COMPTEL is significantly determined by the instrumental background. A substantial suppression is achieved by the combination of effective charged particle shield detectors, time-of-flight measurement techniques, pulse-shape discrimination, Earth horizon angle cuts and proper event selections in energy and $\bar{\varphi}$-space. Table 1 summarizes the presently used event selections.

Two typical background spectra are shown in Figure 3. The top spectrum was taken in phase-1, the bottom spectrum in phase-6. Both spectra show two prominent features: one is at 2.223 MeV (from thermal neutron capture in D1) and the other one at 1.465 MeV. In phase-1 this second feature is mainly caused by ^{40}K-decay in the D1 photomultiplier glasses. In VP 6 the build-up of radioactive ^{22}Na and ^{24}Na is so large that cascade events from their decay products cause an even more intense feature at practically the same energy. These cascade events are also responsible for the broader bumps in the spectrum between 2.5 and 4.0 MeV.

The application of the above mentioned imaging techniques require an accu-

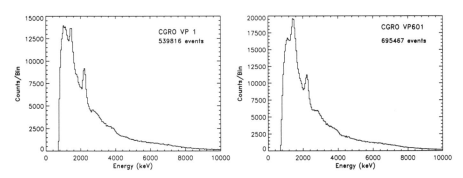

FIGURE 3. COMPTEL background spectra. Left: phase-1, Right: phase-6.

rate knowledge of the COMPTEL background (instrumental and cosmic). A variety of background models has been investigated and is being used. In one method, the background is derived from averaging high-latitude observations. This assumes that the background has a constant shape in the instrument coordinate system for all observations, and it also assumes that the extragalactic source contribution is small and smeared out by the averaging process. A second method derives the background from the cosmic source data that are being studied. This is accomplished by applying a low-pass filter to the 3-D data that smooths the photon distribution and eliminates (in the first approximation) the source signatures (e.g. Bloemen et al. 1994). For line studies, we estimate the background below an underlying cosmic gamma-ray line by averaging the count rate from neighboring energy intervals (Diehl et al. 1994).

In addition to the normal double scatter mode of operation, two of the NaI crystals in the lower D2-detector assembly of COMPTEL are also operated simultaneously as burst detectors. These two modules are used to measure the time history and energy spectra of cosmic gamma-ray bursts and solar flares. Hence, solar flares and cosmic gamma-ray bursts can be measured in the telescope mode (provided the event was within the field-of-view of the instrument), and in the burst mode. During the last few years, a rapid burst response program has been developed, which allows to locate bursts observed in the telescope mode within a few minutes confidence.

Table 2 summarizes the COMPTEL characteristics:

TABLE 2. COMPTEL Characteristics

Energy range [MeV]	0.8 to 30
Energy resolution [FWHM]	5 to 8 %
Effective area [cm^2]	25.8 at 1.27 MeV
	29.3 at 2.75 MeV
	29.4 at 4.43 MeV
Field of view [FWHM]	$\sim 64°$
Angular resolution [FWHM]	1.7° to 4.4°
Position localization at 20% of Crab intensity	0.5 to 1.0°°[D (90 % confidence)

Table 3 summarizes the actual achieved sensitivities for a standard 2-week observation time ($t_{eff} \sim 3.5 \cdot 10^5$ sec) in phase-1 and for the cases, when all data from a certain source in the Galactic Center or anticenter direction are added from either phase-1 to 3 ($\sim 2.6 \cdot 10^6$ sec) or phase-1 to 5 ($\sim 6 \cdot 10^6$ sec).

Six years after the launch of the Observatory, COMPTEL is operating still under very good conditions. Only few of the, in total, 252 photo-multipliers have failed. The high voltage of one of the 14 D2-modules has been switched off. All other modules remain operating.

TABLE 3. 3σ Sensitivity limits of COMPTEL

$E_\gamma [MeV]$	3σ Flux Limits 10^{-5} photons $[cm^{-2}sec^{-1}]$		
	$3.5 \cdot 10^5$ sec	$2.6 \cdot 10^6$ sec	$6.0 \cdot 10^6$ sec
0.75 – 1	20.1	7.4	3.7
1 – 3	16.8	5.5	3.8
3 – 10	7.3	2.8	1.7
10 – 30	2.8	1.0	0.8
1.156	6.2	2.0	1.6
1.809	6.6	2.2	1.6

It, therefore, can be expected that COMPTEL will continue its excellent performance for many years to come. This will give us the opportunity to increase the exposure of the all-sky survey and to reach the ultimate sensitivity that can be obtained with this instrument. In addition, we can expect valuable results from the next solar maximum cycle!

EGRET

The Energetic Gamma-Ray Experiment Telescope (EGRET) images gamma rays with energy from 30 MeV to over 30 GeV. Detection is achieved when the incident gamma ray pair-converts in one of the thin Ta foils that interleave the spark chamber planes and the secondary positron or electron satisfy the coincidence required to trigger the spark chamber. Two segmented scintillator planes form a coincidence and time-of-flight system that defines the aperture, and an anti-coincidence dome covers the upper portion of the instrument to eliminate triggers from incident charged particles. An energy calorimeter called TASC (Total Absorption Shower Counter) measures the total energy of the triggered event. The arrival time to a resolution of 7 μs is stored with the image and energy.

Ground processing of events identifies the two tracks of the secondary electron and positron. The tracks must issue from an inverted 'V' that starts within the spark chamber volume. Multiple scattering and the total energy measured in the TASC are used to estimate the energy of each secondary. That information is used to construct the arrival direction based on an energy-weighted bisector fit. At a later stage of processing when skymaps are constructed, events are accepted only if they are outside the Earth's limb and are within a specified cone about the pointing direction, typically of 30° half angle. A more detailed description of the instrument can be found in Kanbach et al. (1988) and Nolan et al. (1992), and in the calibration paper, Thompson et al. (1993).

EGRET's aperture is comprised of 9 sub-telescopes that point vertically and in 8 directions off-axis. These are dynamically enabled and disabled depending

TABLE 4. EGRET Instrument Characteristics

Energy Range:	20 MeV to over 30 GeV
Energy Resolution:	$\sim 20\%$, over central energy range
Total Area:	$6,400\,cm^2$
Point Source Sensitivity:	$6 \times 10^{-8}\,cm^{-2}s^{-1}$ for E> 100 MeV
Timing Accuracy:	$0.10\,ms$ absolute, $7\,\mu s$ relative
Source Position Location:	5 to 30 $arcmin$, depending on location and spectrum
Wide Field Mode:	
Effective Area[†]:	$\sim 1,500\,cm^2$ between 200 and 1,000 MeV,
Field of View	Approximately Gaussian, $\sim 20°$ HWHM
Point Source Sensitivity:	$6 \times 10^{-8}\,cm^{-2}s^{-1}$ for E> 100 MeV
Narrow Field Mode:	
Effective Area[†]:	$\sim 1,200\,cm^2$ between 200 and 1,000 MeV,
Field of View	Approximately Gaussian, $\sim 12°$ HWHM
Point Source Sensitivity:	$8 \times 10^{-8}\,cm^{-2}s^{-1}$ for E> 100 MeV

†See Figure 1 for more detail on the energy and angle dependence, and for the wide and narrow angle modes of operation.

on the Earth occultation. The nominal, wide field mode was used during the first 4.5 years of the mission when survey and diffuse studies were a high priority. Since the start of Phase 5, the narrow angle mode has been used almost exclusively to conserve spark chamber gas. Table 4 summarizes the characteristics for these two modes of EGRET. The effective area depends on energy and arrival angle with respect to the instrument axis. Figure 4 shows this variation.

The high voltage discharge of the spark chamber slowly deteriorates the neon/ethane spark chamber gas. EGRET has a gas exchange system that provided five refills in orbit, based on a two year mission. Each gas fill was extended as long as practical and this resulted in the need to apply correction factors to the calibrated effective area (shown in Fig. 4). These factors are determined by a lengthy process of comparing overlapping fields of view, and by occasional views of pulsars. The GRO Science Support Center has a matrix of these corrections based on time and energy. The last available fill was done on 09/08/95. A small reserve of about 40% of a fill remains, and at a future date, a partial exchange will be done. By operating in the narrow field mode in about 50% of the viewing periods, about 6.5 million triggers have occurred in the 21 months since the last fill. The previous fills tallied $\sim$ 15 million triggers. Consequently, there is reason to believe that the EGRET spark chamber will be able to function throughout Phase 7 at this same level, and that some capability might still remain for interesting targets of opportunity beyond that.

The spark chamber performance has also degraded apart from the gas qual-

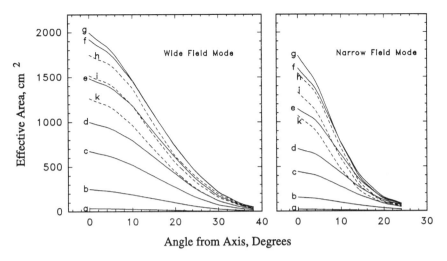

FIGURE 4. Comparison of the effective area of EGRET in the wide and narrow field-of-view modes. The solid curves are for the energy bands (in MeV) of a: 30-50, b: 50-70, c: 70-100, d: 100-150, e: 150-300, f: 300-500, g: 500-1000. The dashed curves similarly are for the high energies where the area decreases, h: 1000-2000, i: 2000-4000, and k: 4000-10000.

ity issue due to the aging of the high voltage pulsers and perhaps from contamination buildup. The scaling factors mentioned above compensate for this change in performance. One of the coincidence photo-tubes failed. This resulted in the loss of a few percent in sensitivity that was corrected by a change in calibration tables. Otherwise, all of EGRET's subsystems are operating well.

Following the discovery that bursts sometimes extend into the EGRET energy regime – in fact to 10 GeV (Hurley et al. 1994) a new mode of operation was incorporated when EGRET is operating in narrow field mode, or is disabled entirely. When the BATSE on-board processor determines that the direction is within about 50° of EGRET's axis and exceeds a threshold, BATSE automatically signals the GRO processor that in turn activates a stored sequence to configure EGRET to a wide field of view for 3 hours and then return to its pre-burst state. These events are infrequent, occurring about twice per month. To date, no burst has been observed in the spark chamber as a result of this operation. It might be particularly useful for solar activity as solar maximum approaches.

The EGRET calorimeter (TASC) provides the energy measurements for events that trigger the spark chamber. However, it can operate independently as an omni-directional detector. As such, it will continue to be useful beyond the lifetime of the spark chamber. This system is designed to study bursts and transient solar events in the energy range from 1 to 200 MeV. The time resolution is normally 32.6 s. When a burst trigger from BATSE is received, a collection four spectra with typical accumulation times of 1, 2, 4, and 16 s are taken. The spectrum mode of the TASC is independent of the energy measurements that are made in response to a spark chamber trigger. The times when triggered processing are occurring are accumulated as dead time in the burst and solar spectrum accumulations. The TASC is not shielded, and has a significant background that has to be removed. When the arrival direction is known, say from a burst or from the Sun, corrections to the spectra due to surrounding spacecraft material must also be made.

The TASC analysis proved to be very important in studying the solar flares that occurred in June 1991. Spectra were obtained during ~15-min time periods following each of 4 flares. Line emission was seen in all 4 events, and time profiles of line emission from neutron capture (2.2 MeV) and the carbon-oxygen complex (4.4 MeV) were determined to a limiting sensitivity of $5 \times 10^{-4} \, cm^{-2} s^{-1}$ (Schncid et al., 1996). The TASC capabilities remain, and it will be an important tool in the time of the next solar maximum.

OSSE

The Oriented Scintillation Spectrometer Experiment (OSSE) covers the 50 keV-10 MeV energy range with four nearly identical actively-shielded scintillation detectors (Johnson et al. 1993). This energy region covers the highest temperature thermal emissions associated with hard X-ray emission from compact objects, non-thermal emissions from a variety of galactic and extragalactic sources, and is also the region where most lines are observed. Each OSSE detector has a 3.8° x 11.4° field-of-view (FWHM) defined by a passive tungsten collimator. Observations are undertaken by alternately viewing source and background regions by offset pointing the detectors on a 2-min time scale to just remove the source in the background FOV's (typical offsets are ± 4.5°). For each 2-min source spectrum, the background is estimated by fitting a quadratic function to 3 or 4 background observations within 6 minutes of the source observations. This has proven to be a reliable approach for estimating the background in the intense and time variable radiation environment. Any associated systematic effects are reduced to levels that are not significant at the limiting OSSE sensitivity for 10^6 sec observations. For non-standard observations (e.g. galactic plane observations which might require large background offsets, mapping observations, and one-sided backgrounds) a small scan-angle dependent background is observed that may require consideration (Kurfess et al. 1997).

Table 5 lists the key instrumental characteristics of the OSSE instrument.

The energy calibration and gain are maintained with the use of an onboard gain control system which maintains the gain of each phoswich PMT

TABLE 5. OSSE Instrument Characteristics

Detector Type:	4 NaI(Tl)-CsI(Na) phoswiches with active NaI shields, passive W collimators.
Energy Range:	50 keV to 10 MeV
	10-200 MeV solar gamma-ray and neutron mode
Energy Resolution:	$\sim$ 8.5% at 661 keV
	$\sim$ 4.5% at 4.4 MeV
Effective Area:	$\sim 2000\, cm^2$ at 0.5 MeV
	$\sim 600\, cm^2$ at 5 MeV
Timing Resolution:	16.38 sec in normal spectral mode
	0.125 ms event by event mode
	4 ms in burst mode
Field-of-View:	3.8° × 11.4° FWHM (0.05-10.0 MeV)
Narrow Line Sensitivity:	$5 \times 10^{-5} cm^{-2} s^{-1}$ at 0.5 MeV
(10^6 s observation)	$8 \times 10^{-5} cm^{-2} s^{-1}$ at 4.4 MeV
Continuum Sensitivity:	$10^{-3} cm^{-2} s^{-1} MeV^{-1}$ for 50-150 keV
(10^6 s observation)	$5 \times 10^{-5} cm^{-2} s^{-1} MeV^{-1}$ for 2-10 MeV

to within ±0.1 %. The continuum and line sensitivities are determined by the background environment, which is largely due to activation of the large OSSE detectors and the surrounding spacecraft. The recent re-boost from 430 km to 515 km altitude has increased the proton dose acquired each day during passages through the South Atlantic Anomaly (SAA) by about a factor of 4. Although the daily summed background spectra has increased only about 50% below 5 MeV, the substantially higher background during SAA orbits may increase the systematic effects for selected observations during these orbits. This is shown in Figure 5 which shows one-day rate histories in the 300-650 keV energy region for three altitudes: 350 km, 430 km, and 510 km. The dramatic increase in the background during SAA orbits is evident. In the worst case scenario, the user may wish to not include these orbits in the data analysis, which will result in about a factor of 1.4 loss of sensitivity during this period immediately following the re-boost. With the onset of the maximum of solar cycle 23 within the next year, the radiation dose acquired during SAA orbits will decrease. The dosage will further decrease as the orbital altitude of CGRO decays. Assuming the CGRO remains operational well beyond the next solar maximum, the OSSE sensitivity should return to pre-reboost levels for most of the remainder of the mission.

Several new capabilities have been added over the past few years. Following the failure of the S/C tape recorders, an on-board storage mode was implemented which enables the key 2-minute detector spectra to be accumulated during periods of TDRSS outage. This has resulted in minimal loss of primary OSSE data due to the tape recorder failures. OSSE also utilizes a burst

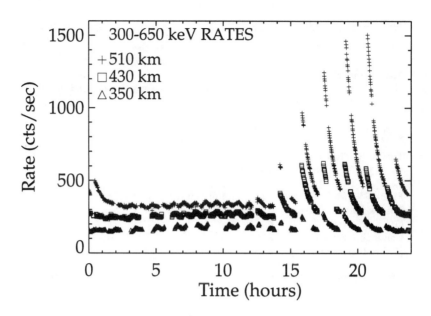

FIGURE 5. One-day rate history in the 300-650 keV energy region for three different altitudes. The 430 km and 510 km data were obtained just prior to and after the recent reboost.

trigger signal from BATSE to acquire shield rates with 16-msec time resolution following each BATSE burst. Recently two new capabilities have been added. Working with the BATSE team, the approximate scan angle location of a burst is encoded in the length of the burst signal, permitting OSSE to slew to the approximate burst location within about 60 seconds. (Matz et al. 1995). Although few bursts lie along the OSSE scan plane, this capability should enable OSSE to observe the fading hard X-ray emission of several strong bursts during the remainder of the mission. Also, a new capability to acquire shield spectra with high time resolution, rather that just rate data in a broad energy band, has been recently implemented.

OSSE and CGRO provide unique capabilities for observing transient phenomena, such as supernovae, novae, outbursts in both galactic and extragalactic sources, and solar flare activity. In addition, OSSE provides a unique capability to map extended line and continuum emission from the Galaxy, as evidenced by the recent discovery of a positron annihilation cloud above the center of the Galaxy (Purcell et al. 1997). These capabilities will provide opportunities for major new discoveries with the extension of the CGRO mission!

CONCLUSIONS

CGRO continues to perform very well, both operationally and scientifically, more that six years after launch. Both the spacecraft and the four instruments are operating near full capability. Following the successful reboost, the observatory is postured to continue operations for many years into the next millennium. In addition, several new capabilities have been added to the instruments which were not available early in the mission. The observatory provides unique scientific capabilities, energy band covered, sensitivity, operational flexibility, and full sky monitoring, that will not be duplicated in the foreseeable future. In view of the dynamic nature of the gamma ray sky, and CGRO's capability to respond to new scientific opportunities, it is imperative that this billion dollar resource be maintained well into the next century.

REFERENCES

1. Barret, D., et al. 1997, *these proceedings.*
2. Barthelmy, S.D., et al. 1996 *AIP Conf. Proc.* **384** 580.
3. Bloemen, H., et al. 1994 *Astrophys. J. Supp.* **92**, 419.
4. de Boer, H., et al., 1992, in "Data Analysis in Astronomy IV", Plenum, N.Y., p. 241.
5. Diehl, R., et al. 1994 *Astrophys. J. Supp.* **92**, 429.
6. Fishman, G.J., et al. 1989 *Proc. GRO Science Workshop*, 2-39.
7. Hurley, K., et al. 1994, *Nature* **372**, 652.
8. Johnson, W.N., et al., 1993 *Astrophys. J. Supp.* **86** 693.
9. Kanbach, G., et al. 1988, *Space Science Rev.* **46**, 69.
10. Kurfess, J.D., et al., 1997, *these proceedings.*
11. Matz, S.M., et al., *AIP Conf. Proc.* **384** 846.
12. Nolan, P., et al. 1992, *IEEE Trans. Nucl. Sci.* **39**, 993
13. Purcell. W.R., et al., 1997 accepted for publication in *Astrophys. J.*
14. Schneid, E.J., et al. 1996, *Astron. Astrophys. Supp.* **120**, 4, 299
15. Schönfelder, V., et al., 1993 *Astrophys. J. Supp.* **86** 657
16. Strong, A.W., et al., 1992, in "Data Analysis in Astronomy IV", Plenum, N.Y., p. 55.
17. Thompson, D.J., et al., 1993, *Astrophys. J. Supp.* **86**, 629

The COMPTON Observatory: Reflections on its Origins and History

D.A. Kniffen[1] and N. Gehrels[2]

[1] Hampden-Sydney College, Hampden-Sydney, Virginia 23943
[2] NASA Goddard Space Flight Center, Greenbelt, Maryland 20771

Abstract. The Gamma Ray Observatory (GRO) was planned as a mission to make the first comprehensive study of the gamma-ray sky over the full gamma-ray spectrum. It followed on the heels of two successful small astronomy satellites with high energy gamma-ray telescopes, SAS-2 and COS-B, and the High Energy Astronomical Observatory (HEAO) with low-energy gamma-ray detectors. The guideline was to cover the spectrum from the high energy limit of the x-ray domain to the highest energies for which a detectable flux might be reasonably observed. The objective was to include transients as well as persistent sources. The efforts to get such a mission approved stands as a model for developing a consensus of the scientific community for a mission involving a relatively small community of observers. A broad and vigorous guest investigator program was a key element in this effort. The results of the guest investigator program are testimony to the wisdom of this plan. The great success of GRO (now the *Compton* Observatory) was a result of the dedicated efforts of a few individuals and a project management team that understood how to work with scientists. In spite of increasing budgetary pressure and long delays beyond their control, the project management brought GRO to its successful launch with a minimum of cost growth resulting from the delays. This development period set the stage for what has been one of the most successful scientific programs in NASA's history. The very exciting results presented at this conference by a very broad community of scientists is a testimony to the dedicated efforts of a small group of visionaries with the foresight to see the potential of such a mission.

INTRODUCTION

Beginning in the late 1950s, space science began to thrive until it reached a crescendo in the early 1970s. Following the prolific and highly successful Explorer series, and early deep space planetary probes, came the High Energy Astronomy Observatory series. Frank McDonald of NASA's Goddard Space Flight Center had recognized in the late 1960s that the time and the technology were ripe for a set of major space observatories to attack the entire spectrum of unsolved problems in astrophysics. It was clear that the path to such a grand scheme would be long and difficult. Gathering support from key administrators at NASA headquarters and the endorsement of NASA's major scientific advisory committees, McDonald sought a broad consensus of support. With this strategy he was successful in convincing NASA to introduce a set of missions, the High Energy Astronomy Observatories (HEAO), into its budget. In early 1970 a call was issued for proposals for x-ray, cosmic-ray and gamma-ray experiments for these missions. On the basis of the proposals a complement of instruments was selected for four missions. The first two were to contain a mix of x-ray, gamma-ray and cosmic-ray experiments. The third was to emphasize focusing x-ray telescopes and the fourth a complement of cosmic-ray experiments. This was destined to usher in the long sought second golden age in man's attempts to penetrate the world beyond his local environment.

HEAO was planned to be a major program attacking a great spectrum of open questions. In 1971 Congress approved funding and the work began. This was one of the greatest pure science efforts ever launched by such a diverse international community in the history of mankind. The momentum was all in the forward direction. This is why scientists everywhere were in a state of shock when, on one devastating day, dubbed the Friday Night Massacre, January 5, 1973, NASA headquarters suspended the HEAO program. In retrospect, this began a new era, or perhaps ended an old one. The days of unconstrained budgets were over. The suspension of the program was purely and simply a result of funding limitations, with no regard for the considerable amount of effort that had already been expended. This scenario has been oft repeated; the Superconducting Super Collider (SSC) is another case in point. In the case of HEAO, constrained budgets accompanied by large cost overruns on the Viking Mars Lander program could only be remedied by an infusion of funds from other programs. HEAO was the only program with enough resources to make up the difference. Politically, HEAO was expendable since it did not have the public appeal of the Mars Lander.

In an earlier day, NASA would have gone to Capital Hill and received the additional funding required for the Viking program. Only heroic efforts by interested NASA managers Drs. Alois Schardt and Albert Opp and Richard Halpern prevented the program from being cancelled outright. Progress in astrophysics was severely set back, but not destroyed. HEAO would survive with a much reduced scope. Most importantly the momentum in space astrophysics would continue its march forward, but at a slower pace. A year later, the HEAO scientists were informed of the details of a scaled down version of the program. Reduced to three satellites, the payload capacity was insufficient to accommodate the larger and heavier high-energy gamma-ray and cosmic-ray instruments. The gamma-ray instruments that did survive included a high-energy x-ray, low energy gamma-ray experiment with Laurence Peterson of the University of California at San Diego and Walter Lewin of MIT as Principal Investigators and a new type of instrument to measure energies with the precision necessary to study the gamma-ray lines. Alan Jacobson of JPL was the Principal Investigator on this cooled germanium detector.

THE SELLING OF THE PROGRAM

The importance of these earlier missions is not only in the scientific observations they achieved, but most importantly that they showed the promise of gamma-ray astronomy. NASA's second Small Astronomy Satellite (SAS-2) had confirmed the earlier results obtained both with balloon-borne detectors and with a small gamma-ray detector flown aboard the third Orbiting Solar Observatory (OSO-3). With remarkab le longetivity, the European COS-B satellite had explored the galactic diffuse gamma-ray emission in detail and established the first catalog of 25 high-energy gamma-ray sources. However the sensitivities of these early missions were still marginal and the very important medium-energy region was left unexplored. There was a great need for another opportunity for the observations in the gamma-ray window to the cosmos. Judging that a dedicated explorer would be the best path, Carl Fichtel, the Principal Investigator of the high-energy gamma-ray experiment that was removed from the HEAO program because it exceeded the weight and size constraints of the de-scoped mission, campaigned for a high energy gamma-ray explorer. The major advantage of an Explorer is that it would not require Congressional approval, a major hurdle. The problem is that there was an inadequate constituency to promote in Congress a mission devoted to just a portion of the gamma-ray band and it was problematical that such an experiment could be accomplished within the constrained resources of an Explorer. Like the original HEAO program, this Gamma Ray Explorer (GRE) was also a budget victim.

Recognizing the considerable scientific advantage of broader spectral coverage, McDonald and Fichtel began to study a much broader mission concept. Congressional approval would be needed for this more ambitious initiative. The arguments for a major thrust in this relatively unexplored portion of the electromagnetic spectrum were compelling. History has shown this to be an accurate assessment. The middle part of the decade of the 1970s was spent studying the possible configurations for a comprehensive mission covering the range from about 100 thousand to about 30 million electron volts of energy. What an ambitious objective! No single instrument could possibly cover this range. In fact at least three, and probably more would be required, each optimized to take advantage of one of the major interactions which could be used to detect and identify gamma rays. To probe this vast energy range with unprecedented sensitivity – at least a factor of ten better than any previous experiment – would require a major observatory. McDonald and Fichtel developed a strategy to pursue Congressional approval for such an undertaking.

The first step was to develop a scientific rationale and a spacecraft concept that could provide the necessary platform on orbit. At the same time, a study team was organized at the Goddard Space Flight Center with McDonald as the Study Scientist and Jeremiah Madden as the Study Manager. This team had worked together very successfully on several Explorer missions. A concept was put together for a spacecraft with five instruments which would be launched aboard the Shuttle. Gamma-ray astronomy does not push the technological limits of a spacecraft, but in this case it did press the envelope on size, data rate, weight and power. The pointing accuracy and platform alignment requirements were modest.

To advance the planning, NASA headquarters released an Announcement of Opportunity for instruments for the definition study of such a mission on November 15, 1977. The announcement indicated a planned 1983 launch. In August 1978, five instruments were selected for a study phase with the intention, in anticipation of Congressional approval, of proceeding into development in fiscal year 1980. The selected experiments were as follows:

BATSE: Burst and Transient Source Event experiment, with Gerald Fishman of NASA's Marshall Space Flight Center (MSFC) as the Principal Investigator. Co-Investigators included scientists from MSFC and from the University of Alabama at Huntsville.

GROSS: Gamma Ray Oriented Scintillation Spectrometer, later renamed the Oriented Scintillation Spectrometer Experiment or OSSE, with James Kurfess of the Naval Research Laboratory (NRL) in Washington, D.C. as the Principal Investigator. Co-Investigators were from NRL, Northwestern University and Rice University. The latter group has since relocated to Clemson University.

GRSE: Gamma Ray Spectroscopy Experiment, with Laurence Peterson of the University of California in San Diego (UCSD) as Principal Investigator. His colleagues included Co-Investigators from UCSD and JPL, a NASA laboratory operated for NASA by the California Institute of Technology. Other Co-Investigators were located at NASA's Goddard Space Flight Center (GSFC), and at two French Institutions, the Center for Nuclear Studies in Saclay and the Center of Space Studies in Toulouse.

COMPTEL: Compton Telescope, with Volker Schönfelder of the Max Planck Institute for Extraterrestrial Physics (MPE) in Garching, Germany, as the Principal Investigator. His Co-Investigators included scientists from the University of Leiden, the Netherlands, from the European Space Agency (ESA/ESTEC) in Noordwijk, the Netherlands, and from the University of New Hampshire.

EGRET: Energetic Gamma Ray Experiment Telescope, with three Co-Principal Investigators: Carl Fichtel of GSFC, Robert Hofstadter of Stanford University, and Klaus Pinkau of MPE. Co-Investigators came from these institutions and from Grumman Aerospace Corporation, now Northrop-Grumman. This team included the lead author of this article, who was initially at GSFC and is now at Hampden-Sydney College in southern Virginia.

The technical capabilities of this complement of instruments would cover the broadest spectral range of any investigation ever dedicated to astronomy. Sources 10 to 20 times fainter than ever seen before in this wavelength band would now become visible. For stronger emitters, the sources would be positioned far better than ever before possible. This is particularly important since it allows them to be identified with counterparts seen at other wavelengths. This is the eventual goal, since it provides the most information about the dynamics of the source giving rise to the emission. Finally, the new spacecraft would measure the spectrum; that is, the amount of emission at each element of energy within the full range, with the greatest precision ever possible. With these improvements the predicted return of new astrophysical information was enormously exciting. But what could not be predicted was even more exciting.

This next step was to organize a team of experts in theory and observation and to develop a Science Plan. This document would set forth the scientific arguments for such a mission in terms which could be understood by the decision makers. Both the team of scientific experts and their product were impressive. No one could fail to catch the enthusiasm of the specialists who pondered what the high-energy gamma-ray sky might reveal. The cornerstone of the Science Plan was the scientific objectives of the mission, which included:

- The study of the dynamic evolutionary forces in compact objects such as neutron stars and black holes;

- A search for evidence of nucleosynthesis – the fundamental building process in nature – particularly in the environment of supernova;

- The study of γ-ray emitting objects whose nature is not yet understood;

- The exploration of our Galaxy in the γ-ray range (especially with regard to regions difficult to observe at other wavelengths), the origin and dynamic pressure effects of the cosmic rays, and structural features particularly related to high energy particles;

- The study of the nature of other galaxies in the energetic realm of γ-rays, especially radio galaxies, Seyfert galaxies, BL Lacertae objects and quasars;

- The study of cosmological effects through the detailed examination of the diffuse radiation and the search for primordial black hole emission; and

- The study of intense γ-ray bursts of many types whose origins remain a mystery.

As the observing teams were preparing their instrument concepts, the process of gaining the support for the mission went into full swing in late 1978. A group of scientists, which was enormously helpful in generating support, was assembled under the leadership of Professor Donald Clayton, then at Rice University. The list of those who contributed or advised included Professor Kenneth Brecher of Boston University, Dr. Herbert Friedman of the Naval Research Laboratory, Professor Marvin Leventhal of the University of Maryland (formerly from Bell Labs), Professor Richard Lingenfelter of the University of California in San Diego (formerly from UCLA), Dr. Reuven Ramaty, Dr. Floyd Stecker and Dr. David Thompson of NASA/GSFC, and Professor William Webber of the University of New Hampshire. Many others, including the Principal and Co-Investigators, were instrumental in contacting their Congressional representatives and appealing to NASA headquarters.

A major boost to the attempts to attain final approval for the mission resulted from these efforts. In 1979, NASA's highest level advisory committee, the National Academy of Sciences Space Science Board, recommended that GRO be given NASA's highest priority for a new space astronomy mission. Their report appeared as part of the Academy's report "A Strategy for Space Astronomy and Astrophysics in the 1980s." However, the Carter Administration's fight against inflation took precedence and the final confirmation was postponed. The instrument development phase continued, however, and so the impact of this decision to delay was minimized.

THE DEVELOPMENT OF GRO

The instrument definition study was concluded with the realization that all five instruments could not be accommodated within the constraints of the mission. These constraints included funding and spacecraft resources such as space, weight, power and telemetry rate, as well as real estate. Only a limited total size could be contained within the envelope of the Shuttle Bay.

Each instrument team was invited to present the results of its definition study. NASA convened a science review panel to prioritize the five investigations. BATSE, COMPTEL, EGRET and OSSE were selected for the development phase of the mission, but GRSE was not. In 1981, both the Astronomy Survey Committee and the Space Science Advisory Committee issued statements lamenting the programmatic decision, but confirming the high priority of GRO with the selected instruments. However, they issued a strong recommendation that every effort be given to recapturing some of the most important losses. In response to these concerns a spectroscopy monitor was added to supplement BATSE with an ability to measure the energies of gamma-rays coming from strong bursts. This recommendation was adopted with the addition of the BATSE Spectroscopy Monitor.

During the definition phase, as the instrument teams were developing the details of their design concepts, two mission contractors were studying spacecraft concepts to accommodate them. At about the same time that the final instrument selection was made, a mission contractor, TRW, was selected to begin the development of the Observatory. In 1981, the new Reagan administration adopted GRO as their candidate for a new NASA initiative. All seemed well, and the new start was approved by Congress. However, concerns in Congress over inflation and the growing budget deficit again caused the funding to be tightly constrained. Understanding that the spacecraft requirements presented by the scientific objectives were quite modest, the project office made the correct decision to continue to emphasize experiment development with the available funds in order to minimize schedule impacts. The technical challenges resided with the instruments. It was not so much an advanced state-of-the-art technology, but the sheer magnitude of the task that presented the challenge. This was far and away the largest and heaviest set of instruments ever placed on orbit. Managing all the pieces and making them play together was a daunting task.

Other elements of the mission were underway in parallel, under the guidance of the Science Working Team. This group, consisting of the Project Scientist as chairman, the NASA Headquarters Program Scientist, each of the Principal Investigators, and key members of the Project Office, played a role analogous to a Board of Directors. A Systems Team, consisting of scientists and engineers from the instrument teams, was assembled to provide technical oversight during the development of the instruments. An Operations Working Group provided similar advice for the development of flight operations procedures.

The schedule was moving smoothly for an August 1988 launch until the tragic day in January 1986 when the Space Shuttle Challenger with its crew of six and a New England school teacher met a fiery fate as one of the solid rocket boosters failed a little over a minute after the lift-off. This was an event of enormous tragedy. Most importantly of all, seven very valuable lives were lost. But, in a sense, NASA lost its innocence. Morale in the agency reached bottom and the public lost confidence in this once infallible Agency. The Shuttle program was suspended until it was clear that the problem which led to this disaster was understood and corrected. NASA had been committed to the Shuttle as its sole launch vehicle for large payloads. Thus, much of the space program was grounded for over two and a half years. So profound was the effect that NASA would no longer place all of its hopes in any one launch vehicle, and the "mixed fleet" concept was born.

As a result of this national tragedy, the launch of GRO was delayed for over two and a half years. The intervening years were busy ones with the construction of the spacecraft, the calibration of the instruments, the integration of the instruments onto the spacecraft, the functional tests of the integrated system and the development and testing of systems for data analysis and for operations. When the spacecraft was finally launched on April 5, 1991, it was prepared to take its place in history as one of the most flawless missions ever launched by NASA. At this writing, the spacecraft and instruments, designed for two years of orbital life, are still operating at near peak performance after six years on orbit.

The launch was truly a splendid sight for all of those who had worked so hard to bring this mission to fruition. The Space Shuttle Atlantis crew, with Commander Steve Nagel, Pilot Ken Cameron, and Mission Specialists Jerry Ross, Linda Godwin and Jay Apt, was magnificent. During the deployment of the spacecraft on the third day of flight, everything worked to perfection until GRO's main communication antenna failed to release from its cradle for deployment. The Project had recognized this as a potential single point failure and had planned for a contingency extravehicular activity (EVA) to correct any problems. Mission Specialists Jerry Ross and Jay Apt were partially suited for the possibility when the call came to proceed. Once outside in the Shuttle bay, they quickly released the boom which had snarled when its insulation caught on a projecting bolt head in the cradle. They finished cranking out the antenna by hand. Shortly thereafter, the spacecraft was released in top operating condition.

THE GUEST INVESTIGATOR PROGRAM

In the years when GRO was preparing for launch, there were changes occurring at NASA. Budget constraints were impacting the planning for new missions. The keystone of NASA orbital science, the Hubble Space Telescope (HST) was launched in 1990. This was the first of a major initiative to provide important new observing capabilities across the electromagnetic spectrum. Dubbed the *Great Observatories*, the four missions included HST, GRO, the Advanced X-ray Astrophysics Facility (AXAF) and the Space Infrared Telescope Facility (SIRTF). GRO had been designed as a two-year mission, believing that subsequent missions would follow in rapid succession. With tightening budget constraints, the decision was made to extend its operations to several years. Designed as a two-year PI class observatory, meaning the original scientific teams would have proprietary access to the data for two years after receipt of usable data, it now made more sense to adopt a program in which a broader science community would have early data access. This would increase the interest in the observatory, and most importantly, enhance the scientific return.

In December 1988 the first meeting of the GRO User's Committee was held with Professor Tom Prince of the California Institute of Technology its first chairman. Established to define the shape of the GRO Guest Investigator Program, this group represented the broad interests of the PI Teams and the potential Guest Observer community. It contained representatives from users interested in the gamma-ray observations directly, those interested in correlating the gamma-ray data with those at other wavelengths, theorists, and other interested parties. This group worked very hard to develop a program which protected the interests of the PI teams, but allowed for the meaningful participation of others. Because the data analysis systems were designed for a PI mode, a GRO Science Support Center was included in the plan to aid access to the data. The mission was divided into cycles with gradual evolution to more guest investigator involvement as the flight operations progressed until the present mode in which all data are made available annually to proposers from the entire scientific community. There no longer is any PI Team preference other than that inherent in their knowledge of the instrument and the data analysis systems.

The results of the Fourth Compton Science Symposium are testimony to the wisdom of the GI program established by this group, now known as the CGRO User's Committee. The chairmen of this committee, Professor Prince, Professor Jonathan Grindlay of Harvard, and Professor Kevin Hurley of the University of California at Berkeley have been enormously successful in designing, implementing and supporting the interests of the GI program, but more importantly the mission as a whole. It should also be pointed out that Dr. Alan Bunner, the NASA Headquarters Program Scientist for GRO, was instrumental to the success of the GI Program. In a tight budget climate, he insured that the funding necessary for the program and the CGRO Science Support Center was made available. The future of gamma-ray astronomy, still only scratching the surface of its potential, will benefit from this vision for years to come.

ORBITAL OPERATIONS

The operations following the EVA described above have been relatively routine for CGRO. The slow turn-on of the instruments to allow outgassing of the high voltage supplies in the multitude of photomultiplier tubes included several weeks of on-orbit verification of instrument operations. As pre-planned by the Science Working Team, the spacecraft was oriented toward the anticenter region of the Milky Way so that the instruments could make use of "standard candles" for calibration of their response functions. The standards were the relatively stable sources contained in the Crab nebula (The Crab nebula pulsar) and the high energy source Geminga, later also identified as a pulsar. On 16 May 1991, the first phase of the viewing program began with the objective of achieving a relatively uniform survey of the full sky with the two wide-field instruments in the medium-energy (COMPTEL) and high-energy (EGRET) regions. This phase lasted about fifteen months and provides a basis for assessing the general features of a static gamma-ray sky. As subsequent observations have shown, however, one of the major features of the gamma-ray sky is the variability in time of a large number of the observed sources. As a reflection of the major and most dynamic energy transfers in the cosmos, this is not a surprise, but argues for the urgency of more sensitive observations in the future to characterize the nature of such variations. This is a key to understanding the mechanisms involved.

The smoothness of the operations is in no small way the result of efforts of the Flight Operations Team (FOT), initially staffed by TRW, the mission contractor, with the able leadership of Mr. Billy Breshears. The Operations Working Group under Mr. Karl Schauer, of GSFC, had wisely made use of the extended prelaunch schedule time to perfect the ground system. The spacecraft, initially deployed at an altitude of 450 kilometers, has now been reboosted twice using its own hydrazine propulsion system, on occasions when atmospheric drag left the spacecraft at an altitude where the lifetime would be short and adequate pointing accuracy would be difficult to achieve. A recent reboost has placed the spacecraft in a 519 kilometer altitude orbit where its lifetime will carry it to the next solar minimum and well into the new millennium. The only significant instrument degradation is with EGRET, in which the gas supply for the spark chambers has now been largely depleted. To extend the performance, this instrument is now cycled on only for high priority targets, and then only in a narrow field mode. This reduces dramatically the number of spark chamber triggers, which is the main component in gas degradation. The amazing performance of EGRET and the other instruments is truly spectacular, given that they were designed for two years of operation.

The only major spacecraft failure has led to one of its greatest benefits. Early in operations, the two redundant tape recorders failed. These recorders were used to record data for two orbits at 32 kilobits per second, and then they were relayed to ground at 512 kilobits per second through the Tracking and Data Relay Satellite System (TDRSS). Failure of the recorders meant that the data must be recovered in "real time". This resulted in some loss of data as the TDRSS coverage around the globe was not complete. Subsequently, one of the TDRSS satellites was maneuvered over Australia where a ground relay station provided data from the "gap". This "near real time" data mode provided an opportunity for a rapid response delivery of BATSE data to allow counterpart searches for the enigmatic gamma-ray bursts. A system was developed and implemented by Dr. Scott Barthtelmy and his colleagues to provide such data within seconds of the detection of the events. Scientists throughout the world have benefitted from this by-product of the tape recorder failures.

At this point, we enter Cycle Seven of the CGRO orbital operations. Long ago declared a Mission Success, this marvelous spacecraft and its instruments have established a new standard for success by a NASA spacecraft. This has not been the result of the efforts of a few, but the combined work of a group of experienced space scientists, technicians and engineers who worked as a team and adopted this program as their own. There are troubling signs that the lessons of this success are not duly recognized and understood by the new NASA management.

I   RETROSPECTIVE

CGRO has been a remarkably successful mission. This Symposium has duly recorded the tremendous advances in our knowledge of the gamma-ray sky. The number of known gamma-ray sources has increased from tens to hundreds. Class studies are now possible for characterizing sources. However, the goal of identifying counterparts of unknown sources in the pre-GRO catalogs has been illusive. The data on time and spectral variability have been so close to the margin of detectability that it is clear that definitive revelations must await the next generation of more sensitive instruments. Prospects currently look promising for this to occur early in the new millennium. To bring this dream to reality will take an even more dedicated effort by the combined community of scientists who understand the potential significance of the results. These will be highly constrained times, with ever more difficult decisions on how to establish priorities. The results of this Symposium will be key in making the arguments for new missions.

PARTICIPANT LIST
THE FOURTH COMPTON SYMPOSIUM
WILLIAMSBURG, VIRGINIA APRIL 27-30, 1997

Aharonian, Felix
MPI f. Kernphysik
aharon@fel.mpi-hd.mpg.de

Aller, Hugh
University of Michigan
haller@umich.edu

Aller, Margo
University of Michigan
margo@astro.lsa.umich.edu

Arzoumanian, Zaven
Cornell University
arzouman@spacenet.tn.cornell.edu

Bandyopadhyay, Reba
Oxford University
rmb@astro.ox.ac.uk

Baring, Matthew
NASA/GSFC
baring@lheavx.gsfc.nasa.gov

Barnes, Sandy
USRA/NASA/GSFC
barnes@grossc.gsfc.nasa.gov

Barret, Didier
CESR, Toulouse
barret@cesr.cnes.fr

Barrett, Paul
USRA/NASA/GSFC
barrett@compass.gsfc.nasa.gov

Barthelmy, Scott
USRA/NASA/GSFC
scott@lheamail.gsfc.nasa.gov

Bazzano, Angela
IAS/CNR
angela@saturn.ies.fro.cnr.it

Bennett, Kevin
ESTEC/ESA
kbennett@astro.estec.esa.nl

Bertsch, David
NASA/GSFC
dlb@mozart.gsfc.nasa.gov

Bhattacharya, Dipen
Univ. of California, Riverside
dipen@tigre@ucr.edu

Bignami, Giovanni
ASI/Roma
bignami@asizou.asi.it

Blazquez, Raquel
Valencia University

Bloemen, Hans
SRON-Utrecht
H.Bloemen@sron.run.nl

Bloom, Steve
NASA/GSFC
bloom@egret.gsfc.nasa.gov

Boettcher, Markus
MPIfR, Bonn
mboett@mipfr-bonn.mpg.de

Bower, Geoffrey
University of California, Berkeley
gbower@astro.berkeley.edu

Brazier, Karen
University of Durham
karen.brazier@dur.ac.uk

Bridgman, William, T.
USRA/NASA/GSFC
bridgman@grossc.gsfc.nasa.gov

Briggs, Michael
University of Alabama, Huntsville
briggs@gibson.msfg.nasa.gov

Buchholz, James
California Baptist College
jbuchholz@earthlink.net

Bunner, Alan
NASA Headquarters
alan.bunner@hq.nasa.gov

Caravro, Patrizia
IFC/Milano
pat@irctr.mi.cnr.it

Carraminana, Alberto
INAOE, Mexico
alberto@inaoep.mx

Castro-Tirado, Alberto
LAEFF-INTA
ajct@laeff.esa.es

Catanese, Michael
Iowa State University
catanese@egret.sao.arizona.ed

Chakrabarty, Deepto
MIT
deepto@space.mit.edu

Chaty, Sylvain
CEA Saclay
chaty@discovery.saclay.cea.fr

Chen, Andrew
Columbia University
awc@astro.columbia.edu

Chen, Wan
NASA/GSFC
chen@rosserv.gsfc.nasa.gov

Chen, Xingming
Lick Observatory
chen@ucolick.org

Cheng, LingXiang
University of Maryland
lxcheng@astro.umd.edu

Chernenko, Anton
Space Research Institute
anton@cgrsmx.ili.rssi.ru

Chernyakova, Maria
Astro Space Center LPI
masha@sigma.iki.rssi.ru

Cheung, Cynthia
NASA/GSFC
cynthia.cheung@gsfc.nasa.gov

Chiang, James
Naval Research Laboratory
chiang@asse.nrl.navy.mil

Clayton, Donald
Clemson University
clayton@gamma.phys.clemson.edu

Cline, Thomas
NASA/GSFC
cline@lheavx.gsfc.nasa.gov

Collmar, Werner
Max-Planck Inst.
wec@mpe-garching.mpg.de

Connell, Paul Henry
Birmingham University
phc@stor.sr.bham.ac.uk

Coppi, Paolo
Yale University
coppi@blazar.astro.yale.edu

Corbel, Stephane
CEA Saclay
corbel@discovery.saclay.cea.fr

Cordes, James
NAIC/Cornell University
cordes@spacenet.tn.cornell.edu

Crider, Anthony
Rice University
acrider@spacsun.rice.edu

Cui, Wei
MIT
cui@space.mit.edu

Cusumano, Giancarlo
IFCAI CNR
cusumano@ifcai.pa.cnr.it

Dal Fiume, Daniele
TESRE/CNR Bologna
daniele@tesre.bo.cnr.it

Daugherty, Joseph
UNC Asheville
daughtery@cs.unca.edu

Davis, Stanley
NASA/GSFC
davis@grossc.gsfc.nasa.gov

Deal-Giblin, Kim
University of Alabama, Huntsville
deal@gibson.msfc.nasa.gov

Del Sordo, Stefano
IFCAI-CNR
delsordo@ifcai.pa.cnr.it

Dermer, Charles
Naval Research Laboratory
dermer@osse.nrl.navy.mil

Diallo, Nene
CEA Saclay
diallo@integral.saclay.cea.fr

Diehl, Roland
Max-Planck Inst.
rod@mpe-garching.mpg.de

Dieters, Stefan
UAH/MSFC

Digel, Seth
NASA/GSFC
digel@gsfc.nasa.gov

Dingus, Brenda
University of Utah
dingus@mail.physics.utah.edu

Dixon, David
Univ. of California, Riverside
dixon@agouti.ucr.edu

Dogiel, Vladimir
P.N. Lebedev Physical Inst.

Dorman, Lev
IZMIRAN, TECHNION, UNAM

Doron, Ted
Chemical Abstracts Service
tdoron@cas.org

Dubath, Pierre
ISDC
pierre.dubath@obs.unige.ch

Durouchoux, Philippe
Saclav-Service & Astrophysique
durouchouse@sapyzg.saclay.cea.fr

Edwards, Philip
ISAS
pge@vsop.isas.ac.jp

Eiker, **erry**, Stephen
Harvard-Smithsonian
seikenberry@cfa.harvard.edu

Esposito, Joseph
USRA/NASA/GSFC
jae@egret.gsfc.nasa.gov

Fargion, Daniele
University of Italy
Fargion@Roma1.infn.it

Fegan, David
Univ. College, Dublin
djfegan@ferdia.ucd.ie

Fender, Robert
University of Sussex
rpf@star.maps.susx.ac.uk

Feroci, Marco
IAS-CNR
feroci@saturn.ias.fra.cnr.it

Finger, Mark
USRA/NASA/MSFC
finger@batse.msfc.nasa.gov

Finley, John
Purdue University
finley@purdd.physics.purdue.edu

Fishman, Gerald
NASA/MSFC
fishman@ssl.msfc.nasa.gov

Focke, Warren
NASA/GSFC
Warren.B.Focke.1@gsfc.nasa.gov

Ford, Eric
Columbia University
eric@astro.columbia.edu

Frail, Dale
NRAO
dfrail@nrao.edu

Gail, Bill
Ball Aerospace Corp.
bgail@ball.com

Galvan, Edward
New Mexico State University
bmcnamar@nmsu.edu

Gehrels, Neil
NASA/GSFC
gehrels@gsfc.nasa.gov

Georgii, Robert
MPI Extra Terrestrische Physik
rog@mpe-garching.mpg.de

Gierlinski, Marek
Jagiellonian University
gier@camk.edu.pl

Godwin, Linda
NASA/JSC

Goldoni, Paolo
SAP/CEA-Saddy
paolo@discovery.saddy.ced.fr

Gorosabel, Javier
LAEFT-INTA
jgu@laexff.esa.es

Gottbrath, Chris
DePauw University

Gouveia Dal Pino, Elisabete
University of Sao Paulo
dalpino@jet.iagusp.usp.br

Graham, Bradley
Naval Research Laboratory
graham@osse.nrl.navy.mil

Grandi, Paola
IAS/CNR
grandi@alphasax2.ias.fra.cnr.it

Greiner, Jochen
AIP Potsdam
jgreiner@aip.de

Grenier, Isabelle
Centre d'etudes de Saclay
isabelle.grenier@cea.fr

Gros, Maurice
CEA Saclay
maurice.gros@saclay.cca.fr

Grove, J. Eric
Naval Research Lab
grove@osse.nrl.navy.mil

Gruber, Duane
Univ. of California, San Diego
dgruber@ucsd.edu

Gursky, Herbert
Naval Research Laboratory
gursky@ssd0.nrl.navy.mil

Gwinn, Carl
Univ. of California, Santa Barbara
cgwinn@physics.ucsb.edu

Hallum, Jeremy
Boston University
jhallum@bu-ast.bu.edu

Harding, Alice
NASA/GSFC
harding@twinkie.gsfc.nasa.gov

Harris, Michael J.
USRA/NRL
harris@osse.nrl.navy.mil

Harrison, Thomas
New Mexico State University
tharriso@nmsu.edu

Hartman, Bob
NASA/GSFC
rch@egret.gsfc.nasa.gov

Hartmann, Dieter
Clemson University
hartmann@grb.phys.clemson.edu

Heindl, William
Univ. of California, San Deigo
wheindl@mamacass.ucsd.edu

Helfer, Larry
University of Rochester
pany@tsepiot.pas.rochester.edu

Henri, Gilles
Observatoire de Grenoble
henri@obs.ujf-grenoble.fr

Hermsen, Willem
SRON Utrecht
W.Hermsen@sron.ruu.nl

Hernanz, Margarita
IEEC/CSIC
hernanz@ieec.fcr.es

Howard, William
USRA
whoward@usra.edu

Hua, Xin-Min
NASA/GSFC
hua@rosserv.gsfc.nasa.gov

Hui, Li
Los Alamos National Lab
hli@lanl.gov

Hunter, Stanley
NASA/GSFC
sdh@egret.gsfc.nasa.gov

Hurley, Kevin
Univ. of California, Berkeley
kherley@sunspot.ssl.berkeley.edu

Jaffe, Tess
HSTX/GSFC
jaffe@rosserv.gsfc.nasa.gov

Jauncey, David
Australia Telescope
djauncey@atnf.csiro.au

Jean, Pierre
CESR, Toulouse
jean@cesr.cnes.fr

Johnson, Neil
Naval Research Laboratory
johnson@osse.nrl.navy.mil

Jones, Brian
  Stanford University
  bbjones@egret7.stanford.edu
Jung, Gregory
  USRA/NRL
  jung@osse.nrl.navy.mil
Kaaret, Philip
  Columbia University
  kaaret@astro.columbia.edu
Kaluzienski, Lou
  NASA Headquarters

Kanbach, Gottfried
  Max-Planck Inst.
  gok@mpe-garching.mpg.de
Kappadath, S.
  University of New Hampshire
  s.cheenu.kappadath@unh.edu
Kaspi, Victoria
  MIT
  vicky@space.mit.edu
Kawai, Nobuyuki
  RIKEN Institute
  nkawai@postman.riken.go.jp
Kazanas, Demosthenes
  NASA/GSFC
  kazanas@lheavx.gsfc.nasa.gov
Kertzman, Mary
  DePauw University

Kifune, Tadashi
  University of Tokyo
  tkifune@icrr.u-toyko.ac.jp
King, Edward
  Australia Telescope
  eking@atnfcsiro.au
Kinzer, Robert
  Naval Research Laboratory
  kinzer@osse.nrl.navy.mil
Kirk, John
  Max Planck Institut Kernphysik
  kirk@boris.mpi-hd.de
Kotani, Taro
  RIKEN Institute
  kotani@riken.go.jp

Kouveliotou, Chryssa
  USRA/NASA/MSFC
  kouveliotou@batse.msfc.nasa.gov
Kozlovsky, Benzion
  NASA/GSFC
  bzk@pair.gsfc.nasa.gov
Krennrich, Frank
  Iowa State University
  frank@egret.sao.arizona.edu
Kretschmar, Peter
  Integral Science Data Centre
  peter.kretschmar@obs.unige.ch
Krimm, Hans
  Hampden-Sydney College
  hansk@pulsar.hsc.edu
Kroeger, Richard
  Naval Research Laboratory
  kroeger@osse.nrl.navy.mil
Kuiper, Lucien
  SRON Utrecht
  L.Kuiper@sron.ruu.nl
Kurczynski, Peter
  University of Maryland
  kraken@rosserv.gsfc.nasa.gov
Kurfess, James
  Naval Research Laboratory
  kurfess@osse.nrl.navy.mil
Kuulkers, Erik
  University of Oxford
  erik@astro.ox.ac.uk
Lahteenmaki, Anne
  Metsahovi Radio Res. Station
  alien@kurp.fi
Lamb, Richard
  Caltech
  lamb@srl.caltech.edu
Langston, Glen
  NRAO
  glangston@nrao.edu
Leeber, Dawn
  New Mexico State University
  dleeber@nmsu.edu
Leising, Mark
  Clemson University
  leising@nova.phys.clemson.edu

Lessard, Rodney
Purdue University
lessard@purdd.physics.purdue.edu

Leventhal, Marvin
University of Maryland
ml@astro.umd.edu

Lewin, Walter
MIT
lewin@space.mit.edu

Liang, Edison
Rice University
liang@spacsun.rice.edu

Lichti, Giselher
MPI Extra Terrestrische Physik
grl@mpe-garching.mpg.de

Lin, Ying-chi
Stanford University
lin@egret0.stanford.edu

Ling, James
Jet Propulsion Laboratory
jling@jplsp.jpl.nasa.gov

Lorenz, Eckart
MPI-Physics
ecl@hegrat.mppmu.mpg.de

Macomb, David
USRA/NASA/GSFC
macomb@cosmic.gsfc.nasa.gov

Madejski, Greg
NASA/GSFC

Magdziarz, Pawel
Jagiellonian University
pavel@camk.edu.pl

Mahoney, William
Jet Propulsion Laboratory
wam@heag4.jpl.nasa.gov

Majmudar, Deepa
Columbia University
dpm@astro.columbia.edu

Malzac, Julien
CESR (CNRS/UPS)
malzac@cest.cnes.fr

Mandrou, Pierre
CESR-Toulouse
mandrou@cesr.cnes.fr

Martin, Michael
University of Louisville
mmarting@zeno.physics.louisville.edu

Massaro, Enrico
Instituto Astronomico - Roma
massaro@astrm2rm.astro.it]

Matteson, Jim
Univ. of California, San Diego
jmatteson@ucsd.edu

Mattox, John
Boston University
mattox@bu.edu

Matz, Steven
Northwestern University
s-matz@nwu.edu

Mayer-Hasselwander, Hans
Max-Planck Inst.
hrm@mpe-garching.mpg.de

McBreen, Brian
University College Dublin
bmcbreen@ollamh.ucd.ie

McCollough, Michael
USRA/NASA/MSFC
mccollough@bowic.msfc.nasa.gov

McConnell, Mark
University of New Hampshire
mark.mcconnell@unh.edu

McLaughlin, Maura
Cornell University
mclaughl@spacenet.tn.cornell.edu

McNamara, Bernie
New Mexico State University
bmcnamar@nmsu.edu

Medina-Tanco, Gustavo
Royal Greenwich Observatory
gmt@ast.cam.ac.uk

Melia, Fulvio
University of Arizona
melia@physics.arizona.edu

Messina, Daniel
NRL/SFA, Inc.
messina@osse.nrl.navy.mil

Milne, Peter
Clemson University
pmilne@astro.phys.clemson.edu

Mirabel, Felix
CEA Saclay
mirabel@discovery.saclay.cea.fr

Mitra, Abhas
Nuclear Research Laboratory
nrl@magnum.barc.ernet.in

Miyaji, Shigeki
Chiba University
miyaji@c.chiba-u.ac.jp

Mori, Masaki
Miyagi University of Education
m-mori3@ipc.miyakyo-u.ac.jp

Morris, Daniel
University of New Hampshire
dmorris@comptel.sr.unh.edu

Moscoso, Michael
University of Texas
mdm@astro.as.utexas.edu

Moskalenko, Igor
MPE Garching
imos@mpe-garching.mpg.de

Mukherjee, Reshmi
USRA/McGill Univ.
muk@hep.physics.mcgill.ca

Muslimov, Alex
NASA/GSFC
muslimov@lhea1.gsfc.nasa.gov

Nagel, Steve
NASA/JSC

Nandikotkur, Girikhar
Iowa State University
giridhar@iastate.edu

Narayan, Ramesh
Harvard-SAO
rnarayan@cfa.harvard.edu

Naya, Juan
USRA/NASA/GSFC
naya@fgrs2.gsfc.nasa.gov

Nayakshin, Sergei
University of Arizona
serg@physics.arizona.edu

Norris, Jay
NASA/GSFC
norris@grossc.gsfc.nasa.gov

Oberlack, Uwe
Max-Planck Inst.
ugo@mpe-garching.mpg.de

Obrebski, Tina
Naval Research Laboratory
tina@osse.nrl.navy.mil

Olive, Jean-Francois
CESR, Toulouse
olive@sigma-o.cesr.cnes.fr

Orlandini, Mauro
TESRE/CNR Bologna
orlandini@tesre.bo.cnr.it

Paciesas, William
UAH/MSFC
william.paciesas@msfc.nasa.gov

Palmer, David
USRA/NASA/GSFC
palmer@lheamail.gsfc.nasa.gov

Palumbo, Giorgio
Bologna Uniiversity
gqcpalumbo@astbo3.bo.astro.it

Parizot, Etienne
CEA-Seclay
parizot@cea.fr

Parsons, Ann
NASA/GSFC
parsons@lheamail.gsfc.nasa.gov

Peaper, David
DePauw University
dpeaper@depauw.edu

Pentecost, Elizabeth
USRA
lpenteco@usra.edu

Petrucci, Pierre-Olivier
Observatoire de Grenoble
petrucci@obs.ujf-grenoble

Phlips, Bernard
USRA/NRL
phlips@osse.nrl.navy.mil

Pian, Elena
STSci
pian@stsci.edu

Piro, Luigi
IAS/CNR
piro@alpha.ias.fra.cnr.it

Pitts, Karl
University of Louisville
kpitts@zeno.physics.louisville.edu

Pohl, Martin
Max-Planck Inst.
mkp@mpe-garching.mpg.de

Poutanen, Juri
Uppsala Astron. Observ.
juri@astro.uu.se

Prince, Thomas
Caltech
prince@srl.caltech.edu

Purcell, William
Northwestern University
w-purcell@nwu.edu

Ramaty, Reuven
NASA/GSFC
ramaty@pair.gsfc.nasa.gov

Ray, Alak
NASA/GSFC
ark@olegacy.gsfc.nasa.gov

Reimer, Olaf
Max-Planck Inst.
olr@mpe-garching.mpg.de

Rephaeli, Yoel
Stanford University
yoelr@memsch.stanford.edu

Reynolds, John
ATNF/CSIRO
jreynold@atnf.csiro.au

Robinson, Craig
USRA/NASA/MSFC
robinson@ssl.msfc.nasa.gov

Rothschild, Richard
Univ. of California, San Diego
rothschild@ucsd.edu

Rubin, Bradley
RIKEN Institute
rubin@crab.riken.go.jp

Ryan, James
University of New Hampshire
jryan@unh.edu

Ryde, Felix
Stockholm Observatory
felix@astro.su.se

Said, Slassi-Sennou
Univ. of California, Berkeley
slassi@ssi.berkeley.edu

Saito, Yoshitaka
RIKEN Institute
saito@crab.riken.go.jp

Salamon, Michael
University of Utah
salamon@cosmic.physics.utah.edu

Santangelo, Andrea
IFCAI-CNR
andrea@ifcai.pa.cnr.it

Schlickeiser, Reinhard
MPI Radioastronomie
rschlickeiser@rmpifr.bonn.mpg.de

Schonfelder, Volker
Max-Planck Inst.
vos@mpe-garching.mpg.de

Schubnell, Michael
University of Michigan
schubnel@umich.edu

Scott, Matthew
USRA/NASA/MSFC
scott@gibson.msfc.nasa.gov

Seifert, Helmut
USRA/NASA/GSFC
seifert@lheamail.gsfc.nasa.gov

Sembroski, Glenn
Purdue University
sembroski@purdd.physics.purdue.edu

Shapiro, Maurice
University of Maryland

Share, Gerald
Naval Research Laboratory
share@osse.nrl.navy.mil

Shrader, Chris
USRA/NASA/GSFC
shrader@grossc.gsfc.nasa.gov

Sikora, Marek
N. Copernicus Astron. Center
sikora@camk.edu.pl

Silva, Allison
New Mexico State University
bmcnamar@nmsu.edu

Skibo, Jeff
Naval Research Laboratory
skibo@osse.nrl.navy.mil

Skinner, Gerry
University of Birmingham
gks@star.sr.bham.ac.uk

Smith, David
Univ. of California, Berkeley
dsmith@ssl.berkeley.edu

Smith, Ian
Rice University
ian@spacsun.rice.edu

Sreekumar, P.
USRA/NASA/GSFC
sreekumar@gsfc.nasa.gov

Srinivasan, Radhika
Purdue University
srinivasan@purdd.physics.purdue.edu

Stacy, Greg
University of New Hampshire
greg.stacy@unh.edu

Starrfield, S.
Arizona State University
sumner.starrfield@asu.edu

Staubert, Ruediger
University of Tuebingen
staubert@astro.uni-tuebingen.de

Stecker, Floyd
NASA/GSFC
stecker@lheavx.gsfc.nasa.gov

Steinle, Helmut
MPI f. Extrat. Physik
hcs@mpe-garching.mpg.de

Stollberg, Mark
UAH/MSFC
stollberg@gibson.msfc.nasa.gov

Strickman, Mark
Naval Research Laboratory
strickman@osse.nrl.navy.mil

Strohmayer, Tod
USRA/NASA/GSFC
stroh@pcasrv1.gsfc.nasa.gov

Strong, Andrew
Max-Planck Inst.
aws@mpe-garching.mpg.de

Sturner, Steven
USRA/NASA/GSFC
sturner@tgrosf.gsfc.nasa.gov

Subramanian, Prasad
George Mason University
psubrama@gmu.edu

Swank, Jean
NASA/GSFC
swank@pcasun1.gsfc.nasa.gov

Takahashi, Tadayuki
ISAS
takahasi@astro.isas.ac.jp

Tatischeff, Vincent
NASA/GSFC
tatische@lhea1.gsfc.nasa.gov

Tavani, Mario
Columbia University
tavani@astro.columbia.edu

Teegarden, Bonnard
NASA/GSFC
bonnard@lheamail.gsfc.nasa.gov

Terasranta, Harri
Metsahovi Radio Res. Station
hte@alpha.hut.fi

The, Lih-Sin
Clemson University
lishin@astro.phys.clemson.edu

Thompson, Dave
NASA/GSFC
djt@egret.gsfc.nasa.gov

Timmes, Frank
Univ. of California, SC
fxt@burn.uchicago.edu

Tingay, Steven
Jet Propulsion Lab
tingay@hyaa.jpl.nasa.gov

Titarchuk, Lev
NASA/GSFC
titarchuk@lheavx.gsfc.nasa.gov

Tompkins, Bill
Stanford University
billt@leland.stanford.edu

Tueller, Jack
NASA/GSFC
tueller@gsfc.nasa.gov

Turver, Keith
Durham University
k.e.turver@durham.ac.uk

Tzioumis, Tasso
ATNF, CSIRO
atzioumi@atnf.csiro.au

Ubertini, Pietro
IAS-CNR
uberotini@alpha1.ias.fra.cnr.it

Ulmer, Melville
Northwestern University
m-ulmer@nwu.edu

Urry, C. Megan
STSci
cmu@stsci.edu

Valtaoja, Esko
Turku University
esko.valaoja.utu.fi

van der Hooft, Frank
University of Amsterdam
vdhooft@astro.uva.nl

van der Meulen, Roel
SRON Utrecht
vdmeulen@sron.ruu.nl

van Dyk, Rob
ESTEC/ESA
rvdijk@estec.esa.ni

van Paradijs, Jan
Univ. of Alabama, Huntsville
jvp@astro.uva.nl

Varendorff, Martin
Max-Planck Inst.
mgv@mpe-garching.mpg.de

Vedrenne, Gilbus
CESR, Toulouse

Vestrand, Tom
University of New Hampshire
tom.vestrand@unh.edu

Vilhu, Osmi
University of Helsinki
osmi.vilhu@helsinki.fi

von Ballmoos, Peter
CESR, Toulouse
pvb@cesr.cnes.fr

von Montigny, Corinna
LSW Heidelberg
cvmontig@lsw.uni-heidelberg.de

Wagner, R.
Ohio State University
rmw@lowell.edu

Wagner, Stefan
LSW Heidelberg
swagner@lsw.uni-heidelberg.de

Wang, John
University of Maryland
jcwang@astro.umd.edu

Watanabe, Ken
USRA/NASA/GFSC
watanabe@grossc.gsfc.nasa.gov

Weekes, Trevor
Whipple Observatory
tweekes@cfa.harvard.edu

White, Stephen
University of Maryland
white@astro.umd.edu

Wietfeldt, Fred
NIST
few@rrdstrad.nist.gov

Wiik, Kaj
Metsahovi Radio Res. Station
kah.wiik@hut.fi

Williams, Grant
Clemson University
ggwilli@hubcap.clemson.edu

Williams, Owen
ESTEC/ESA
owilliam@estec.esa.nl

Wilms, Jorn
IAA Tubingen
wilms@astro.uni.turbingen.de

Wilson, Robert
NASA/MSFC
wilson@gibson.msfc.nasa.gov

Winkler, Christoph
ESA/ESTEC
cwinkler@estsa2.estec.esa.nl

Woods, Peter
Univ. of Alabama, Huntsville
peter.woods@msfc.nasa.gov

Xu, Wenge
  Caltech
  wx@ipac.caltech.edu

Yu, Wenfei
  Inst. of High Energy Physics
  wenfei@gibson.msfc.nasa.gov

Yusef Zadeh, Farmad
  Northwestern University
  zadeh@nwu.edu

Zdziarski, Andrzej
  N. Copernicus Astron. Center
  aaz@camk.edu.pl

Zhang, S. Nan
  USRA/NASA/MSFC
  zhang@ssl.msfc.nasa.gov

Zychi, Piotr
  University of Durham
  piotr.zycki@durham.ac.uk

Author Index

A

Aharonian, F., 361, 1397
Aharonian (HEGRA collaboration), F. A., 1631
Ahlen, S. P., 1636
Akhperjanian, A., 1397
Akyüz, A., 1137
Albernhe, F., 1535, 1559, 1647
Aldering, G. S., 1417
Allen, G. E., 1089
Aller, H. D., 1417, 1423
Aller, M. C., 1423
Aller, M. F., 1417, 1423
Applegate, J. H., 1167
Arnett, D., 1119
Arocas, I., 1647

B

Backman, D. E., 1417
Bailes, M., 583, 602
Balonek, T. J., 1417, 1423
Bandyopadhyay, R., 892
Baring, M. G., 171, 638, 1157
Barret, D., 75, 697, 724, 734, 952, 1498
Barrio, J., 1397
Bartolini, G., 1516
Bazán, G., 1119
Bazzano, A., 719, 729
Begelman, M. C., 1423
Belolipetskiy, S. V., 1606
Bennett, K., 537, 542, 583, 588, 829, 967, 1084, 1099, 1198, 1218, 1243, 1298, 1341, 1582, 1587
Bernlöhr, K., 1397
Bertsch, D. L., 509, 1248, 1346, 1371, 1592, 1606
Beteta, J., 1397
Bhattacharya, D., 1137, 1147, 1626
Bignami, G. F., 387, 573
Biller, S., 558
Blaes, O., 1288
Blanco, P. R., 897, 1089, 1147
Blázquez, R., 1647

Bloemen, H., 249, 537, 829, 967, 1074, 1084, 1099, 1109, 1114, 1218, 1243, 1298, 1341, 1356
Blom, J. J., 1243, 1341, 1356, 1582
Bloom, S. D., 1257, 1262, 1346, 1371
Bloser, P. F., 697, 724, 734, 952, 1498
Boirin, L., 724
Boltwood, P., 1417
Bonnell, J., 1417
Borrel, V., 957, 1535, 1647
Böttcher, M., 1473
Bouchet, L., 957
Boyle, P. J., 558, 592, 1142, 1376, 1381
Bradbury, S., 1397
Bradbury, S. M., 1142, 1402
Brazier, K. T. S., 588, 597, 1267
Breslin, A. C., 1402
Bridgman, W. T., 977
Briggs, M. S., 687
Brinkmann, W., 922
Buccheri, R., 542
Buckley, J. H., 558, 592, 1142, 1376, 1381, 1402
Buckley, T., 665, 675
Burdett, A. C., 1142
Burdett, A. M., 558, 592, 1376, 1381
Burger, M., 878
Bussóns Gordo, J., 558, 1142, 1376, 1381
Bykov, A. M., 249, 1074

C

Campbell-Wilson, D., 937
Caplinger, J., 1417
Caraveo, P. A., 387, 568, 573, 1535
Carramiñana, A., 583, 588, 597, 1267, 1462
Carter-Lewis, D. A., 558, 592, 1142, 1361, 1376, 1381, 1402
Case, G., 1137
Cassé, M., ·1059
Castro-Tirado, A. J., 1333, 1516
Catanese, M. A., 558, 592, 1142, 1376, 1381, 1402
Catelli, J. R., 1606

Cawley, M. F., 558, 592, 1142, 1376, 1381, 1402
Celotti, A., 1417
Chadwick, P. M., 612, 1621
Chakrabarty, D., 773
Chantell, M. C., 1626
Charles, P. A., 892
Chaty, S., 917
Chavushyan, V., 1462
Chen, X., 995
Cheng, L. X., 902, 1012
Chernenko, A., 653
Chernyakova, M. A., 822
Chiaberge, M., 1423
Chiang, J., 1308
Chiappetti, L., 1412
Churazov, E., 957
Clements, S. D., 1423
Cline, T. L., 1007
Cocchi, M., 719, 729
Collin, M., 1559
Collmar, W., 307, 829, 863, 1243, 1298, 1341, 1356, 1417, 1423, 1587
Colombo, E., 592
Connell, P., 1535
Connell, P. H., 1544
Connors, A., 407, 542, 583
Contreras, J., 1397
Coppi, P., 1626
Corbel, S., 912, 932, 937
Cordes, J. M., 633
Cordier, B., 1535, 1559, 1647
Cortina, J., 1397
Costa, M. E., 1433, 1493, 1516
Covault, C. E., 1626
Crary, D. J., 803, 947
Crawford, H., 1606
Crider, A., 868, 1512
Cui, W., 813, 839, 854
Cusumano, G., 553, 758, 793

D

Dal Fiume, D., 553, 758, 793, 1493, 1516
Dalton, J., 1417
Dame, T. M., 1188
D'Amico, N., 602
Daniels, W. M., 1606

Daum, A., 1397
Dazeley, S. A., 1507
Deal, K., 687
Deckers, T., 1397
De Francesco, G., 1423
de Gouveia Dal Pino, E. M., 1203
Deines-Jones, P., 1606
de Jager, O. C., 171
Del Sordo, S., 758, 793
Dermer, C. D., 271, 307, 977, 1044, 1152, 1275, 1308, 1402
Diallo, N., 1535, 1559
Dickinson, M. R., 612, 1621
Diehl, R., 218, 542, 967, 1074, 1084, 1099, 1104, 1109, 1114, 1198, 1218, 1298, 1535, 1554, 1577
Dieters, S., 617
Digel, S. W., 1188
Dingus, B. L., 407, 1248, 1346, 1371, 1592
Dipper, N. A., 612, 1621
Dixon, D. D., 1039, 1137, 1601
Dobrinskaya, J., 932
Dogiel, V. A., 1069
Done, C., 962
Dorman, L. I., 1172, 1178, 1183
Dotani, T., 844, 922
Dove, J., 849
Dragovan, M., 1626
Drucker, A., 1417
Durouchoux, Ph., 887, 912, 932, 937, 1535
Dyachkov, A., 957

E

Ebisawa, K., 623, 844, 922
Edwards, P. G., 1428, 1433, 1507
Eikenberry, S. S., 547
Ellison, D. C., 1157
Esin, A., 887
Esposito, J. A., 1248, 1257, 1346, 1371, 1606

F

Falomo, R., 1417
Fazio, G. G., 547

Fegan, D. J., 361, 558, 592, 1142, 1376, 1381, 1402
Feigl, E., 1397
Fender, R. P., 798, 813, 932, 937
Fenker, H., 1606
Fernandez, J., 1397
Feroci, M., 758, 1493, 1516
Ferrero, J. L., 1647
Fichtel, C. E., 436, 1417
Filippenko, A. V., 932
Finger, M. H., 57, 617, 739, 748, 773, 778, 803, 947
Finley, J. P., 558, 592, 1142, 1376, 1381, 1402
Fiore, F., 1412
Fishman, G. J., 509, 687, 927, 947, 952, 1012, 1283
Fletcher, S., 1099
Focke, W., 854
Fonseca, V., 1397
Ford, E. C., 697, 703, 734
Foster, R. S., 813, 922, 1253
Fraß, A., 1397
Freudling, W., 1417
Freyberg, M. J., 1069
Frontera, F., 758, 1493, 1516
Funk, B., 1397

G

Gaensler, B. M., 1428
Gaidos, J. A., 558, 592, 1142, 1376, 1381, 1402
Galvan, E., 665, 675
Gautier, T. N., 912
Gear, W. K., 1417, 1423
Geckeler, R. D., 753
Gehrels, N., 3, 524, 902, 1007, 1012, 1283
Georganopoulos, M., 1423
Georgii, R., 1109, 1535, 1554
Gerard, E., 892
Ghigo, F. D., 813
Ghisellini, G., 1293, 1412, 1417, 1423
Giarrusso, S., 553, 793
Gierliński, M., 844
Gilfanov, M., 887, 957
Giommi, P., 1412, 1493
Glass, I. S., 1423

Goldoni, P., 957, 1549
Goldwurm, A., 957, 1549
Gómez-Gomar, J., 1125
Gonzalez, J., 1397
González-Pérez, J. N., 1417, 1423
Gordo, J., 592
Gorosabel, J., 1333, 1516
Gossan, B., 1606
Gottbrath, C. L., 1616
Gotthelf, E., 1027
Graham, B. L., 1642
Grandi, P., 1293
Gregorich, D. T., 1626
Greiner, J., 907
Grenier, I. A., 394, 1157, 1188
Grindlay, J. E., 122, 697, 724, 734, 952, 1498
Grove, J. E., 122, 788, 1079, 1293
Gruber, D. E., 744, 753, 788, 897, 1089
Guainazzi, M., 793
Guarnieri, A., 1516
Guichard, J., 1267, 1462

H

Haardt, F., 1293
Hall, P., 1417
Halpern, J. P., 568, 1253
Hanna, D. S., 1626
Hara, T., 1507
Harding, A. K., 39, 607, 638, 648
Harmon, A., 878, 922, 1283
Harmon, B. A., 122, 141, 687, 697, 713, 734, 773, 813, 834, 839, 873, 927, 947, 952, 1498
Harris, M. J., 418, 1007, 1079, 1094
Harrison, T. E., 665, 670, 675, 679, 942
Hartman, R. C., 307, 1248, 1262, 1346, 1366, 1417, 1423, 1428, 1606
Hartmann, D. H., 1039, 1147, 1223, 1228
Haustein, V., 1397
Hayami, Y., 1507
Heidt, J., 1516
Heikkila, C. W., 665
Heinämäki, P., 1423
Heindl, W. A., 744, 788, 854, 897, 1089
Heinzelmann, G., 1397
Heise, J., 719, 729

Hemberger, M., 1397
Henri, G., 1303, 1313
Hermann, G., 1397
Hermsen, W., 39, 537, 542, 583, 588, 829, 967, 1074, 1084, 1099, 1109, 1114, 1218, 1243, 1253, 1298, 1341, 1356
Hernanz, M., 1125
Herter, M., 1423, 1457
Heß, M., 1397
Heusler, A., 1397
Higdon, J. C., 912, 1089
Hillas, A. M., 558, 592, 1142, 1376, 1381, 1402
Hjellming, R. M., 141, 813
Hofmann, W., 1397
Holder, J., 612, 1621
Holl, I., 1397
Hooper, E. J., 1423
Horns, D., 1397
Hua, X.-M., 122, 858, 982, 987, 1520
Hughes, P. A., 1423
Hunstead, R. W., 1428
Hunter, S. D., 192, 1188, 1213, 1248, 1346, 1371, 1606
Hurley, K. H., 407, 1007
Hutchins, J. B., 1606

I

Illarionov, A. F., 822
Inoue, H., 922, 1417
in't Zand, J., 719, 729
Isern, J., 1125
Iyudin, A., 1109

J

Jahoda, K., 692, 844, 1089
Jauncey, D. L., 1433
Jean, P., 1535
Johnson, W. N., 283, 844, 1079, 1288, 1402, 1417, 1423, 1572
Jones, B. B., 783, 1213, 1248, 1366, 1371
Jones, D. L., 1433
José, J., 1125

Jourdain, E., 881, 957, 1303
Jung, G. V., 708, 1079

K

Kaaret, P., 697, 703, 734
Kamae, T., 628
Kamei, S., 1507
Kanbach, G., 597, 1213, 1248, 1253, 1267, 1346, 1371
Kankanian, R., 1397
Kappadath, S. C., 344, 1218, 1587
Kaspi, V. M., 583, 602
Katajainen, S., 1423
Kaul, R. K., 1271
Kawai, N., 628, 922
Kazanas, D., 122, 982, 987, 1520
Kendall, T. R., 612, 1621
Kendziorra, E., 788
Kertzman, M. P., 1616
Khavenson, N., 957
Kidger, M. R., 1417, 1423
Kifune, T., 361, 1507
Kinzer, R. L., 192, 1079, 1193, 1223, 1572
Kirk, J. G., 1478
Kirstein, O., 1397
Kita, R., 1507
Kniffen, D. A., 524, 1248, 1257, 1346, 1371
Knödlseder, J., 1104, 1109, 1114
Köhler, C., 1397
Kolaczyk, E. D., 1039, 1601
Kollgaard, R. I., 1417
Kommers, J., 687
Kondo, Y., 1417
König, M., 763
Konishi, T., 1507
Konopelko, A., 1397
Kornmayer, H., 1397
Kotani, T., 922
Kouveliotou, C., 96, 687, 947
Kozlovsky, B., 1049
Kranich, D., 1397, 1407
Kraus, A., 1346, 1423
Krawczynski, H., 1397
Krennrich, F., 558, 592, 1142, 1376, 1381, 1402
Krennrich, for the Whipple Collaboration, F., 563, 1391

Kretschmar, P., 788
Kreykenbohm, I., 788
Krichbaum, T. P., 1346
Krizmanic, J. F., 1606
Kroeger, R. A., 141, 1642
Krolik, J. H., 972
Kubo, H., 1467
Kuiper, L., 537, 542, 583, 588, 829
Kuleshova, N., 957
Kunz, M., 744
Kurfess, J. D., 509, 708, 1079, 1147, 1193, 1417, 1572, 1642
Kusunose, M., 868, 1512
Kuulkers, E., 683

Lindenstruth, V., 1606
Lindner, A., 1397
Ling, J. C., 858
Lingenfelter, R. E., 1089
Litterio, M., 643
Lorenz, E., 1397, 1407
Lorenz, E. for the MAGIC Telescope Design Group, 1611
Lovell, J. E. J., 1428, 1433
Lu, F. J., 578
Lund, N., 1516
Luo, C., 878, 992
Lyne, A., 588

L

Lähteenmäki, A., 1452
Lainela, M., 1428
Lamb, R. C., 558, 592, 1142, 1376, 1381, 1402
Lammers, U., 683
Lampeitl, H., 1397
Lanteri, L., 1423
Lattanzi, M. G., 573
Laurent, P., 957, 1549
Lawrence, G. F., 1423
Lawson, A. J., 1417
Leahy, D. A., 748
Lebrun, F., 1549
Leeber, D. M., 942
Leising, M. D., 163, 1017, 1022, 1104, 1147, 1223, 1228
Leleux, P., 1535
Leonard, D. C., 932
Leray, J. P., 1647
Lessard, R. W., 558, 592, 1142, 1376, 1381, 1402
Leventhal, M., 208, 902, 1012
Lewin, W. H. G., 687, 947
Li, T. P., 578, 1351
Liang, E. P., 461, 868, 873, 878, 932, 992, 1512
Lichti, G. G., 542, 1084, 1109, 1298, 1423, 1535, 1554
Lin, D., 868, 1512
Lin, R., 1535
Lin, Y. C., 283, 783, 1248, 1346, 1371, 1423

M

Macomb, D. J., 1402
Madejski, G. M., 283, 1417, 1423, 1467
Magdziarz, P., 1288
Magnussen, N., 1397
Mahoney, W. A., 912, 1346
Maisack, M., 788, 1356
Majmudar, D., 1167
Makarov, V. V., 573
Makino, F., 1417, 1467
Malzac, J., 1303
Manchester, R. N., 583, 602
Mandrou, P., 1535
Mandzhavidze, N., 1054
Maraschi, L., 1293, 1412, 1417
Marscher, A. P., 1346, 1417, 1423, 1437
Martin, M. D., 1606
Martini, P., 892
Masaike, A., 1507
Masetti, N., 1516
Mason, P. A., 665, 670, 675, 679
Massaro, E., 553, 643
Massone, G., 573
Masterson, C., 558, 592, 1142, 1376, 1381
Mastichiadis, A., 1478
Matsubara, Y., 1507
Matsuoka, M., 922
Matsuoka, Y., 1507
Matteson, J., 1535
Mattox, J. R., 568, 1152, 1253, 1636
Matz, S. M., 407, 808
Mause, H., 1473

Mayer-Hasselwander, H. A., 1248, 1346, 1371
McCollough, M. L., 813, 834
McCollum, B., 1417
McComb, T. J. L., 612, 1621
McConnell, M. L., 122, 542, 829, 967, 1084, 1099, 1198, 1218, 1341, 1577, 1587
McCulloch, P. M., 1428, 1433
McEnery, J. E., 558, 592, 1142, 1376, 1381, 1402
McHardy, I. M., 1417
McLaughlin, M. A., 633
McNamara, B. J., 665, 670, 675, 679, 942
McNaron-Brown, K., 1079, 1417, 1423, 1572
Medina Tanco, G. A., 1203
Meegan, C., 407
Meier, D. L., 1433
Melia, F., 1000, 1318, 1323, 1328
Meyer, H., 1397
Michelson, P. F., 783, 1346, 1371
Mignani, R., 573
Milne, P. A., 1017, 1022
Mineo, T., 553
Mioduszewski, A. J., 813
Mirabel, I. F., 141, 892, 902, 917
Mirzoyan, R., 1397
Misra, R., 1000
Mitchell, J. W., 1606
Mitra, A. K., 818, 1271
Mizumoto, Y., 1507
Mohanty, G., 558, 592, 1142, 1376, 1381
Möller, H., 1397
Monnelly, G. P., 1498
Moore, E. M., 1423
Moralejo, A., 1397
Morfill, G. E., 1069
Morgan, E. H., 897, 907
Mori, M., 623, 768, 1507
Moriarty, P., 592, 1402
Morris, D. J., 1084, 1109, 1114, 1298
Moskalenko, I. V., 863, 881, 1162
Moss, M., 932
Much, R. P., 537, 542, 829, 1243, 1582
Mücke, A., 1233, 1346, 1371
Mukherjee, R., 394, 1253, 1346, 1371, 1423, 1626
Muller, J. M., 719, 729
Muraishi, H., 1507
Muraki, Y., 1507
Murphy, D. W., 1433
Murphy, R. J., 17, 1079
Muslimov, A., 648

N

Nagase, F., 922
Nair, A. D., 1423
Nair, D., 1417
Naito, T., 1507
Nandikotkur, G., 1361
Narayan, R., 461, 887
Natalucci, L., 719, 729
Navarro, J., 602
Naya, J., 1535, 1544
Naya, J. E., 1567
Nayakshin, S., 1318, 1323, 1328
Nevalainen, J., 887
Nicastro, L., 553, 793, 1493, 1516
Nicolson, G. D., 1433
Nilsson, K., 1423
Nishijima, K., 1507
Nolan, P. L., 783, 1248, 1346, 1371
Nomoto, K., 1089
Nowak, M., 849

O

Oberlack, U., 1104, 1109, 1114, 1577
Ogio, S., 1507
Olive, J.-F., 724
Ong, R. A., 1626
Oosterbroek, T., 683, 758
Orford, K. J., 612, 1621
Orlandini, M., 758, 793, 1493
Osborne, J. L., 612
Oser, S., 1626
Owens, A., 683

P

Paciesas, W. S., 283, 697, 713, 734, 803, 813, 834, 839, 927, 952, 1283
Padilla, L., 1397
Palazzi, E., 758, 1412, 1493, 1516

Palmer, D., 1007
Panter, M., 1397
Parizot, E. M. G., 1059, 1064
Parmar, A. N., 553, 683, 758
Parsons, A., 1283
Patterson, J. R., 1507
Paul, J., 1059
Peaper, D. R., 1616
Peila, A., 1423
Pelling, M. R., 897, 1089
Penton, S., 1417
Perryman, M. A. C., 573
Pesce, J. E., 1417
Petrucci, P. O., 1303, 1313
Petry, D., 1397
Petry, D. for the HEGRA Collaboration, 1407
Phillips, R. B., 1442
Phlips, B. F., 854, 1642
Pian, E., 1412, 1417
Piccioni, A., 1516
Pierkowski, D. B., 1423
Piraino, S., 793
Piro, L., 793
Piro, L. on behalf of the BeppoSAX team, 1485
Pitts, W. K., 1606
Plaga, R., 1397
Pohl, M. K., 449, 1034, 1213, 1233, 1346, 1366, 1371, 1417, 1423
Pohl, M. for the EGRET collaboration, 1596
Pooley, G. G., 813
Poutanen, J., 887, 972
Prahl, J., 1397
Preston, R. A., 1433
Prince, T. A., 57
Prosch, C., 1397
Pühlhofer, G., 1397
Purcell, W. R., 208, 708, 902, 1027, 1079, 1193, 1572
Pursimo, T., 1423

Q

Quinn, J., 558, 592, 1142, 1376, 1381, 1402

R

Ragan, K., 1626
Raiteri, C. M., 1417, 1423
Ramaty, R., 449, 1007, 1049, 1054
Rauterberg, G., 1397
Ray, A., 607
Rayner, S. M., 612, 1621
Reich, W., 1423
Reimer, O. L., 597, 1248, 1267, 1346, 1371
Remillard, R. A., 907
Renda, M., 1417
Rephaeli, Y., 271
Revnivtsev, M., 957
Reynolds, J. E., 1433
Reynolds, S. J., 1157
Rhode, W., 1397
Rivero, R., 1397
Roberts, I. D., 612, 1621
Roberts, M. D., 1507
Roberts, M. S. E., 783
Robinson, C. R., 713, 734, 813, 834, 922, 927, 1498
Robson, E. I., 1417, 1423
Rodgers, A., 1381
Rodgers, A. J., 558, 592, 1142, 1376, 1402
Rodriguez, L. F., 141, 902
Röhring, A., 1397
Roldán, C., 1647
Roques, J.-P., 957, 1535
Rose, H. J., 558, 592, 1142, 1376, 1381, 1402
Rothschild, R. E., 744, 753, 788, 897, 1089
Rowell, G. P., 1507
Rubin, B. C., 713, 778, 927, 947
Rupen, M. P., 141, 813
Ryan, J. M., 17, 537, 542, 829, 967, 1084, 1099, 1109, 1114, 1198, 1218, 1243, 1356, 1577, 1582
Ryde, F., 972

S

Sahakian, V., 1397
Saito, Y., 628
Sako, T., 1507

Sakurazawa, K., 1507
Salamon, M. H., 1238
Sambruna, R. M., 1412, 1417
Samimi, J., 1039, 1601
Samorski, M., 1397
Samuelson, F. W., 558, 592, 1142, 1376, 1381, 1402
Sanchez, F., 1535
Sanchez, J., 1397
Sandhu, J. S., 602
Santangelo, A., 758
Saunders, M. A., 1601
Schandl, S., 763
Schirmer, A. F., 1417
Schlickeiser, R., 449, 1473
Schmele, D., 1397
Schmidt, T., 1397
Schneid, E. J., 1346, 1371, 1592
Schönfelder, V., 509, 542, 583, 588, 829, 863, 967, 1069, 1074, 1084, 1099, 1109, 1114, 1198, 1218, 1243, 1298, 1341, 1356, 1535, 1554, 1582, 1587
Schubnell, M., 1386
Scott, D. M., 617, 739, 744, 748, 778, 803
Segreto, A., 553, 758, 793
Seifert, H., 1007, 1535, 1544, 1567
Seitz, T., 1516
Sembroski, G. H., 558, 592, 1142, 1376, 1381, 1402, 1616
Shahbaz, T., 892
Share, G. H., 17, 1079, 1223
Shaw, S. E., 612, 1621
Sheth, S., 878
Shibata, S., 628
Shrader, C. R., 3, 328, 892, 1417
Sikora, M., 494, 1417, 1423
Sillanpää, A., 1417, 1423
Silva, A., 665
Simrall, J. H., 1606
Skibo, J. G., 449, 977, 1044, 1152, 1208, 1308
Skinner, G. K., 1535, 1544
Smale, A., 724
Smith, A. G., 1423
Smith, D. A., 962, 1288
Smith, D. M., 208, 902, 1012
Smith, I. A., 110, 868, 932, 1512
Smith, M. J. S., 719, 729
Smith, P. S., 1417

Song, L. M., 578
Sood, R. K., 932, 937
Sparks, W. M., 1130
Spencer, R. E., 937
Sreekumar, P., 436, 768, 1248, 1346, 1361, 1366, 1371, 1606
Srinivasan, R., 558, 592, 1142, 1376, 1381, 1402
Stacy, J. G., 1341, 1356, 1442
Stamm, W., 1397
Stappers, B. W., 602
Stark, M. J., 692
Starrfield, S., 1130
Staubert, R., 744, 753, 763, 788
Stecker, F. W., 344, 1238
Steinle, H., 283, 542, 829, 1298, 1356
Stelzer, B., 753
Steppe, H., 1423
Stevens, J. A., 1417, 1423
Stocke, J., 1417
Stollberg, M. T., 803
Streitmatter, R. E., 1606
Strickman, M., 607
Strohmayer, T. E., 692, 773
Strong, A. W., 192, 537, 542, 829, 1074, 1099, 1109, 1114, 1162, 1198, 1298, 1356, 1544, 1596
Sturner, S. J., 1152, 1535, 1544, 1567
Sun, X. J., 578
Sunyaev, R., 957
Susukita, R., 1507
Suzuki, A., 1507
Suzuki, R., 1507
Swank, J. H., 692, 724, 854, 897, 995, 1089

T

Taam, R. E., 995
Tagliaferri, G., 1412
Takahashi, T., 1467
Takalo, L. O., 1417, 1423
Takeshima, T., 922
Tamura, T., 1507
Tanaka, Y., 922
Tanimori, T., 1507
Tashiro, M., 1467
Tatischeff, V., 1049, 1054
Tavani, M., 75, 697, 734, 922, 1253

Teegarden, B. J., 1007, 1535, 1544, 1567
Templeton, M., 665, 670, 675
Teräsranta, H., 1346, 1417, 1423, 1447, 1452
Thaddeus, P., 1188
The, L.-S., 1022, 1147, 1223, 1228
Thommes, E., 1516
Thompson, D. J., 39, 394, 1248, 1253, 1257, 1262, 1346, 1371, 1417, 1606
Thompson, R., 1417
Thornton, G. J., 1507
Timmes, F. X., 218
Tingay, S. J., 1433
Titarchuk, L. G., 477, 982, 987, 1520
Tompkins, W. F., 783, 1248, 1371
Tornikoski, M., 1346, 1417, 1423, 1428
Tosti, G., 1417
Treves, A., 1412, 1417
Truran, J. W., 1130
Tserenin, N., 957
Tueller, J., 902, 1012
Tümer, O. T., 1626
Turcotte, P., 1417
Turver, K. E., 612, 1621
Tzioumis, A. K., 937, 1433

U

Ubertini, P. on behalf of the IBIS Consor 1527
Ubertini, P., 719, 729
Ueda, Y., 922
Ulmer, M. P., 39, 633
Ulrich, M., 1397
Unwin, S. C., 1417
Urry, C. M., 1293, 1412, 1417

V

v.Montigny, C., 1248
Vacanti, G., 1412
Valtaoja, E., 1346, 1417, 1423, 1452
van der Hooft, F., 947
van der Klis, M., 947
van der Meulen, R. D., 537, 542, 1074
van Dijk, R., 829, 967, 1099, 1577
van Paradijs, J., 96, 617, 687, 947
Varendorff, M., 542, 1218, 1577

Vargas, M., 887, 957
Vaughan, B. A., 849
Vedrenne, G., 1535
Vestrand, W. T., 768, 1442
Vilhu, O., 887
Villata, M., 1417, 1423
Visser, G., 1606
Völk, H., 1397
von Ballmoos, P., 1109, 1114, 1535
von Montigny, C., 307, 1257, 1262, 1346, 1371, 1423, 1457

W

Wagner, R. M., 892
Wagner, S. J., 1346, 1417, 1423, 1457
Wallyn, P., 912
Walsh, K. M., 1606
Waltman, E. B., 813
Wang, J. C. L., 658
Watanabe, K., 1223, 1228
Webb, J. R., 1417
Weekes, T. C., 361, 558, 592, 1142, 1376, 1381, 1402
Wehrle, A. E., 328, 1417, 1437
Weidenspointner, G., 1218, 1577, 1587
Wessolowski, U., 1109, 1587
West, M., 558
Westerhoff, S., 1397
Wheaton, Wm. A., 858
White, N. E., 922
Wichmann, R., 1423
Wiebel-Sooth, B., 1397
Wiedner, C. A., 1397
Wiescher, M. C., 1130
Wiik, K., 1452
Williams, D. A., 1626
Williams, O. R., 1243, 1298, 1341, 1356, 1582
Willmer, M., 1397
Wilms, J., 753, 788, 849
Wilson, C. A., 617, 687, 773, 834, 1283
Wilson, R. B., 739, 748, 773, 778, 803
Winkler, C., 542, 1109, 1114, 1218
Wirth (HEGRA collaboration), H., 1397

Witzel, A., 1346, 1423
Wolf, C., 1516
Woods, P., 687
Woosley, S. E., 1089, 1228
Wu, M., 578, 1351

X

Xu, W., 1417, 1437

Y

Yamaoka, K., 922
Yanagita, S., 1507
Yoshida, T., 1507
Yoshikoshi, T., 1507

Yu, W., 578, 734, 1351
Yusef-Zadeh, F., 1027

Z

Zavattini, G., 1493
Zdziarski, A. A., 283, 844, 1288
Zhang, S., 578, 1351
Zhang, S. N., 141, 697, 713, 734, 813, 834, 839, 873, 878, 922, 927, 952, 1253, 1283, 1498
Zook, A. C., 1417
Zweerink, A., 558
Zweerink, J., 592, 1142, 1376, 1381, 1402
Zych, A., 1137
Życki, P. T., 962